Industrial
Air Pollution
Control Systems

Industrial Air Pollution Control Systems

William L. Heumann

McGraw-Hill

New York   San Francisco   Washington, D.C.   Auckland   Bogotá
Caracas   Lisbon   London   Madrid   Mexico City   Milan
Montreal   New Delhi   San Juan   Singapore
Sydney   Tokyo   Toronto

Library of Congress Cataloging-in-Publication Data

Heumann, William L.
    Industrial air pollution control systems / William L. Heumann.
       p.   cm.
    Includes index.
    ISBN 0-07-031430-6 (alk. paper)
    1. Flue gases—Purification.   2. Particles—Environmental aspects.
  3. Air—Purification—Equipment and supplies.   I. Title.
  TD885.H48   1997
  628.5'3—dc21   97-8938
                                             CIP

McGraw-Hill

A Division of The **McGraw·Hill** *Companies*

1 2 3 4 5 6 7 8 9 0   DOC/DOC   9 0 2 1 0 9 8 7

ISBN 0-07-031430-6

The sponsoring editor for this book was Robert Esposito, the editing supervisor was Bernard Onken, and the production supervisor was Pamela A. Pelton. It was set in Century Schoolbook by Don Feldman of McGraw-Hill's Professional Book Group composition unit.

Printed and bound by R. R. Donnelley & Sons Company.

McGraw-Hill books are available at special quantity discounts to use as premiums and sales promotions, or for use in corporate training programs. For more information, please write to the Director of Special Sales, McGraw-Hill, 11 West 19th Street, New York, NY 10011. Or contact your local bookstore.

This book is printed on recycled, acid-free paper containing a minimum of 50% recycled de-inked fiber.

To my wife, Lynn, whose love, encouragement, and hard work enabled me to survive the multi-year pregnancy that gave birth to this work.

It is my hope that, armed with common sense and this tool, engineers, students, and scientists will more easily find the best solutions to their air pollution control problems.

Contents

Section 1 Basic Knowledge   1

Preface

This book is intended for those who have a vested interest in industrial air pollution control systems and how they work. Throughout my career, I have been astonished at the uniqueness and variety of industrial processes. Two processes that seemingly should be the same, invariably, are not. By the sheer number of variables that are present, not the least of which is the number of people that are involved in the initial design and installation, every system has a personality of its own. Over the years of operation, demands on a given system change and modifications ensue. Its personality of operation grows and changes with the environment surrounding it. As a valuable part of the production or as a regulated requirement, the air pollution control system must exist as part of the industrial process.

As a result of the complex design, many times the pollution control system or the process, or both, will not work properly. This dysfunction may result in an emission standard being missed, lost production, poor operating efficiencies, unreasonably high maintenance, and a host of other problems.

In this text, we have attempted to provide engineers, students, and scientists with the tools and knowledge necessary to understand how to make their air pollution control systems work. Many talented, experienced, and learned individuals have contributed their time to this endeavor. By supplying their efforts to this work, these contributors have transferred some of the knowledge they accumulated over numerous applications and years of experience. I hope this work will provide a shortcut in reaching a high level of proficiency for the reader without requiring the high cost of direct experience and mistakes that many of us have made.

Physics, the theories of operation, and the math that support them are used when necessary to convey an understanding to the reader. We have purposely stayed away from the derivation of most theories and physics involved in an effort to keep this text as simple and useful as possible in the direct application of air pollution control technologies. The equations and formulas presented herein are those that have proven valuable to the authors in practice or in understanding important concepts.

This book is more than an overview of the control technologies and their application. However, it is less than the complete and final word on all the subjects covered herin. Every effort has been made to ensure that this work will provide a compilation of the knowledge and tools needed to assist with a wide variety of air pollution control problems.

William L. Heumann

Introduction

The Reasons of Air Pollution Control

There are two primary motivations behind the utilization of industrial air pollution control technologies. These are

1. They must be used because of legal or regulatory requirements
2. They are integral to the economical operation of an industrial process

Although economists would point out that both of these motivations are really the same, that is, it is less expensive for an industrial user to operate with air pollution control than without, the distinction in application type is an important one. The distinction is important at the beginning of this work because many of the users for equipment, technology, and processes described herein may be for applications that the user does not consider to be "air pollution control." In general, air pollution control is used to describe those applications that are driven by regulations and/or health considerations, while applications that deal with product recovery are considered process applications. Nevertheless, the technical issues, equipment design, operation, etc., will be similar if not identical. In fact, what differs between these uses is that the economics that affect the decision making process will often vary to some degree.

The Methods of Air Pollution Control

There are two general methods of reducing emissions from any system. These are

1. Change, modify, or otherwise optimize the process so that the unwanted emissions are reduced or eliminated

2. Add a control technology or air pollution control device to the system

Although this work deals almost exclusively with the field of add-on control technologies, some mention of system optimization is appropriate.

Industrial process design and engineering is a broad field and one that does not lend itself to simple generalizations. I feel it would be safe to say though that if system modifications that can reduce pollution are available, they should be considered and evaluated prior to beginning the process of evaluation of add-on technologies. Often, not only will the immediate economics (capital expenditures) be more favorable, but the long-term system economics (e.g., reliability, waste disposal, maintenance cost, and energy cost) may be lower.

Once system modifications have been ruled out as a method of air pollution control, the process of selection of the most economical control system begins. Since there are generally two distinct categories of pollutant, particulate or gaseous, we sectionalized this work into these categories. Of course, there are many industrial applications in which there are pollutants from both categories and a combination of technologies may be utilized to effect an acceptable solution. The mechanisms that are generally employed for the removal of industrial pollutants from gas streams are

Particulate control mechanisms

1. Gravity

2. Inertia

3. Centrifugal force

4. Inertial impaction

5. Interception

6. Diffusion and brownian motion

7. Electrostatic precipitation

8. Miscellaneous mechanisms such as growth of particles by condensation, thermophoresis, photophoresis, sonic agglomeration, and magnetism.

Gaseous control mechanisms

1. Diffusion

2. Absorption

3 Adsorption

4. Thermal oxidation

5. Miscellaneous mechanisms such as biofiltration, energy-driven (other than thermal energy) destruction, and condensation.

The ease with which the reader will grasp the operation and characteristics of any given technology will likely relate to his or her background and experience. In general, particulate control technologies are more easily understood by physicists and mechanical engineers, while gaseous control technologies are most easily understood by chemists, chemical engineers, and process engineers. This undoubtedly relates to the nature of the mechanisms listed above and the practical embodiment of these into an operational system.

The order of subjects within this book was designed to follow the thought and learning process that an unfamiliar user would follow from start to end in solving an air pollution control application. Within each broad category of control technologies (particulate or gaseous), the order of technologies presented represents a simple hierarchy of selection. This hierarchy of selection is not appropriate in all cases but is typical of the order of elimination that may be followed. In other words, for a particulate control application, it makes sense to consider cyclones first, media filtration second, etc. As technologies are eliminated for various reasons (not able to practically achieve desired collection efficiency, safety issues, etc.), the next technology to be considered will be more expensive, more complicated, and/or more expensive to operate. Although for any given application there is an optimum selection that represents the most economical achievement of the project goals, most often this is unknown at the beginning of the process. Following this hierarchy of selection and eliminating inappropriate selections will help result in the most economical solution.

Nomenclature

Roman Letters

a	acceleration, ft/s^2 [cm/s^2]
A	area, ft^2 [m^2]
A/C	air-to-cloth ratio
AMW	apparent molecular weight
C	Cunningham correction factor
C^D	drag coefficient
d	diameter, ft [μm]
D	diameter, ft [m]
E_d	collection efficiency at particle size d, % by weight
E_{df}	fractional efficiency at particle size d, % by weight
E_F	collection efficiency as fraction ($E_T/100$)
E_{MID}	collection efficiency at particle size MID, % by weight
E_T	total collection efficiency, % by weight
E^v	evaporation rate, lb/min [kg/m]
F	force, lb [N]
g	gravitation acceleration, ft/s^2 [m/s^2]
G	solids mass flux, lb/s [kg/s]
H	height, ft [m]
H_a	absolute humidity-ratio
H_u	specific humidity-ratio
h	enthalpy, Btu/lb m [cal/g]
h_p	drainpipe submergence below minimum water level, in [cm]
h_n	scrubber minimum elevation above maximum water level, in [cm]
h_u	latent heat of vaporization, Btu [cal]
h_t	total heat, Btu [cal]
i	current, A
I	ionic strength

I	current density, A/ft^2 [A/m^2]
Kn	Knudsen number
l	cyclone exponent, lb [kg]
M	mass flux, lb/s [kg/s]
M	mass, lb [kg]
n	number
N	moles, lb [kg]
$\dot{N}$	mole flux, mol (lb)/min [mol (kg)/h]
N_{av}	Avogadro's number
P	pressure, lb/in^2 [kg/cm^2]
$P(D)$	particle penetration fraction
P_w	partial pressure of water vapor, psia [kg/cm^2]
P_s	saturation pressure of water vapor, psia [kg/cm^2]
Q	volumetric flow rate, ft^3/min [m^3/h]
r	radius, ft [m]
R	universal gas constant, ft · lb/mol (lb) · °R [m · kg/mol (kg) · K]
Re	Reynolds number
RH	relative humidity, %
Sh	specific heat, Btu/lb (mass) m · °F [J/g · °C]
S_t	Stokes number
t	time, s
T	temperature, °R [K]
U	velocity, ft/s [m/s]
V	volume, ft^3 [m^3]
V_r	repulsive energy per unit area
v	specific volume, ft^3/lb m [cm^3/g]
W	dust loading, gr/ft^3 [g/m^3]
w	width, ft [m]
X	displacement, ft [m]

Greek Letters

α	angle
Δ	change
Γ	conductivity
Υ	interfacial tension
η	viscosity, lb(mass)/ft.s [cp]
κ	dielectric constant
λ	mean free path
ξ	viscous drag coefficient
ρ	density, lb/ft^3 [kg/m^3]
σ	standard deviation
Ψ	impaction parameter
ω	migration velocity, ft/s [m/s]

Subscripts

A	absolute
a	aerodynamic
c	centrifugal
b	bulk
d	droplet
f	flow
g	gravity
G	gas
I	index or impaction
l	limiting
L	liquid
p	particle
q	particle charge
Q	collector charge
r	relative
R	residual
s	surface
t	terminal
T	tangential
TR	throat
1	inlet or upstream
2	outlet or downstream
0	original
N	new

Basic Knowledge

Gas Physics

William L. Heumann

Fisher-Klosterman, Inc.
Louisville, KY

1.1 Introduction

A basic understanding of gas and particle physics is vital in understanding the operation and selection of air pollution control equipment. Without a good understanding of gas physics it is often difficult to understand the characteristics of the system in which control equipment is to be used. It is necessary to have an accurate description of the gas flow to select most air handling equipment including simple ductwork, fans, and collection equipment. The volume of gas to be handled is one of the primary criteria that will determine the size of most air pollution equipment.

The gas density will affect the pressure drop (energy consumption) required to move a given volume of a gas through ductwork and most air pollution control equipment. The viscosity of the gas will affect the collection efficiency of inertial collectors (cyclones, venturi scrubbers, etc.) and the pressure drop in media filtration devices (baghouses, cartridge collectors, etc.). For these reasons, it is important that the reader have a good working knowledge of these physical properties before endeavoring to understand equipment operation.

1.2   Boyle's Law, Charles' Law, and the Ideal Gas Law

In most systems utilizing air pollution control equipment, we are dealing with systems that include a ductwork connection to the process to transport the gases, an air moving device such as a fan, and the pollution control device(s). The properties of gases must be known at the collection equipment and fan for proper selection. The gas properties must also be known for proper selection and design of the ductwork.

Gases are compressible. For a given mass flow rate the actual volume of gas passing through the system is not constant within the system due to changes in pressure. This physical property is described by Boyle's law, which states that for a given volume of gas in a system held at constant temperature, *as pressure goes up, volume goes down,* and vice versa. Therefore, in a pneumatic system where gas is caused to move through the system by the fact that gases will flow from a zone of high pressure to that of low pressure, we will always have a greater actual volume of gas at the end of the system than at the beginning assuming temperature has remained constant. The pressure will decrease as the gas progresses downstream in a pneumatic system resulting in an increased volume. Boyle's law may be written as

$$\frac{Q_1}{Q_2} = \frac{P_2}{P_1} \tag{1.1}$$

where Q_1 = original gas volume at pressure P_1
$\quad\quad Q_2$ = new gas volume at pressure P_2
$\quad\quad P_1$ = original pressure (units for pressure must be absolute)
$\quad\quad P_2$ = new pressure (units for pressure must be absolute)

Boyle's Law Example   The gas flow rate is measured as 10,000 actual cubic feet per minute (acfm) [16,992 cubic meters per hour (m³/h)] at −7 inches water column (in wc) [−17 millimeters of water (mmH₂O)] in a pneumatic conveying system at point 1. Downstream at point 2, the pressure is measured as −32 in wc [−813 mmH₂O]. Assuming the temperature remains constant, what is the volumetric flow rate at point 2?

solution

$$\frac{Q_1}{Q_2} = \frac{P_2}{P_1}$$

Q_1 = 10,000 acfm = [16,992 m³/h]

$P_1 = \dfrac{-7}{27.67} + 14.70 = 14.45$ pounds per square inch absolute (psia)
$\quad\quad\quad\quad\quad\quad\quad\quad\quad\quad$ (converted to absolute pressure)

$$= \left[\frac{-178}{10,000} + 1.033 = 1.015 \text{ kilograms per square centimeter (kg/cm}^2) \right]$$

$$P_2 = \frac{-32}{27.67} + 14.70 = 13.54 \text{ psia (converted to absolute pressure)}$$

$$= \left[\frac{-813}{10,000} + 1.033 = 0.952 \text{ kg/cm}^2 \right]$$

$$Q_2 = \frac{P_1}{P_2} \cdot Q_1$$

$$= \frac{14.45}{13.54} \cdot Q_1 = 1.07 \cdot Q_1 = 1.07 \cdot 10,000 = 10,700 \text{ acfm}$$

$$= \left[\frac{1.015}{0.952} \cdot Q_1 = 1.07 \cdot Q_1 = 1.07 \cdot 16,992 = 18,181 \text{ m}^3/\text{h} \right]$$

Charles' law provides us with the relationship between gas volume and temperature as pressure is held constant. Charles' law states that gas volume is directly proportional to the change in the absolute temperature of the gas. In other words, *as temperature goes up, volume goes up,* and vice versa as pressure is held constant. Charles' law may be written as

$$\frac{Q_1}{Q_2} = \frac{T_1}{T_2} \tag{1.2}$$

where Q_1 = original gas volume at temperature T_1
 Q_2 = new gas volume at temperature T_2
 T_1 = original temperature (units for temperature must be absolute)
 T_2 = new temperature (units for temperature must be absolute)

Charles' Law Example   Gas flow rate is measured as 10,000 acfm [16,992 m³/h] at point 1 in a pneumatic system at a temperature of 70 degrees Fahrenheit (°F) [21 degrees Celsius (°C)]. At point 2 in the system the temperature is measured at 600°F [316°C]. Assuming the pressure remains constant, what is the volumetric flow rate at point 2?

solution

$$\frac{Q_1}{Q_2} = \frac{T_1}{T_2}$$

$$Q_2 = \frac{T_2}{T_1} \cdot Q_1$$

$$Q_1 = 10,000 \text{ acfm} = [16,992.4 \text{ m}^3/\text{h}]$$

T_1 = 70.0 + 459.7 = 529.7°R (converted to absolute temperature)

= [21.1 + 273.2 = 294 kelvins (K)]

T_2 = 600 + 459.7 = 1059.7°R (converted to absolute temperature)

= [315.6 + 273.2 = 588.8 K]

$$Q_2 = \frac{1059.7}{529.7} \cdot Q_1 = 2.0 \cdot 10{,}000 = 20{,}000 \text{ acfm}$$

$$= \left[\frac{588.8}{294.3} \cdot Q_1 = 2.0 \cdot 16{,}992.4 = 33{,}984.8 \text{ m}^3/\text{h} \right]$$

Boyle's and Charles' laws may be combined into the form

$$\frac{Q_1}{Q_2} = \frac{T_2}{T_1} \cdot \frac{P_2}{P_1}$$

Combined Form Example   Gas flow is measured as 10,000 acfm [16,992 m³/h] at point 1 in a system at 70°F [21°C] and −7 in wc [−178 mmH$_2$O]. At point 2 of the system the temperature is measured as 600°F [316°C] and the pressure as −32 in wc [−813 mmH$_2$O]. What is the volumetric flow rate at point 2?

solution

$$Q_2 = \frac{T_2}{T_1} \cdot \frac{P_1}{P_2} \cdot Q_1$$

T_1 = 530°R [294 K]

T_2 = 1060°R [589 K]

P_1 = 14.45 psia [1.015 kg/cm²]

P_2 = 13.54 psia [0.952 kg/cm²]

Q_2 = 2 · 1.07 · Q_1 = 21,400 ft³/min = [36,364 m³/h]

Most of the gases we encounter in industrial pollution control may be reasonably described by the ideal gas law. *Ideal* gases exhibit a very useful property: A mole of any ideal gas occupies the same volume as a mole of any other ideal gas at the same temperature and pressure. This property is more precisely defined by Avogadro's law which states that equal volumes of gases contain equal numbers of molecules.[1]

In English units:

2.73 · 10^{26} molecules = 1 mole (mol) [pound (lb)] any ideal gas = 386.7 ft³ @ 70°F and 29.92 inHg.

In Système International (SI) units:

$$6.02 \cdot 10^{23} \text{ molecules} = 1 \text{ mol (g) any ideal gas} = 22.4 \text{ liters (L) @ } 0°C$$
$$\text{and 760 millimeters of mercury (mmHg)}$$

From this fixed relationship we can see that if we know the volume of a gas, we immediately know how many moles of that gas are present. For example, if we have 386.7 ft^3 of an ideal gas at standard temperature and pressure (STP), then we have 386.7/386.7 or 1 mol (lb) of that gas. Within certain limitations, we do not need to know the specific gas composition to reach this conclusion. It is true for all ideal gases.

When dealing with gases of known composition, we can use this property to determine the weight of a given volume of gas. More specifically, since we are most frequently concerned with flowing gases, this property is useful in determining the density of a gas within the system.

To illustrate the use of this principle let us assume we have a flowing gas at STP (70°F and 29.92 inHg) [21°C and 760 mmHg] which is comprised by volume of 80% nitrogen (N_2) and 20% oxygen (O_2). If we examine any block of this gas that is 386.7 ft^3 [10.95 m^3] in volume, it will contain 1 mol (lb) of the mixture. We know that the weight of each gas is equal to its percentage of the total volume multiplied by its molecular weight (MW). In other words, if we had 100% nitrogen and 386.7 ft^3 [10.95 m^3] of volume at STP, we would have 28 lb [12.7 kg] of nitrogen since the molecular weight of nitrogen is 28. Unless the gas is one of the noble elements of the periodic table (argon, helium, etc.) or is already combined with other elements (H_2O, SO_2, etc.), it is likely that it will be diatomic. Diatomic means that the normal state in which it may be found in nature is as two atoms. Therefore, the gaseous form of nitrogen is N_2. The molecular weight for gaseous nitrogen is two times the atomic weight of 14 which is indicated in the periodic table or in App. 1A: Properties of the Elements.

In the example given, our gas is made up of not 100% N_2 but 80%. This means the weight of N_2 in our gas mixture is 0.8 × 28 = 22.4 lb [10.2 kg]. The weight of O_2 in our mixture is 0.2 × 32 = 6.4 lb [2.9 kg]. Therefore, the total weight of this gas mixture is 28.8 lb [13.1 kg]. This value is referred to as the apparent molecular weight (AMW) of the mixture. Density is defined as mass per unit volume. So for this gas mixture our gas density $\rho_G = 28.8/386.7 = 0.0745$ lb/ft^3 [1.19 kg/m^3]. The determination of gas density is crucial for many of the calculations required in the selection and design of pollution control equipment and systems. It is apparent from the above discussion that one must know the molecular weight of a gas before its density may be calculated. For mixtures of gases such as combustible gases, air with water vapor, and other process gases, it is recommended that the AMW be calculated based upon the actual gas mixture.

AMW Example 1 (Gas Composition by Weight) A gas has been analyzed to have the following composition:

Nitrogen (N_2)   74% by weight
Oxygen (O_2)   12% by weight
Water (H_2O)   14% by weight

What is the AMW for this gas?

solution

Gas	Weight fraction M		AMW		$F = M/\text{AMW}$
N_2	0.74	÷	28	=	0.0264
O_2	0.12	÷	32	=	0.0038
H_2O	0.16	÷	18	=	0.0089
	$\Sigma M = 1.0$				$\Sigma F = 0.0391$

$$\text{AMW} = \frac{\Sigma M}{\Sigma F} = \frac{1.0}{0.0391} = 23.87$$

AMW Example 2 (Gas Composition by Volume) A gas has been analyzed to have the following composition:

N_2 74% by volume
O_2 12% by volume
H_2O 16% by volume

What is the AMW for this gas?

solution

Gas	Volume fraction V		AMW		$G = V \times \text{AMW}$
N_2	0.74	×	28	=	20.72
O_2	0.12	×	32	=	3.84
H_2O	0.16	×	18	=	2.88
	$\Sigma V = 1.0$				$\Sigma G = 27.44$

$$\text{AMW} = \frac{\Sigma G}{\Sigma V} = \frac{27.44}{1.0} = 27.44$$

In this example, we infer a fixed volume of gas since we know the weight or volume fraction. The logic and method are identical for flowing gases since we may analyze any portion of that flowing gas to determine the properties of the rest assuming the gas is well-mixed. For convenience we may select a molar volume as defined by the ideal gas law out of the flowing mass. By Charles' law we know that as we increase temperature, the volume of gas increases. As the volume of that fixed amount of gas increases, its mass remains constant. The density of the gas, which is defined as the mass per unit volume, goes down. By Boyle's law we know that as pressure increases, volume decreases; the mass remains constant and subsequently the density increases. The opposite is also true. The general rules to remember are

Increased temperature	Volume increases, density decreases
Decreased temperature	Volume decreases, density increases
Increased pressure	Volume decreases, density increases
Decreased pressure	Volume increases, density decreases

If we incorporate Boyle's and Charles' laws into the ideal gas law, we have

In English units:

$$\rho_G = \frac{AMW}{386.7} \cdot \frac{530}{T} \cdot \frac{P}{14.7} \tag{1.3a}$$

In SI units:

$$\rho_G = \frac{AMW}{22.4} \cdot \frac{273}{T} \cdot \frac{P}{1.033} \tag{1.3b}$$

where ρ_G = gas density, lb/ft^3 [kg/m^3]
P = absolute pressure, lb/in^2 [kg/cm^2]
T = absolute temperature, °R [K]

Density Example 1 A gas with an AMW of 18 is at a temperature of 1000°F and a pressure of 40 psig (pounds per square inch gauge). What is the density of this gas?

solution

$$\rho_G = \frac{AMW}{386.7} \frac{530}{T} \frac{P}{14.7} = \left[\frac{AMW}{22.4} \frac{273}{T} \frac{P}{1.033} \right]$$

$$AMW = 18$$

$$T = 1000 + 460 = 1460°R = [538 + 273 = 811 \text{ K}]$$

$$P = 40 + 14.7 = 54.7 \text{ psia} = [2.812 + 1.033 = 3.845 \text{ kg/cm}^2]$$

$$\rho_G = \frac{18}{386.7} \times \frac{530}{1460} \times \frac{54.7}{14.7} = 0.0629 \text{ lb/ft}^3$$

$$= \left[\frac{18}{22.4} \times \frac{273}{811} \times \frac{3.845}{1.033} = 1.01 \text{ kg/m}^3 \right]$$

Density Example 2   A gas has been analyzed to have the following composition:

N_2 80% by volume

O_2 19% by volume

At a temperature of 146°F [63°C] and a pressure of −12 in wc [−305 mmH$_2$O], what is the gas density?

solution

Gas	Volume fraction, V		MW		G
N_2	0.80	×	28	=	22.4
O_2	0.19	×	32	=	6.08
	$\Sigma V = 0.99$				$\Sigma G = 28.48$

$$\text{AMW} = \frac{28.48}{0.99} = 28.77$$

$$\rho_G = \frac{\text{AMW}}{386.7}\ \frac{530}{T}\ \frac{P}{14.7} = \left[\frac{\text{AMW}}{22.4}\ \frac{273}{T}\ \frac{P}{1.033}\right]$$

$$T = 460 + 146 = 606°R = [273 + 63 = 336\ \text{K}]$$

$$P = \frac{-12}{27.67} + 14.7 = 14.27\ \text{psia} = \left[\frac{-305}{10,000} + 1.033 = 1.003\ \text{kg/cm}^2\right]$$

$$\rho_G = \frac{28.77}{386.7} \times \frac{530}{606} \times \frac{14.27}{14.7} = 0.0632\ \text{lb/ft}^3$$

$$= \left[\frac{28.77}{22.4} \times \frac{273}{336} \times \frac{1.003}{1.033} = 1.01\ \text{kg/m}^3\right]$$

Another useful form of the ideal gas law and one that is familiar to many people is

$$PV = NRT \tag{1.4}$$

where P = absolute pressure, lb/ft^2 [kg/m^2]
   T = absolute temperature, °R [K]
   V = gas volume, ft^3 [m^3]
   R = universal gas constant, 1545.32 ft · lb/[mol (lb) · °R] [847.82 m · kg/{mol (kg) · K}]
   N = moles of gas, mol (lb) [mol (kg)]

Ideal Gas Law Example   A combustion process produces 421 mol (lb) [190.9 mol (kg)] of an ideal gas per hour at 1850°F [1010°C] and −6 in wc [−152 mmH$_2$O] static pressure. What is the volumetric flow rate? *Note:* In this example, we are dealing with a generation of a volume of gas per unit of time. Therefore, we may substitute Q (volumetric flow rate) for V (volume) into Eq. (1.4):

$$PQ = \dot{N}RT \tag{1.5}$$

where P = pressure, lb/ft^3 [kg/m^2]
Q = volumetric flow rate, ft^3/min [m^3/h]
$\dot{N}$ = mol (lb)/min [mol (kg)/h]
R = universal gas constant, 1545.32 ft · lb/[mol (lb) · °R] [847.82 m · kg/{mol (kg) · K}]
T = temperature, °R [K]

solution

$$Q = \frac{\dot{N}RT}{P}$$

$$\dot{N} = \frac{421}{60} = 7.0 \text{ mol (lb)/min} = [190.9 \text{ mol (kg)/h}]$$

$$T = 1840 + 460 = 2310°R = [1010 + 273 = 1283 \text{ K}]$$

$$P = \left(\frac{-6}{27.67} + 14.7\right) \cdot 144 = 2085.6 \text{ lb/ft}^2$$

$$= \left[\left(\frac{-152}{10,000} + 1.033\right) \cdot 10,000 = 10,178 \text{ kg/m}^2\right]$$

$$Q = \frac{7.0 \cdot 1545.32 \cdot 2310}{2085.6} = 11,981 \text{ ft}^3/\text{min}$$

$$= \left[\frac{190.9 \cdot 847.82 \cdot 1283}{10,178} = 20,402 \text{ m}^3/\text{h}\right]$$

Before leaving the subject of the ideal gas law, it is important to discuss the validity of these laws since no gas behaves exactly as an ideal gas. No gas is truly an ideal gas. For most pollution control applications, use of the ideal gas laws is reasonably accurate for air, water vapor, nitrogen, oxygen, and carbon dioxide; virtually all other normal gases; and combinations thereof. As a gas approaches the liquid state, deviations from an ideal gas increase. In most pollution control applications the possibility of a gas approaching its liquid state is usually limited to the water vapor or acid gases reaching their dew points in a gas mixture. In these cases, the ideal gas law is reasonably accurate since the condensing vapor comprises a small part of the gas mixture. If very high pressures are present, resulting in the majority of gases approaching the liquid state, a more detailed calculation should be performed.[2]

1.3 Dalton's Law

Dalton's law is one of the most simple to remember although its implications and applications are sometimes not so apparent. Dalton's law states that the partial pressure exerted by any single gas in a mixture is equal to its percentage of the total. In other words, if we are examining a volume of gas that is 20% by volume O_2 and 80% by volume N_2 at a pressure of 100 psia [7.031 kg/cm^2], the partial pressure of oxygen (PO_2) is 20 psia [1.406 kg/cm^2] and PN_2 is 80 psia [5.625 kg/cm^2]. There are many interesting properties and phenomena of gases that relate to partial pressures. For instance, divers must beware of the intoxicating effect of nitrogen (nitrogen narcosis) at partial pressures greater than about 80 psia [5.625 kg/cm^2], since this can occur when breathing air at depths greater than 160 ft [50 m]. Divers performing deep dives may breathe specially mixed gases to avoid this problem. They also must contend with and prevent the possibility of oxygen poisoning at PO_2 greater than 30 psia [2.109 kg/cm^2]. These divers may breathe a mixture of oxygen and nitrogen and other inert gas(es) such as helium. The human body assimilates O_2 and reacts to CO_2 in our blood based upon partial pressures, not by the percentage composition. At extreme elevations people suffer from a shortness of breath. The air at these high elevations still has approximately 20% O_2 by volume, but the total air pressure is reduced so that the PO_2 is reduced. The normal PO_2 at sea level is about 2.9 psia [0.204 kg/cm^2], while at 14,000 ft [4361 m] elevation above sea level the PO_2 drops to about 1.9 psia [0.134 kg/cm^2] or 65% of its sea level value. It is useful to apply an understanding of partial pressures to understand phase states of gases as discussed in the following section.

1.4   The Thermodynamics of Gases and Gaseous Mixtures

To better explain how partial pressures help describe other important characteristics of gases let us look at the properties of water. Most of us are familiar with water in its three common states: solid, liquid, and gas. At normal atmospheric pressure, water is a solid below 32°F [0°C]. As the temperature of water increases, it changes from a solid state (ice) into a liquid state. It remains in a liquid state until it reaches 212°F [100°C], when the water changes to a gaseous state (steam). These temperatures are known as the freezing point and boiling point of water, respectively. Pressure also affects the points at which a substance changes state. As pressure increases, the temperature at which the freezing or boiling occurs increases. The description of these characteristics for water readily available as "steam tables" or saturation tables which are included as Appendix 1B.

In most industrial applications we are concerned with mixtures of gases. Usually these mixtures consist of air and water or are mixtures in which the air is replaced by the products of combustion. In designing equipment and systems that convey and filter these gases, it is vital to understand and properly apply an understanding of water as described above. For example, if we have a mixture of air (80% by volume) and water (20% by volume) at atmospheric pressure, 14.7 psia [1.034 kg/cm^2], and a temperature of 212°F [100°C], then the partial pressures of air and water are approximately 11.8 and 2.9 psia [0.830 and 0.204 kg/cm^2], respectively.

By examination of the saturation pressures table (see App. 1B), we can determine that at 212°F [100°C] the saturation pressure for water is 14.7 psia [1.034 kg/cm^2]. In other words, our mixture would have to be 100% water to change from a gaseous to a liquid state at these conditions. If the temperature of our gas is 100°F [38°C] though, we have a completely different condition. From the saturation temperatures table (see App. 1B), one can see that the saturation pressure for water is 0.9503 psia [0.067 kg/cm^2] at 100°F [38°C], which is less than the partial pressure of water in our mixture of 2.9 psia [0.204 kg/cm^2]. Therefore, some of the water must change state from gaseous form to liquid at these conditions. The volume of water that will undergo this phase change is an amount that will leave a remaining partial pressure of water in the gas mixture of 0.9503 psia [0.067 kg/cm^2]. If this was an actual application, the system would need to be designed to deal with the condensation that occurs or prevent its occurrence by keeping the gas temperature above the point where condensation (the change in state from a gas to a liquid) will occur. This temperature is called the *dew point*. From the saturation pressures table (see App. 1B), we can see that the saturation temperature for water at 2.9 psia [0.204 kg/cm^2] is 139.4°F (interpolate between 1 and 3 psia) [59.7°C] which for this example is the dew point. Above that temperature, condensation will not occur. If the temperature of this gas drops below 139.4°F [59.7°C], condensation will occur.

There are easily collected measurements that are useful in determining the properties of an unknown gas mixture. The most common of these measurements are the dry bulb and wet bulb temperatures. Dry bulb temperature is the temperature as measured with a conventional temperature-sensing device (thermometer, thermocouple, etc.). Wet bulb temperature is the temperature that will be measured if the sensing portion of the device is covered by a liquid film. If the gas surrounding the temperature sensor is not saturated, water will be evaporated from the surface of the sensing device, causing a reduction in the measured temperature. The wet bulb temperature is the saturation temperature for that gas mixture. From the wet bulb temperature, it is possible to determine the amount of water vapor in the gas mixture and the temperature at which condensation will occur.

When describing gas mixtures that are primarily humid air, the term *relative humidity* is often used. Relative humidity is the ratio between the partial pressure of water vapor in a gas stream and the saturation partial pressure of water vapor at the same temperature.

Relative Humidity Example Assume a gas mixture of 80% air by volume and 20% water by volume at 180°F [82°C] at an absolute pressure of 14.2 psia [0.998 kg/cm²].

$$RH = \frac{D_W}{P_{WS}} \cdot 100 \qquad (1.6)$$

where RH = relative humidity
D_W = partial pressure of water vapor, psia [kg/cm²]
P_{WS} = saturation pressure of water vapor at same temperature as D_W, psia [kg/cm²]

solution

$$D_W = 0.20 \cdot 14.2 = 2.84$$

$$P_{WS} = 7.52$$

$$RH = \frac{2.84}{7.52} \cdot 100 = 37.78\%$$

There are several other descriptions of humidity that are used within various engineering calculation tables and charts. Since it is impossible to foresee which format(s) the reader might desire or be required to use, it is important to have some familiarity with each. The absolute humidity ratio is the mass of water vapor divided by the mass of dry gas, expressed as[2]

$$H_a = \frac{M_L}{M_G} \qquad (1.7)$$

where H_a = absolute humidity ratio
M_L = mass of liquid vapor, lb [kg]
M_G = mass of dry gas, lb [kg]

The specific humidity ratio is the mass of water vapor divided by the total mass of the mixture, expressed as

$$H_u = \frac{M_L}{M_L + M_G} \qquad (1.8)$$

where H_u = specific humidity ratio
M_L = mass of liquid vapor, lb [kg]
M_G = mass of dry gas, lb [kg]

Humidity Example Assume a gas stream has the following mass flow rate:

Nitrogen (N_2)   1456 lb/h [660.4 kg/h]
Oxygen (O_2)   210 lb/h [kg/h]
Water (H_2O)   150 lb/h [68.0 kg/h]

What are the absolute and specific humidity ratios?

Solution

$$M_L = 150 \text{ lb/h} = [68.0 \text{ kg/h}]$$

$$M_G = 1666 \text{ lb/h} = [755.7 \text{ kg/h}]$$

$$H_a = \frac{150}{1666} = 0.09 = \left[\frac{68.0}{755.7} = 0.09 \right]$$

$$H_u = \frac{150}{1666 + 150} = 0.08 = \left[\frac{68}{756.7 + 68} = 0.08 \right]$$

For most substances there are conditions at which they may exist as any of three forms or states, that is, solid, liquid, or vapor. Although we are most frequently concerned with the properties of water, it is important to realize the significance of this statement to many general industrial applications.

Phase changes that occur in materials may affect the design, operation, and performance of many pollution control systems. Some common cases where such phase changes occur are systems that melt, incinerate, or vaporize a solid product. Above the molten pools of glass or metals in furnaces there are molecules of that substance in vaporized gaseous form. As the gases are ventilated from the process, they cool and condense from vapor to liquid and then to solid states.

The testing and design of systems where phase changes occur must take these properties into account. In the cases of metal and glass furnaces, we cannot measure or collect these materials with conventional particle-gas separation devices while they are in a vapor state. They will simply pass through the separators as do the rest of the gas molecules. Since particles that form from condensation and freezing of a substance previously in the vapor state may be extremely small and difficult to collect, many collection systems and devices optimize this process. By cooling the substance below the melting temperature well ahead of the collection point, it will agglomerate and form into larger particles which are easier to collect.

In pollution control systems, cooling of the gas is a technique frequently implemented to allow for the use or optimization of certain technologies. For this reason it is important that those engineers who are involved with their design and operation have a good working knowledge of the basic thermodynamic properties and characteristics of gases. The most basic of these properties is specific heat. *Specific heat* is the term given to the heat required to raise one unit of mass of a substance one degree. In English units the common unit of heat is the British thermal unit (Btu), and the most common SI unit is the calorie.

Specific heat is then defined as the amount of heat required to raise 1 lb of a substance 1°F in English units. In SI units, specific heat is defined as the mount of heat required to raise 1 g of water 1°C. Specific-heat data for various substances is readily available and may be used easily to approximate the amount of heat to be added or removed from a system to warm or cool a system. It is somewhat more difficult to use specific-heat data to calculate very accurate heat balances since the specific heat for a given substance is not constant over a range of temperatures. For example, the specific heats for air (see App. 1C) at 100°F and 1000°F are 0.2392 Btu/(lbm · °F) and 0.2486 Btu/(lbm · °F) [.2392 cal/g-°C and .2486 cal/g-°C], respectively.

Specific-Heat Example   A flowing gas stream of 1000 lb/min [453.6 kg/min] is 90% air and 10% water (by weight). The process gas flow is initially at a temperature of 1000°F [537.8°C]. By using an indirect heat exchanger, how much heat must be removed from the gas to ensure the gas temperature entering a baghouse is less than 500°F [260°C]? *Note:* Since we are looking for a *minimum* performance specification for the heat exchanger, use the maximum specific-heat values for the calculation.

solution

$$\Delta h = h_g \cdot M \tag{1.9}$$

$$h_g = \Delta T \cdot S_h$$

$$S_h = (0.9 \cdot 0.2486) + (0.1 \cdot 0.4748) = 0.2712 \text{ Btu/(lbm} \cdot °F)$$
$$[(.9 \cdot 248.6) + (.1 \cdot 474.8) = 271.2 \text{ cal/kg} - °C]$$

$$\Delta T = 1000°F - 500°F = 500°F \text{ [277.8°C]}$$

$$\Delta h = 0.2712 \text{ Btu/(lbm} \cdot °F) \cdot (500°F)(1000 \text{ lb/min}) = 135,600 \text{ Btu/min}$$
$$[271.2 \text{ cal/(kg} - °C) \cdot (277.8) \cdot (453.6 \text{ kg/min}) = 34,173,934 \text{ cal/min}]$$

Cooling, as required in the previous example, may be accomplished by a number of methods such as quenching with water, direct heat exchange, and indirect heat exchange. Another method of cooling this gas would be to mix it with a lower-temperature gas. This method is called dilution cooling.

Water has numerous valuable and interesting thermodynamic properties, but one of the most useful to air pollution control engineers is its high latent heat of vaporization. Latent heat of vaporization is the name given to the heat required to cause a substance to change state from a liquid to a gas as opposed to that required to change temperature without changing state. At atmospheric pressures water will begin to boil or rapidly vaporize at 212°F [100°C]. The temperature of the liquid water will not exceed 212°F [100°C] until all the water has been evaporated, and yet we must add significant heat to the process to keep the water boiling. The heat we are adding is providing for heat losses from our system and to vaporize the water from the liquid state.

The latent heat of vaporization of water is 971.2 Btu/lb or 539.6 calories/g (cal/g). For example, if we have 1 lb [453.6 g] of water at 60°F [156°C] that we wish to evaporate, we may calculate the heat required as follows. The heat h_L required to raise the water temperature from 60 to 212°F [15.6 to 100°C] is

$$h_L = M \cdot S_h \cdot (T_2 - T_1)$$

$$= 1 \text{ lb} \cdot 1 \text{ Btu/(lb} \cdot \text{°F)} \cdot (212\text{°F} - 60\text{°F}) = 152 \text{ Btu}$$

$$= [453.6 \text{ g} \cdot 1 \text{ cal/g} \cdot (100\text{°C} - 15.6\text{°C}) = 38{,}283.8 \text{ cal}] \qquad (1.10)$$

The latent heat of vaporization L_h is 971.2 Btu/lb [539.6 cal/g].

$$h_v = L_h M = 971.2 \text{ Btu/lb} \times 1 \text{ lb} = 971.2 \text{ Btu}$$

$$= [539.6 \text{ cal/g} \cdot 453.6 \text{ g} = 244{,}762.6 \text{ cal}]$$

and

$$h_t = h_L + h_U = 1123.2 \text{ Btu} = [283{,}046.4 \text{ cal/g}] \qquad (1.11)$$

where M = mass, lb [g]
S_h = specific heat of substance, Btu/(lb · °F) [cal/(g · °C)]
T_1 = initial temperature, °F [°C]
T_2 = final temperature, °F [°C]
h_v = latent heat required to vaporize water, Btu/lb [cal/g]
h_t = total heat, Btu [cal]

If we wish to reverse the process, the system must absorb or remove the same amount of heat from the vapor.

In the above example we used a constant specific heat for water of 1.0 Btu/lb. In fact the specific heats for most substances are not constant with temperature as may be seen from App. 1C. If a very precise calculation of thermodynamic properties is desired, we recommend the use of the enthalpy value. *Enthalpy,* or heat content, is the sum of the internal energy of a system plus the product of the pressure-volume work done on the system or

$$H = E + PV \qquad (1.12)$$

where H = enthalpy or heat content
   E = internal energy of the system
   P = pressure
   V = volume

Many tables and references are available listing enthalpy values for air, air-water mixtures, and other substances at atmospheric and nonatmospheric conditions. When available, this method of heat calculation is more accurate and is often easier to use. Using Properties of Dry Air, Water Vapor, and Saturated Air-Water Mixtures (App. 1D), we can determine the heat that must be removed from a system to cool 1 lb [453.6 g] of dry air at 155°F [68.3°C] to 120°F [48.9°C] at 14.7 psia [1.033 kg/cm²] by

$$\Delta H = M \cdot (H_{155} - H_{120})$$

$$= 1 \cdot (29.581 - 21.153) = 8.428 \text{ Btu}$$

In other words, for each pound of gas we wish to cool, we will need to remove 8.428 Btu of heat from the system.[2]

The practical value of cooling with water becomes readily apparent by example. Assume a gas flow of 30,000 acfm [50,977 m³/h] (dry air) at 1400°F [760°C] and 14.7 psia [1.033 kg/cm²]. We wish to cool this gas to the saturation temperature prior to scrubbing the gas stream. From our psychrometric charts (see App. 1E) we can see that this gas will saturate at about 156°F at an absolute humidity ratio of 0.26 lb/lb dry air. From the ideal gas law, Eq. (1.3a and b), our dry gas density is

$$\rho_G = \frac{29}{386.7} \cdot \frac{530}{1400 + 460} \cdot \frac{14.7}{14.7} = 0.0214 \text{ lb/ft}^3$$

$$= \left[\frac{29}{22.4} \cdot \frac{273}{273 + 760} \cdot \frac{1.033}{1.033} = 0.3421 \text{ kg/m}^3 \right]$$

Our mass flow rate is

$$M = (30,000 \text{ ft}^3/\text{m}) \cdot (0.0214 \text{ lb/ft}^3) = 642 \text{ lb/min}$$

$$= [(50,997 \text{ m}^3/\text{h}) \cdot (0.3421 \text{ kg/m}^3) = 17,439 \text{ kg/h}]$$

Our evaporation losses or the amount of water that will be evaporated in this process is

$$E_V = M \cdot (H_{a2} - H_{a1}) = 642 \cdot (0.26 - 0) \ 166.9 \text{ lb/min} = [4541.5 \text{ kg/h}]$$

where E_V = water evaporated, lb/min [kg/h]
   M = mass of gas to be cooled, lb/min [kg/h]
   H_{a2} = absolute humidity ratio at saturation, lb/lb dry air [kg/kg dry air]
   H_{a1} = absolute humidity ratio at initial temperature, lb/lb dry air [kg/kg dry air]

Although this may seem like a large amount of water [about 20 gallons per minute (gal/min) or 91 L/min] keep in mind that in this example we are removing about 190,000 Btu/min or 11.4 million Btu/h [2.88×10^9 cal/h] from the gas stream to drop the temperature from 1400 to 156°F [760 to 68.8°C].

1.5 Gas Viscosity

Many aspects of the behavior in multiphase systems in which one of the phases is gaseous are dependent on the viscosity of the gas. To cite two common examples, it is necessary to have a value for gas viscosity prior to determining a Reynolds number for a given set of conditions or for utilizing Stokes' law to calculate terminal velocity for particles within the system.

Viscosity is the property of a fluid or gas that describes its resistance to changes in shape or form. More specifically, viscosity is the ratio between the rate of shear and the shear stress.[3,4]

For most gases, the effect of temperature on viscosity may be expressed by Sutherland's equation

$$\eta = \eta_0 \cdot \left(\frac{T_0 + C}{T + C} \cdot \frac{T}{T_0} \right)^{1.5} \tag{1.13}$$

where η = gas viscosity at T
   η_0 = gas viscosity at T_0
   T = new gas temperature, K
   T_0 = original gas temperature, K
   C = Sutherland's constant

For most applications, it is safe to assume that gas viscosity is independent of pressure. To determine the viscosity of a gas mixture, we may arrive at a reasonable estimate by multiplying each of the component viscosities by the volume (or mole) fraction of that component and summing the products.

Since viscosity is a measured physical property, it is advisable to use accurately measured results as opposed to calculated values, when possible. An example of calculating gas viscosity for a mixture is shown below (see App. 1F).

Oxygen	20% by volume
Nitrogen	80% by volume
Temperature	23°C

Gas	Volume fraction	$\times$	viscosity [micropoise (μP)]	=	viscosity fraction
O_2	0.20	$\times$	203.9	=	40.78
N_2	0.80	$\times$	176.5	=	141.20
	$\Sigma = 1.0$				$\Sigma = 181.98$

Thus, $\eta = 181.98/1.0 = 181.98 \ \mu P$.

References

1. David Holliday and Robert Resnick, *Fundamentals of Physics,* 3d ed., John Wiley and Sons, New York, 1988, p. 485.
2. Robert Jorgensen, *Fan Engineering,* Buffalo Forge Company, Buffalo, NY, 1983, p. 1–7 (Section 1, p. 7).
3. Edward W. Washburn (ed.), *International Critical Tables of Numerical Data, Physics, Chemistry and Technology,* National Research Council and National Academy of Sciences, New York, McGraw-Hill, 1929, vol. 5, p. 1.
4. *Handbook of Chemistry and Physics,* 57th Edition, Robert C. Weast, Ph.D., (ed.), CRC Press, Cleveland, OH, 1976, p. F-124.

Properties of the Elements

Element	Symbol	Atomic Number	Atomic Weight	Density g/cm³ at 20°C	Melting point, °C	Boiling point, °C
Actinium	Ac	89	227	10.07	1051	(3200)
Aluminum	Al	13	26.98	2.699	660	2520
Americium	Am	95	243	11.7	1541	–
Antimony	Sb	51	121.76	6.69	630.8	1587
Argon	Ar	18	39.95	1.7837×10^{-3}	-189.3	-185.9
Arsenic	As	33	74.92	5.73	817 (28 at.)	613
Astatine	At	85	210	–	(302)	(337)
Barium	Ba	56	137.3	3.5	729	1805
Berkelium	Bk	97	247	–	–	–
Beryllium	Be	4	9.01	1.848	1289	2472
Bismuth	Bi	83	208.98	9.747	271.37	1564
Boron	B	5	10.81	2.34	2092	4002
Bromine	Br	35	79.90	3.12 (liquid @ 20° C)	-7.2	58.78
Cadmium	Cd	48	112.41	8.65	321.03	767
Calcium	Ca	20	40.08	1.55	842	1494
Californium	Cf	98	251	–	–	–
Carbon	C	6	12.01	2.25	(3550)	4830
Cerium	Ce	58	140.11	6.768	798	3443
Cesium	Cs	55	132.91	1.873	28.39	671
Chlorine	Cl	17	35.45	3.214×10^{-3} (0°C)	-101	-33.9
Chromium	Cr	24	52.00	7.19	1863	2672
Cobalt	Co	27	58.93	8.85	1495	2928
Copper	Cu	29	63.55	8.96	1084.87	2563
Curium	Cm	96	247	(13.51)	(1345)	–
Dysprosium	Dy	66	162.5	8.551	1412	2567
Einsteinium	Es	99	252	–	–	–
Erbium	Er	68	167.26	9.066	1529	2868
Europium	Eu	63	151.97	5.244	822	1527
Fermium	Fm	100	257	–	–	–
Fluorine	F	9	19.00	1.696×10^{-3} (0° C)	-219.62	-188.13
Francium	Fr	87	223	–	27	677
Gadolinium	Gd	64	157.25	7.901	1313	3273
Gallium	Ga	31	69.72	5.904	29.77	2205
Germanium	Ge	32	72.61	5.323	938.35	2834
Gold	Au	79	196.97	19.32	1064.43	2857
Hafnium	Hf	72	178.49	13.31	2231	4603
Helium	He	2	4.00	0.1785×10^{-3}	-272.2	-268.9
Holmium	Ho	67	164.93	8.795	1474	2700

Element	Symbol	Atomic Number	Atomic Weight	Density g/cm^3 at 20°C	Melting point, °C	Boiling point, °C
Hydrogen	H	1	1.01	0.08988 x 10^{-3}	-259.34	-252.87
Indium	In	49	114.82	7.31	156.634	2073
Iodine	I	53	126.90	4.93	113.5	184.35
Iridium	Ir	77	192.22	22.42	2447	4428
Iron	Fe	26	55.85	7.874	15381	2862
Krypton	Kr	36	83.80	3.733 x 10^{-3}	-157.37	-153.23
Lanthanum	La	57	138.91	6.145	918	3464
Lawrencium	Lw	103	(262)	–	–	–
Lead	Pb	82	207.2	11.35	327.50	1750
Lithium	Li	3	6.941	0.534	180.55	1342
Lutetium	Lu	71	174.967	9.841	1663	3402
Magnesium	Mg	12	24.31	1.74	650	1090
Manganese	Mn	25	54.94	7.43	1246	2062
Mendelevium	Md	101	258	–	–	–
Mercury	Hg	80	200.59	13.546	-38.842	357
Molybed- enum	Mo	42	95.94	10.22 (20° C)	2623	4639
Neodymium	Nd	60	144.2	7.01	1021	3074
Neon	Ne	10	20.18	0.8999 x 10^{-3}	-248.597	-246.08
Neptunium	Np	93	237	20.25	639	(3902)
Nickel	Ni	28	58.69	8.902	1455	2914
Niobium	Nb	41	92.91	8.57 (20° C)	2469	4744
Nitrogen	N	7	14.01	1.2506 x 10^{-3}	-210	-195.8
Nobelium	No	102	259	–	–	–
Osmium	Os	76	190.2	22.57	3033	5012
Oxygen	O	8	16.00	1.3318 x 10^{-3}	-218.79	-182.97
Palladium	Pd	46	106.42	12.02	1555	2964
Phosphorus	P	15	30.97	1.82	44.14	280
Platinum	Pt	78	195.08	21.45 (20° C)	1769	3827
Plutonium	Pu	94	244	19.84 (25° C)	640	3230
Polonium	Po	84	209	9.32	254	962
Potassium	K	19	39.10	0.862	63.71	759
Praseo- dymium	Pr	59	140.91	6.773	931	3520
Promethium	Pm	61	(145)	7.264	1042	–
Protactinium	Pa	91	231.04	(15.37)	1572	–
Radium	Ra	88	(226)	5.0	700	1140
Radon	Rn	86	(222)	9.73 x 10^{-3} (0° C)	(-71)	-61.7
Rhenium	Re	75	186.21	21.02	3186	5596
Rhodium	Rh	45	102.91	12.41	1963	3697
Rubidium	Rb	37	85.47	1.532	39.48	688

Element	Symbol	Atomic Number	Atomic Weight	Density g/cm^3 at 20°C	Melting point, °C	Boiling point, °C
Ruthenium	Ru	44	101.07	12.41	2334	4150
Samarium	Sm	62	150.36	7.52	1074	1794
Scandium	Sc	21	44.96	2.989	1541	2836
Selenium	Se	34	78.96	4.79	221	685
Silicon	Si	14	28.09	2.33	1414	3267
Silver	Ag	47	107.87	10.50	961.93	2163
Sodium	Na	11	22.99	0.9712	97.81	882.9
Strontium	Sr	38	87.62	2.54	769	1384
Sulfur	S	16	32.07	2.07	115.22	444.674
Tantalum	Ta	73	180.95	16.654	3020	5458
Technetium	Tc	43	98	11.50	22.04	4265
Tellurium	Te	52	127.60	6.24	449.57	988
Terbium	Tb	65	158.93	8.23	1356	3230
Thallium	Tl	81	204.38	11.85	304	1473
Thorium	Th	90	232.04	11.72	1755	4788
Thulium	Tm	69	168.93	9.321	1545	1950
Tin	Sn	50	118.71	7.31	231.9	2603
Titanium	Ti	22	47.88	4.541	1670	3289
Tungsten	W	74	183.85	19.3	3422	5555
Uranium	U	92	238.03	(18.95)	1135	4134
Vanadium	V	23	50.94	6.11	1910	3409
Xenon	Xe	54	131.29	5.887×10^{-3}	-111.76	-108
Ytterbium	Yb	70	173.04	6.903	819	1196
Zinc	Zn	30	65.39	7.133	419.58	907
Zirconium	Zr	40	91.22	6.506	1855	4409

Values in parentheses are uncertain or estimated values.

All the physical properties are given for a pressure of one atmosphere except where otherwise specified.

The data for gases is valid only when these are in their usual molecular state, such as H_2, He, O_2, Ne, etc.

Sources: 1. *Fundamentals of Physics,* David Halliday and Robert Resnick, John Wiley and Sons, New York, NY, 1988.

2. *Handbook of Chemistry and Physics,* 73rd edition, David R. Lide, CRC Press, Boca Raton, FL, 1992.

Saturation:
Pressures English Units

Pressure (psia) p	Temperature (° F) t	Specific Volume		Enthalpy		
		Saturated Liquid ft³/lbm v_f	Saturated Vapor ft³ /lbm v_g	Saturated Liquid BTU/lbm h_f	Evaporated BTU/lbm h_{fg}	Saturated Vapor BTU/lbm h_g
0.08866	32.02	.016022	3302.0	0.01	1075.4	1075.4
0.10	35.02	.016021	2946.0	3.02	1073.7	1076.7
0.20	53.15	.016027	1526.3	21.22	1063.5	1084.7
0.50	79.56	.016071	641.5	47.65	1048.6	1096.2
1.0	101.70	.016136	333.6	69.74	1036.0	1105.8
3.0	141.43	.016300	118.72	109.39	1013.1	1122.5
5.0	162.21	.016407	73.53	130.17	1000.9	1131.0
10	193.19	.016590	38.42	161.23	982.1	1143.3
14.696	211.99	.016715	26.80	180.15	970.4	1150.5
20	227.96	.016830	20.09	196.26	960.1	1156.4
30	250.34	.017004	13.748	218.93	945.4	1164.3
40	267.26	.017146	10.501	236.16	933.8	1170.0
50	281.03	.017269	8.518	250.24	924.2	1174.4
60	292.73	.017378	7.177	262.25	915.8	1178.0
70	302.96	.017478	6.209	272.79	908.3	1181.0
80	312.07	.017570	5.474	282.21	901.4	1183.6
90	320.31	.017655	4.898	290.76	895.1	1185.9
100	327.86	.017736	4.434	298.61	889.2	1187.8
120	341.30	.017886	3.730	312.67	878.5	1191.1
140	353.08	.018024	3.221	325.05	868.7	1193.8
160	363.60	.018152	2.836	336.16	859.8	1196.0
180	373.13	.018273	2.533	346.29	851.5	1197.8
200	381.86	.018387	2.289	355.6	843.7	1199.3
250	401.04	.018653	1.8448	376.2	825.8	1202.1
300	417.43	.018896	1.5442	394.1	809.8	1203.9
400	444.70	.019340	1.1620	424.2	781.2	1205.5
500	467.13	.019748	.9283	449.5	755.8	1205.3
600	486.33	.02013	.7702	471.7	732.4	1204.1
800	518.36	.02087	.5691	509.7	689.6	1199.3
1000	544.75	.02159	.4459	542.4	650.0	1192.4
1500	596.39	.02346	.2769	611.5	557.2	1168.7
2000	636.00	.02565	.18813	671.9	464.4	1136.3
2500	668.31	.02860	.13059	730.9	360.5	1091.4
3000	695.52	.03431	.08404	802.5	213.0	1015.5
3203.6	705.44	.05053	.05053	902.5	0	902.5

Pressure (bars) p	Temperature (° C) t	Specific Volume		Enthalpy		
		Saturated Liquid cm³/gm v_f	Saturated Vapor cm³/gm v_g	Saturated Liquid j/gm h_f	Evaporated j/gm h_{fg}	Saturated Vapor j/gm h_g
0.006113	0.01	1.0002	206.136	0.001	2501.3	2501.4
0.010	6.98	1.0002	129.208	29.30	2484.9	2514.2
0.020	17.50	1.0013	67.004	73.48	2460.0	2533.5
0.060	36.16	1.0064	23.739	151.53	2415.9	2567.4
0.10	45.81	1.0102	14.674	191.83	2392.8	2584.7
0.20	60.06	1.0172	7649.0	251.40	2358.3	2609.7
0.60	85.94	1.0331	2732.0	359.86	2293.6	2653.5
1.00	99.63	1.0432	1694.0	417.46	2258.0	2675.5
1.50	111.37	1.0528	1159.3	467.11	2226.5	2693.6
2.00	120.23	1.0605	885.7	504.70	2201.9	2706.7
2.50	127.44	1.0672	718.7	535.37	2181.5	2716.9
3.00	133.55	1.0732	605.8	561.47	2163.8	2725.3
3.50	138.88	1.0786	524.3	584.33	2148.1	2732.4
4.0	143.63	1.0836	462.5	604.74	2133.8	2738.6
6.0	158.85	1.1006	315.7	670.56	2086.3	2756.8
8.0	170.43	1.1148	240.4	721.11	2048.0	2769.1
10.0	179.91	1.1273	194.44	762.81	2015.3	2778.1
12.0	187.99	1.1385	163.33	798.65	1986.2	2784.8
14.0	195.07	1.1489	140.84	830.30	1959.7	2790.0
16.0	201.41	1.1587	123.80	858.79	1935.2	2794.0
18.0	207.15	1.1679	110.42	884.79	1912.4	2797.1
20	212.42	1.1767	99.63	908.79	1890.7	2799.5
30	233.90	1.2165	66.68	1008.42	1795.7	2804.2
40	250.40	1.2522	49.78	1087.31	1714.1	2801.4
60	275.64	1.3187	32.44	1213.35	1571.0	2784.3
80	295.06	1.3842	23.52	1316.64	1441.3	2758.0
100	311.06	1.4524	18.026	1407.56	1317.1	2724.7
120	324.75	1.5267	14.263	1491.3	1193.6	2684.9
160	347.44	1.7107	9.306	1650.1	930.6	2580.6
200	365.81	2.036	5.834	1826.3	583.4	2409.7
220	373.80	2.742	3.568	2022.2	143.4	2165.6
220.9	374.14	3.155	3.155	2099.3	0	2099.3

Source: Thermodynamics 1: An Introduction to Energy, John R. Dixon, Prentice-Hall, Inc., Englewood Cliffs, NJ, 1975.

Saturation:
Temperatures English Units

Temperature (° F) t	Pressure (psia) p	Specific Volume		Enthalpy		
		Saturated Liquid ft³/lbm v_f	Saturated Vapor ft³ /lbm v_g	Saturated Liquid BTU/lbm h_f	Evaporated BTU/lbm h_{fg}	Saturated Vapor BTU/lbm h_g
32.018	.08866	.016022	3302	.01	1075.4	1075.4
40	.12166	.016020	2445	8.03	1070.9	1078.9
50	.17803	.016024	1704.2	18.06	1065.2	1083.3
60	.2563	.016035	1206.9	28.08	1059.6	1087.7
70	.3632	.016051	867.7	38.09	1054.0	1092.0
80	.5073	.016073	632.8	48.09	1048.3	1096.4
90	.6988	.016099	467.7	58.07	1042.7	1100.7
100	.9503	.016130	350.0	68.05	1037.0	1105.0
110	1.2763	.016166	265.1	78.02	1031.3	1109.3
120	1.6945	.016205	203.0	88.00	1025.5	1113.5
130	2.225	.016247	157.17	97.98	1019.8	1117.8
140	2.892	.016293	122.88	107.96	1014.0	1121.9
150	3.722	.016343	96.99	117.96	1008.1	1126.1
160	4.745	.016395	77.23	127.96	1002.2	1130.1
170	5.996	.016450	62.02	137.97	996.2	1134.2
180	7.515	.016509	50.20	147.99	990.2	1138.2
190	9.343	.016570	40.95	158.03	984.1	1142.1
200	11.529	.016634	33.63	168.07	977.9	1145.9
210	14.125	.016702	27.82	178.14	971.6	1149.7
220	17.188	.016772	23.15	188.22	965.3	1153.5
240	24.97	.016922	16.327	208.44	952.3	1160.7
260	35.42	.017084	11.768	228.76	938.8	1167.6
280	49.18	.017259	8.650	249.18	924.9	1174.1
300	66.98	.017448	6.742	269.73	910.4	1180.2
400	247.1	.018638	1.8661	375.12	826.8	1202.0
450	422.1	.019433	1.1011	430.2	775.4	1205.6
500	680.0	.02043	0.6761	487.7	714.8	1202.5
550	1044.0	.02175	0.4249	549.1	641.6	1190.6
600	1541.0	.02363	0.2677	616.7	549.7	1166.4
650	2205.0	.02673	0.16206	695.9	423.9	1119.8
700	3090.0	.03666	0.07438	822.7	167.5	990.2
705.44	3204.0	.05053	0.05053	902.5	0	902.5

Temperature (°C) t	Pressure (bars) p	Specific Volume		Enthalpy		
		Saturated Liquid cm³/gm v_f	Saturated Vapor cm³/gm v_g	Saturated Liquid j/gm h_f	Evaporated j/gm h_{fg}	Saturated Vapor j/gm h_g
0.01	0.006113	1.0002	206.136	0.01	2501.3	2501.4
1	0.006567	1.0002	192.577	4.16	2499.0	2503.2
10	0.012276	1.0004	106.379	42.01	2477.7	2519.8
20	0.02339	1.0018	57.791	83.96	2454.1	2538.1
30	0.04246	1.0043	32.894	125.79	2430.5	2556.3
40	0.07384	1.0078	19.523	167.57	2406.7	2574.3
50	0.12349	1.0121	12.032	209.33	2382.7	2592.1
60	0.19940	1.0172	7671.0	251.13	2358.5	2609.6
70	0.3119	1.0228	5042.0	292.98	2333.8	2626.8
80	0.4739	1.0291	3407.0	334.91	2308.8	2643.7
90	0.7014	1.0360	2361.0	376.92	2283.2	2660.1
100	1.0135	1.0435	1672.9	419.04	2257.0	2676.1
110	1.4327	1.0516	1210.2	461.30	2230.2	2691.5
120	1.9853	1.0603	891.9	503.71	2202.6	2706.3
130	2.701	1.0697	668.5	546.31	2174.2	2720.5
140	3.613	1.0797	508.9	589.13	2144.7	2733.9
150	4.758	1.0905	392.8	632.20	2114.3	2746.5
160	6.178	1.1020	307.1	675.55	2082.6	2758.1
170	7.917	1.1143	242.8	719.21	2049.5	2768.7
180	10.021	1.1274	194.05	763.22	2015.0	2778.2
190	12.544	1.1414	156.54	807.62	1978.8	2786.4
200	15.538	1.1565	127.36	852.45	1940.7	2793.2
220	23.18	1.1900	86.19	943.62	1858.5	2802.1
240	33.44	1.2291	59.76	1037.32	1766.5	2803.8
260	46.88	1.2755	42.21	1134.37	1662.5	2796.9
280	64.12	1.3321	30.17	1235.99	1543.6	2779.6
300	85.81	1.4036	21.67	1344.0	1404.9	2749.0
320	112.74	1.4988	15.488	1461.5	1238.6	2700.1
340	145.86	1.6379	10.797	1594.2	1027.9	2622.0
360	186.51	1.8925	6.945	1760.5	720.5	1.1379
374.136	220.9	3.155	3.155	2099.3	0	2099.3

Source: Thermodynamics 1: An Introduction to Energy, John R. Dixon, Prentice-Hall, Inc., Englewood Cliffs, NJ, 1975.

Specific Heats of Various Gases (Btu/lbm−°R)

TEMPERATURE		AIR	N₂	O₂	H₂O	CO₂	CO	H₂
°R	°F							
100.0	-359.7	0.2392	-	-	-	-	-	-
200.0	-259.7	0.2392	0.2480	0.2173	0.4411	0.1589	0.2481	2.7599
300.0	-159.7	0.2392	0.2480	0.2173	0.4415	0.1674	0.2481	3.0957
400.0	-59.7	0.2393	0.2481	0.2174	0.4421	0.1815	0.2482	3.2961
500.0	40.3	0.2396	0.2481	0.2184	0.4439	0.1964	0.2483	3.3948
600.0	140.3	0.2403	0.2485	0.2206	0.4473	0.2100	0.2487	3.4355
700.0	240.3	0.2416	0.2491	0.2239	0.4527	0.2221	0.2498	3.4509
800.0	340.3	0.2434	0.2503	0.2278	0.4592	0.2326	0.2517	3.4598
900.0	440.3	0.2458	0.2521	0.2321	0.4667	0.2421	0.2541	3.4648
1000.0	540.3	0.2486	0.2546	0.2363	0.4748	0.2507	0.2570	3.4692
1100.0	640.3	0.2516	0.2573	0.2404	0.4833	0.2584	0.2603	3.4742
1200.0	740.3	0.2547	0.2603	0.2442	0.4919	0.2654	0.2638	3.4811
1300.0	840.3	0.2579	0.2635	0.2476	0.5008	0.2716	0.2673	3.4906
1400.0	940.3	0.2611	0.2668	0.2507	0.5099	0.2773	0.2707	3.5025
1500.0	1040.3	0.2642	0.2700	0.2534	0.5191	0.2824	0.2741	3.5169
1600.0	1140.3	0.2671	0.2731	0.2560	0.5285	0.2870	0.2773	3.5352
1700.0	1240.3	0.2698	0.2760	0.2583	0.5380	0.2913	0.2803	3.5560
1800.0	1340.3	0.2725	0.2789	0.2604	0.5476	0.2951	0.2830	3.5789
1900.0	1440.3	0.2750	0.2816	0.2622	0.5570	0.2986	0.2856	3.6032
2000.0	1540.3	0.2773	0.2842	0.2638	0.5663	0.3017	0.2881	3.6290

Sources:
1. Fan Engineering, Robert Jorgensen, Buffalo Forge Co., Buffalo, NY., 1982.
2. Illustration courtesy of Fisher-Klosterman, Inc., Louisville, KY.

Properties of Dry Air, Water Vapor, and Saturated Air-Water Vapor Mixtures; Temperature Range, −10 to + 211°F; Pressure, 14.7 psia

Temp. °F	Saturation Pressure	Saturation humidity, Wt. water vapor/lb. dry air	Saturation moisture content, Wt. water vapor/cu. ft. sat. mixture	Saturation density Wt. air plus water vapor/cu. ft. sat. mixture	Volume			Enthalpy		
	Lb./sq. in	Pounds	Pounds	Pounds	Dry air cu. ft./lb.	Water vapor, cu. ft./lb.	Saturated mixture cu. ft./lb. dry air	Dry air B.T.U. /lb.	Water vapor, B.T.U. /lb.	Saturated mixture, B.T.U./lb. dry air
-10	0.013595	.0005759	.00005080	0.08826	11.326	18.228	11.336	-10.086	1056.5	-9.478
-9	0.014281	.0006050	.00005324	0.08806	11.351	18.269	11.362	-9.846	1056.9	-9.207
-8	0,014997	.0006353	.00005579	0.08787	11.376	18.309	11.388	-9.606	1057.4	-8.934
-7	0.015745	.0006670	.00005844	0.08767	11.402	18.350	11.414	-9.366	1057.8	-8.660
-6	0.016526	.0007002	.00006121	0.08747	11.427	18.390	11.440	-9.126	1058.3	-8.385
-5	0.017343	.0007348	.00006409	0.08727	11.452	18.431	11.466	-8.886	1058.7	-8.108
-4	0.018196	.0007710	.00006709	0.08709	11.478	18.471	11.492	-8.646	1059.2	-7.829
-3	0.019085	.0008088	.00007022	0.08689	11.503	18.512	11.518	-8.406	1059.6	-7.549
-2	0.020014	.0008482	.00007347	0.08670	11.528	18.552	11.544	-8.165	1060.0	-7.266
-1	0.020982	.0008893	.00007686	0.08651	11.553	18.593	11.570	-7.925	1060.5	-6.982
0	0.021994	.0009322	.00008039	0.08632	11.579	18.634	11.596	-7.685	1060.9	-6.696
1	0.023048	.0009769	.00008406	0.08613	11.604	18.674	11.622	-7.445	1061.4	-6.408
2	0.024148	.0010236	.00008788	0.08594	11.629	18.715	11.649	-7.205	1061.8	-6.118
3	0.025294	.0010723	.00009185	0.08575	11.655	18.755	11.675	-6.965	1062.3	-5.826
4	0.026488	.0011230	.00009598	0.08556	11.680	18.796	11.701	-6.725	1062.7	-5.531
5	0.027733	.001176	.0001003	0.08537	11.705	18.836	11.727	-6.485	1063.1	-5.234
6	0.029039	.001231	.0001047	0.08518	11.731	18.877	11.754	-6.244	1063.6	-4.935
7	0.030384	.001289	.0001094	0.08500	11.756	18.917	11.780	-6.004	1064.0	-4.633
8	0.031790	.001348	.0001142	0.08481	11.781	18.958	11.807	-5.764	1064.5	-4.329
9	0.033253	.001411	.0001192	0.08463	11.806	18.998	11.833	-5.524	1064.9	-4.022
10	0.034779	.001475	.0001244	0.08444	11.832	19.039	11.860	-5.284	1065.4	-3.712
11	0.036367	.001543	.0001298	0.08426	11.857	19.079	11.886	-5.044	1065.8	-3.399
12	0.038026	.001613	.0001354	0.08405	11.882	19.120	11.913	-4.804	1066.2	-3.083
13	0.039735	.001686	.0001412	0.08389	11.908	19.161	11.940	-4.563	1066.7	-2.765
14	0.041522	.001762	.0001473	0.08371	11.933	19.201	11.967	-4.323	1067.3	-2.443

Temp. °F	Saturation Pressure	Saturation humidity, Wt. water vapor/lb. dry air	Saturation moisture content, Wt. water vapor/cu. ft. sat. mixture	Saturation density Wt. air plus water vapor/cu. ft. sat. mixture	Volume			Enthalpy		
	Lb./sq. in.	Pounds	Pounds	Pounds	Dry air cu. ft./lb.	Water vapor cu. ft./lb.	Saturated mixture cu. ft./lb. dry air	Dry air B.T.U. /lb.	Water vapor, B.T.U. /lb.	Saturated mixture, B.T.U./lb. dry air
15	0.43384	.001842	.0001537	0.08360	11.958	19.242	11.984	-4.083	1067.6	-2.1171
16	0.045316	.001924	.0001600	0.08335	11.983	19.282	12.021	-3.843	1088.0	-1.7882
17	0.047325	.002009	.0001668	0.08317	12.009	19.322	12.048	-3.603	1068.5	-1.4558
18	0.049411	.002098	.0001738	0.08299	12.034	19.363	12.075	-3.363	1068.9	-1.1197
19	0.051583	.002191	.0001810	0.08281	12.059	19.403	12.102	-3.122	1069.3	-0.7797
20	0.053838	.002287	.0001885	0.08264	12.085	19.444	12.129	-2.882	1069.8	-0.43572
21	0.056179	.002387	.0001963	0.08246	12.110	19.485	12.156	-2.642	1079.2	-0.08768
22	0.058606	.002490	0002044	0.08228	12.135	19.525	12.184	-2.402	1070.7	0.26436
23	0.061133	.002598	.0002128	0.08210	12.160	19.566	12.211	-2.162	1071.1	0.62115
24	0.063751	.002710	.0002214	0.08194	12.186	19.606	12.239	-1.922	1071.6	0.98221
25	0.066474	.002826	.0002304	0.08175	12.211	19.647	12.267	-1.6813	1072.0	1.348
26	0.069295	.002947	.0002397	0.08158	12.236	19.687	12.294	-1.4411	1072.5	1.719
27	0.072225	.003072	.0002493	0.08140	12.262	19.728	12.322	-1.2010	1072.9	20.95
28	0.075263	.003202	.0002592	0.08123	12.287	19.769	12.350	-0.9608	1073.3	2.476
29	0.078417	.003337	.0002695	0.08106	12.312	19.809	12.378	-0.7206	1073.8	20862
30	0.081684	.003476	.0002802	0.08089	12.337	19.850	12.405	-04804	1074.2	3.254
31	0.085072	.003621	.0002912	0.08071	12.363	19.890	12.435	-0.2402	1074.7	3.651
32	0.088579	.003787	.0003038	0.08055	12.388	19.918	12.463	0.0000	1075.1	4.071
33	0.092218	.003943	.0003157	0.08037	12.413	19.959	12.492	0.2402	1075.5	4.481
34	0.096000	.004106	.0003280	0.08020	12.439	20.000	12.520	0.4804	1076.0	4.898
35	0.09991	.004274	.0003406	0.08003	12.464	20.040	12.549	0.7206	1076.4	5.321
36	0.10396	.004449	.0003537	0.07986	12.489	20.080	12.578	0.9608	1076.9	5.751
37	0.10815	.004629	.0003672	0.07969	12.514	20.121	12.607	1.2010	1077.3	6.188
38	0.11250	.004817	.0003812	0.07969	12.540	20.162	12.637	1.4412	1077.8	6.633
39	0.11699	.005011	.0003956	0.07935	12.565	20.203	12.666	1.6814	1078.2	7.084

Temp. °F	Saturation Pressure	Saturation humidity, Wt. water vapor/lb. dry air	Saturation moisture content, Wt. water vapor/cu. ft. sat. mixture	Saturation density Wt. air plus water vapor/cu. ft. sat. mixture	Volume			Enthalpy		
	Lb./sq. in.	Pounds	Pounds	Pounds	Dry air cu. ft./lb.	Water vapor cu. ft./lb.	Saturated mixture cu. ft./lb. dry air	Dry air B.T.U. /lb.	Water vapor, B.T.U. /lb.	Saturated mixture, B.T.U./lb. dry air
40	0.12164	.005212	.0004105	0.07918	12.590	20.243	12.695	1.9216	1078.7	7.544
41	0.12646	.005420	.0004259	0.07901	12.615	20.284	12.725	2.1618	1079.1	8.010
42	0.13144	.005635	.0004418	0.07884	12.641	20.324	12.755	2.4020	1079.5	8.485
43	0.13660	.005859	.0004582	0.07867	12.666	20.365	12.785	2.6422	1080.0	8.969
44	0.014194	.006090	.0004752	0.07851	12.691	20.404	12.815	2.8824	1080.4	9.462
45	0.14745	.006329	.0004927	0.07834	12.717	20.445	12.846	3.1226	1080.9	9.963
46	0.15316	.006576	.0005107	0.07817	12.742	20.485	12.876	3.3628	1081.3	10.474
47	0.15906	.006833	.0005294	0.07801	12.767	20.525	12.907	3.6030	1081.8	10.995
48	0.16516	.007098	.0005486	0.07784	12.792	20.565	12.938	3.8432	1082.2	11.505
49	0.17146	.007369	.0005682	0.07767	12.818	20.605	12.970	4.0834	1082.7	12.062
50	0.17798	.007655	.0005888	0.07750	12.843	20.646	13.002	4.3238	1083.1	12.615
51	0.18471	.007948	.0006099	0.07734	12.868	20.686	13.033	4.5639	1083.5	13.176
52	0.19167	.008252	.0006316	0.07717	12.894	20.726	13.065	4.8041	1084.0	13.749
53	0.19885	.008565	.0006540	0.07701	12.919	20.766	13.097	5.0444	1084.4	14.332
54	0.20627	.008889	.0006771	0.07684	12.944	20.807	13.129	5.2846	1084.9	14.929
55	0.21394	.009225	.0007009	0.07668	12.969	20.847	13.162	5.5248	1085.3	15.536
56	0.22185	.009571	.0007254	0.07651	12.995	20.887	13.195	5.7650	1085.8	16.157
57	0.23002	.009929	.0007506	0.07635	13.020	20.927	13.228	6.0053	1086.2	16.790
58	0.23845	.010299	.0007766	0.07618	13.045	20.967	13.262	6.2456	1086.7	17.438
59	0.24716	.010681	.0008034	0.07602	13.070	21.006	13.295	6.4858	1087.1	18.097
60	0.25614	.01108	.0008310	0.07586	13.096	21.046	13.329	6.7260	1087.5	18.771
61	0.26541	.01149	.0008595	0.07569	13.121	21.087	13.363	6.9664	1087.9	19.461
62	0.27497	.01191	.0008886	0.07553	13.146	21.126	13.398	7.2067	1088.3	20.164
63	0.28483	.01234	.0009188	0.07536	13.172	21.166	13.433	7.4469	1088.8	20.885
64	0.29500	.01279	.0009498	0.07520	13.197	21.208	13.468	7.6872	1089.2	21.620

Temp. °F	Saturation Pressure	Saturation humidity, Wt. water vapor/lb. dry air	Saturation moisture content, Wt. water vapor/cu. ft. sat. mixture	Saturation density Wt. air plus water vapor/cu. ft. sat. mixture	Volume			Enthalpy		
	Lb./sq. in.	Pounds	Pounds	Pounds	Dry air cu. ft./lb.	Water vapor cu. ft./lb.	Saturated mixture cu. ft./lb. dry air	Dry air B.T.U. /lb.	Water vapor, B.T.U. /lb.	Saturated mixture, B.T.U./lb . dry air
65	0.30549	.01326	.0009816	0.07503	13.222	21.247	13.504	7.9275	1089.7	22.373
66	0.31630	.01374	.0010144	0.07487	13.247	21.287	13.540	8.1679	1090.1	23.140
67	0.32744	.01423	.0010479	0.07469	13.273	21.327	13.580	8.4081	1090.5	23.926
68	0.33893	.01423	.0010829	0.07454	13.298	21.367	13.613	8.6484	1090.9	24.729
69	0.35077	.01527	.0011186	0.07438	13.323	21.407	13.650	8.8888	1091.3	25.552
70	0.36297	.01581	.001155	0.07421	13.348	21.447	13.688	9.1290	1091.7	26.392
71	0.37554	.01638	.001193	0.07405	13.374	21.486	13.726	9.3694	1092.2	27.255
72	0.38848	.01700	.001232	0.07389	13.399	21.527	13.764	9.6098	1092.7	28.137
73	0.40182	.01755	.001272	0.07372	13.424	21.566	13.803	9.8500	1093.1	29.037
74	0.41556	.01817	.001313	0.07356	13.450	21.606	13.842	10.0904	1093.6	29.962
75	0.42969	.01881	.001355	0.07339	13.475	21.646	13.882	10.331	1094.0	30.907
76	0.44425	.01946	.001398	0.07323	13.500	21.685	13.922	10.571	1094.4	31.865
77	0.45923	.02014	.001443	0.07306	13.525	21.725	13.963	10.812	1094.9	32.865
78	0.47467	.02084	.001488	0.07290	13.551	21.765	14.004	11.052	1095.3	33.880
79	0.49055	.02156	.001535	0.07273	13.576	21.804	14.046	11.292	1095.7	34.920
80	0.50689	.02231	.001583	0.07257	13.601	21.844	14.088	11.533	1096.2	35.985
81	0.52370	.02308	.001645	0.07240	13.626	21.884	14.131	11.773	1096.6	37.077
82	0.54099	.02387	.001684	0.07223	13.652	21.923	14.175	12.013	1097.1	38.197
83	0.55878	.02468	.001736	0.07206	13.677	21.964	14.219	12.254	1097.5	39.342
84	0.57707	.02552	.001789	0.07190	13.702	22.003	14.264	12.494	1097.9	40.515
85	0.59588	.02639	.001844	0.07173	13.728	22.043	14.309	12.735	1098.3	41.718
86	0.61522	.02728	.001901	0.07156	13.753	22.082	14.355	12.975	1098.7	42.952
87	0.63510	.02821	.001958	0.07139	13.778	22.122	14.402	13.217	1099.1	44.216
88	0.65555	.02916	.002018	0.07123	13.803	22.161	14.449	13.457	1099.6	45.515
89	0.67656	.03014	.002079	0.07106	13.829	22.200	14.497	13.696	1100.0	46.845

Temp. °F	Saturation Pressure	Saturation humidity, Wt. water vapor/lb. dry air	Saturation moisture content, Wt. water vapor/cu. ft. sat. mixture	Saturation density Wt. air plus water vapor/cu. ft. sat. mixture	Volume			Enthalpy		
	Lb./sq. in.	Pounds	Pounds	Pounds	Dry air cu. ft./lb.	Water vapor cu. ft./lb.	Saturated mixture cu. ft./lb. dry air	Dry air B.T.U. /lb.	Water vapor, B.T.U. /lb.	Saturated mixture, B.T.U./lb. dry air
90	0.69816	.03115	.002141	0.07088	13.854	22.240	14.547	13.938	1100.5	48.212
91	0.72036	.03219	.002205	0.07071	13.879	22.279	14.597	14.177	1100.9	49.612
92	0.74316	.03326	.002271	0.07054	13.904	22.319	14.647	14.418	1101.4	51.050
93	0.76659	.03437	.002338	0.07037	13.930	22.358	14.699	14.658	1101.8	52.522
94	0.79065	.03551	.002407	0.07020	13.955	22.398	14.751	14.899	1102.3	54.037
95	0.81537	.03668	.002478	0.07003	13.980	22.437	14.804	15.139	1102.7	55.586
96	0.84074	.03789	.002550	0.06985	14.006	22.476	14.854	15.380	1103.1	57.179
97	0.86681	.03914	.002624	0.06968	14.031	22.515	14.913	15.620	1103.5	58.810
98	0.89358	.04043	.002701	0.06951	14.056	22.555	14.968	15.860	1103.9	60.486
99	0.92105	.04175	.002779	0.06933	14.081	22.594	15.025	16.101	1104.3	62.209
100	0.94926	.04312	.002859	0.06916	14.107	22.633	15.083	16.341	1104.8	63.980
101	0.97821	.04453	.002941	0.06898	14.132	22.673	15.142	16.582	1105.2	65.794
102	1.00792	.04498	.003025	0.06880	14.157	22.712	15.202	16.822	1105.6	67.657
103	1.03842	.04748	.003111	0.06863	14.182	22.751	15.263	17.063	1106.1	69.577
104	1.06965	.04902	.003198	0.06845	14.208	22.790	15.325	17.304	1160.5	71.541
105	1.1018	.05061	.003289	0.06827	14.233	22.829	15.389	17.544	1106.9	73.563
106	1.1347	.05225	.003381	0.06809	14.258	22.868	15.453	17.785	1107.3	75.639
107	1.1685	.05394	.003476	0.06791	14.283	22.907	15.519	18.025	1107.7	77.771
108	1.2031	.05568	.003572	0.06773	14.309	22.946	15.587	18.266	1108.1	79.967
109	1.2386	.05747	.003671	0.06755	14.334	22.985	15.655	18.506	1108.5	82.215
110	1.2750	.05932	.003772	0.06737	14.359	23.024	15.725	18.747	1109.0	84.535
111	1.3123	.06123	.003876	0.06718	14.384	23.063	15.796	18.987	1109.4	86.915
112	1.3506	.06319	.003982	0.06700	14.410	23.101	15.869	19.228	1109.8	89.360
113	1.3897	.06522	.004090	0.06681	14.435	23.140	15.944	19.469	1110.2	91.873
114	1.4300	.06731	.004202	0.06662	14.460	23.178	18.020	19.709	1110.6	94.461

Temp. °F	Saturation Pressure	Saturation humidity, Wt. water vapor/lb. dry air	Saturation moisture content, Wt. water vapor/cu. ft. sat. mixture	Saturation density Wt. air plus water vapor/cu. ft. sat. mixture	Volume			Enthalpy		
	Lb./sq. in.	Pounds	Pounds	Pounds	Dry air cu. ft./lb.	Water vapor cu. ft./lb.	Saturated mixture cu. ft./lb. dry air	Dry air B.T.U. /lb.	Water vapor, B.T.U. /lb.	Saturated mixture, B.T.U./lb. dry air
115	1.4711	.06946	.004315	0.06643	14.486	23.218	16.098	19.950	1111.1	97.128
116	1.5133	.07168	.004431	0.06624	14.516	23.256	16.178	20.190	1111.5	99.866
117	1.5566	.07397	.004550	0.06805	14.536	23.295	16.259	20.431	1112.0	102.688
118	1.6008	.07633	.004671	0.06586	14.561	23.333	16.343	20.672	1112.4	105.586
119	1.6462	.07877	.004795	0.06567	14.587	23.372	16.428	20.912	1112.8	108.571
120	1.6927	.08128	.004922	0.06547	14.612	23.408	16.515	21.153	1113.3	111.65
121	1.7403	.08388	.005052	0.06528	14.637	23.448	16.603	21.394	1113.7	114.81
122	1.7890	.08655	.005184	0.06508	14.662	23.487	16.695	21.634	1114.2	118.07
123	1.8389	.08931	.005320	0.06488	14.688	23.526	16.789	21.875	1114.6	121.42
124	1.8900	.09216	.005458	0.06468	14.713	23.565	16.885	22.116	1115.0	124.88
125	1.9423	.09511	.005600	0.06448	14.738	23.602	16.983	22.356	1115.4	128.44
126	1.9958	.09815	.005745	0.06428	14.763	23.641	17.084	22.597	1115.8	132.11
127	2.0506	.10129	.005893	0.06408	14.789	23.679	17.187	22.838	1116.3	135.91
128	2.1066	.10453	.006045	0.6387	14.814	23.718	17.293	23.079	1116.7	139.81
129	2.1640	.10788	.006199	0.06366	14.839	23.756	17.402	23.319	1117.1	143.83
130	2.2227	.1113	.006357	0.06345	14.864	23.794	17.514	23.560	1117.6	147.99
131	2.2828	.1149	.006519	0.06325	14.890	23.833	17.628	23.801	1118.0	152.27
132	2.3442	.1186	.006683	0.06303	14.915	23.871	17.746	24.041	1118.4	156.64
133	2.4072	.1224	.006851	0.06282	14.941	23.910	17.867	24.282	1118.8	161.23
134	2.4715	.1264	.007024	0.06261	14.965	23.948	17.991	24.523	1119.2	165.95
135	2.5373	.1304	.007199	0.06239	14.991	23.987	18.119	24.764	1119.6	170.79
136	2.6045	.1346	.007377	0.06217	15.016	24.025	18.251	25.005	1120.1	175.81
137	2.6733	.1390	.007561	0.06195	15.041	24.063	18.386	25.245	1120.5	181.01
138	2.7436	.1435	.007747	0.06173	15.067	24.103	18.525	25.486	1120.9	186.35
139	2.8155	.1482	.007937	0.06150	15.092	24.139	18.669	25.727	1121.3	191.88

Temp. °F	Saturation Pressure	Saturation humidity, Wt. water vapor/lb. dry air	Saturation moisture content, Wt. water vapor/cu. ft. sat. mixture	Saturation density Wt. air plus water vapor/cu. ft. sat. mixture	Volume			Enthalpy		
	Lb./sq. in.	Pounds	Pounds	Pounds	Dry air cu. ft./lb.	Water vapor cu. ft./lb.	Saturated mixture cu. ft./lb. dry air	Dry air B.T.U. /lb.	Water vapor, B.T.U. /lb.	Saturated mixture, B.T.U./lb. dry air
140	2.8890	.1530	.008131	0.06128	15.117	24.178	18.816	25.968	1121.7	197.59
141	2.9641	.1580	.008329	0.06105	15.142	24.215	18.969	26.209	1122.1	203.50
142	3.0409	.1632	.008532	0.06082	15.168	24.253	19.126	26.449	1122.5	209.62
143	3.1193	.1685	.008738	0.06058	15.193	24.290	19.288	26.690	1122.9	215.94
144	3.1915	.1741	.008949	0.06035	15.218	24.329	19.454	26.931	1123.3	222.49
145	3.2814	.1798	.009163	0.06012	15.243	24.367	19.626	27.172	1123.7	229.26
146	3.3651	.1858	.009383	0.05988	15.269	24.405	19.804	27.413	1124.1	236.29
147	3.4506	.1920	.009607	0.05964	15.294	24.443	19.987	27.654	1124.6	243.59
148	3.5379	.1984	.009835	0.05940	15.320	24.481	20.176	27.895	1125.0	251.12
149	3.6271	.2051	.010066	0.05915	15.344	24.518	20.374	28.135	1125.4	258.94
150	3.7182	.2120	.01030	0.05890	15.370	24.555	20.576	28.376	1125.8	267.06
151	3.8113	.2192	.01055	0.05866	15.395	24.593	20.786	28.617	1126.2	275.48
152	3.9063	.2267	.01079	0.05840	15.420	24.630	21.004	28.858	1126.6	284.22
153	4.0033	.2334	.01104	0.05815	15.445	24.668	21.229	29.099	1127.0	293.30
154	4.1023	.2425	.01130	0.05789	15.471	24.705	21.462	29.340	1127.4	302.73
155	4.2034	.2509	.01156	0.05763	15.496	24.742	21.704	29.581	1127.8	312.55
156	4.3066	.2596	.01183	0.05737	15.521	24.780	21.955	29.822	1128.2	322.75
157	4.4120	.2688	.01210	0.05711	15.546	24.818	22.216	30.063	1128.7	333.41
158	4.5195	.2782	.01237	0.05684	15.572	24.855	22.487	30.304	1129.1	344.46
159	4.6292	.2881	.01265	0.05657	15.597	24.892	22.769	30.545	1129.5	356.00
160	4.7412	.2985	.01294	0.05630	15.622	24.929	23.063	30.786	1129.9	368.13
161	4.8554	.3092	.01323	0.05603	15.647	24.966	23.368	31.027	1130.3	380.56
162	4.9720	.3205	.01353	0.05575	15.673	25.003	23.685	31.268	1130.7	393.67
163	5.0909	.3323	.01384	0.05547	15.698	25.040	24.017	31.509	1131.1	407.35
164	5.2122	.3446	.01414	0.05519	15.723	25.077	24.365	31.750	1131.5	421.66

Source: Psychrometric Tables and Charts, O. T. Zimmerman and Irvin Lavine, Industrial Research Services, Inc., Dover, NH, 1964.

Temp. °F	Saturation Pressure	Saturation humidity, Wt. water vapor/lb. dry air	Saturation moisture content, Wt. water vapor/cu. ft. sat. mixture	Saturation density Wt. air plus water vapor/cu. ft. sat. mixture	Volume			Enthalpy		
	Lb./sq. in.	Pounds	Pounds	Pounds	Dry air cu. ft./lb.	Water vapor cu. ft./lb.	Saturated mixture cu. ft./lb. dry air	Dry air B.T.U. /lb.	Water vapor, B.T.U. /lb.	Saturated mixture, B.T.U./lb. dry air
165	5.3358	.3575	.01446	0.05490	15.748	25.114	24.725	31.991	1131.9	436.61
166	5.4621	.3710	.01478	0.05462	15.744	25.151	25.102	32.232	1132.3	452.30
167	5.5908	.3851	.01511	0.05434	15.799	25.188	25.492	32.473	1132.7	468.72
168	5.7220	.4000	.01543	0.05402	15.824	25.225	25.914	32.714	1133.1	485.95
169	5.8558	.4156	.01577	0.05373	15.849	25.261	26.347	32.955	1133.5	504.05
170	5.9923	.4320	.01612	0.05343	15.875	25.298	26.804	33.196	1133.9	523.06
171	6.1314	.4493	.01647	0.05314	15.900	25.335	27.272	33.437	1134.3	543.08
172	6.2733	.4675	.01682	0.05281	15.925	25.372	27.787	33.678	1134.7	564.15
173	6.4179	.4867	.01719	0.05251	15.950	25.408	28.315	33.919	1135.1	586.38
174	6.5653	.5070	.01756	0.05219	15.976	25.445	28.876	34.161	1135.5	609.84
175	6.7156	.5284	.01793	0.05187	16.001	25.481	29.465	34.402	1135.9	634.63
176	6.8687	.5511	.01832	0.05155	16.026	25.518	30.089	34.643	1136.3	660.88
177	7.0247	.5752	.01871	0.05123	16.051	25.554	30.749	34.884	1136.7	688.69
178	7.1838	.6008	.01910	0.05090	16.077	25.591	31.449	35.125	1137.1	718.25
179	7.3458	.6279	.01950	0.05057	16.102	25.627	32.193	35.366	1137.5	749.63
180	7.5109	.6569	.01992	0.05023	16.127	25.664	32.984	35.607	1137.9	783.08
181	7.6791	.6878	.02033	0.04989	16.152	25.700	33.829	35.848	1138.3	818.78
182	7.8504	.7209	.02076	0.04955	16.178	25.736	34.731	36.090	1138.7	856.97
183	8.0247	.7563	.02119	0.04920	16.203	25.772	35.694	36.331	1139.1	897.79
184	8.2027	.7943	.02163	0.04885	16.228	25.809	36.728	36.572	1139.5	941.68
185	8.3836	.8352	.02207	0.04850	16.253	25.844	37.839	36.813	1139.9	988.88
186	8.5678	.8794	.02253	0.04814	16.279	25.881	39.037	37.054	1140.3	1039.79
187	8.7554	.9271	.02299	0.04778	16.304	25.916	40.332	37.296	1140.7	1094.88
188	8.9465	.9790	.02346	0.04742	16.329	25.953	41.737	37.537	1141.1	1154.70
189	9.1411	1.0355	.02393	0.04705	16.354	25.988	43.265	37.778	1141.5	1219.80

Temp. °F	Saturation Pressure	Saturation humidity, Wt. water vapor/lb. dry air	Saturation moisture content, Wt. water vapor/cu. ft. sat. mixture	Saturation density Wt. air plus water vapor/cu. ft. sat. mixture	Volume			Enthalpy		
	Lb./sq. in.	Pounds	Pounds	Pounds	Dry air cu. ft./lb.	Water vapor cu. ft./lb.	Saturated mixture cu. ft./lb. dry air	Dry air B.T.U. /lb.	Water vapor, B.T.U. /lb.	Saturated mixture, B.T.U./lb. dry air
190	9.3392	1.097	.02442	0.04667	16.380	26.024	44.935	38.019	1141.9	1291.0
191	9.5409	1.165	.02491	0.04630	16.405	26.060	46.764	38.261	1142.3	1369.5
192	9.7463	1.240	.02542	0.04592	16.430	26.093	48.780	38.502	1142.7	1455.1
193	9.9553	1.322	.02592	0.04553	16.455	26.131	51.011	38.743	1143.1	1550.4
194	10.1684	1.414	.02644	0.04514	16.481	26.167	53.488	38.985	1143.5	1656.2
195	10.385	1.517	.02697	0.04474	16.506	26.202	56.265	39.226	1143.9	1775.0
196	10.605	1.633	.02750	0.04434	16.530	26.238	59.381	39.467	1144.3	1908.3
197	10.829	1.765	.02805	0.04394	16.556	26.273	62.918	39.709	1144.7	2059.6
198	11.057	1.915	.02860	0.04353	16.582	26.309	66.963	39.950	1145.1	2232.8
199	11.289	2.089	.02916	0.04312	16.607	26.344	71.630	40.191	1145.5	2432.7
200	11.526	2.292	.02973	0.04270	16.632	26.380	77.102	40.433	1145.9	2667.2
201	11.766	2.532	.03031	0.04228	16.657	26.415	83.543	40.674	1146.2	2943.0
202	12.010	2.820	.03090	0.04185	16.683	26.450	921.270	40.916	1146.6	3274.2
203	12.259	3.173	.03149	0.04142	16.708	26.486	100.750	41.158	1146.0	3679.6
204	12.512	3.614	.03210	0.04098	16.733	26.521	112.590	41.399	1147.4	4188.6
205	12.769	4.181	.03272	0.04054	16.758	26.556	127.80	41.640	1147.7	4840.5
206	13.031	4.939	.03346	0.04024	16.784	26.591	147.60	41.882	1148.1	5712.8
207	13.297	6.000	.03398	0.03965	16.809	26.626	176.56	42.124	1148.4	6932.3
208	13.568	7.594	.03463	0.03919	16.834	26.661	219.30	42.366	1148.8	8766.8
209	13.843	10.248	.03528	0.03873	16.859	26.696	290.44	42.608	1149.2	11820.3
210	14.122	15.54	.03595	0.03826	16.885	26.731	432.25	42.849	1149.6	17906.0
211	14.407	31.49	.03667	0.03779	16.910	26.765	859.82	43.090	1150.0	36260.0

Enthalpy of Dry Air and Water Vapor; Temperature Range, 212 to 1000°F; Pressure, 14.7 psia

Temp. °F	Enthalphy, B.T.U./lb		Temp. °F	Enthalphy, B.T.U./lb		Temp. °F	Enthalphy, B.T.U./lb	
	Dry Air	Water Vapor		Dry Air	Water Vapor		Dry Air	Water Vapor
			240	50.099	1164.2	290	62.215	1188.1
			241	50.340	1164.7	292	62.700	1189.0
212	43.331	1150.4	242	50.581	1165.2	294	63.186	1190.0
213	43.573	1150.9	243	50.824	1165.7	296	63.671	1190.9
214	43.815	1151.4	244	51.066	1166.2	298	64.157	1191.9
215	44.057	1151.9	245	51.308	1166.6	300	64.643	1192.8
216	44.399	1152.4	246	51.540	1167.1	302	65.129	1193.7
217	44.640	1152.9	247	51.781	1167.6	304	65.616	1194.7
218	44.481	1153.4	248	52.023	1168.1	306	66.101	1195.6
219	45.123	1153.9	249	52.265	1168.6	308	66.588	1196.6
220	45.265	1154.4	250	52.518	1169.0	310	67.074	1197.5
221	45.507	1154.8	252	53.001	1170.0	312	67.560	1198.4
222	45.748	1155.3	254	53.486	1170.9	314	68.047	1199.4
223	45.990	1155.8	256	53.970	1171.9	316	68.533	1200.3
224	46.231	1156.3	258	54.455	1172.8	318	69.020	1201.3
225	46.475	1156.8	260	54.939	1173.8	320	69.507	1202.2
226	46.715	1157.3	262	55.424	1174.8	322	69.993	1203.2
227	46.956	1157.8	264	55.909	1175.7	324	70.479	1204.1
228	47.198	1158.3	266	56.393	1176.7	326	70.966	1205.1
229	47.439	1158.8	268	56.878	1177.6	328	71.453	1206.0
230	47.680	1159.3	270	57.362	1178.6	330	71.940	1207.0
231	47.923	1159.8	272	57.848	1179.5	332	72.427	1207.9
232	48.164	1160.3	274	58.332	1180.5	334	72.915	1208.9
233	48.406	1160.8	276	58.818	1181.5	336	73.401	1209.8
234	48.648	1161.3	278	59.303	1182.4	338	73.889	1210.8
235	48.890	1161.8	280	59.788	1183.3	340	74.375	1211.7
236	49.131	1162.2	282	60.273	1184.3	342	74.863	1212.6
237	49.372	1162.7	284	60.759	1185.2	344	75.350	1213.6
238	49.615	1163.2	286	61.244	1186.2	346	75.838	1214.5
239	49.857	1163.7	288	61.729	1187.1	348	76.326	1215.5

Temp. °F	Enthalphy, B.T.U./lb		Temp. °F	Enthalphy, B.T.U./lb		Temp. °F	Enthalphy, B.T.U./lb	
	Dry Air	Water Vapor		Dry Air	Water Vapor		Dry Air	Water Vapor
350	76.813	1216.4	420	93.947	1249.3	490	111.22	1282.4
352	77.301	1217.3	422	94.439	1250.3	492	111.71	1283.3
354	77.790	1218.3	424	94.930	1251.2	494	112.21	1284.3
356	78.278	1219.2	426	95.421	1252.2	496	112.70	1285.2
358	78.767	1220.2	428	95.913	1253.1	498	113.20	1286.2
360	79.255	1221.1	430	96.405	1254.1	500	113.69	1287.1
362	79.744	1222.0	432	96.898	1255.0	505	114.93	1289.5
364	80.233	1223.0	434	97.390	1256.0	510	116.17	1291.9
366	80.721	1223.9	436	97.882	1256.9	515	117.41	1294.3
368	81.210	1224.9	438	98.374	1257.9	520	118.05	1296.6
370	81.699	1225.8	440	98.866	1258.8	525	119.89	1299.0
372	82.188	1226.7	442	99.359	1259.7	530	121.14	1301.4
374	82.677	1227.7	444	99.852	1260.7	535	122.38	1303.8
376	83.165	1228.6	446	100.344	1261.6	540	123.63	1306.2
378	83.651	1229.6	448	100.837	1262.6	545	124.88	1308.6
380	84.143	1230.5	450	101.33	1263.5	550	126.12	1311.0
382	84.632	1231.4	452	101.82	1264.4	555	127.37	1313.3
384	85.121	1232.4	454	102.32	1265.7	560	128.62	1315.7
386	85.611	1233.3	456	102.81	1266.3	565	129.87	1318.1
388	86.100	1234.3	458	103.31	1267.3	570	131.12	1320.5
390	86.590	1235.2	460	103.80	1268.2	575	132.37	1322.9
392	87.079	1236.1	462	104.29	1269.1	580	133.62	1325.3
394	87.569	1237.1	464	104.79	1270.1	585	134.88	1327.6
396	88.059	1238.0	466	105.28	1271.0	590	136.13	1330.0
398	88.548	1239.0	468	105.77	1272.0	595	137.38	1332.4
400	89.039	1239.9	470	106.27	1272.9	600	138.63	1334.8
402	89.529	1240.8	472	106.76	1273.8	605	139.90	1337.2
404	90.020	1241.8	474	107.26	1274.8	610	141.14	1339.6
406	90.510	1242.7	476	107.75	1275.7	615	142.41	1342.1
408	91.000	1243.7	478	108.25	1276.7	620	143.66	1344.5
410	91.491	1244.6	480	108.74	1277.6	625	144.92	1346.9
412	91.981	1245.5	482	109.24	1278.6	630	145.18	1349.3
414	92.474	1246.5	484	109.73	1279.5	635	147.43	1351.7
416	92.964	1247.4	486	110.23	1280.5	640	148.70	1354.2
418	93.455	1248.4	488	110.72	1281.4	645	149.96	1356.6

Temp. °F	Enthalphy, B.T.U./lb		Temp. °F	Enthalphy, B.T.U./lb		Temp. °F	Enthalphy, B.T.U./lb	
	Dry Air	Water Vapor		Dry Air	Water Vapor		Dry Air	Water Vapor
650	151.22	1359.0	800	189.44	1432.3	950	228.35	1507.7
655	152.48	1361.4	805	190.73	1434.8	955	229.66	1510.2
660	153.73	1363.8	810	192.02	1437.3	960	230.97	1512.8
665	155.01	1366.3	815	193.30	1439.8	965	232.28	1515.3
670	156.27	1368.7	820	194.59	1442.3	970	233.60	1517.9
675	157.53	1371.1	825	195.88	1444.8	975	234.91	1520.4
680	158.81	1373.5	830	197.17	1447.3	980	236.22	1522.9
685	160.07	1375.9	835	198.46	1449.8	985	237.54	1525.5
690	161.33	1378.4	840	199.75	1452.3	990	238.86	1528.0
695	162.61	1380.8	845	201.06	1454.8	995	240.17	1530.6
700	163.88	1383.2	850	202.33	1457.3	1000	241.48	1533.1
705	165.14	1385.7	855	203.63	1459.8			
710	166.42	1388.1	860	204.96	1462.3			
715	167.69	1390.6	865	206.22	1464.8			
720	168.96	1393.0	870	207.52	1467.3			
725	170.24	1395.5	875	208.82	1469.8			
730	171.52	1397.9	880	210.12	1472.3			
735	172.79	1400.4	885	211.42	1474.8			
740	174.07	1402.8	890	212.72	1477.3			
745	175.34	1405.3	895	214.01	1479.8			
750	176.62	1407.8	900	215.31	1482.3			
755	177.90	1410.2	905	216.61	1484.8			
760	179.18	1412.7	910	217.91	1487.4			
765	180.47	1445.1	915	219.21	1489.9			
770	181.74	1417.6	920	220.52	1492.5			
775	183.02	1420.0	925	221.82	1495.0			
780	184.31	1422.5	930	223.12	1497.5			
785	185.59	1424.9	935	224.43	1500.1			
790	186.87	1427.4	940	225.73	1502.6			
795	188.15	1429.8	945	227.04	1505.2			

Source: *Psychrometric Tables and Charts,* O. T. Zimmerman and Irvin Lavine, Industrial Research Services, Inc., Dover, NH, 1964.

Psychometric Charts

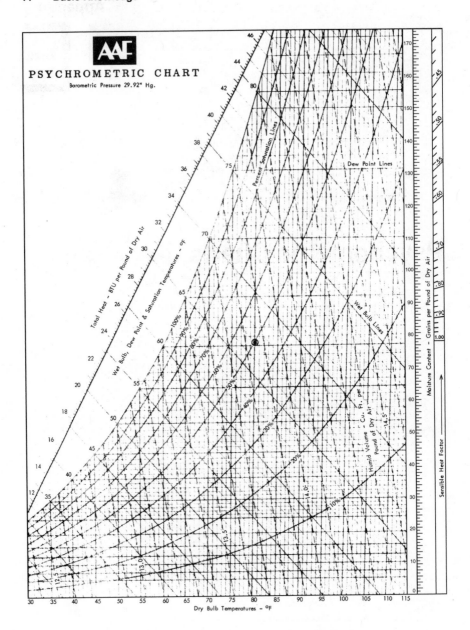

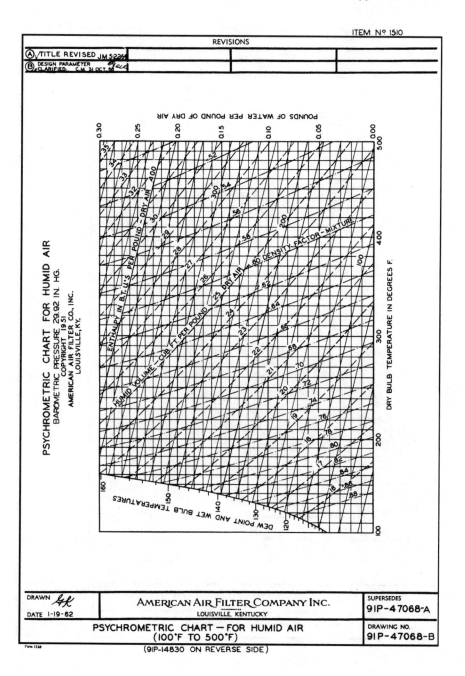

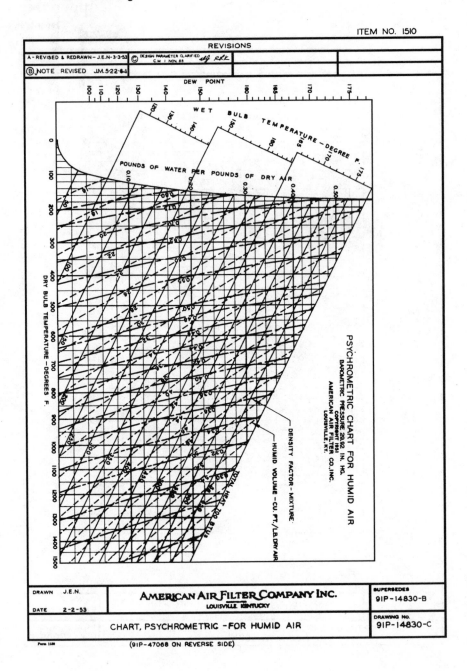

Viscosity of Some Common Cases

VISCOSITY OF ACETONE VAPOR

| TEMPERATURE | VISCOSITY (MICROPOISE) | |
(°C)	CALCULATED	REPORTED
-20.0	58.8	
-15.0	60.2	
-10.0	61.6	
-5.0	63.1	
0.0	64.5	
25.0	71.6	
50.0	78.8	
75.0	86.0	
100.0	93.1	93.1
119.0	98.5	99.1
150.0	107.3	
190.4	118.6	118.6
212.5	124.8	124.0
247.7	134.5	133.4
306.4	150.4	148.1
350.0	162.1	
400.0	175.2	
500.0	200.7	
600.0	225.3	
700.0	248.9	
800.0	271.8	
900.0	293.8	
1000.0	315.0	
1100.0	335.6	
1200.0	355.5	

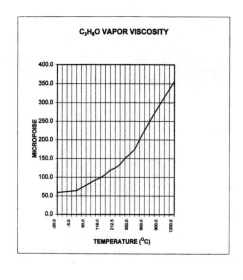

NOTES:

1) CALCULATED VALUES ARE BASED UPON THE EQUATION

$V=V_0*(T_0+C)/(T+C)*(T/T_0)^{1.5}$, WHERE C=670

2) SOURCES OF DATA ARE THE INTERNATIONAL CRITICAL TABLES AND HANDBOOK
   OF CHEMISTRY AND PHYSICS, CRC PRESS, INC., CLEVELAND, OHIO (1976)

3) ILLUSTRATION COURTESY OF FISHER-KLOSTERMAN, INC.

VISCOSITY OF AIR

TEMPERATURE ($^\circ$C)	VISCOSITY (MICROPOISE)	
	CALCULATED	REPORTED
-194.2	52.4	55.1
-183.1	60.5	62.7
-104.0	113.2	113.0
-69.4	133.7	133.3
-31.6	154.5	153.9
0.0	170.9	170.8
18.0	179.8	182.7
40.0	190.4	190.4
54.0	196.9	195.8
74.0	206.1	210.2
229.0	269.2	263.8
334.0	306.2	312.3
357.0	313.9	317.5
409.0	330.6	341.3
466.0	348.1	350.1
481.0	352.6	358.3
537.0	369.0	368.6
565.0	376.9	375.0
620.0	392.1	391.6
638.0	397.0	401.4
750.0	426.1	426.3
810.0	441.0	441.9
923.0	467.8	464.3
1034.0	492.9	490.6
1134.0	514.5	520.6

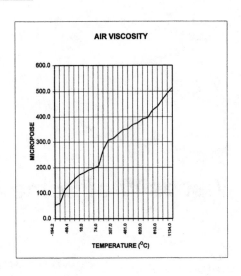

NOTES:
1) CALCULATED VALUES ARE BASED UPON THE EQUATION

$V = V_o \ast (T_o + C)/(T + C) \ast (T/T_o)^{1.5}$, WHERE C=120
2) SOURCES OF DATA ARE THE INTERNATIONAL CRITICAL TABLES AND HANDBOOK
   OF CHEMISTRY AND PHYSICS, CRC PRESS, INC., CLEVELAND, OHIO (1976)
3) ILLUSTRATION COURTESY OF FISHER-KLOSTERMAN, INC.

VISCOSITY OF AMMONIA VAPORS

TEMPERATURE ($^{\circ}$C)	VISCOSITY (MICROPOISE) CALCULATED	REPORTED
-109.1	51.4	
-90.0	58.6	
-78.5	62.9	67.2
-50.0	73.5	
0.0	91.8	91.8
10.0	95.4	
20.0	99.0	98.2
50.0	109.6	109.2
100.0	126.9	127.9
132.9	137.9	139.9
150.0	143.5	146.3
200.0	159.7	164.6
250.0	175.2	181.4
300.0	190.3	198.7
400.0	219.0	
450.0	232.7	
500.0	246.0	
550.0	258.9	
600.0	271.5	
700.0	295.6	
800.0	318.6	
900.0	340.6	
1000.0	361.6	
1100.0	381.8	
1200.0	401.3	

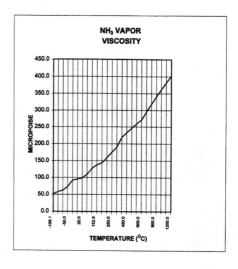

NOTES:
1) CALCULATED VALUES ARE BASED UPON THE EQUATION
$V=V_o*(T_o+C)/(T+C)*(T/T_o)^{1.5}$, WHERE C=370
2) SOURCES OF DATA ARE THE INTERNATIONAL CRITICAL TABLES AND HANDBOOK
OF CHEMISTRY AND PHYSICS, CRC PRESS, INC., CLEVELAND, OHIO (1976)
3) ILLUSTRATION COURTESY OF FISHER-KLOSTERMAN, INC.

VISCOSITY OF BENZENE VAPOR

TEMPERATURE ($^{\circ}$C)	VISCOSITY (MICROPOISE)	
	CALCULATED	REPORTED
-20.0	57.7	
-15.0	59.2	
-10.0	60.6	
-5.0	62.0	
0.0	63.4	
14.2	67.4	73.8
50.0	77.6	
75.0	84.7	
100.0	91.8	91.8
131.2	100.6	103.1
150.0	105.9	
194.6	118.4	119.8
212.5	123.4	123.0
252.5	134.4	134.3
312.8	150.8	148.4
350.0	160.7	
400.0	173.8	
500.0	199.5	
600.0	224.2	
700.0	248.0	
800.0	271.0	
900.0	293.2	
1000.0	314.7	
1100.0	335.5	
1200.0	355.6	

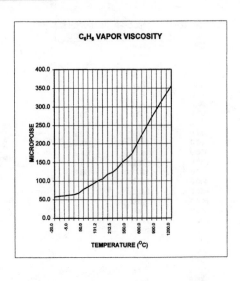

NOTES:
1) CALCULATED VALUES ARE BASED UPON THE EQUATION

$V = V_o * (T_o + C)/(T + C) * (T/T_o)^{1.5}$, WHERE C=700
2) SOURCES OF DATA ARE THE <u>INTERNATIONAL CRITICAL TABLES</u> AND <u>HANDBOOK OF CHEMISTRY AND PHYSICS</u>, CRC PRESS, INC., CLEVELAND, OHIO (1976)
3) ILLUSTRATION COURTESY OF FISHER-KLOSTERMAN, INC.

VISCOSITY OF BROMINE VAPOR

TEMPERATURE ($^{\circ}$C)	VISCOSITY (MICROPOISE)	
	CALCULATED	REPORTED
-48.7	106.8	
-30.0	117.3	
-20.0	122.8	
-10.0	128.4	
0.0	133.9	
12.8	141.0	151.0
25.0	147.7	
65.7	169.8	170.0
99.7	188.0	188.0
139.7	209.0	208.0
179.7	229.6	227.0
220.3	250.0	248.0
300.0	288.9	
350.0	312.4	
400.0	335.2	
450.0	357.5	
500.0	379.2	
550.0	400.3	
600.0	420.9	
700.0	460.7	
800.0	498.7	
900.0	535.1	
1000.0	570.1	
1100.0	603.7	
1200.0	636.2	

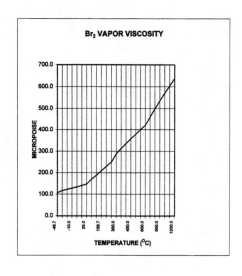

NOTES:
1) CALCULATED VALUES ARE BASED UPON THE EQUATION

$V=V_o*(T_o+C)/(T+C)*(T/T_o)^{1.5}$, WHERE C=460

2) SOURCES OF DATA ARE THE INTERNATIONAL CRITICAL TABLES AND HANDBOOK OF CHEMISTRY AND PHYSICS, CRC PRESS, INC., CLEVELAND, OHIO (1976)

3) ILLUSTRATION COURTESY OF FISHER-KLOSTERMAN, INC.

VISCOSITY OF BUTANE VAPOR

TEMPERATURE (°C)	VISCOSITY (MICROPOISE)	
	CALCULATED	REPORTED
-20.0	72.6	
-15.0	74.1	
-10.0	75.7	
-5.0	77.2	
0.0	78.7	
14.7	83.2	83.2
50.0	93.7	
75.0	101.0	
100.0	108.2	108.2
125.0	115.2	
150.0	122.2	
175.0	128.9	
200.0	135.6	
250.0	148.6	
300.0	161.1	
350.0	173.2	
400.0	184.9	
500.0	207.3	
600.0	228.3	
700.0	248.3	
800.0	267.3	
900.0	285.3	
1000.0	302.7	
1100.0	319.3	
1200.0	335.3	

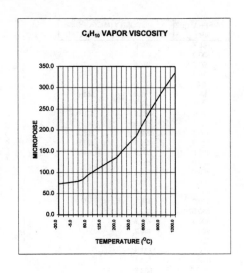

C_4H_{10} VAPOR VISCOSITY

NOTES:
1) CALCULATED VALUES ARE BASED UPON THE EQUATION
   $V = V_o * (T_o + C)/(T + C) * (T/T_o)^{1.5}$, WHERE C=345
2) SOURCES OF DATA ARE THE INTERNATIONAL CRITICAL TABLES AND HANDBOOK
   OF CHEMISTRY AND PHYSICS, CRC PRESS, INC., CLEVELAND, OHIO (1976)
3) ILLUSTRATION COURTESY OF FISHER-KLOSTERMAN, INC.

VISCOSITY OF CARBON DIOXIDE VAPOR

TEMPERATURE (°C)	VISCOSITY (MICROPOISE)	
	CALCULATED	REPORTED
-134.3	65.2	
-125.0	70.3	
-97.8	85.0	89.6
-78.2	95.5	97.2
0.0	135.8	
15.0	143.2	145.7
23.0	147.1	147.1
30.0	150.5	153.0
99.1	182.9	186.1
182.4	219.3	222.1
200.0	226.7	
235.0	241.0	241.5
302.0	267.3	268.2
350.0	285.4	
402.0	304.2	
450.0	321.0	
490.0	334.5	330.0
550.0	354.3	
608.0	372.7	
685.0	396.2	380.0
850.0	443.4	435.8
900.0	457.0	
1000.0	483.3	
1052.0	496.5	478.6
1200.0	532.6	

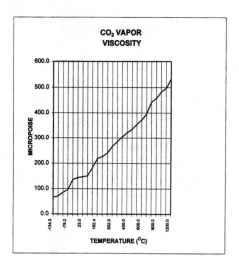

CO$_2$ VAPOR VISCOSITY

NOTES:

1) CALCULATED VALUES ARE BASED UPON THE EQUATION

   $V = V_o*(T_o+C)/(T+C)*(T/T_o)^{1.5}$, WHERE C=274

2) SOURCES OF DATA ARE THE <u>INTERNATIONAL CRITICAL TABLES</u> AND <u>HANDBOOK OF CHEMISTRY AND PHYSICS</u>, CRC PRESS, INC., CLEVELAND, OHIO (1976)

3) ILLUSTRATION COURTESY OF FISHER-KLOSTERMAN, INC.

VISCOSITY OF CARBON MONOXIDE VAPOR

TEMPERATURE (°C)	VISCOSITY (MICROPOISE)	
	CALCULATED	REPORTED
-222.0	31.1	
-191.5	53.1	56.1
-78.5	124.9	127.0
0.0	166.0	166.0
15.0	173.2	172.0
21.7	176.4	175.3
100.0	211.1	210.0
150.0	231.4	
200.0	250.4	
250.0	268.5	
300.0	285.6	
350.0	301.9	
400.0	317.5	
450.0	332.6	
500.0	347.0	
550.0	361.0	
600.0	374.4	
650.0	387.5	
700.0	400.2	
750.0	412.5	
800.0	424.5	
900.0	447.7	
1000.0	469.7	
1100.0	490.9	
1200.0	511.2	

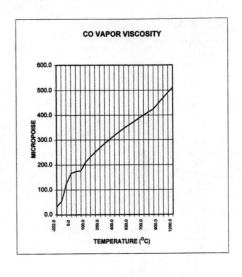

NOTES:
1) CALCULATED VALUES ARE BASED UPON THE EQUATION

$V = V_o * (T_o + C)/(T + C) * (T/T_o)^{1.5}$, WHERE C=118
2) SOURCES OF DATA ARE THE INTERNATIONAL CRITICAL TABLES AND HANDBOOK OF CHEMISTRY AND PHYSICS, CRC PRESS, INC., CLEVELAND, OHIO (1976)
3) ILLUSTRATION COURTESY OF FISHER-KLOSTERMAN, INC.

VISCOSITY OF CHLORINE VAPOR

TEMPERATURE	VISCOSITY (MICROPOISE)	
(°C)	CALCULATED	REPORTED
-118.0	65.6	
-100.0	74.6	
-50.0	99.2	
-25.0	111.2	
0.0	123.1	
12.7	129.0	129.0
20.0	132.4	132.7
50.0	146.2	146.9
99.1	168.0	168.0
150.0	189.8	187.5
200.0	210.3	208.5
250.0	230.1	227.6
300.0	249.2	
350.0	267.6	
400.0	285.4	
450.0	302.6	
500.0	319.3	
550.0	335.5	
600.0	351.2	
700.0	381.4	
800.0	410.1	
900.0	437.5	
1000.0	463.7	
1100.0	488.8	
1200.0	512.9	

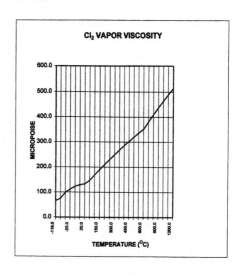

NOTES:

1) CALCULATED VALUES ARE BASED UPON THE EQUATION

$V = V_o \cdot (T_o + C)/(T + C) \cdot (T/T_o)^{1.5}$, WHERE C=325

2) SOURCES OF DATA ARE THE INTERNATIONAL CRITICAL TABLES AND HANDBOOK OF CHEMISTRY AND PHYSICS, CRC PRESS, INC., CLEVELAND, OHIO (1976)

3) ILLUSTRATION COURTESY OF FISHER-KLOSTERMAN, INC.

VISCOSITY OF CHLOROFORM VAPOR

TEMPERATURE ($^\circ$C)	VISCOSITY (MICROPOISE)	
	CALCULATED	REPORTED
-58.0	74.1	
-30.0	84.3	
-10.0	91.5	
0.0	95.1	93.6
14.2	100.1	98.9
50.0	112.4	
100.0	129.0	129.0
121.3	135.9	135.7
189.1	156.9	157.9
212.5	163.8	164.0
250.0	174.7	177.6
307.5	190.9	194.7
350.0	202.4	
400.0	215.4	
450.0	228.0	
500.0	240.3	
550.0	252.1	
600.0	263.6	
650.0	274.8	
700.0	285.7	
800.0	306.6	
900.0	326.5	
1000.0	345.5	
1100.0	363.8	
1200.0	381.3	

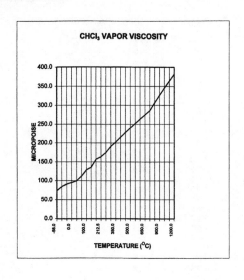

CHCl$_3$ VAPOR VISCOSITY

NOTES:
1) CALCULATED VALUES ARE BASED UPON THE EQUATION
   $V = V_o*(T_o+C)/(T+C)*(T/T_o)^{1.5}$, WHERE C=292
2) SOURCES OF DATA ARE THE INTERNATIONAL CRITICAL TABLES AND HANDBOOK
   OF CHEMISTRY AND PHYSICS,CRC PRESS, INC., CLEVELAND, OHIO (1976)
3) ILLUSTRATION COURTESY OF FISHER-KLOSTERMAN, INC.

VISCOSITY OF EHTHANE VAPOR

TEMPERATURE	VISCOSITY (MICROPOISE)	
(°C)	CALCULATED	REPORTED
-159.5	32.2	
-78.5	59.6	63.4
-50.0	69.0	
-25.0	77.0	
0.0	84.8	84.8
17.2	90.1	90.1
50.8	100.2	100.1
75.0	107.2	
100.4	114.5	114.3
125.0	121.3	
150.0	128.1	
200.3	141.4	140.9
250.0	153.9	
300.0	166.1	
350.0	177.7	
400.0	189.0	
450.0	199.8	
500.0	210.3	
600.0	230.3	
700.0	249.2	
800.0	267.1	
900.0	284.1	
1000.0	300.4	
1100.0	316.0	
1200.0	331.0	

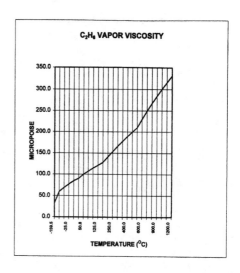

NOTES:
1) CALCULATED VALUES ARE BASED UPON THE EQUATION
   $V=V_0*(T_0+C)/(T+C)*(T/T_0)^{1.5}$, WHERE C=270
2) SOURCES OF DATA ARE THE INTERNATIONAL CRITICAL TABLES AND HANDBOOK
   OF CHEMISTRY AND PHYSICS, CRC PRESS, INC., CLEVELAND, OHIO (1976)
3) ILLUSTRATION COURTESY OF FISHER-KLOSTERMAN, INC.

VISCOSITY OF EHTHYLENE VAPOR

TEMPERATURE (^{o}C)	VISCOSITY (MICROPOISE)	
	CALCULATED	REPORTED
-168.3	33.8	
-75.7	68.2	69.9
-44.1	79.3	76.9
-38.6	81.2	78.5
0.0	94.1	90.7
13.8	98.6	95.4
15.0	99.0	99.0
20.0	100.6	100.8
50.0	110.1	110.3
100.0	125.2	125.7
150.0	139.5	140.3
200.0	153.2	154.1
250.0	166.2	166.6
302.0	179.2	180.0
350.0	190.7	
400.0	202.2	
450.0	213.2	
500.0	223.9	
600.0	244.3	
700.0	263.5	
800.0	281.6	
900.0	298.9	
1000.0	315.4	
1100.0	331.2	
1200.0	346.3	

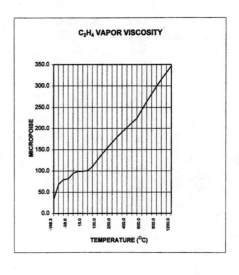

C_2H_4 VAPOR VISCOSITY

NOTES:
1) CALCULATED VALUES ARE BASED UPON THE EQUATION
   $V=V_o*(T_o+C)/(T+C)*(T/T_o)^{1.5}$, WHERE C=226
2) SOURCES OF DATA ARE THE INTERNATIONAL CRITICAL TABLES AND HANDBOOK
   OF CHEMISTRY AND PHYSICS,CRC PRESS, INC., CLEVELAND, OHIO (1976)
3) ILLUSTRATION COURTESY OF FISHER-KLOSTERMAN, INC.

VISCOSITY OF ETHYL ALCOHOL VAPOR

TEMPERATURE ($^{\circ}$C)	VISCOSITY (MICROPOISE)	
	CALCULATED	REPORTED
-115.7	38.9	
-75.0	51.9	
-50.0	60.0	
-25.0	68.0	
0.0	76.1	
25.0	84.2	
50.0	92.2	
75.0	100.1	
100.0	108.0	108.0
130.2	117.4	117.3
170.7	129.9	129.3
191.8	136.3	135.5
212.5	142.5	140.0
251.7	154.1	151.9
308.7	170.6	167.0
350.0	182.3	
400.0	196.2	
500.0	222.9	
600.0	248.4	
700.0	272.7	
800.0	296.1	
900.0	318.5	
1000.0	340.0	
1100.0	360.8	
1200.0	380.8	

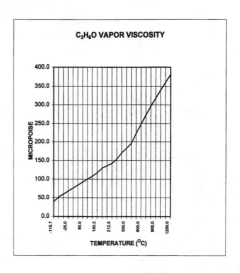

NOTES:
1) CALCULATED VALUES ARE BASED UPON THE EQUATION
   $V = V_0 \cdot (T_0 + C)/(T + C) \cdot (T/T_0)^{1.5}$, WHERE C=525
2) SOURCES OF DATA ARE THE INTERNATIONAL CRITICAL TABLES AND HANDBOOK
   OF CHEMISTRY AND PHYSICS, CRC PRESS, INC., CLEVELAND, OHIO (1976)
3) ILLUSTRATION COURTESY OF FISHER-KLOSTERMAN, INC.

VISCOSITY OF HELIUM VAPOR

TEMPERATURE (°C)	VISCOSITY (MICROPOISE) CALCULATED	REPORTED
-271.7	0.3	
-191.6	69.5	87.1
-100.0	134.6	
-50.0	163.5	
0.0	189.3	189.3
13.7	195.9	195.4
23.0	200.3	
27.4	202.3	
100.1	234.2	234.1
150.0	254.2	
200.0	273.0	267.2
250.0	290.7	285.3
282.0	301.6	299.2
350.0	323.5	
407.0	340.9	343.6
450.0	353.5	
486.0	363.8	370.6
550.0	381.4	
606.0	396.1	408.7
676.0	413.9	430.3
817.0	447.6	471.3
900.0	466.4	
1000.0	488.1	
1100.0	508.8	
1200.0	528.8	

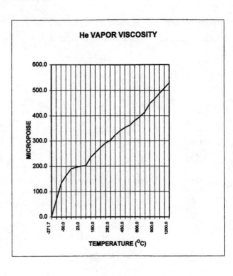

NOTES:
1) CALCULATED VALUES ARE BASED UPON THE EQUATION
   $V=V_o*(T_o+C)/(T+C)*(T/T_o)^{1.5}$, WHERE C=71.4
2) SOURCES OF DATA ARE THE INTERNATIONAL CRITICAL TABLES AND HANDBOOK
   OF CHEMISTRY AND PHYSICS,CRC PRESS, INC., CLEVELAND, OHIO (1976)
3) ILLUSTRATION COURTESY OF FISHER-KLOSTERMAN, INC.

VISCOSITY OF HYDROGEN VAPOR

TEMPERATURE ($^{\circ}$C)	VISCOSITY (MICROPOISE)	
	CALCULATED	REPORTED
-263.3	2.2	
-257.7	4.1	5.7
-97.5	59.5	61.5
-31.6	76.5	76.7
0.0	83.8	
14.8	87.1	87.7
20.7	88.4	87.6
28.1	90.0	89.2
100.5	104.6	104.6
150.0	113.7	
229.1	127.2	126.0
250.0	130.6	
299.0	138.2	138.1
350.0	145.7	
412.0	154.4	155.4
450.0	159.5	
490.0	164.8	167.2
550.0	172.4	
601.0	178.6	182.9
713.0	191.6	198.2
825.0	203.8	213.7
900.0	211.6	
1000.0	221.5	
1100.0	231.1	
1200.0	240.3	

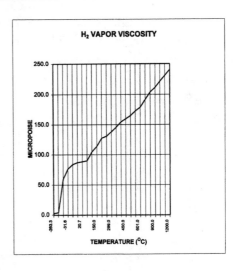

NOTES:
1) CALCULATED VALUES ARE BASED UPON THE EQUATION

$V=V_o*(T_o+C)/(T+C)*(T/T_o)^{1.5}$, WHERE C=83.0

2) SOURCES OF DATA ARE THE <u>INTERNATIONAL CRITICAL TABLES</u> AND <u>HANDBOOK OF CHEMISTRY AND PHYSICS</u>,CRC PRESS, INC., CLEVELAND, OHIO (1976)

3) ILLUSTRATION COURTESY OF FISHER-KLOSTERMAN, INC.

VISCOSITY OF HYDROGEN BROMIDE VAPORS

TEMPERATURE (°C)	VISCOSITY (MICROPOISE)	
	CALCULATED	REPORTED
-138.8	74.4	
-100.0	101.1	
-78.5	116.0	
-50.0	135.6	
0.0	169.4	
10.0	176.1	
18.7	181.9	181.9
50.0	202.4	
100.2	234.5	234.4
132.9	254.9	
150.0	265.3	
200.0	295.2	
250.0	324.2	
300.0	352.1	
400.0	405.4	
450.0	430.9	
500.0	455.6	
550.0	479.6	
600.0	503.0	
700.0	547.9	
800.0	590.7	
900.0	631.5	
1000.0	670.7	
1100.0	708.3	
1200.0	744.4	

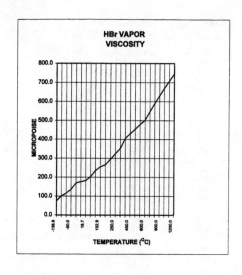

HBr VAPOR VISCOSITY

NOTES:
1) CALCULATED VALUES ARE BASED UPON THE EQUATION

$V=V_o*(T_o+C)/(T+C)*(T/T_o)^{1.5}$, WHERE C=375

2) SOURCES OF DATA ARE THE INTERNATIONAL CRITICAL TABLES AND HANDBOOK
OF CHEMISTRY AND PHYSICS,CRC PRESS, INC., CLEVELAND, OHIO (1976)

3) ILLUSTRATION COURTESY OF FISHER-KLOSTERMAN, INC.

VISCOSITY OF HYDROGEN CHLORIDE VAPORS

TEMPERATURE ($^\circ$C)	VISCOSITY (MICROPOISE)	
	CALCULATED	REPORTED
-150.8	52.0	
-125.0	65.8	
-100.0	79.2	
-50.0	105.9	
0.0	132.1	
12.5	138.5	138.5
25.0	144.9	
50.0	157.5	
100.3	182.2	182.2
150.0	205.7	
200.0	228.6	
250.0	250.7	
300.0	272.0	
350.0	292.6	
400.0	312.6	
450.0	332.0	
500.0	350.7	
550.0	369.0	
600.0	386.7	
700.0	420.8	
800.0	453.3	
900.0	484.2	
1000.0	513.8	
1100.0	542.3	
1200.0	569.7	

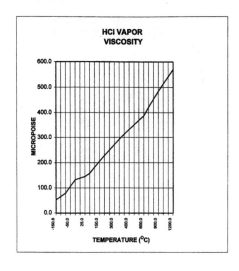

NOTES:
1) CALCULATED VALUES ARE BASED UPON THE EQUATION

$V=V_o*(T_o+C)/(T+C)*(T/T_o)^{1.5}$, WHERE C=357
2) SOURCES OF DATA ARE THE INTERNATIONAL CRITICAL TABLES AND HANDBOOK
   OF CHEMISTRY AND PHYSICS, CRC PRESS, INC., CLEVELAND, OHIO (1976)
3) ILLUSTRATION COURTESY OF FISHER-KLOSTERMAN, INC.

VISCOSITY OF HYDROGEN IODIDE VAPORS

TEMPERATURE (°C)	VISCOSITY (MICROPOISE)	
	CALCULATED	REPORTED
-123.3	85.7	
-100.0	102.0	
-78.5	117.1	
-50.0	137.1	
0.0	171.7	
10.0	178.5	
20.6	185.7	185.7
50.0	205.4	201.8
100.2	238.3	238.3
132.9	259.2	
150.0	270.0	262.7
200.0	300.7	292.4
250.0	330.5	318.9
300.0	359.3	
400.0	414.3	
450.0	440.6	
500.0	466.2	
550.0	491.0	
600.0	515.2	
700.0	561.7	
800.0	606.0	
900.0	648.4	
1000.0	689.0	
1100.0	727.9	
1200.0	765.5	

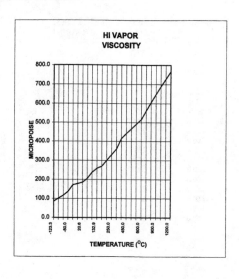

HI VAPOR VISCOSITY

NOTES:
1) CALCULATED VALUES ARE BASED UPON THE EQUATION
$V=V_o*(T_o+C)/(T+C)*(T/T_o)^{1.5}$, WHERE C=390
2) SOURCES OF DATA ARE THE INTERNATIONAL CRITICAL TABLES AND HANDBOOK OF CHEMISTRY AND PHYSICS,CRC PRESS, INC., CLEVELAND, OHIO (1976)
3) ILLUSTRATION COURTESY OF FISHER-KLOSTERMAN, INC.

VISCOSITY OF HYDROGEN SULFIDE VAPOR

TEMPERATURE (^{o}C)	VISCOSITY (MICROPOISE)	
	CALCULATED	REPORTED
-134.0	54.5	
-120.0	61.1	
-100.0	70.5	
-50.0	93.9	
0.0	116.6	116.6
10.0	121.1	
17.0	124.2	124.1
50.0	138.6	
100.0	159.7	158.7
150.0	180.1	
200.0	199.7	
250.0	218.6	
300.0	236.8	
350.0	254.4	
400.0	271.4	
450.0	287.9	
500.0	303.9	
550.0	319.3	
600.0	334.4	
700.0	363.3	
800.0	390.7	
900.0	416.9	
1000.0	442.0	
1100.0	466.0	
1200.0	489.1	

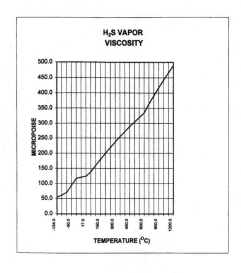

NOTES:
1) CALCULATED VALUES ARE BASED UPON THE EQUATION

$V=V_o*(T_o+C)/(T+C)*(T/T_o)^{1.5}$, WHERE C=331
2) SOURCES OF DATA ARE THE <u>INTERNATIONAL CRITICAL TABLES</u> AND <u>HANDBOOK</u>
   <u>OF CHEMISTRY AND PHYSICS</u>,CRC PRESS, INC., CLEVELAND, OHIO (1976)
3) ILLUSTRATION COURTESY OF FISHER-KLOSTERMAN, INC.

VISCOSITY OF IODINE VAPOR

TEMPERATURE	VISCOSITY (MICROPOISE)	
(^{o}C)	CALCULATED	REPORTED
38.7	138.0	
50.0	143.7	
75.0	156.4	
100.0	169.1	
124.0	181.1	184.0
170.0	204.0	204.0
205.4	221.3	220.0
247.1	241.5	240.0
300.0	266.5	
350.0	289.7	
400.0	312.4	
450.0	334.6	
500.0	356.3	
550.0	377.6	
600.0	398.4	
650.0	418.8	
700.0	438.8	
750.0	458.4	
800.0	477.6	
850.0	496.4	
900.0	514.9	
950.0	533.1	
1000.0	550.9	
1100.0	585.6	
1200.0	619.2	

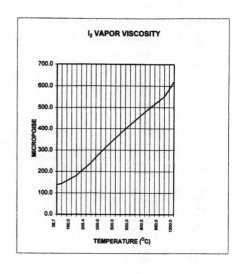

NOTES:
1) CALCULATED VALUES ARE BASED UPON THE EQUATION
   $V = V_o*(T_o+C)/(T+C)*(T/T_o)^{1.5}$, WHERE C=590
2) SOURCES OF DATA ARE THE INTERNATIONAL CRITICAL TABLES AND HANDBOOK
   OF CHEMISTRY AND PHYSICS,CRC PRESS, INC., CLEVELAND, OHIO (1976)
3) ILLUSTRATION COURTESY OF FISHER-KLOSTERMAN, INC.

VISCOSITY OF ISOPENTANE VAPOR

TEMPERATURE (^oC)	VISCOSITY (MICROPOISE)	
	CALCULATED	REPORTED
-20.0	56.6	
-15.0	57.9	
-10.0	59.2	
-5.0	60.5	
0.0	61.8	
25.0	68.3	
50.0	74.7	
75.0	81.1	
100.0	87.4	87.4
125.0	93.6	
150.0	99.8	
190.4	109.7	
212.5	115.0	115.0
250.0	123.8	
300.0	135.4	
350.0	146.6	
400.0	157.6	
500.0	178.8	
600.0	198.9	
700.0	218.2	
800.0	236.6	
900.0	254.3	
1000.0	271.3	
1100.0	287.6	
1200.0	303.4	

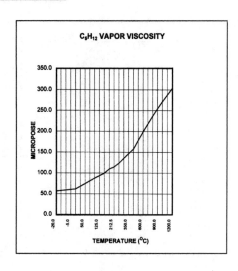

C_5H_{12} VAPOR VISCOSITY

NOTES:
1) CALCULATED VALUES ARE BASED UPON THE EQUATION

$V=V_o*(T_o+C)/(T+C)*(T/T_o)^{1.5}$, WHERE C=500
2) SOURCES OF DATA ARE THE <u>INTERNATIONAL CRITICAL TABLES</u> AND <u>HANDBOOK OF CHEMISTRY AND PHYSICS</u>,CRC PRESS, INC., CLEVELAND, OHIO (1976)
3) ILLUSTRATION COURTESY OF FISHER-KLOSTERMAN, INC.

VISCOSITY OF METHANE VAPOR

TEMPERATURE ($^{\circ}$C)	VISCOSITY (MICROPOISE)	
	CALCULATED	REPORTED
-205.9	22.3	
-181.6	32.4	34.8
-78.5	74.1	76.0
0.0	102.7	102.7
20.0	109.5	108.7
50.0	119.5	
100.0	135.3	133.1
150.0	150.2	
200.5	164.5	160.5
250.0	177.9	
284.0	186.7	181.3
300.0	190.7	
350.0	203.1	
380.0	210.2	202.6
450.0	226.3	
499.0	237.1	226.4
550.0	247.9	
600.0	258.2	
650.0	268.2	
700.0	277.9	
800.0	296.5	
900.0	314.1	
1000.0	331.0	
1100.0	347.2	
1200.0	362.7	

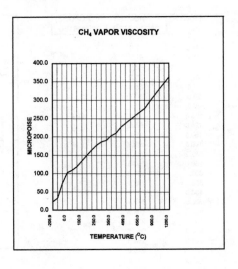

NOTES:
1) CALCULATED VALUES ARE BASED UPON THE EQUATION
   $V = V_0 * (T_0 + C)/(T + C) * (T/T_0)^{1.5}$, WHERE C=198
2) SOURCES OF DATA ARE THE <u>INTERNATIONAL CRITICAL TABLES</u> AND <u>HANDBOOK OF CHEMISTRY AND PHYSICS</u>, CRC PRESS, INC., CLEVELAND, OHIO (1976)
3) ILLUSTRATION COURTESY OF FISHER-KLOSTERMAN, INC.

VISCOSITY OF METHYL CHLORIDE VAPOR

TEMPERATURE (^oC)	VISCOSITY (MICROPOISE) CALCULATED	REPORTED
-92.4	60.3	
-50.0	77.6	
-15.3	91.7	92.0
0.0	97.8	96.9
15.0	103.9	104.0
50.0	117.8	
99.1	137.0	137.0
121.3	145.5	
182.4	168.5	168.0
212.5	179.5	
250.0	193.0	
302.0	211.3	211.0
350.0	227.6	
400.0	244.2	
450.0	260.4	
500.0	276.1	
550.0	291.5	
600.0	306.4	
650.0	321.0	
700.0	335.3	
800.0	362.9	
900.0	389.3	
1000.0	414.6	
1100.0	439.0	
1200.0	462.5	

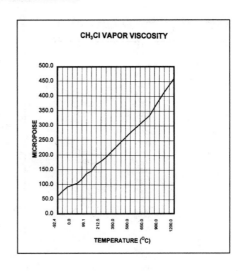

CH$_3$Cl VAPOR VISCOSITY

NOTES:
1) CALCULATED VALUES ARE BASED UPON THE EQUATION

$V=V_o*(T_o+C)/(T+C)*(T/T_o)^{1.5}$, WHERE C=454
2) SOURCES OF DATA ARE THE INTERNATIONAL CRITICAL TABLES AND HANDBOOK
   OF CHEMISTRY AND PHYSICS, CRC PRESS, INC., CLEVELAND, OHIO (1976)
3) ILLUSTRATION COURTESY OF FISHER-KLOSTERMAN, INC.

VISCOSITY OF NITROGEN VAPOR

TEMPERATURE (°C)	VISCOSITY (MICROPOISE)	
	CALCULATED	REPORTED
-226.1	28.8	
-150.0	82.3	
-100.0	113.1	
-21.5	155.2	156.3
0.0	165.7	
10.9	170.9	170.7
23.0	176.5	176.5
27.4	178.5	178.1
127.2	220.9	219.1
150.0	229.7	
200.0	248.4	
226.7	257.9	255.9
299.0	282.4	279.7
350.0	298.7	
400.0	313.9	
450.0	328.6	
490.0	339.9	337.4
550.0	356.3	
600.0	369.5	
700.0	394.6	
825.0	424.1	419.2
900.0	441.0	
1000.0	462.5	
1100.0	483.1	
1200.0	503.0	

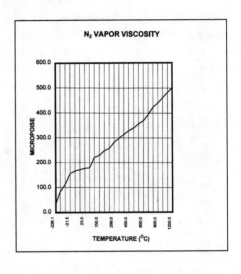

NOTES:
1) CALCULATED VALUES ARE BASED UPON THE EQUATION

$V = V_0 * (T_0 + C)/(T + C) * (T/T_0)^{1.5}$, WHERE C=110.6
2) SOURCES OF DATA ARE THE <u>INTERNATIONAL CRITICAL TABLES</u> AND <u>HANDBOOK</u>
   <u>OF CHEMISTRY AND PHYSICS</u>, CRC PRESS, INC., CLEVELAND, OHIO (1976)
3) ILLUSTRATION COURTESY OF FISHER-KLOSTERMAN, INC.

VISCOSITY OF OXYGEN VAPOR

TEMPERATURE ($^{\circ}$C)	VISCOSITY (MICROPOISE) CALCULATED	REPORTED
-219.1	37.1	
-150.0	92.5	
-100.0	128.5	
-21.5	178.5	
0.0	191.0	189.0
19.1	201.7	201.8
23.0	203.9	203.9
27.4	206.3	
127.7	257.4	256.8
150.0	267.9	
200.0	290.3	
227.0	302.0	301.7
283.0	325.0	323.3
350.0	351.1	
402.0	370.3	369.3
450.0	387.3	
496.0	403.0	401.3
550.0	420.8	
608.0	439.3	437.0
690.0	464.2	461.2
829.0	504.0	501.2
900.0	523.3	
1000.0	549.3	
1100.0	574.3	
1200.0	598.3	

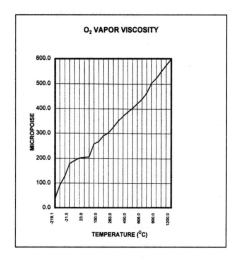

NOTES:
1) CALCULATED VALUES ARE BASED UPON THE EQUATION
   $V = V_o*(T_o+C)/(T+C)*(T/T_o)^{1.5}$, WHERE C=127
2) SOURCES OF DATA ARE THE <u>INTERNATIONAL CRITICAL TABLES</u> AND <u>HANDBOOK OF CHEMISTRY AND PHYSICS</u>,CRC PRESS, INC., CLEVELAND, OHIO (1976)
3) ILLUSTRATION COURTESY OF FISHER-KLOSTERMAN, INC.

VISCOSITY OF PROPANE VAPOR

TEMPERATURE (°C)	VISCOSITY (MICROPOISE)	
	CALCULATED	REPORTED
-100.0	45.6	
-50.0	60.2	
-25.0	67.4	
-5.0	73.0	
0.0	74.3	
17.9	79.2	79.5
50.0	87.9	
75.0	94.4	
100.4	100.9	100.9
125.0	107.1	
150.0	113.2	
175.0	119.2	
199.3	124.9	125.1
250.0	136.5	
300.0	147.4	
350.0	158.0	
400.0	168.2	
500.0	187.5	
600.0	205.7	
700.0	222.9	
800.0	239.2	
900.0	254.7	
1000.0	269.5	
1100.0	283.7	
1200.0	297.4	

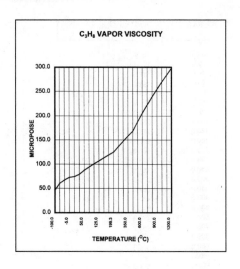

NOTES:
1) CALCULATED VALUES ARE BASED UPON THE EQUATION
$$V = V_o * (T_o + C)/(T + C) * (T/T_o)^{1.5}, \text{ WHERE } C = 290$$
2) SOURCES OF DATA ARE THE INTERNATIONAL CRITICAL TABLES AND HANDBOOK OF CHEMISTRY AND PHYSICS, CRC PRESS, INC., CLEVELAND, OHIO (1976)
3) ILLUSTRATION COURTESY OF FISHER-KLOSTERMAN, INC.

VISCOSITY OF PROPYLENE VAPOR

TEMPERATURE (°C)	VISCOSITY (MICROPOISE)	
	CALCULATED	REPORTED
-100.0	47.5	
-50.0	63.2	
-25.0	70.9	
-5.0	77.0	
0.0	78.5	
16.7	83.5	83.4
49.9	93.3	93.5
75.0	100.5	
100.1	107.6	107.6
125.0	114.5	
150.0	121.3	
175.0	127.9	
199.4	134.3	133.8
250.0	147.2	
300.0	159.4	
350.0	171.3	
400.0	182.7	
500.0	204.5	
600.0	225.0	
700.0	244.5	
800.0	262.9	
900.0	280.5	
1000.0	297.4	
1100.0	313.5	
1200.0	329.1	

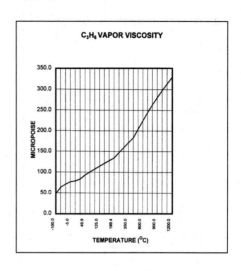

NOTES:
1) CALCULATED VALUES ARE BASED UPON THE EQUATION

$V = V_o \cdot (T_o + C)/(T + C) \cdot (T/T_o)^{1.5}$, WHERE C=330
2) SOURCES OF DATA ARE THE <u>INTERNATIONAL CRITICAL TABLES</u> AND <u>HANDBOOK OF CHEMISTRY AND PHYSICS</u>, CRC PRESS, INC., CLEVELAND, OHIO (1976)
3) ILLUSTRATION COURTESY OF FISHER-KLOSTERMAN, INC.

VISCOSITY OF SULFUR DIOXIDE VAPORS

TEMPERATURE (°C)	VISCOSITY (MICROPOISE)	
	CALCULATED	REPORTED
-95.5	71.2	
-90.0	73.9	
-75.0	81.1	85.8
-20.0	107.5	107.8
0.0	117.0	117.0
18.0	125.5	124.2
25.0	128.8	
50.0	140.4	
100.4	163.3	161.2
150.0	185.3	
199.4	206.5	203.8
250.0	227.6	
293.0	245.0	244.7
350.0	267.4	
400.0	286.4	
450.0	304.9	
490.0	319.4	311.5
550.0	340.5	
600.0	357.5	
700.0	390.4	
800.0	421.7	
900.0	451.7	
1000.0	480.4	
1100.0	508.0	
1200.0	534.7	

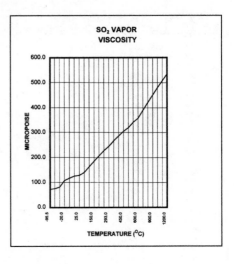

NOTES:

1) CALCULATED VALUES ARE BASED UPON THE EQUATION

$V = V_o \cdot (T_o + C)/(T + C) \cdot (T/T_o)^{1.5}$, WHERE C=416

2) SOURCES OF DATA ARE THE INTERNATIONAL CRITICAL TABLES AND HANDBOOK OF CHEMISTRY AND PHYSICS, CRC PRESS, INC., CLEVELAND, OHIO (1976)

3) ILLUSTRATION COURTESY OF FISHER-KLOSTERMAN, INC.

VISCOSITY OF WATER VAPOR

TEMPERATURE (°C)	VISCOSITY (MICROPOISE) CALCULATED	REPORTED
-17.3	83.5	
-15.0	84.4	
-10.0	86.4	
-5.0	88.4	
0.0	90.4	90.4
15.0	96.4	97.5
20.6	98.6	97.5
28.9	101.9	100.6
100.0	130.2	132.0
150.0	150.0	144.5
200.0	169.4	163.5
250.0	188.6	182.7
300.0	207.4	202.4
350.0	225.9	221.8
400.0	244.0	241.2
450.0	261.8	
500.0	279.3	
550.0	296.4	
600.0	313.2	
700.0	345.8	
800.0	377.2	
900.0	407.5	
1000.0	436.7	
1100.0	465.0	
1200.0	492.4	

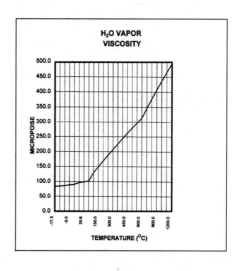

NOTES:
1) CALCULATED VALUES ARE BASED UPON THE EQUATION

$V=V_0*(T_0+C)/(T+C)*(T/T_0)^{1.5}$, WHERE C=650

2) SOURCES OF DATA ARE THE INTERNATIONAL CRITICAL TABLES AND HANDBOOK
   OF CHEMISTRY AND PHYSICS,CRC PRESS, INC., CLEVELAND, OHIO (1976)
3) ILLUSTRATION COURTESY OF FISHER-KLOSTERMAN, INC.

Guide to Densities and Specific Gravities of Some Common Bulk Solids

Material Name	Alternate Name	Particle Specific Gravity Range	Avg. S.G.	Bulk Density Range LB/FT^3	Avg. B.D. LB/FT
Acetate Powder	-	1.4	1.4	-	-
Adipic Acid	-	1.4	1.4	-	-
Alfalfa, Ground	-	-	-	19.8	19.8
Alfalfa, Meal	-	-	-	28.1	28.1
Aluminum Fluoride	-	2.9	2.9	70.1	70.1
Aluminum, Ground	-	2.7	2.7	34.5	34.5
Aluminum Oxide	Corundum	3.9 - 4.0	4.0	54.6 - 64.5	59.6
Ammonium Oxalate	-	1.5	1.5	21.4	21.4
Ammonium Phosphate	-	1.7	1.7	21.0	21.0
Ammonium Sulfate		1.7 - 1.8	1.8	48.0 - 56.0	52.0
Asbestos	Amosite	1.8 - 2.2	2.0	5.0 - 8.9	7.0
Asbestos	Brake Lining	-	-	36.2	36.2
Avicel	Crystalline Cellulose	1.7	1.7	18.7	18.7
Bark Dust	-	.80 - 1.16	.98	-	
Barley	-	-		39.3	39.3
Bastnaesite	-	4.9 - 5.2	5.1	-	
Beet Pulp	-	1.5	1.5	26.0	26.0
Bone Meal	-	3.0	3.0	52.6	52.6
Borax	Tincal	1.7	1.7	51.6 - 53.0	52.0
Boric Acid	-	1.4	1.4	-	
Bran Dust	-	-		26.8 - 28.9	27.9
Brick Dust		1.4 - 2.2	1.8	57.0	57.0
Cadmium Oxide		7.0 - 8.1	7.6		
Calcium Carbonate		2.7 - 3.0	2.9		
Calcium Fluoride	Fluorspar	3.0 - 3.3	3.2	101.0 - 105.0	103.0

Material Name	Alternate Name	Particle Specific Gravity Range	Avg. S.G.	Bulk Density Range LB/FT^3	Avg. B.D. LB/FT
Calcium Phosphate	-	2.2	2.2	62.5 - 65.0	63.8
Calcium Propionate	-	1.2	1.2		
Calcium Sucaryl	Calcium Cyclamate	-	-	25.5	25.5
Carbon, Activated	-	2.0	2.0	39.9	39.9
Carbon, Black	Furnace Black	1.8 - 2.1	2.0	34.8	34.8
Catalyst, FCC		1.8 - 2.6	2.3	44.5	44.5
Celite	Diatomaceous Earth	.24 - .34	.29	11.0 - 12.5	11.8
Cement	Portland Cement	2.6 - 2.8	2.7	94.0 - 95.0	94.5
Ceramic Dust				45.0	45.0
Charcoal Dust		1.6	1.6	25.0	25.0
Chicken Meal				25.9	25.9
Chocolate Fudge				37.5	37.5
Chrome Ore		7.2	7.2	85.0	85.0
Cinnabar	Natural Vermilion	8.1	8.1		
Clay		1.8 - 2.6	2.2	58.0 - 59.0	58.5
Clay, Baked	Porcelain	2.3 - 2.5	2.4	56.0	56.0
Clay, Kaolin		2.2 - 2.6	2.4		
Coal Dust		1.2 - 1.8	1.5	25.9	25.9
Cobaltic Oxide		4.8 - 5.6	5.2	-	-
Cocoa Mix		1.5	1.5	-	-
Coffee Instant		1.5	1.5	15.0	15.0
Coke		1.0 - 1.7	1.4	28.4 - 46.8	37.6
Copper		8.9	8.9		

Material Name	Alternate Name	Particle Specific Gravity Range	Avg. S.G.	Bulk Density Range LB/FT^3	Avg. B.D. LB/FT
Copper Cones		3.1 - 4.1	3.5		
Copper Oxide		5.7 - 6.3	6.0		
Copper Sulfate	Blue Vitriol	2.3	2.3	50.2	50.2
Cork		.22 - .26	.24	14.9 - 15.2	15.0
Corn Gluten		1.0 - 1.1	1.1	26.0	26.0
Corn, Ground		-	-	38.0 - 45.0	41.5
Corn Hulls		-	-	4.6	4.6
Corn Meal		.92 - .93	.93	32.7	32.7
Cotton Seed Meal		.92	.92	48.0	48.0
Cryolite	Ice Stone	2.9 - 3.0	3.0	-	-
Diammonium Phosphate	Ammonium Phosphate	1.6	1.6	-	-
Diatomaceous Earth	Celite	.24 - .34	.29	11.0 - 12.5	11.8
Diatomaceous Earth	Diatomite	1.9 - 2.4	2.2	10.9	10.9
Diazo Salts	Sodium Nitrate	2.3	2.3	-	-
Dicalcium Phosphate		2.3	2.3	103.0	103.0
Dolomite	Pearl Spar	2.9	2.9	87.5	87.5
Egg White Dehyd.		-	-	22.4	22.4
Feldspar	Potassium Alumino Silicate	1.5	1.5	73.0	73.0
Filtrol	Clay	-	-	44.0	44.0
Flax	-	-	-	7.0	7.0
Flour	-	-	-	23.7	23.7
Fluorspar	Fluorite	3.0 - 3.3	3.2	101.0 - 105.0	103
Flyash	-	1.3 - 2.3	1.8	49.0 - 49.7	49.3
Fumaric Acid	Boletic Acid	1.6	1.6	45.7	45.7
Gelatin		1.3	1.3	40.5 - 43.5	42.0
Hydrite MP	Clay	2.6	2.6	45.6	45.6

Material Name	Alternate Name	Particle Specific Gravity Range	Avg. S.G.	Bulk Density Range LB/FT^3	Avg. B.D. LB/FT
Iron Ore	-	3.0 - 5.2	4.6	193.5	193.5
Iron Oxide		3.3 - 5.1	4.2	-	-
Jello	-	1.3	1.3	28.5	28.5
Kool Aid	-	-	-	34.8	34.8
Kralac Plastic	-	1.0	1.0	42.7	42.7
Lacquer	-	-	-	25.5	25.5
Lead Borate	-	5.6	5.6	-	-
Lead Dust	-	11.4	11.4	184.0	184.0
Lead Oxide	-	8.3 - 9.2	8.8	-	-
Lignite	Coal	1.2	1.2	75.0 - 94.0	83.5
Lime, Granular	-	-		39.2	39.2
Lime, Hydrated	Slacked Lime	1.3 - 1.4	1.4	28.8 - 36.4	31.6
Limestone		2.7 - 2.8	2.8	167.0 - 171.0	169.0
Liver Meal	-	-	-	38.8	38.8
Magnesium Carbonate	Magnesia Alba	3.0 - 3.1	3.1	-	-
Magnesium Oxide	Magnesia	3.7	3.7	-	-
Magnesium Silicate	-	2.6 - 2.8	2.7	19.2	19.2
Milk, Powdered	-	1.2	1.2	30.2	30.2
Molasses	-	1.5	1.5	48.5	48.5
Molybdenum Disulfide	-	4.8	4.8	-	-
Molybdenum Trioxide	-	4.4	4.4	-	-
Molybdenum	-	10.2	10.2	-	-
Nickel Carbonate	-	2.6	2.6	19.0	19.0
Nickel Dust	-	8.6	8.6	218.0	218.0
Niobium Concentrate	-	8.6	8.6	-	-
Nitrophenide		-	-	25.8	25.8
Oats, Ground	-	-	-	26.8	26.8

Material Name	Alternate Name	Particle Specific Gravity Range	Avg. S.G.	Bulk Density Range LB/FT^3	Avg. B.D. LB/FT
Oat Hulls	-	-	-	17.8	17.8
Paper Dust	-	0.7 - 1.2	1.0	6.1	6.1
Pea Meal	-	-	-	35.0	35.0
Pectin	-	-	-	38.5	38.5
Polystryrene	-	1.1	1.1	27.5	27.5
Polyurethane	Isocyanate Resin	-	-	2.0 - 30.0	16.0
Polyvinyl Alcohol	PVA	1.21 - 1.31	1.26	17.0 - 23.0	20.0
Polyvinyl Butyral	-	1.1	1.1	-	-
Polyvinyl Chloride	PVC	1.0 - 1.5	1.3	27.0 - 38.0	32.5
Porcelain	-	2.3 - 2.5	2.4	56.0	56.0
Potash	-	2.0	2.0	-	-
Potato Granules	-	-	-	56.0	56.0
Potatoes Instant	-	-	-	45.0	45.0
Protein Hydrolyste	-	1.4	1.4	-	-
Pumice	Igneous Rock	-	-	49.5	49.5
Pyrite	Fools Gold	4.9 - 5.2	5.1	52.0	52.0
Redwood Sawdust	-	-	-	17.7	17.7
Resin, Pathic	-	-	-	16.6	16.6
Resin, Phenolic	-	1.07	1.07	21.6	21.6
Resin, Saran	Thermoplastic Resin	-	-	33.2	33.2
Resin, Vinyl	Vinyl Plastic	-	-	19.3 - 22.2	20.8
Rubber, Synthetic	Butyl Rubber	-	-	18.0	18.0
Rubber, Ground	-	-	-	20.4 - 24.0	22.2
Rubber, Shredded	-	-	-	6.5 - 10.5	8.5

Material Name	Alternate Name	Particle Specific Gravity Range	Avg. S.G.	Bulk Density Range LB/FT^3	Avg. B.D. LB/FT
Sand Foundry	-	2.6	2.6		
Santobrite	Sodium Penta Chlorophenate	2.1	2.2	48.6	48.6
Santocel	Silica Aerogel	-	-	1.4 - 4.8	3.1
Shale	-	1.6 - 2.7	2.2	63.6	63.6
Silica		2.2 - 2.6	2.4	16.7	16.7
Silica Flour	Ajax	1.2	1.2	57.0	57.0
Silica Floor	Comet	-	-	67.7	67.7
Silica Gel		2.1 - 2.3	2.2	19.0	19.0
Silicon Carbide		3.17	3.17	66.7	66.7
Slate, Powdered		2.6 - 3.3	3.0	46.8 - 50.7	48.8
Sodium, Bicarbonate	Baking Soda	2.2	2.2	-	-
Sodium Carbonate	Soda Ash	2.5	2.5	-	-
Sodium Chromate		1.5	1.5	30.8	30.8
Sodium Fluoride		2.8	2.8	82.5	82.5
Sodium Phosphate		2.0	2.0	24.0	24.0
Sodium Stannate	Preparing Salt	3.2	3.2	-	-
Sodium Sulfate	-	2.7	2.7	78.0	78.0
Sodium Sulfite		2.6	2.6	80.0	80.0
Sodium Tripolyphosphate		2.5	2.5	41.0	41.0
Soy Bean Dust		1.7	1.7	-	-
Soy Bean Meal		-	-	43.2	43.2
Starch, Corn		1.5	1.5	33.0 - 38.0	35.5
Starch, Potato		1.5	1.5	36.0	36.0
Starch, Wheat		1.5	1.5	36.8	36.8
Stearic Acid	Fatty Acid	.84	.84	30.5	30.5
Sugar		1.6	1.6	35.1	35.1
Taconite	Iron Oxide	4.0	4.0	108.5	108.5
Talc		2.7 - 2.8	2.8	24.5 - 26.8	25.7
Titanium Dioxide		3.8	3.8	-	-
Tobacco Dust		-	-	20.6	20.6
TPA	Terephaltic Acid	1.5 - 1.6	1.6	19.3	19.3
Trap Rock		2.6 - 2.9	2.8	-	-
Urea	Carbamide	1.3	1.3	44.6	44.6

Material Name	Alternate Name	Particle Specific Gravity Range	Avg. S.G.	Bulk Density Range LB/FT^3	Avg. B.D. LB/FT
Vanillin	-	-	-	43.8	43.8
Vegetable Fibers	-	-	-	15.7	15.7
Vitamin	-	-	-	15.5	15.5
Wheat Flour	-	-	-	34.6	34.6
Wheat Germ	-	-		26.9	26.9
Wood Flour	Wood Meal	-	-	13.9 - 16.6	15.3
Wool Floss	-	-	-	10.8	10.8
Zinc Dust	-	7.1	7.1	178.0	178.0
Zinc Oxide	Chinese White	5.5	5.5	-	-
Zinc Sulphide	-	Greater Than 4.0		-	-
Zirconium Oxide	Zirconia	5.7	5.7	79.7	79.7

Particle Physics

Albert J. Heber

Purdue University
West Lafayette, IN

Nomenclature

a	acceleration, ft/s^2 [m/s^2]
d	particle diameter, ft [μm, m]
e	elementary charge, one electron
g	gravitational acceleration, ft/s^2 [m/s^2]
k	Boltzmann constant = 1.38×10^{-16} lb$_f \cdot$ ft/°R [1.38×10^{-23} N $\cdot$ m/K]
n	number
t	time, s
x	displacement, ft [m]
C	Cunningham slip correction factor
D	characteristic length, ft [m]
E	electric field strength, V/ft [V/m]
F	force, lb [N]
Kn	Knudsen number
λ	path length, ft [m]
Re	Reynolds number

S_t Stokes number

T temperature, °R [K]

U velocity, ft/s [m/s]

Z characteristic length, ft [m]

σ diffusivity, ft^2/s [m^2/s]

ρ density, lb/ft^3 [kg/m^3]

η gas viscosity, lb$_f$ · s/ft^3 or lb$_m$/ft · s [kg/m · s]

χ particle shape factor

τ relaxation time, s

Subscripts

c centrifugal

e equivalent

f force

i initial

m mass

n normal

p particle

rms   root mean square

t terminal

l streamline

D drag

E electrical

G gas

S sphere

T tangential

ct   terminal centrifugal

TS   terminal settling

2.1 Introduction

An aerosol, or airborne particles, includes dusts, fumes, mists, and smoke. The principal aerodynamic behavior of these particles is determined by the characteristic properties of each particle and any external forces. Physical forces and characteristic properties include gravitational settling, curvilinear motion, brownian motion, thermophoresis, impaction, interception, coagulation, condensation, evaporation, electric charge, and light scattering and extinction. These forces and characteristics of particles will be discussed in this chapter.

2.2   Gas Properties and Motion

Knowledge of certain aspects of the kinetic theory of gases is necessary to understand the interaction of particles with the surrounding gas. This kinetic theory explains temperature, pressure, mean free path, viscosity, and diffusion in regard to the motion of gas molecules.[1] The theory assumes gases contain a large number of molecules which are small enough so that the relevant distances between them are discontinuous. The theory also assumes that the molecules are rigid spheres that travel in straight lines and experience perfectly elastic collisions.

Air molecules travel an average of 1519 ft/s [463 m/s] at standard conditions. Increased molecule weight will decrease speed. Molecular velocity increases as the square root of absolute temperature increases. Temperature, then, is an indication of the kinetic energy of gas molecules. Pressure arises with molecular impacts on a surface and is directly related to concentration. The transfer of momentum by randomly moving molecules from a faster moving layer of gas to an adjacent slower moving layer of gas is represented by gas viscosity. Viscosity of a gas is independent of pressure but will increase as temperature increases. Finally, diffusion is the transfer of molecular mass without any fluid flow.[1] Diffusion transfer of gas molecules is from a higher to a lower concentration. Movement of gas molecules by diffusion is directly proportional to the concentration gradient, inversely proportional to concentration, and proportional to the square root of absolute temperature.

The average distance a molecule travels in a gas between collisions with other molecules is called the *mean free path*. It is the kinetic theory's most critical quantity. The mean free path increases with increasing temperature and decreases with increasing pressure.[1] Mean free path and other properties of air at standard conditions are given in Table 2.1.

TABLE 2.1   Properties of Air at Atmospheric Pressure and 68°F (20°C)

Molecular weight	0.0637 lb/mol [0.0289 kg/mol]
Average molecular velocity	1519 ft/s [463 m/s]
Molecular diameter	1.21×10^{-9} ft [0.00037 μm]
Mean free path	2.16×10^{-7} ft [0.066 μm]
Viscosity	3.78×10^{-7} $lb_f \cdot$ s/ft^2 or 1.22×10^{-5} lb_m/ft $\cdot$ s [1.81×10^{-5} $\dfrac{N \cdot s}{m^2}$ or 1.81×10^{-5} kg/m $\cdot$ s]
Density	0.075 lb_m/ft^3 [1.2 kg/m^3]
Diffusion coefficient	2.04×10^{-4} ft^3/s [1.9×10^{-5} m^3/s]

Gas flow is characterized by the Reynolds number, which is a dimensionless index that describes the flow regime. The Reynolds number for gas is the following ratio of inertial to frictional forces:

$$Re = \frac{\rho \, U_G D}{\eta} \qquad (2.1)$$

where   Re = Reynolds number
 ρ = gas density, lb/ft³ [kg/m³]
 U_G = gas velocity, ft/s [m/s]
 D = characteristic length, ft [m]
 η = gas viscosity, lb$_m$/ft · s [kg/m · s]

It helps to determine the flow regime, the application of certain equations, and geometric similarity.[2] The flow is laminar (flow streamlines remain distinct as very little mixing occurs) at low Reynolds numbers, and viscous forces predominate. Inertial forces dominate the flow at high Reynolds numbers when mixing causes the streamlines to disappear. Laminar and turbulent flows are characterized by Re < 1 and Re > 1000, respectively, for flow of a sphere in still air and by Re < 2100 and Re > 4000, respectively, for flow through a circular duct.

2.3   Gravitational Settling

The Reynolds number for a particle characterizes its motion through a gas and is

$$Re_p = \frac{\rho_G \, (U_p - U_G)d}{\eta_G} \qquad (2.2)$$

where Re_p = Reynolds number for a particle
 ρ_G = gas density, lb/ft³ [kg/m³]
 U_p = particle velocity, ft/s [m/s]
 U_G = gas velocity, ft/s [m/s]
 d = particle diameter, ft [m]
 η_G = gas viscosity, lb/ft · s [kg/m · s]

For example, a particle with a diameter of 2.62×10^{-5} ft [8.0 μm] and a density of 93.6 lb/ft³ [1500 kg/m³] traveling at 0.328 ft/s [0.1 m/s] in standard air has a particle Reynolds number of

$$Re_p = \frac{(2.62 \times 10^{-5}\ \text{ft}) \times 0.328\ \text{ft/s} \times 0.075\ \text{lb}_m/\text{ft}^3}{1.22 \times 10^{-5}\ \text{lb}_m/\text{ft} \cdot \text{s} = 0.053}$$

$$= \left[\frac{(8.0 \times 10^{-6} \text{ m}) \times 0.10 \text{ m/s} \times 1.203 \text{ kg/m}^3}{1.81 \times 10^{-5} \text{ kg/m} \cdot \text{s}} = 0.053 \right]$$

The drag force acting on a spherical particle in a still gas assuming laminar flow is predicted by the Stokes equation[3]:

$$F_D = \frac{3 \, \Pi \, \eta \, U_p \, d \, \chi}{C} \qquad (2.3)$$

where F_D = drag force, lb_f [N]
  U_p = particle velocity, ft/s [m/s]
  d = particle diameter, ft [m]
  η = gas viscosity, $\text{lb}_f \cdot \text{s/ft}$ [N $\cdot$ s/m^2]
  χ = dynamic particle shape factor
  C = Cunningham slip correction factor

The assumptions of the Stokes equation include a continuous, incompressible, viscous, and infinite gas with rigid and spherical particles.[4] Assuming unit values of χ and C (both described later), $\eta = 3.81 \times 10^{-7} \text{ lb}_f \cdot \text{s/ft}^2$, $u = 0.328$ ft/s, $d = 2.62 \times 10^{-5}$ ft (8.0 μm), and the drag force on the particle is

$$F_D = 3 \, \Pi \, (3.81 \times 10^{-7} \text{ lb}_f \cdot \text{s/ft}^2) \times 0.328 \text{ ft/s} \times (2.62 \times 10^{-5} \text{ ft})$$
$$= 3.086 \times 10^{-11} \text{ lb}_f \qquad (2.3)$$

$$= [3 \, \Pi \, (1.82 \times 10^{-5} \text{ N} \cdot \text{s/m}) \times 0.1 \text{ m/s} \times (8 \times 10^{-6} \text{ m})$$
$$= 1.372 \times 10^{-10} \text{ N}]$$

If $\text{Re}_p > 1000$, Newton's drag force applies and the drag force on the particle is proportional to particle diameter d^2 and U^2 (see Ref. 5). If $\text{RE}_p < 1$, the particle drag force is in the Stokes region, as illustrated in Fig. 2.1, and is proportional to d and U [see Eq. (2.5)]. Most airborne particles are in the Stokes region as they settle. For example, the Stokes equation applies for quartz particles below 2.78×10^{-4} ft [85 μm] in diameter.[6] Setting the Stokes drag force equal to the gravitational force, terminal settling velocity may be solved with the following:

$$U_{TS} = \frac{(\rho_p - \rho_G) \, d^2 \, g \, C}{18 \, \eta \, \chi} \qquad (2.4)$$

where U_{TS} = terminal settling velocity, ft/s [m/s]
  ρ_p = particle density, lb_m/ft^3 [kg/m^3]
  ρ_G = gas density, lb_m/ft^3 [kg/m^3]
  d = particle diameter, ft [m]

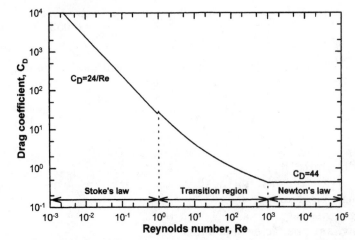

Figure 2.1 Drag coefficient versus Reynolds number for spheres. (*From Ref. 1.*)

g = gravitational acceleration, ft/s^2 [m/s^2]
C = Cunningham slip correction factor
η = gas viscosity, lb$_m$/ft · s [kg/m · s]
χ = dynamic particle shape factor

It is apparent from this equation that the settling velocity is proportional to the particle's cross-sectional area and the ratio of its density to the gas viscosity. The terminal settling velocity of the 2.62×10^{-5} ft [8 µm] particle where $C = 1.0165$ (explained later) in still air is

$$U_{\mathrm{TS}} = \frac{(93.6 - 0.075 \text{ lb}_m/\text{ft}^3)(2.62 \times 10^{-5} \text{ ft})^2 (32.2 \text{ ft/s}^2)(1.017)}{18 (1.214 \times 10^{-5} \text{ lb}_m/\text{ft} \cdot s)(1)}$$

$$= 0.0096 \text{ ft/s}$$

$$= \left[\frac{(1500 - 1.2 \text{ kg/m}^3)(8 \times 10^{-6} \text{ m})^2 (9.8 \text{ m/s}^2)(1.017)}{18 (1.81 \times 10^{-5} \text{ kg/m} \cdot s)(1)} \right.$$

$$\left. = 0.0029 \text{ m/s} \right] \tag{2.4}$$

The buoyancy effect ρ_G is usually neglected in this equation because the density of particles is usually about 1000 times greater than the density of air.

Equation (2.5) gives an accuracy of 1 percent for Cunningham slip coefficient when Re$_p$ is between 10^{-7} and 10^{-6}, and 10 percent for Re$_p$ as low as 4×10^{-11} and as high as 10^{-5} corresponding to Knudsen number (Kn) values from about 0.3 to 13. Particles with Kn greater than 10 in air at standard conditions are smaller than 3.28×10^{-8} ft [0.01 μm]. These particles are extremely small and remain suspended in still air.[7] The settling velocity at high Re is given in Refs. 2 and 4.

The terminal settling velocity can also be calculated as the acceleration of gravity times the relaxation time. The time it takes for a particle to reach its terminal velocity is 92 ms for a particle with a diameter of 3.28×10^{-4} ft [100 μm] and less than 1.0 ms for particles with diameters smaller than 3.28×10^{-5} ft [10 μm]. For most problems in pollution control, one can assume that particles come instantly to the terminal velocity and therefore neglect calculating velocities during acceleration or deceleration.[1]

A nonspherical particle settles more slowly than a sphere of the same volume. The dynamic shape factor χ accounts for this phenomenon in Eqs. (2.5) and (2.8). Dynamic shape factors for quartz, sand, and talc are 1.36, 1.57, and 2.04, respectively, and 1.05 to 1.11 for bituminous coal.[8]

The Knudsen number is determined by dividing the mean free path of gas molecules by particle radius and is used to divide the motion of aerosol particles into the following regimes:

$0 < \mathrm{Kn} < 0.01$	Continuum
$0.01 < \mathrm{Kn} < 0.2$	Slip flow
$0.2 < \mathrm{Kn} < 10$	Transition
$10 < \mathrm{Kn}$	Free molecule

Particle size is so large compared to the mean free path of gas molecules in the continuum regime that it "feels" like a continuum to the particle. At the other extreme, in the free molecule regime, the particles approach the size of gas molecules and behave just like them.

Figure 2.2 shows the relative size and spacing of air molecules at standard conditions as compared to a particle with a diameter of 3.28×10^{-7} ft [0.1 μm]. Small particles tend to slip between air molecules. The degree of slip is related to the mean free path of gas molecules and decreases with higher densities as molecules move closer together. The Cunningham slip correction factor C is used to predict behavior of small particles (Kn > 0.1). For air, and $d > 3.28 \times 10^{-7}$ ft [0.1 μm],

$$C = 1 + \frac{2.52\lambda}{d} \qquad (2.5)$$

where d = particle diameter, ft [m]
   C = Cunningham slip correction factor
   λ = mean free path of the gas, ft [m]

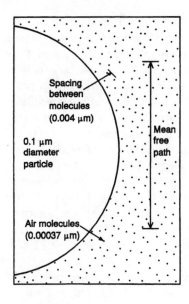

Figure 2.2 Relative size and spacing of air molecules at standard conditions.

This correction is negligible for particles with diameters larger than 3.28×10^{-6} ft [1 μm]. It is 1.02, 1.15, and 2.90 for particles with diameters of 3.28×10^{-5}, 3.28×10^{-6}, and 3.28×10^{-7} ft [10, 1.0, and 0.1 μm], respectively, and increases rapidly for particles with smaller diameters. Equation (2.8) modified for $d < 3.28 \times 10^{-8}$ ft [0.1 μm] is given in Eq. (10.9).[4] The correction also increases with temperature. For example, for a particle with a diameter of 5.25×10^{-6} ft [1.6 μm], C is 1.1 at standard conditions and 1.37 at 120°C.

Free-falling or sedimentation diameter and Stokes diameter refer to the diameter of a sphere having the same terminal settling velocity and density as a nonspherical or irregular particle.[7] They should be used in the behavioral equations presented above. Stokes diameter applies specifically to laminar flow (Re < 0.2). The equivalent volume diameter d_e of a particle is the diameter of a sphere having the same volume as an irregular particle. The dynamic particle shape factor χ is a proportionality constant accounting for the effect of shape on particle motion. It is defined as the ratio of the actual resistance force of the nonspherical particle to the resistance force of a sphere having the same volume and velocity.[1]

The aerodynamic diameter of a nonspherical particle is the diameter of a unit density sphere having the same resistance to motion as the particle in question. An aerodynamic diameter is a key property in filtration, air cleaning, and aerosol sampling. It is used in describing a device's collection efficiency as a function of particle size.[8]

2.4 Relaxation Time and Stopping Distance

The velocity of moving particles must be changed before capturing them in collection devices. Sometimes, to increase collection efficiency, the collection device accelerates the particles first before decelerating them.[7]

The time required for a particle to adjust from one steady-state velocity to another is characterized by its relaxation time.[7] Relaxation time is the period during which most of the motion change occurs when a force on the particle changes, e.g., time for a falling particle to reach terminal velocity. Relaxation time of a particle is independent of the type of force applied to it and is defined as

$$\tau = \frac{d^2 \, \rho_p \, C}{18\eta} \qquad (2.6)$$

where τ = relaxation time, s
d = particle diameter, ft [m]
ρ_p = particle density, lb_m/ft^3 [kg/m^3]
C = Cunningham slip correction factor
η = gas viscosity, $lb_m/ft \cdot s$ [$kg/m \cdot s$]

Relaxation time is proportional to the square of the particle diameter and is also affected by the temperature and pressure of the gas [see Eq. (2.9)]. The relaxation times for particles with diameters of 3.28×10^{-6}, 3.28×10^{-5}, and 3.28×10^{-4} ft [1.0, 10, and 100 μm] at standard conditions are 3.6×10^{-6}, 3.1×10^{-4}, and 3.1×10^{-2} s, respectively. For the example particle with a diameter of 3.28×10^{-6} ft [8 μm], the relaxation time is 2.99×10^{-4} s.

Stopping distance is the maximum distance a moving particle will travel in still air after all external forces are removed. For particles in the Stokes region, it is calculated as velocity times relaxation time. Applications of this might be an electrical force suddenly turned off or a particle moving in an airstream that abruptly turns 90°. Stopping distances are small and particle motion is dominated by the motion of the gas. For example, the stopping distance of the example particle with a diameter of 2.62×10^{-7} ft [8 μm] and initially traveling 32.8 ft/s [10 m/s] (Re_p = 5.3) in standard air is only 0.01 ft [0.003 m].

2.5 Curvilinear Particle Motion

Curvilinear particle motion causes a centrifugal force and is used in impaction, interception, and cyclone separation of particles.[7] Impaction is the collection of particles by inertia, interception, or brownian diffusion. When a gas stream changes direction while parti-

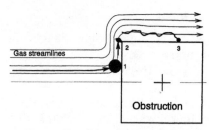

Figure 2.3 Possible particle movement near an obstruction in a moving gas stream.

cles are moving with it, the resulting centrifugal force causes particles to deviate from streamlines because of their inertia. This occurs when obstacles in the gas stream are used to collect particles (particle 1 in Fig. 2.3). Inertial impaction is the predominant mechanism occurring in particle collection devices. Interception occurs when a particle following a gas streamline flows around a collector at a distance less than the particle radius (particle 2). Interception increases with larger particle sizes relative to the collector. Particles can also impact on the surface because of diffusion (particle 3). Collectors may also be larger particles or droplets that are moving and overtaking smaller particles. This process is called scavenging.

Particle removal becomes more efficient with increases in the Stokes number, a parameter that characterizes inertial impaction. The Stokes number S_t is a dimensionless ratio of particle stopping distance to a characteristic distance of the collector. It represents the ratio of drag forces to viscous forces and is defined as

$$S_t = \frac{d^2 \, \rho_p \, (U_p - U_G) \, C}{9 \eta L} \tag{2.7}$$

where S_t = Stokes number
   d = particle diameter, ft [m]
   ρ_p = particle density, lb/ft³ [kg/m³]
   U_p = particle velocity, ft/s [m/s]
   U_G = gas velocity, ft/s [m/s]
   C = Cunningham slip correction factor
   η = gas viscosity, lb$_m$/ft · s [kg/m · s]
   L = characteristic length of the collector, ft [m]

The Stokes number increases as particles resist changing their directions as gas streamlines change. Critical Stokes numbers above which collection by impaction occurs are 0.30 for a stream approaching a cylinder, 0.20 for a stream approaching a circular disc, 0.16 for a jet perpendicular to a plane, and 0.10 for a stream approaching a sphere.[7] Inertial impingement is the process of collecting particles from gas streams on relatively large surfaces. Such surfaces can be flat plates, large spheres, or large cylinders. Collection efficiencies have been collected for each type.[7]

The terminal centrifugal velocity for spherical particles in the continuum and slip flow regimes is

$$U_{ct} = \frac{d^2 \, \chi \rho_p U_t^2 C}{18 \eta r_1} \tag{2.8}$$

where U_{ct} = terminal centrifugal velocity, ft/s [m/s]
  d = particle diameter, ft [m]
  χ = dynamic particle shape factor
  ρ_p = particle density, lb/ft^3 [kg/m^3]
  U_t = terminal velocity, ft/s [m/s]
  C = Cunningham slip correction factor
  η = gas viscosity, lb$_m$/ft · s [kg/m · s]
  r_1 = radius of collector at inlet, ft [m]

The velocity of a small accelerating spherical particle at time t is predicted by

$$U_p = \tau \, C \, a(1 - e^{-t/\tau C}) \tag{2.9}$$

where U_p = particle velocity, ft/s [m/s]
  τ = relaxation time, s
  C = Cunningham slip correction factor
  a = acceleration, ft/s^2 [m/s^2]
  t = time, s

The velocity of a small decelerating spherical particle is predicted by

$$U_p = U_{pi} e^{-t/\tau C} \tag{2.10}$$

where U_p = particle velocity, ft/s [m/s]
  U_{pi} = initial particle velocity, ft/s [m/s]
  τ = relaxation time, s
  C = Cunningham slip correction factor
  t = time, s

While accelerating from a standstill, the distance a particle in the continuum or slip flow regimes travels is

$$x = \tau C \, (at - U_p) \tag{2.11}$$

where x = displacement, ft [m]
   τ = relaxation time, s
   C = Cunningham slip correction factor
   a = acceleration, ft/s^2 [m/s^2]
   t = time, s
   U_p = particle velocity, ft/s [m/s]

While decelerating from velocity U_{pi}, the distance traveled is

$$x = U_{pi}\,\tau\, C(1-e^{-t/\tau C}) \qquad (2.12)$$

where x = displacement, ft [m]
   U_{pi} = initial particle velocity, ft/s [m/s]
   τ = relaxation time, s
   C = Cunningham slip correction factor
   t = time, s

Graphical methods are required to determine the velocity or distance traveled by an accelerating or decelerating particle in the Newton regime for Re_p from 0.1 to 400.[7]

2.6   Brownian Motion and Diffusion

Particles with diameters smaller than 3.28×10^{-7} ft [0.1 μm] appear increasingly like gas molecules in their movement, in their kinetic energy, and in the diffusion force caused by their random motion. This random motion, called *Brownian motion,* is a result of thermal energy as fluid molecules bombard the tiny particle. A particle will agitate more slowly than a gas molecule of the same size because of its much heavier mass. Whereas gas molecules move in straight lines between collisions, particle movement is characterized by smooth curves. Brownian motion also causes particles to rotate. For example, a particle with a diameter of 1.64×10^{-5} ft [5 μm] rotates an average of 1.3 revolutions per minute (r/min) in standard air.[4]

The net transfer of particles through air by brownian motion is the diffusion coefficient times the concentration gradient (Fick's First Law of Diffusion). The diffusion coefficient is

$$\sigma = \frac{kTC}{3\,\Pi\,\eta\,d} \qquad (2.13)$$

where σ = diffusivity, ft^2/s [m^2/s]
   k = Boltzmann constant, 1.38×10^{-16} lb$_f \cdot$ ft/°R [1.38×10^{-23} N $\cdot$ m/K]

T = absolute temperature, °R [K]
C = Cunningham slip correction factor
η = gas viscosity, $\mathrm{lb_f} \cdot \mathrm{s/ft^3}$ [kg/m · s]
d = particle diameter, ft [m]

It does not depend on particle density but is affected by gas viscosity and temperature. The diffusion coefficient is inversely proportional to particle size (see Fig. 2.4). The slip correction factor C amplifies this inverse dependence on size.

The diffusional transport of a particle with a diameter of 3.28×10^{-8} ft [0.01 μm] is 20,000 times as fast as a particle with a diameter of 3.28×10^{-4} ft [10 μm],[1] and its diffusional velocity is nearly 4000 times faster than its terminal settling velocity.[8] Diffusion is the primary transport mechanism for small particles (diameters $< 3.28 \times 10^{-7}$ ft [0.1 μm]) unless there is convection. The root mean square distance that particles travel in three-dimensional motion is $x_{rms} = (4D_t/\Pi)^{0.5}$ (see Ref. 5) and is only 3.9×10^{-4} ft [0.000 12 m] for our particle with a diameter of 2.62×10^{-5} ft [8 μm].

The concentration gradient needed for particle transfer by diffusion is always present near solid surfaces since particles adhere to them. In other words, a solid surface is a sink for particles. Extremely close to the surface, the concentration is zero if all striking particles are removed, and the movement of particles to the surface decreases the ambient concentration.[1] Diffusive deposition from flowing gas streams decreases as the flow rate increases.[7]

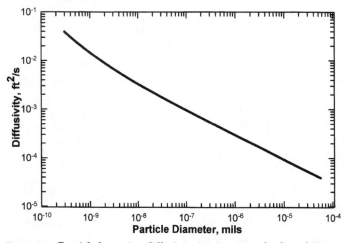

Figure 2.4   Particle brownian diffusivity in air at standard conditions.

2.7 Coagulation

Brownian motion causes particles to collide and adhere to each other in a process called thermal coagulation. Coagulation increases particle size and decreases particle number. Coagulation depends on n^2 where n is the number of particles in a given volume. When the particle concentration is relatively high, such as in cigarette smoke and condensation vapors,[9] coagulation occurs quickly but then slows down as the particle concentration decreases.

There are other types of coagulation besides thermal coagulation. Scavenging of small particles by large particles moving faster because of external forces (differential settling, rain droplets, spray nozzles) is called kinematic coagulation. Turbulent coagulation is caused by eddies that produce relative motion between particles. Acoustic coagulation occurs when intense sound waves create relative motion between particles.[10] Shear coagulation occurs as particles in slightly different streamlines travel at different velocities, e.g., in air velocity gradients.

2.8 Condensation and Evaporation

Condensation and evaporation of particles occur all the time in the atmosphere. Condensation occurring without any foreign nuclei (dust particles) is called homogeneous condensation,[11] and supersaturation of around 500 percent is required before it will occur.[4] In heterogeneous condensation, a gas such as water vapor condenses on soluble and insoluble nuclei, e.g., aerosol particles. Since liquid vapor pressure is affected by the curvature of the surface to which it adheres, irregular particles condense moisture better than spherical particles.[6] Vapor pressure of tiny water droplets (diameters $< 3.28 \times 10^{-7}$ ft [0.1 μm]) increases and surface tension decreases exponentially with smaller sizes (Fig. 2.5).[12] It is impossible to condense moisture onto very small particles (diameters < 0.01 μm) unless the vapor density is very high.

Figure 2.6 shows the condensation growth and evaporation of a salt particle with a diameter of 6.89×10^{-7} ft [0.21 μm] at various relative humidities. Notice the hysteresis between droplet formation and recrystallization. Soluble nuclei in the air grow from a size which scatters very little light to a size which scatters a lot of light, i.e., 3.28×10^{-7} to 3.28×10^{-6} ft [0.1 to 1.0 μm]. This is how summertime hazes are developed.

Condensation on particles is enhanced if the particles carry a charge because the charge reduces the vapor pressure at its surface.[6] It is also enhanced if the particle has a chemical affinity for the vapor. For example, hygroscopic substances have an affinity for water vapor.

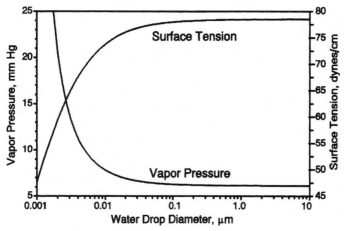

Figure 2.5 Surface effects as related to particle size of water droplets.

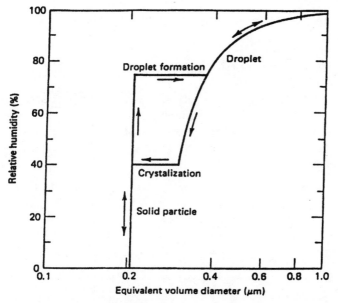

Figure 2.6 Relative humidity versus equivalent volume particle diameter showing the transition from a 10^{-14}-g sodium chloride particle to a droplet with increasing humidity and recrystallization with decreasing humidity. Transitions are approximate. (*From Ref. 1.*)

Particles that are porous allow additional condensation on internal surfaces.

Condensation can be used to make submicron particles grow to a size where they can be captured by industrial particle collection systems. Steam injection, subcooling of a saturated gas stream, and adiabatic expansion of gas streams to promote condensation have been used to improve scrubber performance.[11]

Evaporation is the reverse process of condensation by which more adsorbed liquid vapor molecules leave the particle surface than those that adsorb onto the surface. Calculations of the time for the droplet to evaporate should take evaporative cooling, Stephan flow, and the Kelvin effect into account.[11] The Kelvin effect relates to the increase in surface vapor pressure resulting from the curvature of very small particles.[10] Because of the Kelvin effect, the lifetime of a water droplet with a diameter of 3.28×10^{-7} ft [0.1 μm] in saturated water vapor at 68°F [20°C] is only 8.92×10^{-5} s.[4] Tiny droplets of pure water evaporate in a few milliseconds whereas droplets with a solid nucleus evaporate more slowly.[1] Droplet velocity increases the evaporation of droplets with diameters larger than 1.31×10^{-4} ft [40 μm].[4]

2.9 Electrical Properties

Charged particles subjected to an electric field move due to the force induced on the particle and do so without any current flow. The electrostatic force can be thousands of times greater than the force of gravity for highly charged particles[1] and is used effectively in air cleaning equipment.

A particle may have a positive, negative, or zero charge.[7] Particles less than 3.28×10^{-7} ft [0.1 μm] in diameter do not contain a charge naturally. The number of natural charges increases as particle diameter increases (Fig. 2.7).[4] The median number of natural charges for most particles is one electron.

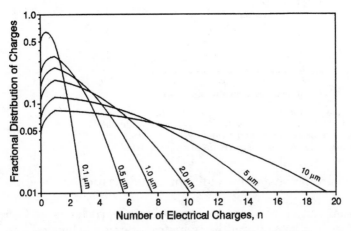

Figure 2.7 Fractional distribution of natural charges on particles.

Particles can be charged by gaseous ion diffusion, static electrification, and field charging. Gaseous ion diffusion occurs when uncharged particles acquire charge by diffusion of charged unipolar gaseous ions to their surface through random collisions (Brownian motion) between ions and particles. An aerosol particle rarely has a zero charge since the ion concentration in the atmosphere is so high. The equilibrium charge is called the Boltzmann equilibrium.[1] Corona discharges, radiation, and other energy releases can produce gaseous ions.[13] High-energy electric fields create large concentrations of charged ions by corona discharge and therefore large numbers of charges on particles. They are used in electrostatic precipitators.

The process of negative corona discharge and particle charging is illustrated in Fig. 2.8. Positive ions move toward the negative electrode in the active zone within the corona discharge. Negative ions produced as gas molecules acquire free electrons and move toward the positive collector. Particles become negatively charged because of the negative ions in the passive zones. This negative charge causes the particles to deposit onto the positively charged collector. Field charging dominates for particles with diameters $> 2.62 \times 10^{-4}$ ft [0.8 μm], and gaseous ion diffusion charging dominates for particles with diameters $< 0.66 \times 10^{-4}$ ft [0.2 μm]. The minimum charge occurs for particles with a diameter of 9.84×10^{-7} ft [0.3 μm].[7]

Static electrification occurs as particles separate from material or other surfaces during generation.[13] Three mechanisms of static elec-

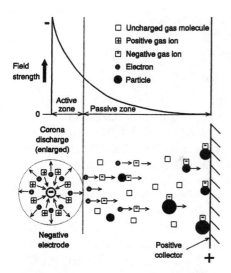

Figure 2.8 Processes of corona discharge and particle charging.

trification are electrolytic charging, spray electrification, and mechanical contact charging. Electrolytic charging occurs when high dielectric liquids such as water are separated from solid surfaces. Spray electrification results when the surfaces of charged liquids are disrupted by atomization or bubbling. Mechanical contact charging occurs as dry nonmetallic particles separate from solid surfaces, e.g., grinding, or resuspension of dry powders. Particle charges can also be developed by friction and contact between particles of different size and mass, e.g., dust moving in a duct system.[6] The charges in an ungrounded duct can reach 20,000 volts (V) for a coal dust cloud moving at 13.2 miles per second (mi/s), and up to 6000 V for wheat, starch, and lycopodium dusts traveling at 1.6 mil/s. The voltage increases with temperature and decreases with increasing humidity.[6]

The migration velocity for a charged particle in an electric field is

$$U_E = \frac{n_E eEC}{3\Pi \eta d} \qquad (2.14)$$

where U_E = electrophoretic particle velocity, ft/s [m/s]
  n_E = number of electric charges on the particle
  e = elementary charge, one electron
  E = electric field strength, V/ft [V/m]
  C = Cunningham slip correction factor
  η = gas viscosity, lb_m/ft · s [kg/m · s]
  d = particle diameter, ft [m]

Electrical migration velocities U_e are typically 0.5 to 1.0 ft/s [0.15 to 0.30 m/s] in electrostatic precipitators.[7]

2.10 Light Scattering and Extinction

Particles absorb, reflect, refract, and scatter radiation including visible light. The visual impact of dust is governed by particle diameters in the size range of 3.28×10^{-7} to 6.56×10^{-6} ft [0.1 to 2.0 μm]. Scattering of light by a particle is governed primarily by the ratio of the particle size to radiation wavelength. Scattering of light by particles less than 1.64×10^{-7} ft [0.05 μm] in diameter follows the theory of molecular scattering. Scattering of light by large particles (diameter $> 3.28 \times 10^{-4}$ ft [100 μm]) is analyzed by geometric optics. Scattering is more complicated between these sizes, the Mie scattering region, because the wavelengths of light and particle size are similar. The intensity of scattered light also depends on the angle of the scattered light, e.g., forward versus backward scattering.

Light absorption depends on the particle material. A beam of light is diminished by particles absorbing and scattering the light in a process called extinction. Light beam attenuation by extinction is described by the Lambert-Beer law.[1] Opacity meters in smoke stacks measure extinction.

2.11   Adhesion

Small airborne particles experience adhesion forces that cause them to attach very firmly to surfaces. Many particle collection methods such as filtration depend on adhesion forces. Adhesion forces include the London–van der Waals force, electrostatic force, and surface tension of adsorbed liquid films.

The adhesive force can vary by three orders of magnitude for materials ranging from soft plastic to quartz. It is influenced by particle shape, surface characteristics, environmental conditions, contact duration, and initial contact velocity.[1]

The higher the speed with which a particle strikes a surface, the better the adhesion due to deformation of the particle. If the size, hardness, and velocity of the particle are large enough, the particle may bounce. Particle bounce can be minimized by coating the surface with oil or grease.

A higher relative humidity may produce a tiny pool of liquid at the particle-surface interface. This increases the adhesion force due to surface tension.[13] Charged particles and surfaces have a much higher adhesive force.

Adhesive forces are generally inversely proportional to diameter, whereas particle removal forces are proportional to d^3 for vibration and centrifugal forces, and to d^2 for air currents. Therefore, it is more difficult to remove small particles from surfaces, especially those with diameters smaller than 3.28×10^{-5} ft [10 μm].[1] However, large agglomerates or chunks of small particles are more easily removed.

2.12   Filtration

Aerosol filters do not work as microscopic sieves like liquid filters but rather by collision and attachment to the fiber surfaces.[1] After the first particles adhere, subsequent particles are captured by the initial particles to form chains. Filtration is thus dependent upon properties of both the aerosol particles and the gas. Since the stated pore size for a filter (e.g., 2.62×10^{-6} ft [0.8 μm]) is based on liquid filtration, the high collection efficiencies of aerosol filters extend to much smaller particles.

Particles collide with a filter fiber by five basic mechanisms: (1) inertial impaction, (2) interception, (3) diffusion, (4) gravitational settling, and (5) electrostatic attraction. A filter fiber forces a change in the flow direction or streamlines of a gas flowing through the filter. If the inertia of the particle is sufficient, i.e., large enough Stokes number, the particle will depart from its streamline, collide with the fiber, and be collected by impaction. Very fine particles may strictly follow gas streamlines, but because of their finite size, the side of the particles nearest the fibers are intercepted by them. Small particles moving by brownian motion will move from nonintercepting streamlines to intercepting streamlines thus increasing their chances of being collected.[1] Collection of large particles by gravitational settling can be significant at low gas velocities through a horizontal filter. Finally, charged particles will be attracted to oppositely charged filter fibers.

Filter efficiency as influenced by size and face velocity is shown in Fig. 2.9. Note the particle sizes with the minimum efficiency where they are too small for effective impaction to take place and too large for significant deposition by diffusion. It is often necessary to use experimentally derived efficiency data rather than theoretical data because actual collection depends on the fraction of particles that remain attached to the filter.[13]

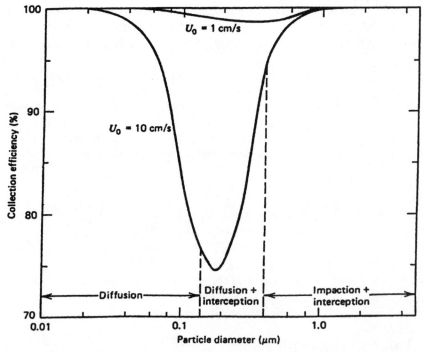

Figure 2.9 Filter efficiency versus particle size for face velocities of 0.03 and 3.3 ft/s [1 and 10 cm/s]; $t = 1$ mm, alpha = 0.05 and $d_f = 6.56 \times 10^{-6}$ ft [2 μm]. (*From Ref. 1.*)

2.13   Dispersion of Airborne Aggregates

Airborne aggregates are dispersed by changes in velocity (acceleration or deceleration), bending and shear stresses caused by air velocity gradients and turbulence, and impacts on obstacles in the airstream. Maximum stresses occur just as the particles are put into a flow field and accelerated. The impact of aggregates on obstacles in the airstream is the only effective dispersion mechanism for small spherical particles. Acceleration or deceleration is also effective if the particles are irregular and nonspherical.[11]

Dust particles deposited on fibers, plates, and other dust collecting media may be resuspended due to fluid drag over the particles. The resistance to resuspension is dominated by adhesion for small particles.[11] Resuspension of an aggregate particle from a plane surface will occur when the external force on the particle exceeds the adhesive force of the particle to the surface or exceeds the strength of the aggregate particle. The external force is principally due to bending stress that is much higher than the shear stress on the aggregate particle.

2.14   Explosiveness

Aerosols can explode violently because small particulates provide extremely large surface areas. One gram or 0.0022 lb of unit density material has a surface area of 646 ft^2 [60 m^2] when divided into particles that are 3.28×10^{-7} ft [0.1 μm] in diameter. This large surface area provides larger numbers of adsorption sites that increase the speed of oxidation reactions in air and make explosions possible. These reactions can occur between substances within the particle, between particles themselves, or between particles of a gaseous chemical species.[10] Explosiveness of dust therefore increases with smaller particle size. The temperature of nearby particles increases rapidly since oxidation reactions produce heat. The large surface area of the particles encourages more reaction and more heat. Runaway reactions are the result.[4]

Organic materials such as sugars, starches, and coal and many inorganic substances that are normally noninflammable will explode violently as an aerosol.[6] For example, the explosiveness of atomized aluminized dust can be greater than grain dust. Many metals such as iron, zinc, magnesium, and lead are relatively inactive chemically as aggregate material. However, they are extremely active as aerosol particles.

Dust and oxygen must be present at the proper concentration and there must be a source of ignition for an explosion to occur.[6] A minimum aerosol concentration of 0.0013 to 0.004 lb$_m$/ft^3 [0.02 to 0.060

kg/m^3] in a confined space is required for an explosion, depending on the type of particles. The maximum concentration is about 0.125 to 0.375 lb$_m$/ft^3 [2 to 6 kg/m^3] as particles too close together inhibit their exposure to oxygen.[4] After ignition occurs, the flame propagates throughout the space containing explosive concentrations. Propagation velocity depends on dust characteristics but is generally less than 984 ft/s [300 m/s].

2.15   Radiometric Forces

Radiometric forces cause particle motion and include diffusiophoresis, thermophoresis, and photophoresis. These forces are weak and have less practical application for particle collection than gravitational, electrical, and centrifugal forces.

2.15.1   Diffusiophoresis

Particle motion due to gas density gradients is called diffusiophoresis. Differences in molecular impacts on opposite sides of the particle cause the motion. It includes Stephan flow, which is caused by the flow of molecules toward or away from the surface of a volatile liquid because of condensation or evaporation, respectively. Stephan flow causes a particle to flow toward a liquid during condensation and away during evaporation.[7]

2.15.2   Thermophoresis

Thermophoresis is the motion of a particle in the direction of cooler gas temperatures. This motion occurs because gas molecules on the hot side of the particle exert more force as compared to those on the cooler side.[5] Thus the cold surfaces of heat exchangers pick up dust from the warm airstream and hot-water pipes remain particlefree. Also, cigarette smoke is deposited on the inside surface of windows during the winter.[1] The velocity of particles in a temperature field depends relatively little on their size, unlike their velocity under gravitational and inertial forces which drops rapidly with smaller sizes.[3] Thermophoretic velocity is independent of size for particles with Kn > 1.[14]

For sampling of submicron particles, thermal precipitators have high efficiency and are charge- and size-independent. Most thermal precipitators operate at nearly 100 percent efficiency for particles with diameters smaller than 3.28×10^{-5} ft [10 μm], but their volumetric flow rates range from only 0.01 to 3.5 ft^3/min [0.012 to 6.0 m^3/h]. About 10 times as much energy is required for their use as a dust collector as compared to an electrostatic precipitator.[14] Their application as a dust collector is more feasible when heat is available in the dirty gas to be cleaned. The use of a cooled packed bed rather than an open pipe also improves performance. Promise is also seen where thermal deposition can supplement and enhance another collection mechanism in an air cleaner,[14] especially when finer and finer particles need to be removed.

2.15.3 Photophoresis

Photophoresis is particle motion in the direction of radiation due to absorbed radiation warming one side of the particle more than the other. Transparent particles may move toward the radiation source as they absorb radiation on the opposite side of the particle.[7] Light can also move a particle by the pressure of the light itself. Particles have been lifted against gravity by high-intensity laser beams.[15]

2.16 Acoustic Force

Particles in a sound field can (1) vibrate with the gas, (2) drift translationally, and (3) circulate in the vibrating medium. Drift can be exploited to remove particles. Drift velocity increases with particle size and sound frequency.[13] Drift occurs from (1) radiation pressure, (2) periodic change in viscosity of vibrating gas, (3) difference in phase of vibrating particles and air, (4) sound wave distortion, and (5) asymmetry in standing wave field. Acoustic fields have been used to enhance particle coagulation, evaporation, or condensation.[2] Particle velocities due to acoustic forces can be 0.023 to 1.3 ft/s [0.007 to 0.4 m/s] for small to large particles, respectively.[13]

References

1. W. C. Hinds, *Aerosol Technology: Properties, Behavior, and Measurement of Airborne Particles,* John Wiley and Sons, New York, 1986, pp. 13–324.
2. P. A. Baron and K. Willeke, "Gas and Particle Motion," in K. Willeke and P. A. Baron (eds.), *Aerosol Measurement: Principles, Techniques and Applications,* Van Nostrand Reinhold, New York, 1993, pp. 23–39.
3. N. A. Fuchs, *The Mechanics of Aerosols,* Pergamon Press, New York, 1964, pp. 23–59.
4. P. C. Reist, *Aerosol Science and Technology,* McGraw-Hill, 1993, pp. 59–328.
5. G. M. Hidy, *Aerosols: An Industrial and Environmental Science,* Academic Press, New York, 1984, pp. 19–22.
6. J. M. Dallavalle, *Micromeritics: The Technology of Fine Particles,* Pitman Publishing Corp., New York, 1948, pp. 21–259.
7. H. E. Hesketh, *Air Pollution Control: Traditional and Hazardous Pollutants,* Technomic Publishing Co., Lancaster, PA, 1991, pp. 3–195.
8. C. N. Davies, "Particle Fluid Interaction," *J. Aerosol Science,* vol. 10, 1979, pp. 477–513.

9. T. T. Mercer, "Brownian Coagulation: Experimental Methods and Results," in D. T. Shaw (ed.), *Fundamentals of Aerosol Science,* John Wiley and Sons, New York, 1978, pp. 86–87.
10. W. C. Hinds, "Physical and Chemical Changes in the Particulate Phase," in K. Willeke and P. A. Baron (eds.), *Aerosol Measurement: Principles, Techniques and Applications,* Van Nostrand Reinhold, New York, 1993, pp. 43–51.
11. T. Yoshida, Y. Kousaka, and K. Okuyama, *Aerosol Science for Engineers,* Power Co., LTD, Tokyo, 1979, pp. 36–178.
12. R. Defay, I. Prigogine, A. Bellemans, and D. H. Everett, *Surface Tension and Adsorption,* John Wiley and Sons, New York, 1966, pp. 217–258.
13. H. E. Hesketh, *Fine Particles in Gaseous Media,* Ann Arbor Science Publishers, An Arbor, MI, 1977, pp. 29–143.
14. J. A. Gieseke, "Thermal Deposition of Aerosols," in W. Strauss (ed.), *Air Pollution Control,* Wiley-Interscience, New York, 1972, pp. 218–248.
15. A. Ashkin, "Applications of Laser Radiation Pressure," *Science,* vol. 210, 1980, pp. 1081–1088.

3

Defining the Pollution Problem

Sherril Wayne Norman

Air Techniques, Inc.
Marietta, GA

3.1 Introduction

For centuries, humans have sought to improve their living and working environment by controlling the waste products from both their working and living surroundings. Simple forms of air pollution control can be traced back to the stone age when fires were built in such a manner as to exhaust particulate and waste gases away from living areas. Early civilizations learned how to construct fireplaces and smoke stacks to vent waste gas and particulate. With the advent of the industrial revolution these simple technologies were transported from the home into the workplace. Our level of technology and construction increased, which resulted in an increased capability of building stacks and ventilation systems that would control and disperse pollutants.

From this very basic foundation, advanced stacks with concomitant air pollution control equipment have now been developed which clean up waste gas streams. Scrubbers, baghouses, electrostatic precipitators, cyclones, thermal oxidizers, and various other control devices are applied to waste particulate and gas streams to purify emissions. This equipment is applied for a variety of reasons:

Personal health.   Pollutants are controlled to protect our health and the health of others. This is achieved in the plant work area, for example, by placing an exhaust hood over a contaminated area and channeling the contaminants outside or to a control device. This is done to protect our work environment and the health of workers. The fumes and particulate from various processes are exhausted to high elevations from smokestacks to disperse the pollutants or are directed to control devices that remove the pollutants. These measures are taken to control the emissions which are vented to our own atmosphere and environment. This is done to keep the air that we breathe cleaner and our environs protected.

Environmental protection.   Pollution control equipment is applied to keep the land, the plants, and the ambient air that we breathe protected from particulate and gases that can destroy the environment. We control many air pollutants because they are toxic and can destroy not only our environment but also the health of each of us and our loved ones.

Product or material recovery.   Many processes would normally generate particulate matter and other pollutants into the atmosphere if it were not for pollution control devices. Some devices are used to capture particulate in the dry state so that they can be reused. Some manufacturing processes create valuable waste products which can be recycled, used as a by-product, or used in the manufacture of other products. This recovery has helped pay for many pollution control systems, while protecting our environment.

3.2   Data Acquisition and Test Methods

In any pollution control project the first step is to identify the problem. Obviously, there must first be recognition that a problem exists and that the problem warrants consideration of pollution control technologies. There are many sources of pollution in the United States that utilize ventilation to the atmosphere as the sole control system. This ventilation allows the pollution to disperse and fall out across the environment. There are many processes today that have operated for years within state and federal regulations while continuing to emit certain levels of pollution. These levels may be within the allowable emission code or state law even though they disperse pollutants throughout our atmosphere. Dispersion of gaseous and particulate matter is recognized as one form of pollution control. Our nation's scientists have condoned this form of control because they have concluded that our environment can accept this regulated level of emission. An example of this form of control is the very basic wood-fired stove or

fireplace in a personal residence. In addition to the products of combustion resulting from the conversion that takes place in the fireplace,
particulate is usually emitted in the form of unburned carbon or ash as
well. Most of us do not consider this level of pollution to be a problem.

The identity of an air pollution problem has to be recognized and
quantified based upon some standard. For environmental applications,
the standards are usually state or federally imposed regulations for a
certain process. When these do not apply, other standards may come
into play. For personal health or industrial hygiene, these standards
may be in-house or Occupational Safety and Health Administration
(OSHA) standards which control the workplace, requiring the fumes or
emissions generated inside to be controlled or emitted outside. Other
standards can be self-imposed by "good neighbor" industries that
admit they have a problem and want to portray a good image to their
community. Many times the situations these industries seek to correct
are not problems relative to state code, federal code, or emission level.

In the United States, laws and regulations are established by each
state and are subsequently reviewed, modified, and adopted by the
federal government. The federal government controls emissions standards on existing industrial processes and new processes. The federal
agency that governs these air emission regulations is called the
Environmental Protection Agency (EPA). There is a range of state
agencies with a variety of names and purview, such as the
Department of Natural Resources (DNR) and Department of Health
and Environmental Control (DHEC). The state and federal agencies
work together and in many cases overlap. Most other countries develop and maintain environmental regulations in a similar manner.

Existing and new industries must have a permit to operate any
process that creates emissions. This permit identifies allowable emissions and states the conditions under which the process must be run.
Enforcement of this permit is both a state and federal enforcement
function. New processes that industry desires to construct must go
through a permitting procedure that will result in an operating permit. The procedure protects the environment from an industry desir-
ing to pollute more than our laws will allow for a specific process and
pollutant.

The United States (U.S.) government passed its first Clean Air Act
legislation in 1970. Those laws were the first requirements placed on
industry, imposing a permitting process and a requirement for identification of pollutants through a regulating and enforcement process.
For 20 years the U.S. government modified the Clean Air Act with
various amendments making laws more stringent and reducing the
levels of emission in existing or new processes. Legislation such as
the 1990 Clean Air Act Amendments was enacted to identify and con-

trol large sources of pollution such as our utility industry. These legislations were targeted against pollutants like SO_x which causes acid rain. Simultaneously, these acts tighten and strengthen the laws regulating existing pollutants and toxins. In addition, the Clean Air Act has a financial impact on those who emit, by charging industry for the pollution they generate on a per-pound or per-ton basis. Those who violate this law can be criminally charged even though they may be an officer or employee of a company.

3.3 Identifying the Problem

Once it is determined that a potential pollution problem exists, it must be determined to what extent the problem exists and how it can be corrected. This process usually requires data acquisition. The identity of the problem typically falls into one of two categories. If a problem exists and the process is physically capable of operating, the path of data acquisition will be quite different from a new and nonfunctioning process. On existing processes we have an excellent opportunity for physically gathering information through field testing. Data acquisition is possible in the majority of the applications that exist.

There are very few processes in existence that would prevent the gathering of field data through the developed test methods. Examples of those processes that inhibit obtaining data are those that operate at very high pressures and those in explosive environments with highly toxic or radioactive components, making it difficult or dangerous for such field data acquisition. There are also existing applications that present difficult test locations because of their physical layout. In most cases however, the process is amenable to the gathering of data by utilization of EPA test methods or modifications thereof. To identify the air pollution control problem in the majority of applications, we must obtain samples of the waste particulate along with other physical data, such as temperatures, flows, pressures, gas density, and moisture.

This data acquisition is called *engineering testing,* as compared with *compliance testing,* which will be discussed later. Engineering testing follows the best common practice, such as OSHA and EPA guidelines, procedures given in the *Industrial Ventilation Guide,* or other established acquisition methods. Data is gathered so that a solution to a problem can be engineered or a benchmark can be established for pollution control equipment or ventilation design purposes. Information about the physical operating condition of the process gas stream is very important to the engineer designing the equipment for the system for various reasons. Obviously, performance of the equipment hinges upon gathering this data. In most cases, performance guarantees for the operation of control equipment are formulated using predicted equipment performance from this established baseline data. The more accurate this data, the better the chance that the guarantee will be met typically. The more accurate the data, the more precise the cost estimates become. Where a risk of inaccurate data exists, manufacturers of equipment tend to be more conservative and include larger safety factors for protection, thus resulting in more expensive systems.

The gathering of field data is based upon the concept of simulating the conditions that exist in the process through the test train. Samples are extracted through various types of mechanisms that will render a physical catch of either a gaseous or particulate sample that can be analyzed. Where possible, most firms use procedures established by the EPA, OSHA, or American Society of Mechanical Engineers (ASME). The established procedures provide quality assurance in the data collection process. In most cases the samples are extracted isokinetically. Isokinetic means that the velocity at the entrance to the sample probe is the same as the velocity at a given point in the duct or stack at a particular given time. Minimal testing usually includes volumetric flows, pressure, and temperature.

The second category into which an environmental problem may fall is that in which testing or physical data cannot be accurately obtained because the process to be tested is physically nonexistent or cannot be tested due to reasons stated previously. Design data for processes falling into this category then must be estimated through some means. In many cases historical data may be used for these purposes. Published manuals provide a point of reference, such as the EPA data source *AP 42,* a manual giving emission criteria for various processes. This manual tends to be very general and contains some common processes. For unique processes one must find other sources of process data. Many times test data from other applications can be used with safety-factor or emission-factor multipliers to gauge what one might expect for emissions from a new or untestable process. Great care must be used when applying assumed data and safety factors, as often the safety factor is not adequate or is grossly conservative. Both may have unacceptable consequences.

3.4 Test Methods

In the early 1970s the U.S. EPA established various test methods for compliance with the regulations that had been established by the Clean Air Act. Those test methods have been refined, modified, expanded, and amended over the years. For particulate testing we have a well-defined method known as method 5 which includes in its format methods 1, 2, and 3. These methods include establishing the volumetric flow of the process, the gas composition, and other physical data that one needs in evaluating a system. Method 5 is a dry particulate test train that will remove particulate from a process duct or stack and catch it on a filter so that it can later be analyzed. The probe and probe tip are heated to keep the gases above the dew point until they reach the filter. The gas, once exhausted from the filter, goes through a wet condensing section where moisture is determined. This is one of the most basic methods utilized for proving EPA compliance for particulate.

Likewise, this method is used extensively to identify baseline data or for engineering testing in identifying the design data for a process that has an emission problem. Methods that have been derived subsequently for sampling gaseous emissions such as SO_2, NO_x, and volatile organic compounds include some form of the methods implemented in method 5 when used to identify certain physical properties of the gas stream. All the methods have specific procedures for capturing gas samples that can be utilized to determine the total concentration of the pollutant(s). As with method 5, these sample methods are utilized extensively in an engineering mode to help identify concentrations for design purposes.

In engineering testing, when establishing criteria for design, it is often better to obtain some data than no data at all, although it may not adhere strictly to EPA test standards. Most engineers and test firms will take liberties with test procedures when necessary so that some physical information can be obtained. The process may not allow strict adherence to the test method format for sampling. For example, there may be obstructions too close to the sampling location, the temperature may dictate utilization of special equipment, or the loading of particulate or the concentration of the gas may make it unreasonable to run the test for the specified period of time that the EPA methods dictate. Whatever the case, obtaining some information about the process and physical properties is much more important than following the test method exactly. Physical testing and the information acquired in the field are considerably better than any information that can be obtained from reference or resource material, even when it exists.

Some of the greatest challenges often come from the field-testing standpoint. One must consider how to test large, high-temperature processes with elevated test locations during unfavorable ambient conditions. These are typical challenges that field engineers and testers must face at some time.

In addition to the guidelines established by EPA methods, test procedures are outlined in other publications that can be useful in data acquisition. A good source of air test procedures is the American Conference of Governmental Hygienists' publication *Industrial Ventilation Guide*. One section in this guide describes various airflow testing procedures. This focuses primarily upon gaining volumetric flows, temperatures, and gas composition. The publication is written in simple terms and is easy to follow.

Another publication that is quite good as a guide for measuring physical properties in a system is the Air Moving and Condensing Association's (AMCA) publication called *Testing and Trouble Shooting*. This document is also written in basic, simple to understand language.

3.5   Airflow Measurements

There are a wide variety of techniques to be utilized to determine airflow in a system. Most all the techniques are identified in the publications referenced previously. This section will deal with airflow measurements in a confined duct or stack. The intent is to cover several of the most frequently utilized techniques that an engineer will find useful, present consistency and provide convenience for a wide variety of applications.

Prior to taking any airflow readings one must find a good location in the stack or duct system for obtaining that measurement. The measurements need to be taken in a straight run of duct or stack, away from any obstruction. Generally speaking, one needs to be at least five equivalent duct diameters downstream of any obstruction or disturbance and at least one to two equivalent diameters upstream of a similar disturbance. A disturbance or obstruction in a system such as a damper, an elbow, transition, fan, process inlet or outlet, or any other similar devices will cause the flow to be turbulent and less laminar. A prerequisite of any airflow measurement is to pick a location where the airflow readings can be consistent and give good results. It should always be remembered that airflow measurements are performed on fluids that are compressible. The gas flow will react in areas of obstruction. For example, the gas being measured will hug the outside wall of an elbow. Velocity pressures measured in an elbow will be higher against the outside wall than the inside wall.

Once a good location has been chosen, three basic pieces of information must be obtained. One must first obtain the cross-sectional area of the duct or stack where the measurement is being taken. This is accomplished by utilizing a Pitot tube or some other marked object that can be inserted into the ductwork to measure the internal dimension. If the duct can be measured from the outside, the cross-sectional area can be calculated as well assuming the wall thicknesses are known. The temperature of the gas stream must be measured. It is always a good idea to obtain both the wet bulb and dry bulb temperature, particularly if moisture is significant in the gas stream. The three most common methods of obtaining the temperature are with a mercury glass thermometer, a bimetal dial thermometer, or a thermocouple attached to an electronic sensor. Any of these can be utilized accurately to read a dry bulb temperature. The temperature should be taken by inserting the device into the duct or stack and closing off the area around the device to prevent any leakage in of outside air. If a temperature differential exists, the process temperature may be influenced by leakage of ambient air. Once a dry bulb temperature is obtained and recorded, a wet bulb temperature should be measured as well. This is done most simply by taking a cotton shoelace and cutting it into strips of 2 or 3 inches (in). This will make an open-ended sock or wick that can be placed over the end of the thermometer or the thermocouple. The tip of the temperature-sensing device that is going to be inserted into the system duct or stack should be covered. The wick can be held in place by using tape on low-temperature applications. On high-temperature applications, glass tape must be utilized for measurements requiring lengthy duration. With the wick in place, the tip of the temperature-measuring instrument is saturated with water. The tip can then be inserted into the gas stream for making a wet bulb measurement. One must carefully watch the thermometer or digital readout as the temperature reading increases. Hesitation will take place at the wet bulb reading. The hesitation can be for a fraction of a second to several seconds based upon the conditions in the duct system. Several readings may be required to accurately reach a wet bulb determination. With both the wet bulb and dry bulb readings, the density can be calculated by reading the actual humidity from psychometric tables or charts (see Appendix 1E) or by calculation.

The last of the three major items obtained prior to making velocity flow measurements is a static pressure reading. This is important for any measurement of gas flow rates and velocities since it is necessary in calculating the gas density. Any high negative- or positive-pressure condition can be significant in affecting the results of any airflow measurements. The most common device used to read static pressure is a U-tube manometer utilizing water. Most standard U-tubes read in a range up to 20 in wg. Manometers reading several times this range are also available. To obtain higher pressure readings, fluids other than water can be utilized in the manometer as an alternative to the U-tube with water. A common substitute is mercury. Figure 3.1 shows an example of a U-tube manometer.

Another common device for measuring static pressure is the utilization of the static leg of an oil slant gauge (manometer). This device is

Figure 3.1 U-tube manometer. (*Courtesy Dwyer Instruments, Inc.*)

more commonly used in obtaining velocity pressure readings. Care should be taken with the use of these devices, since many of the slant gauges are limited in their pressure range and are much more sensitive than U-tube manometers utilizing water or other liquids. Figure 3.2 shows an example of a slant gauge manometer.

Lastly, a magnahelic dial gauge may be utilized and is an easy way to obtain static pressure. Care should be taken to attach to the correct port when using either the slant gauge utilizing oil or the magnahelic gauge. The magnahelic gauge must be calibrated as well and checked periodically for accuracy. The static pressure measurement can be taken at the same location as the temperature measurement. Usually a hole with a diameter between $\frac{1}{4}$ and $\frac{3}{4}$ in is sufficient to obtain many of the readings discussed in this section. As in the case of temperature, one must be sure that the area around any large opening in a duct is closed off temporarily while the measurement is being made. Any other flexible material may be used that will sufficiently seal the hole temporarily. To make any measurement on an ambient- or low-temperature system, flexible tubing can be utilized from the reading point to the device used to make the reading. If a high-temperature system is being analyzed, a short piece of stainless-steel tubing can be attached to a flexible hose. This will reduce the hazards while making

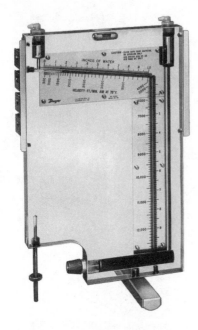

Figure 3.2 Slant gauge manometer. (*Courtesy Dwyer Instruments, Inc.*)

the measurement. This extension to the measurement system will
allow the measurement to be made without damaging the tubing or
the individual making the reading. The length of the stainless-steel
tubing should be determined by the temperature in the duct or stack.
The higher the temperature, the longer the tubing. Usually a tube
between 2 to 5 ft is normal.

An alternative technique for static pressure measurement is to utilize only the static leg of a Pitot tube with the static pressure leg of
the tube attached to the static pressure port of the measurement
device. The velocity pressure leg is disconnected from the instrument.

Once the above three pieces of data are obtained, one is now ready
to measure velocity pressures in the duct or stack. As in the case of
measuring static pressure, the three most common techniques utilized today are a slant manometer utilizing oil, a magnahelic gauge,
or an electronic device. The most common device is probably the slant
manometer using oil. Many engineers think that this is the most
accurate and consistent for the investment. As with any instrument,
care should be taken in its handling. Once a test location is chosen,
the instrument should be set up, leveled, and zeroed. The magnahelic
gauge is more convenient than the oil manometer and is easily moved
from site to site. Its accuracy, however, is often questioned. In many
cases just moving the device from site to site may cause it to lose calibration. It must be adjusted and recalibrated when this occurs. The
magnahelic gauge is popular because it is easy to use, provides a dial

readout, and does not require adjustment of or replacement of oil. However, it is also vulnerable to fine particles of dust, should the gas stream being measured prove to be very dirty. Again, care and handling can provide useful results with a magnahelic gauge on clean or dirty gas streams.

Presently, most advanced velocity pressure measuring devices for field engineers are electronic devices that give digital readouts. The devices available today incorporate features that will allow temperature, static pressure, and velocity pressure to be made with one basic instrument. Accessories or attachments to the basic instrument offer the above features. In measuring velocity pressure and static pressure a transducer is utilized with the instrument. These devices give not only quick, accurate results but also offer features that include averaging, conversion, and totaling, incorporating convenience for the field engineer. The major drawback is their expense, relative to the other tools that are mentioned above. In addition to their expense, most of the units have moderately low temperature limitations, usually in the range of 300 to 350°F [149 to 177°C].

Regardless of the three types of units mentioned above for measuring flow in a confined system, the technique used is identical. One must utilize a Pitot tube inserted into the duct or stack to obtain the readings. The tube can be a standard or S-type Pitot tube. A standard Pitot tube has a velocity port (total pressure) at the tip of the tube and static ports along the outside circumference of the tube perpendicular to the velocity port. The S-type Pitot tube has the velocity port (total pressure) opposite the static port. An illustration of two types of Pitot tubes can be seen in Fig. 3.3. The Pitot tube works by measuring total pressure, which is the velocity and static pressure combined, on one leg of the tube. The other leg measures static pressure. The static pressure is internally subtracted from the total pressure thereby giving the resultant velocity reading. The correction factor or calibration factor for a standard Pitot tube is 0.99. In most cases, this is ignored in velocity calculations. With the S-type Pitot tube, however, it can be in the range of 0.85 and the accuracy can vary as much as 15 percent relative to the standard Pitot tube. The correction factor can be obtained by having the S-type Pitot tube calibrated in a wind tunnel or duct system against a standard Pitot tube. Normally speaking, a standard Pitot tube is used in measuring flows in clean gas systems or with very low particulate stream concentration. The S-type Pitot tube is used where dust loading is heavier, which could result in plugging of a standard Pitot tube.

The connection between the Pitot tube and the device giving the velocity pressure readout is usually flexible hose. Care should be taken to make sure that the hose is of good quality and that no leaks

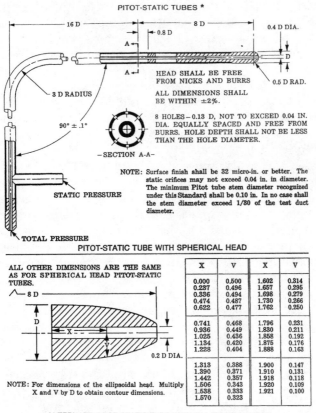

PITOT-STATIC TUBES *

16 D

8 D

0.8 D

0.4 D DIA.

A

D

A

HEAD SHALL BE FREE
FROM NICKS AND BURRS

0.5 D RAD.

3 D RADIUS

ALL DIMENSIONS SHALL
BE WITHIN ±2%.

90° ± .1°

8 HOLES – 0.13 D, NOT TO EXCEED 0.04 IN.
DIA. EQUALLY SPACED AND FREE FROM
BURRS. HOLE DEPTH SHALL NOT BE LESS
THAN THE HOLE DIAMETER.

– SECTION A-A –

NOTE: Surface finish shall be 32 micro-in. or better. The
static orifices may not exceed 0.04 in. in diameter.
The minimum Pitot tube stem diameter recognized
under this Standard shall be 0.10 in. In no case shall
the stem diameter exceed 1/30 of the test duct
diameter.

STATIC PRESSURE

TOTAL PRESSURE

PITOT-STATIC TUBE WITH SPHERICAL HEAD

ALL OTHER DIMENSIONS ARE THE SAME
AS FOR SPHERICAL HEAD PITOT-STATIC
TUBES.

8 D

D

X

0.2 D DIA.

NOTE: For dimensions of the ellipsoidal head. Multiply
X and V by D to obtain contour dimensions.

X	V	X	V
0.000	0.500	1.602	0.314
0.237	0.496	1.657	0.295
0.336	0.494	1.698	0.279
0.474	0.487	1.730	0.266
0.622	0.477	1.762	0.250
0.741	0.468	1.796	0.231
0.936	0.449	1.830	0.211
1.025	0.436	1.858	0.192
1.134	0.420	1.875	0.176
1.228	0.404	1.888	0.163
1.313	0.388	1.900	0.147
1.390	0.371	1.910	0.131
1.442	0.357	1.918	0.118
1.506	0.343	1.920	0.109
1.538	0.333	1.921	0.100
1.570	0.323		

ALTERNATE PITOT-STATIC TUBE WITH ELLIPSOIDAL HEAD

Figure 3.3 Pitot static heads.

are present in the hose or at the connecting points. It is important to
mark both ends of one tube to identify it as the static or velocity side
so that connections to the tube and unit can be made quickly and
accurately. Prior to connecting the instruments, the Pitot tube must
be marked for insertion into the duct. This will vary based upon the
area of the duct and whether it is round, rectangular, or square.

The Pitot tube must be marked prior to testing so that the engineer
can insert the tube to the correct locations. The Pitot tube can be
marked with any type of marker tape for low-temperature systems or
with high-temperature glass tape for high-temperature systems. The
tape is used to locate the point in the duct where the tip of the Pitot
tube should be located for obtaining the reading. Care should be
taken when inserting the tube to protect the instrument against high
positive or negative readings as the flexible hose is being inserted.
Usually crimping or pinching the tube is the best procedure.

The correct location for Pitot tube insertion is dependent on the duct diameter and slope at the test location. The procedure for Pitot test locations varies between round and rectangular ducts and between small and large sizes of each slope. In an effort to correctly calculate volumetric flow rates from velocity pressure readings, it is important to measure velocity pressure at the center of equal-area sections. This means that within a rectangular duct, the internal cross section is divided into a grid of equal-area small rectangles. We then measure the velocity pressure at the center of each. Likewise, circular ducts are divided into concentric rings of equal area. General recommendations for test locations and traverse points are shown in Figs. 3.4 to 3.6. When calculating gas flow rates by velocity pressure measurements from a traverse, it is important to calculate the gas velocity within each section of equal area prior to averaging or summarizing the results.

Once the tube is fully in the duct system and stabilized, readings can begin. One moves from the first point to subsequent points of measurement, taking care to hold the Pitot tube in the correct position, relative to gas flow. This is called *transversing* the duct or stack. The correct position for the Pitot tube is so that the velocity pressure port is pointed directly upstream and so that the tube itself is perpendicular to the gas flow. Readings tend to be higher nearer the centers of the ducts for well-developed flow profiles or higher near the duct walls for turbulent or cyclonic flows. Once the readings have been

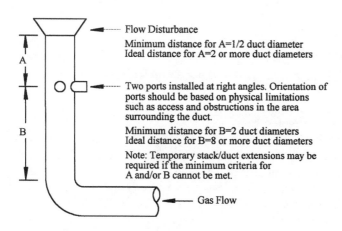

A flow disturbance may be any of the following
fan, bend, expansion or contraction, circular-rectangular
transition, stack exit, or open flame.

Figure 3.4 EPA method-1 test port requirements for circular stacks and ducts. (*Illustration by Rhonda White.*)

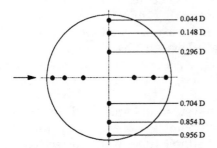

Pitot traverse points in a rectangular duct.
Centers of 16 to 64 equal areas. Located not
more than 6" apart.

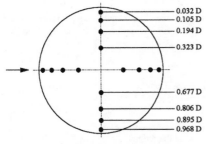

Pitot traverse points in a circular duct.
6 locations in centers of equal annular area.

Pitot traverse points in a circular duct.
8 locations in centers of equal annular area.

Figure 3.5 Pitot traverse locations. (*Illustration by Rhonda White.*)

obtained, they may be used for calculating gas flow. In measuring velocity pressure at various points, the goal is to integrate the multiple points into an average velocity pressure reading. That velocity pressure reading then can be utilized to calculate velocity. The velocity is utilized in the airflow measurement formula

$$Q = A \cdot V \tag{3.1}$$

where Q = gas volume
A = duct or stack area
V = average gas velocity

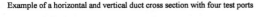

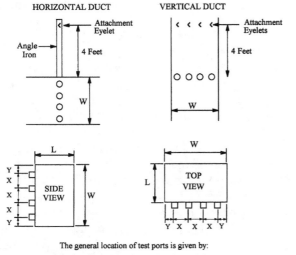

The general location of test ports is given by:

X = W/(# of test ports)   Y = X/2

The number of test ports required is dependent on the distance from up and downstream flow disturbances. This determination must be made on a case by case basis.

The location of the test ports with respect to up and downstream disturbances is the same as for circular ducts using the equivalent diameter for rectangular ducts.

$$\text{equivalent diameter} = \frac{2 \times L \times W}{L + W}$$

Figure 3.6 Test port requirements for rectangular stacks and ducts. (*Courtesy Air Techniques.*)

For measuring gas streams that are not standard air, or for measuring other gas influences which could affect a gas stream molecular weight, a sample gas flow rate calculation sheet is also included in App. 3A.

In addition to measuring flows with Pitot tubes, other devices can be used, based upon the location and accuracy desired. Simple orifice plates are utilized in many instances where the gas stream may be very dirty and where continuous measurements are required. These devices give moderate results as they are not easily calibrated over a broad range. Other devices such as vane or hot wire anemometers work well in clean gas streams or on open inlets where the duct or stack does not confine the gas flow. More information on these methods can be obtained from the *Industrial Ventilation Guide*.

3.6   Compliance Testing

After control equipment is installed and in many cases for engineering or data gathering purposes, tests utilizing EPA methods as guidelines will be administered. Compliance with state and federal codes requires compliance testing or isokinetic sampling. This method describes procedures which adhere strictly to EPA methods, as contrasted with engineering testing discussed previously. In any form of particulate or aerosol testing or sampling it is important that the information obtained be accurate and representative of the process conditions encountered. The accuracy is determined by isokinetic sampling procedures. In isokinetic sampling, one is trying to reproduce in the test train the same conditions that exist in the duct system at a specific point and at the same time. Isokinetic sampling in practice involves a process whereby the test train pump is adjusted to pull the sample volume at the same velocity as is being measured at that point inside the duct. An attempt is made to reproduce temperature, velocity, and concentration by utilizing instruments to measure the gas constituents while adjusting to equal the flow at the measurement point in the stack or duct. This is done by a variable flow sampling system utilizing a pressure pump and flow controls. Whether the stream is being sampled for gases or particulate, the goal is to produce a representative sample. Generally speaking, most particulate sampling work requires isokinetic procedures. However, most gas analysis work does not have to be isokinetic, since concentrations are assumed to be uniform. Gases may contain only submicron particulate matter (less than 1 μm in size) which does not exhibit the same effects of mass and inertia as larger particulate, thus making isokinetic sampling less important. It is important that gas flow measurements accompany any gas analysis. The U.S. EPA has outlined procedures for both sampling isokinetically and for obtaining results that fall within acceptable isokinetic standards. The acceptance is within plus or minus 10 percent of the attempted 100 percent accuracy. Throughout the remaining part of this chapter we will refer to U.S. EPA methods. The EPA sampling procedures and methods can be reviewed in a copy of *40CFR60*. In App. 3C we include a list of more frequently used EPA test methods. Here, isokinetic procedures and guidelines are outlined in detail for many test methods.

3.7   Particulate Concentration Measurement

It has always been useful to know the concentration of particulate in a gas stream both on the inlet and outlet sides of an emission control device. Prior to development of the EPA methods, various techniques were created by engineers to obtain this information. In most cases, a test system included a probe, a filter, a pump or vacuum device, and a method of monitoring pressure drop or flow through the system. Although early test methods revealed results that allowed engineers to design and apply equipment, the results were frequently inconsistent due to variables in the test methods. Much progress has been made in the development of the EPA test methods. Method 5 (particulate concentration method) utilizes all the above named devices and more. The EPA methods improved the original procedure and added techniques that would allow the engineer to easily obtain information with good repeatability. A description of the method-5 test train follows. Figure 3.7 illustrates the schematic for a method-5 test train. In many cases it is of interest to the engineer to obtain particulate results in the stack. Conditions may dictate utilization of an in-stack filter or thimble because the dust loading is expected to be very high. Very high dust loading might plug the standard method-5 probe or the standard filter. The in-stack technique is EPA method 17. This method is illustrated schematically in Fig. 3.8.

Method 5 utilizes a probe that is usually housed in a configuration which uses a thermocouple alongside an S-type Pitot tube, all mounted to a probe assembly. The probe itself is usually heated with resis-

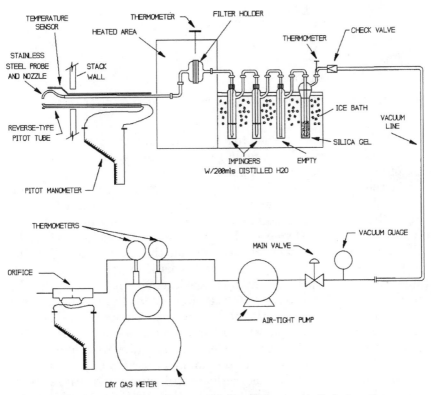

Figure 3.7 EPA method-5 sample train schematic. (*Courtesy Air Techniques.*)

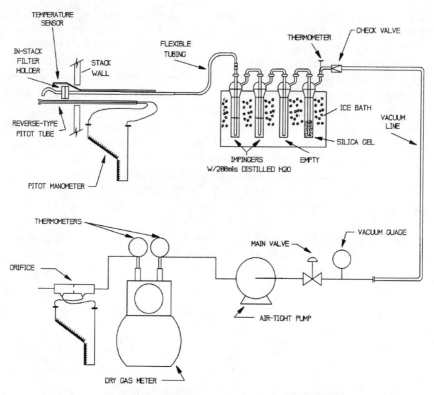

Figure 3.8   EPA method-17 sample train schematic. (*Courtesy Air Techniques.*)

tance heating and is attached to a filter box containing a filter holder, which is also heated. The sample gas is then extracted through a section of glass impingers. The impingers are maintained in an ice bath during the testing. Here the water vapor from the sample gas is condensed along with the other condensables. Lastly, there is a moisture trap utilizing silica gel desiccant to extract remaining moisture from the stream. The gas then goes through a pump and meter. Method 17 utilizes a similar arrangement except that the filter is placed in the stack and is kept warm from the stack gases. Particulates are separated at the filter prior to the impinger section. In EPA testing, the terms *front half* and *back half* are often used. Front half to most testers means the catch on the filter or the particulate that is actually separated from the sample gas stream, which dries on the filter. The back half would be the condensables or very fine particulate matter that would actually condense in the impinger section or pass through the particulate filter. Various methods will record and utilize front-half and back-half information for evaluation. Variations in the equipment and plumbing used on the front and back halves will vary with the method and the desired results of the method. Some examples of this are illustrated in Figs. 3.9 and 3.10 showing EPA method 202 and EPA method 5E, respectively. Note the difference in the appearance of the impingers in method 202 as opposed to those illustrated in method 5E. Also note that method 5E shows a solution of sodium hydroxide, while method 202 shows distilled water. This is just one example of how variations to the basic method can produce results for specific targets.

The main goal for any particulate testing is to get a gravimetric dry weight of the material collected. The filter is one of the areas to be analyzed. In addition, the probe and the probe nozzle are cleaned on the site with solutions. Those solutions are retained in the custody of the test team until the solids can be recovered at the laboratory. Various techniques for recovery are utilized, but usually the probe wash is evaporated and dried to yield the additional particulate from the field test. The totals of this particulate together with the filter

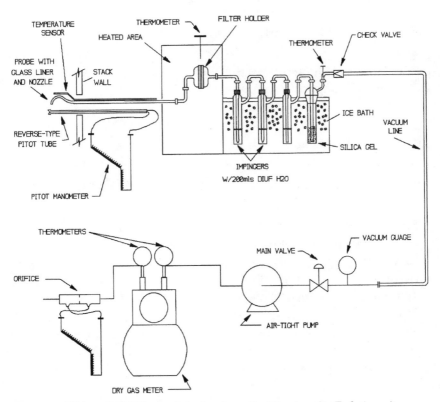

Figure 3.9   EPA method-202 sample train schematic. (*Courtesy Air Techniques.*)

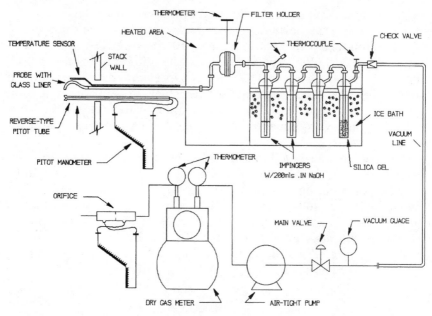

Figure 3.10   EPA method-5E sample train schematic. (*Courtesy Air Techniques.*)

particulate yield the total catch for a test. This catch divided by the gas volume through the sample train will give the concentration of dust load.

3.8 Particulate Method 5

A description of method 5 which includes methods 1, 2, 3, and 4 is given below as an example of procedures to be followed to produce method-5 results.

EPA method 1 is used to determine the number and location of the sampling points within the duct or stack.

EPA method 2 is used to determine stack gas velocity and volumetric flow rate. An S-type Pitot tube and inclined gauge manometer and temperature-measuring device are used to measure stack conditions.

EPA method 3 is used to determine the dry molecular weight and stack gas composition.

EPA method 4 is used to determine the stack gas moisture composition.

EPA method 5, Determination of Particulate Emissions from Stationary Sources, is used to determine particulate concentrations. The sampling consists of a calibrated nozzle, probe with heated liner, glass fiber filter and filter holder, four impingers, 50-ft umbilical cord, pump, dry gas meter, and orifice. A detailed description and schematic of the train can be found in previous pages.

Before each test, the probe and filter holder assembly are secured in the filter box. In each of the first two impingers 100 milliliters (mL) of distilled water is placed. The third impinger is left empty, and the fourth impinger is loaded with 200 g of silica gel. Before each repetition, an optional leak check is performed to ensure all connections are secure.

The nozzle and filter holder assembly are placed in the stack to warm up to the stack temperature prior to commencing sampling. The probe is inserted such that the nozzle is located at the first sample traverse point. The stack gas parameters are recorded on the field data sheet (Figs. 3.11 and 3.12), the pump turned on, and the sampling rate set at the isokinetic rate. At the end of the sampling period for the first point, the probe is moved to the next traverse point and the sampling rate is adjusted to maintain the isokinetic rate for the measured gas parameters at that point.

This procedure is followed until all the traverse points have been sampled. At each point, the following information is measured and recorded on the field data sheet: dry gas meter volume, stack gas velocity pressure differential, orifice meter pressure differential, stack gas temperature, impinger train exit temperature, dry gas meter inlet and outlet temperatures, and sample train system vacuum. After all the points are sampled, the pump can be turned off and the probe removed from the stack.

Immediately following each repetition, a leak check is performed on the sample train and must be less than the allowable 0.02 ft^3/min. The train is then disassembled. The nozzle and filter holder assembly must first be removed and immediately sealed to ensure no particulate matter is lost. Both ends of the probe assembly should be sealed to ensure no particulate matter is lost from that component. Next, the moisture catch from impingers 1, 2, and 3 should be measured with a graduated cylinder, and the silica gel from the fourth impinger returned to its container and sealed. The impingers are then reloaded as previously described.

The filter holder and nozzle should be disconnected to recover the particulate sample. The filter holder is capped, labeled, and stored for transport to a laboratory. The nozzle should be brushed and rinsed with reagent grade acetone until all particulate matter is removed. The acetone wash is then placed in a labeled polyethylene container.

After the first test is completed, the sealed filter holder assembly should be returned to the laboratory. The filter should be removed

AIR TECHNIQUES, INC.   Run #_____   Method #_____
Field Data Sheet

Plant_____ Source_____
Date____/____/____ Start_____ am / pm Stop_____ am / pm City_____ State_____
Operators_____/_____/_____ Probe I.D.#_____
Console #_____ EPA Box #_____ Nozzle #_____ Diameter_____ Probe Liner_____
Console Km (-) _____ K-Factor (-) _____ Barometer #_____
Ambient Conditions: Temp. (°F) _____ Pressure (in. Hg) _____
Stack Conditions: Static Press. (in. H_2O)_____ Pressure (in. Hg) _____ Moisture (%) _____

Leak Checks:	**Pre-Test**	**Post-Test**
Sample Train:	_____CFM @_____in. Hg	_____CFM @_____in. Hg
Pitot Dynamic:	_____in. H_2O for 15 sec.	_____in. H_2O for 15 sec.
Pitot Static:	_____in. H_2O for 15 sec.	_____in. H_2O for 15 sec.
Method #3 Train:	_____in. Hg for 30 sec.	_____in. Hg for 30 sec.

Fyrite Analysis: %CO2_____/_____/_____   %O2_____/_____/_____
Water Collected (ml.):   Tare _____ Gross _____ Net _____
Moisture Collected in Silica Gel (g.):   Tare _____ Gross _____ Net _____
Minutes per Point: _____

POINT #	VOLUME	PRESSURE			TEMPERATURE					PUMP
	GAS METER	PITOT DEL P	ORIFICE		STACK	GAS METER		FILTER	SILICA GEL	VACUUM
			DESIRED	ACTUAL		IN	OUT			
	CUBIC FEET	IN. H2O	IN. H20	IN. H20	°F.	°F.	°F.	°F.	°F.	IN. HG

Comments: _____
_____

QA checked by: _____

Figure 3.11 Sample field data sheet. (*Courtesy Air Techniques.*)

AIR TECHNIQUES, INC.   Run # _1_   Method # _5_
Field Data Sheet

Run #1 Method 5

Plant - _UGA_   Source _Boiler Exhaust Stack_
Date _10 / 4 / 94_ Start _9:45_ am/ pm Stop _11:12_ am/ pm City _Athens_   State _GA_
Operators _Taylor_ / _Parker_ /   Probe I.D.# _5'#4_
Console # _12_ EPA Box # _#16_ Nozzle # _5/16E_ Diameter _.310_ Probe Liner _Stainless Steel_
Console Km (-) _.75_ K-Factor (-) _5.3_   Barometer # _C_
Ambient Conditions: Temp. (°F) _68_ Pressure (in. Hg) _29.43_
Stack Conditions: Static Press. (in. H_2O) _-.60_ Pressure (in. Hg) _____ Moisture (%) _5%_

Leak Checks:	Pre-Test	Post-Test
Sample Train:	_.002_ CFM @ _7_ in. Hg	_.019_ CFM @ _5_ in. Hg
Pitot Dynamic:	_4.6_ in. H_2O for 15 sec.	_3.5_ in. H_2O for 15 sec.
Pitot Static:	_7.9_ in. H_2O for 15 sec.	_6.1_ in. H_2O for 15 sec.
Method #3 Train:	_Good_ in. Hg for 30 sec.	_Good_ in. Hg for 30 sec.

orsat   Fyrite Analysis: %CO_2 _10.4 / 10.4 / 10.4_   %O_2 _8.2 / 8.3 / 8.4_
Water Collected (ml.):   Tare _200_   Gross _238_   Net _38_ ⟩48
Moisture Collected in Silica Gel (g.): Tare _200_   Gross _210_   Net _10_
Minutes per Point: _~~3.5~~ 3.0_

POINT #	VOLUME	PRESSURE			TEMPERATURE					SILICA GEL	PUMP VACUUM
	GAS METER	PITOT DEL P	ORIFICE		STACK	GAS METER					
			DESIRED	ACTUAL		IN	OUT	FILTER			
	CUBIC FEET	IN. H2O	IN. H20	IN. H20	°F.	°F.	°F.	°F.	°F.		IN. HG
1	610.723	.26	---1.4	1.4	414	64	63	230	62		1.0
2	612.950	.25	1.3	1.3	418	69	63	233	60		1.0
3	615.285	.22	1.1	1.1	419	71	63	232	60		1.0
4	617.030	.18	.95	.95	418	73	64	235	59		1.0
5	618.900	.14	.74	.74	398	74	64	234	60		1.0
1	620.385	.22	1.1	1.1	412	74	65	233	59		1.0
2	622.370	.23	1.2	1.2	424	76	65	235	59		1.0
3	624. —	.17	.90	.90	424	77	66	241	58		1.0
4	626. —	.12	.64	.64	421	76	66	255	57		1.0
5	627.600	.08	.42	.42	415	77	66	256	58		1.0
1	628.845	.23	1.2	1.2	421	72	67	256	60		1.0
2	630.910	.18	.95	.95	426	77	67	257	58		1.0
3	632.700	.12	.64	.64	423	78	68	260	58		1.0
4	634.200	.08	.42	.42	419	78	68	265	59		1.0
5	635.400	.07	.37	.37	410	78	68	263	59		1.0
1	636.500	.23	1.2	1.2	416	75	69	252	62		1.0
2	638. —	.17	.90	.90	425	79	69	250	59		1.0
3	640.400	.15	.79	.79	424	80	69	249	59		1.0
4	642.050	.10	.53	.53	423	80	69	254	58		1.0
5	643.330	.07	.37	.37	409	79	70	260	58		1.0
1	644.465	.18	.95	.95	416	73	70	258	60		1.0
2	646.240	.20	1.0	1.0	421	77	70	255	58		1.0
3	648.070	.20	1.0	1.0	423	79	70	254	57		1.0
4	649.850	.18	.95	.95	424	81	70	252	58		1.0
5	651.415	.12	.64	.64	412	82	71	250	58		1.0
END	653.083										

Comments: • Soot Blowers open from 10:00 — 10:10 during this run
42.360   ISO = 101 %   BWS = 5.4%

_,401_   _.866_   _39.00_   QA checked by: _____

(Point group markers at left margin: A — 3,6,9,12; B — 15; C; D; E)

Figure 3.12 Example of completed field data sheet. (_Courtesy Air Techniques._)

from the holder assembly in a controlled environment to ensure that no particulate matter is lost during the recovery process. The filter is placed in a labeled container and desiccated for a minimum of 24 h and is then weighed to a constant weight. The filter holder assembly is rinsed with acetone and the wash added to the labeled nozzle wash container of the same repetition. If any filter fibers remain on the holder gasket, they should be carefully removed and added to the probe wash container. A sample of the acetone used for cleanups should be saved as a blank in a separate container labeled *Acetone Blank*. After the test, the probe wash containers from each test repetition, along with the acetone blank, are returned to the laboratory. The contents of each sample bottle are transferred to tared beakers. The containers are then rinsed with acetone to ensure that all the particulate matter is recovered. The volume of the acetone is recorded and then evaporated by placing the beakers in a low-temperature oven. The evaporation should be closely supervised to prevent bumping and subsequent loss of the sample. Each beaker, with the residue, is desiccated and weighed to the nearest 0.1 mg. The preweighed silica gel samples are returned to the laboratory in their original containers. The samples are weighed to the nearest 0.5 g. The above procedures are repeated for three test runs to complete a field and lab portion of method 5.

Once the laboratory completes its procedures on the weights for the particulate catch, test calculations must be performed and the data reduced to be able to calculate the concentration or the total weight gain. Appendix 3C is an example of standard nomenclature used in the stack sampling industry and is widely accepted by most testing firms performing EPA methods. Appendix 3A shows test calculations for determining moisture in the stack gas, stack velocity and volume, particulate concentration, and emission rate. Lastly, there is a calculation to determine isokinetics or the determination for acceptability of the results. In Fig. 3.11, a blank field data sheet is illustrated. This is the type of format utilized by many test firms for logging data from field test work. Figure 3.12 shows a completed field data sheet with actual results from a test. All the data and information shown are important as the data is reduced and the report completed. With the data gathered from method 5 the engineer has the information needed to evaluate the process particulate levels at the test location and source tested.

3.9   Particle Sizing Techniques

As one works with particulate emission control equipment, sooner or later the words *particle size* become a topic of discussion. Particle size information is important to the air pollution control system manufacturer, particularly on equipment where fractional efficiency is considered (i.e., devices such as wet scrubbers, cyclones, and other mechanical collectors). Most equipment performance will vary with the particle size distribution of the particulate in the airstream to be controlled. Therefore, particle size measurements have become increasingly important to the environmental engineering field.

Within any sample of particulate, individual particles are of unique sizes, shapes, densities, composition, porosity, etc. Therefore, as with any population within nature, no two members of that population are exactly alike. In this case, we are specifically concerned with the sizes of the particles within the gas stream.

To accurately describe a given particulate sample, we use the terms *less than* and *greater than,* as appropriate. In other words, we may accurately measure and describe how much of the particulate is below or above a given particle size without ever determining the exact size of each particle. The result is a size distribution which expresses how much of the sample is below or above various specified particle sizes.

To obtain a particle size distribution, one must first collect a dry sample in the state in which it exists in the gas stream. The methods discussed above, and particularly methods 5 and 17, are most commonly utilized for this purpose. Method 17 becomes a favorite in this case because a large quantity of sampling can be obtained on heavy concentrations. Method 5 can be utilized when the sample can be run longer than normal, so that enough material can be obtained on the filter for a size distribution to be performed.

There is also a particle size technique utilizing an attachment to the EPA sample train for in-stack particle sizing. Basically, this method utilizes a device that attaches to the end of the probe much like the method-17 filter. The attachment used is a cascade impactor which uses internal plates with various size openings and filters to separate particles in the gas stream. This technique requires little lab work other than filter weights and calculations. The in-stack sample technique has limitations because it does not work well on gas streams where the moisture is extremely high or where the loading is high. Because of the small openings in the impactor plates, the gas stream can be sampled for only a short duration in heavily loaded particulate gas streams. This makes this method questionable for some processes. Moreover, the accuracy is better for some applications than for others.

The majority of the particle size techniques utilized by the environmental engineering field employ taking the dry sample from method 5 or 17 to the laboratory for particle size analysis of the dry sample. The most important thing for any engineer to remember is that the technique must take into account physical properties of the dust to be sampled as it exists within the process. One must select a technique that will not skew the results or damage the sample. In addition, one must also discuss the results of particle size techniques with the equipment manufacturers with whom one is working. Most equipment firms have based their performance curves and other efficiency data on specific particle sizing techniques. It is important that a correlation between the various techniques be obtained so that the cyclone or scrubber efficiency curves can easily relate to the fractional efficiency curves that result from the particle size data. In choosing a laboratory technique, one must consider the density of the particulate and its behavior in the fluid used as a settling medium as integral factors. Many of the above techniques give accurate results up to 1 μm. Other techniques have problems in the submicron ranges. This submicron range is important for meeting some of the strict air pollution laws of today and is the important range or zone for equipment manufacturers to study. If the particle size technique cannot deliver results in the micron range being considered, then it becomes of less value to the engineer.

The laboratory methods used must also be selected so that the technique will not agglomerate, deagglomerate, or fracture particles unlike that which is occurring in the process and will not otherwise impart skewed results. The Bahco analyzer (ASME Power Test Code #28), for instance, has the potential to fracture particles where wet methods could result in agglomeration or deagglomeration of some materials. By utilizing the particle size distribution and the control equipment's fractional efficiency curve, the engineer can properly select equipment with removal efficiencies to meet the goals and guarantees established by all parties.

Some methods use water and others use air as the separation fluid. Listed below are some of the techniques and the various unique features of those techniques. Particle sizing methods generally fall into two broad categories, those that measure the physical size and dimensions of the particles and those that measure the aerodynamic properties of the particles. In the situation where air pollution control solutions are sought that use inertial forces (cyclones and scrubbers), it is generally advisable to use aerodynamic particle size distributions. It is generally more appropriate to utilize physical size methods to describe particulate if media filtration (baghouses or cartridge collectors) is the anticipated selection. In any case, it is vital to ensure that the particle sizing method is consistent with the requirements of the equipment manufacturer(s) to accurately predict and warrant equipment performance.

Common methods of *physical* particle size measurements are

Sieve	A mesh or screen which allows smaller particles to pass through the sieve and larger particles to be retained.
Light scattering	Elyzone and Microtrac are common proprietary methods.
Variable orifice	Coulter Counter is the most common proprietary method. The Coulter Counter uses an electrical resistance technique that measures the average size of macromolecules, emulsions, and particles in suspension using a microscopic scan technique.
Electron microscopy	Nondestructive physical sizing technique. This non-aerodynamic method uses an electronic microscope and reference size particles to evaluate the sample. This has an added advantage of chemical analysis by x-ray diffraction or fluorescence.
Visual microscopy	A basic visual technique using reference particles against the sample.

Common methods of *aerodynamic* particle size measurements are

Liquid sedimentation	Andreasen Pipette, Whitby Centrifuge, Micromeretics Sedigraph, Horiba, and Shimadzu are common methods or manufacturers.
	Andreasen Pipette. Uses a pipette, dispersant fluid, and weighings over time.
	Whitby Centrifuge. Is an extension of the Andreasen technique and employs the use of centrifuges.
	Micromeretics. Provides results in Stokes equivalent spherical diameters from 0.01 to 100 μm; an x-ray beam measures particle concentration of materials.
	Horiba and Shimadzu. Provides results in Stokes equivalent diameters from 1 to about 200 μm; a light beam measures particle concentration in conjunction with a centrifuge.
Air classification	Bahco Micro-particle Classifier (ASME PTC #28) and Cascade Impactor are common methods.
	Bahco Centrifuge. A laboratory rotary classifier provides results in Stokes equivalent diameters from about 1.5 to 40 μm. Sample is introduced into a spiral-shaped air current flowing toward the center; this method ensures good deagglomeration and dispersion.

Air sedimentation

Cascade Impactor. Is an in-stack particle classification method that separates particles based on slotted offset orifices in stages of the impactor. Provides results in Stokes equivalent diameters from about 0.3 to 15 μm. It uses verifying velocity of the particle and slot size to classify the particulate.

Micromerograph is the most common method.

Micromerograph. Uses Stokes' Law of Fall as a basis, which equates the measured terminal velocity of particles falling through a gas to provide particle size in Stokes equivalent diameters from about 0.0 to 100 μm.

3.10   Gaseous Testing

Like particulate sampling, test methods for gaseous emissions have been developed and improved with the advent of the EPA methods. Today, most sampling for gases is conducted utilizing an EPA, NIOSH, or OSHA method or variations that use compliance and noncompliance instrumentation. There is such a wide variety of techniques for measuring various gaseous emissions that entire books have been written about them. An attempt to highlight some of the common gas emission sampling techniques and their variations is made here.

Generally speaking, more than one method exists for obtaining similar results for gaseous emission tests. Sampling firms classify two categories of gaseous testing techniques as *manual* EPA methods and *instrumental* methods. Manual methods use a combination of field and lab techniques usually employing a container to house the gas sample. Instrumental methods use gas analyzers to measure concentrations on-line or in real time. Most common gas emission tests can be conducted using real-time analyzers or compliance analyzers. The EPA has established these as acceptable for compliance testing. These analyzers meet the strict guidelines that the EPA has established for its specifications to give repeatability and accuracy. Some of the more common gas emission analyzers today are for SO_2, NO_x, CO, CO_2, and hydrocarbons.

3.11   Sulfur Dioxides

Like method 5 for particulate, method 6 (Fig. 3.13) was derived by the EPA for determining sulfur dioxide emissions. The following is a procedure for sulfur dioxide sampling. It resembles a method-5 train in some respects. The particulate filter is removed because gas sampling does not require particulate measurements.

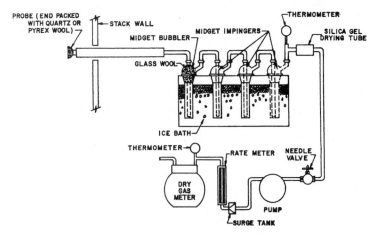

Figure 3.13 EPA method-6 sample train schematic. (*Courtesy Air Techniques.*)

EPA method 6, Determination of Sulfur Dioxide Emissions from Stationary Sources, is used to measure sulfur dioxide emissions. The sample train consists of a glass probe, four midget impingers, umbilical cord, flow control valves, pump, rotameter, and a dry gas meter. Approximately 20 mL of 80% isopropanol is placed in the first impinger to remove sulfur trioxide and acid mist. Fifteen milliliters of 3% hydrogen peroxide solution is placed in between the first and second impingers to prevent any solution carryover. After the train is assembled, a leak check should be conducted to ensure that all connections are secure. The probe is positioned at the sampling point within the stack, the initial gas meter reading is recorded, and the pump is turned on. A constant sampling rate of approximately 1 L/min is maintained during the sampling period. The dry gas meter volume, meter temperature, fourth impinger outlet temperature, and sampling flow rate are recorded every 5 min during sampling.

At the end of the sampling period, the pump is turned off and the probe removed from the stack. The final dry gas meter volume is recorded and another leak check should be conducted. The train is purged with ambient air for 15 min. After each test, the contents of impinger 1 are discarded and the contents of impingers 2 and 3 are saved in a labeled polyethylene container. Impingers 2 and 3 are then rinsed with distilled water and the rinsings added to the same container. In the laboratory, each sample is placed in a 100-mL volumetric flask and diluted to 100 mL with deionized, distilled water. A 2-mL aliquot of this solution is placed in a 250-mL Erlenmeyer flask. Eighty milliliters of isopropanol is added followed by two to four drops of thorin indicator. The solution is titrated to a pink endpoint using

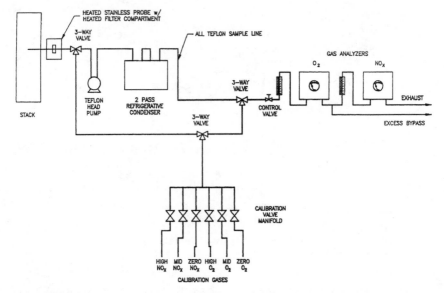

Figure 3.14   EPA method-7E NO$_x$ measuring system. (*Courtesy Air Techniques.*)

0.0100 N barium perchlorate. Each titration is repeated to ensure repeatability of results. Results are reduced and transferred to the report.

EPA method 6C is an instrumental analyzer method for determining sulfur dioxide. This method is similar to the NO$_x$ analyzer method (method 7E) which is illustrated in Fig. 3.14. The analyzer techniques utilize analyzers that will draw a suitable gas sample from the process duct or stack to an analyzer that is calibrated for the specific gas range to be detected. Along with the analyzer, calibration gases are required as well as the appropriate recording or data-logging equipment needed to produce real-time, on-line results. Many of these real-time analyzers are utilized not only for sampling specific sources for compliance or for engineering but are also applied in continuous monitoring systems as well. The term *continuous emission monitor* (CEM) has become commonly used. When several analyzers to obtain data in real-time fashion for continuous recording are incorporated into a system and packaged with calibration gases and recording equipment, they comprise a CEM system. They can be stationary or they can be portable and utilized for mobile field data gathering.

3.12   Nitrous Oxides

Like method 6 for sulfur dioxide sampling, there are several techniques for determining nitrogen oxide (NO$_x$) emissions. Method 7 is

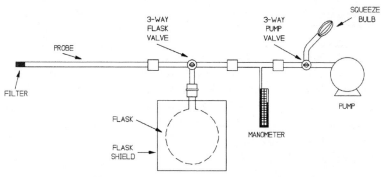

Figure 3.15   EPA method-7 sample train schematic. (*Courtesy Air Techniques.*)

the EPA manual method. A method-7 test train is illustrated in Fig. 3.15. A written procedure for method 7 follows. Method 7E is the instrumental analyzer method that produces continuous real-time results for nitrogen oxides. EPA method 7, Determination of Nitrogen Oxide Emissions from Stationary Sources, is used to determine oxide emission concentrations. Before sampling, 25 mL of a dilute sulfuric acid–hydrogen peroxide absorbing solution is pipetted into each of twelve 2-L flasks. Immediately before sampling, each flask is evacuated to an absolute pressure of less than 3 in Hg. After observing the vacuum for any leaks for 60 s with the pump off, the flask is connected to the ¼-in outer-diameter (OD) stainless-steel sampling probe. The flask valve is turned to the purge position and sampling is commenced. The flask valve is closed after pressure equilibrium between the stack and flask is reached (about 15 s). The probe is then disconnected and the flask is shaken for 5 min. The flask is then stored for transport. Four samples are taken for each test repetition. After the flasks set for more than 16 h, they are shaken again for about 2 min. Each is then connected to a mercury manometer and the internal pressure of the flask measured. The contents of the flask are then transferred to a sample bottle, and the flask is rinsed twice with 5-mL portions of distilled water. These rinses are also added to the sample bottle. Finally, the pH of the sample is adjusted to between 9 and 12 by adding drops of a normal solution of sodium hydroxide. The sample bottle is sealed for shipment to the laboratory.

In the laboratory, the content of each sample storage bottle is placed in a 50-mL volumetric flask. The storage bottle is rinsed twice with 5-mL portions of deionized, distilled water. A 25-mL aliquot from each flask is evaporated to dryness on a steam bath in a porcelain evaporating dish. After cooling, 2 mL of phenodisulfonic acid solution is added to the dried residue and titrated thoroughly. One milliliter of deionized, distilled water is added followed by four drops of concentrated sulfuric acid. The solutions are heated on a steam bath for 3 min and allowed to cool. Twenty milliliters of deionized, distilled water is added and mixed well with the solution. The pH is adjusted to 10 using concentrated ammonium hydroxide. In the absence of solids, the solution is transferred to a 100-mL volumetric flask and diluted to the mark with deionized, distilled water. The absorbency of each solution is measured at the optimum wavelength and used for standard solutions, using a blank solution as zero reference. The results are reduced and transferred to a report.

3.13   Other EPA Methods

Each year new methods for sampling technology are promulgated and approved by the U.S. EPA. Extensive research and development and refining are required for each of these methods before they are approved. As previously mentioned, App. 3B lists some of the more common EPA-approved methods. This will give the reader some idea of the wide range of techniques that have been developed since 1970. There is not a method for every source or pollutant, but generally speaking, by adapting similar techniques to a defined similar source, one can usually find a method that will apply to help gather field data. In addition to this list, there are illustrations of other common methods. Included in these methods are examples of sampling for volatile, semivolatile, formaldehyde, and volatile organic compounds.

Volatile organic compounds (VOCs) are another class of compounds for which there are test methods being utilized today. Again, as with other classes of compounds, there are various methods that can be utilized to obtain VOC concentrations. The most commonly referenced method for determining VOC concentrations is EPA method 25 for compliance testing. Most of the current engineering and much of the compliance testing for VOCs, however, is performed using EPA method 25A utilizing the hydrocarbon analyzer. Schematics of both test trains are included in this section as Figs. 3.16 and 3.17. Additional test train schematics are shown in Figs. 3.18 to 3.24.

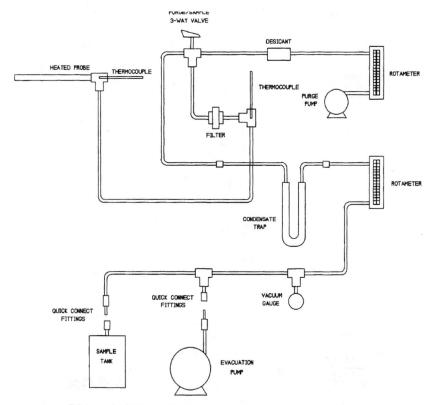

Figure 3.16 EPA method-25 sample train schematic. (*Courtesy Air Techniques.*)

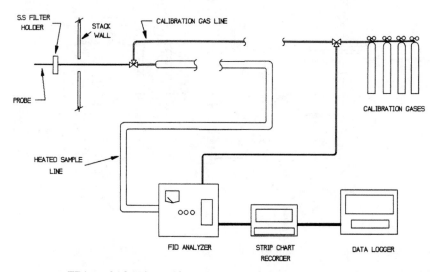

Figure 3.17 EPA method-25A sample train schematic. (*Courtesy Air Techniques.*)

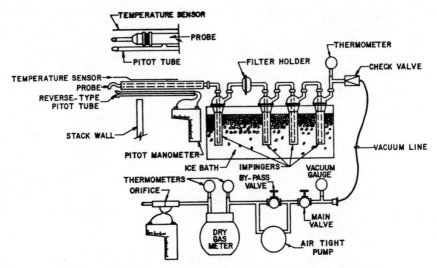

Figure 3.18 EPA method-8 sample train schematic. (*Courtesy Air Techniques.*)

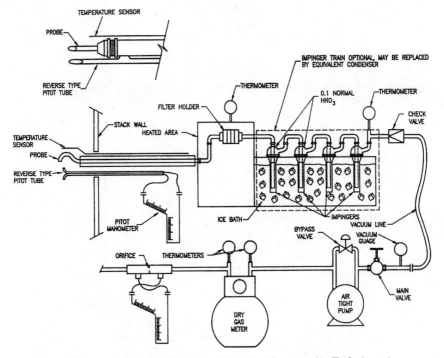

Figure 3.19 EPA method-12 sample train schematic. (*Courtesy Air Techniques.*)

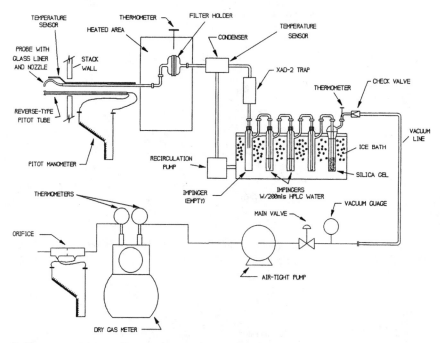

Figure 3.20   EPA method-23/semi-vost sample train schematic. (*Courtesy Air Techniques.*)

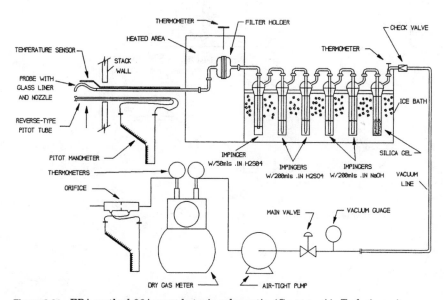

Figure 3.21   EPA method-26A sample train schematic. (*Courtesy Air Techniques.*)

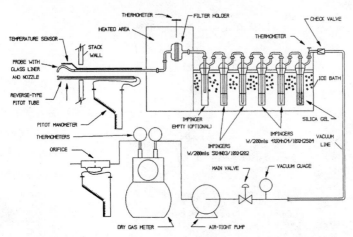

Figure 3.22 Trace metal emissions sample train schematic. (*Courtesy Air Techniques.*)

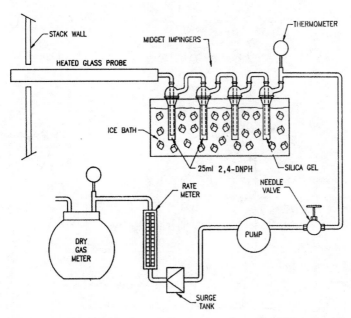

Figure 3.23 Formaldehyde sampling train schematic. (*Courtesy Air Techniques.*)

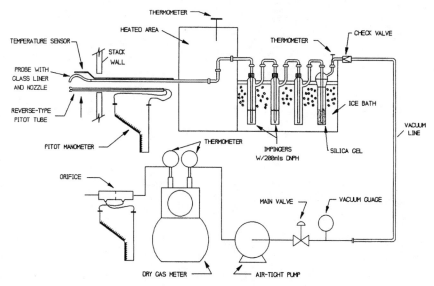

Figure 3.24 EPA method 0011 sample train schematic. (*Courtesy Air Techniques.*)

Test Calculations

I. Determination of moisture in stack gases

   A. Volume of water vapor collected (cubic feet)

$$V_{wstd} = 0.04707 \times V_{lc} \tag{3.2}$$

   B. Dry gas volume through meter (cubic feet)

$$V_{mstd} = 17.64 \times V_m \times Y \times \frac{P_{bar} + (\Delta H/13.6)}{t_m + 460} \tag{3.3}$$

   C. Moisture content (B_{ws})

$$B_{ws} = \frac{V_{wstd}}{V_{wstd} + V_{mstd}} \tag{3.4}$$

   D. Wet molecular weight (M_s)

$$M_s = [M_d \times (1 - B_{ws})] + (18.0 \times B_{ws}) \tag{3.5}$$

II. Actual stack gas volume sampled (cubic feet)

$$V_{ma} = \frac{V_{mstd} \times (ts + 460) \times P_{std}}{(1 - B_{ws}) \times T_{std} \times P_s} \tag{3.6}$$

III. Determination of stack gas velocity and volumetric flow rate

A. Stack gas velocity (feet per second)

$$v_s = K_p \times C_p \times v_s = K_p \times C_p \times \sqrt{\Delta P} \times \sqrt{\frac{t_s + 460}{P_s \times M_s}} \qquad (3.7)$$

B. Stack volumetric flow rate (cubic feet per minute)

1. Dry standard conditions (Qs)

$$Q_s = 60 \times (1 - B_{ws}) \times v_s \times A_s \times \frac{T_{std}}{t_s + 460} \times \frac{P_s}{P_{std}} \qquad (3.8)$$

2. Actual conditions (Q_a)

$$Q_a = v_s \times A_s \times 60 \qquad (3.9)$$

IV. Determination of particulate concentration (grainloading)

A. Dry standard conditions (c_s)

$$c_s = 0.01543 \times \frac{m_n}{V_{mstd}} \qquad (3.10)$$

B. Actual conditions (c_{sl})

$$c_{sl} = 0.01543 \times \frac{m_n}{Vm_a} \qquad (3.11)$$

V. Emission rate (pounds per hour)

$$E = 60 \times Q_s \times \frac{c_s}{7000} \qquad (3.12)$$

VI. Pollutant emission rate (pounds per million Btus)

$$E = F_d \times \left(\frac{c_s}{7000} \times \frac{20.9}{20.9 - \%O_2} \right) \qquad (3.13)$$

VII. Heat input (MMBtu per hour)

$$Q_h = 60 \times \frac{Q_s}{F_d} \times \frac{20.9 - \%O_2}{20.9} \qquad (3.14)$$

VIII. Determination of acceptability of sampling results (I)

$$I = \frac{(t_s + 460) \times \{(0.00267 \times V_{lc}) + [(V_m \times Y)/(t_m + 460)]\} \times [P_{bar} + (\Delta H/13.6)]}{0.599 \times \theta \times v_s \times P_s A_n} \qquad (3.15)$$

EPA Test Methods

Method 5	Determination of Particulate Emissions from Stationary Sources
Method 5A	Determination of Particulate Emissions from the Asphalt Processing and Asphalt Roofing Industry
Method 5D	Determination of Particulate Matter Emissions from Positive Pressure Fabric Filters
Method 5E	Determination of Particulate Emissions from the Wool Fiberglass Insulation Manufacturing Industry
Method 5F	Determination of Nonsulfate Particulate Matter from Stationary Sources
Method 6	Determination of Sulfur Dioxide Emissions from Stationary Sources
Method 6A	Determination of Sulfur Dioxide, Moisture and Carbon Dioxide Emissions from Fossil Fuel Combustion Sources
Method 6B	Determination of Sulfur Dioxide and Carbon Dioxide Daily Average Emissions from Fossil Fuel Combustion Sources
Method 6C	Determination of Sulfur Dioxide Emissions from Stationary Sources (Instrumental Analyzer Procedure)
Method 7	Determination of Nitrogen Oxide Emissions from Stationary Sources
Method 7A	Determination of Nitrogen Oxide Emissions from Stationary Sources (Ion Chromatography)
Method 7B	Determination of Nitrogen Oxide Emissions from Stationary Sources (Ultraviolet Spectrophotometry)
Method 7C	Determination of Nitrogen Oxide Emissions from Stationary Sources (Alkaline-Permanganate/Ion Colorimetric Method)
Method 7D	Determination of Nitrogen Oxide Emissions from Stationary Sources (Alkaline-Permanganate/Ion Chromatographic Method)
Method 7E	Determination of Nitrogen Oxide Emissions from Stationary Sources (Instrumental Analyzer Procedure)
Method 8	Determination of Sulfuric Acid Mist and Sulfur Dioxide Emissions from Stationary Sources
Method 9	Visual Determination of the Opacity of Emissions from Stationary Sources
Method 10	Determination of Carbon Monoxide Emissions from Stationary Sources
Method 10A	Determination of Carbon Monoxide Emissions in Certifying Continuous Emission Monitoring Systems at Petroleum Refineries
Method 10B	Determination of Carbon Monoxide Emissions from Stationary Sources

Method 11	Determination of Hydrogen Sulfide Content of Fuel Gas Streams in Petroleum Refineries
Method 12	Determination of Inorganic Lead Emissions from Stationary Sources
Method 13A	Determination of Total Fluoride Emissions from Stationary Sources—SPADNS Zirconium Lake Method
Method 13B	Determination of Total Fluoride Emissions from Stationary Sources—Specific Ion Electrode Method
Method 14	Determination of Fluoride Emissions from Potroom Roof Monitors for Primary Aluminum Plants
Method 15	Determination of Hydrogen Sulfide, Carbonyl Sulfide, and Carbon Disulfide Emissions from Stationary Sources
Method 15A	Determination of Total Reduced Sulfur Emissions from Sulfur Recovery Plants in Petroleum Refineries
Method 16	Semicontinuous Determination of Sulfur Emissions from Stationary Sources
Method 16A	Determination of Total Reduced Sulfur Emissions from Stationary Sources (Impinger Technique)
Method 16B	Determination of Total Reduced Sulfur Emissions from Stationary Sources
Method 17	Determination of Particulate Emissions from Stationary Sources (In-Stack Filtration Method)
Method 19	Determination of Sulfur Dioxide Removal Efficiency and Particulate Matter, Sulfur Dioxide and Nitrogen Oxides Emission Rates
Method 20	Determination of Nitrogen Oxides, Sulfur Dioxide and Oxygen Emissions from Stationary Gas Turbines
Method 21	Determination of Volatile Organic Compounds Leaks
Method 22	Visual Determination of Fugitive Emissions from Material Sources and Smoke Emissions from Flares
Method 24	Determination of Volatile Matter Content, Water Content, Density, Volume Solids and Weight Solids of Surface Coatings
Method 24A	Determination of Volatile Matter Content and Density of Printing Inks and Related Coatings
Method 25	Determination of Total Gaseous Nonmethane Organic Emissions as Carbon
Method 25A	Determination of Total Gaseous Organic Concentration Using a Flame Ionization Analyzer
Method 25B	Determination of Total Gaseous Organic Concentration Using a Nondispersive Infrared Analyzer
Method 27	Determination of Vapor Tightness of Gasoline Delivery Tank Using Pressure-Vacuum Test

Method 101   Determination of Particulate and Gaseous Mercury
                    Emissions from Chlor-Alkali Plants—Air Streams

Method 101A   Determination of Particulate and Gaseous Mercury
                    Emissions from Sewage Sludge Incinerators

Method 102   Determination of Particulate and Gaseous Mercury
                    Emissions from Chlor-Alkali Plants—Hydrogen Streams

Method 103   Beryllium Screening Method

Method 104   Determination of Beryllium Emissions from Stationary
                    Sources

Method 105   Determination of Mercury in Wastewater Treatment Plant
                    Sewage Sludge

Method 106   Determination of Vinyl Chloride from Stationary Sources

Method 107   Determination of Vinyl Chloride Content of Process
                    Stream Samples

Method 107A   Determination of Vinyl Chloride Content of Solvents,
                    Resin-Solvent Solution, Polyvinyl Chloride Resin, Resin
                    Slurry, Wet Resin and Latex Samples

Method 108   Determination of Particulate and Gaseous Arsenic
                    Emissions

Method 108A   Determination of Arsenic Content in Ore Samples from
                    Nonferrous Smelters

Method 108B   Determination of Arsenic Content in Ore Samples from
                    Nonferrous Smelters

Method 108C   Determination of Arsenic Content in Ore Samples from
                    Nonferrous Smelters

Method 200   Determination of Total Particulate, Free Chloride and
                    Total Chlorides from Secondary Aluminum Smelters and
                    Other Stationary Sources

Method 201A   Determination of Total Particulate less than 10 Micron
                    Emissions from Stationary Sources

Method 202   Determination of Condensable Particulate Emissions from
                    Stationary Sources

Nomenclature for Chap. 3

Sherril Wayne Norman
Special Note For Chapter 3

The nomenclature utilized within this chapter posed a unique problem. That is we could make the nomenclature match that which is used throughout the rest of this work and not match that which is commonly used within the testing industry. Conversely, we could match that used within the testing industry and within the referenced EPA documents and subsequently not match other sections of this book. Since we have endeavored to define nomenclature at each usage, we felt the latter approach was most appropriate. I apologize to the reader for any inconvenience this causes.

A_s Cross-sectional area of stack, ft^2

A_n Cross-sectional area of nozzle, ft^2

ACF   Actual cubic feet of gas at stack conditions

ACFM   Actual cubic feet of gas per minute at stack conditions

B_{ws} Proportion by volume of water vapor in gas stream

C_a Acetone blank residue concentration, mg/g

C_p Pitot tube coefficient

c_{sl}	Particulate concentration in stack gas, g/ACF
c_s	Particulate concentration in stack gas, g/dscf
ΔH	Pressure drop across orifice meter, in H_2O
d_p	Nozzle diameter, in
Δp	Velocity head of stack gas, in H_2O
dscf	Cubic feet of dry gas corrected to standard conditions
E	Particulate emission rate, lb/h
F_d	Ratio of combustion gas volumes to heat input, dscf/MMBtu
I	Percentage of isokinetic sampling, %
K_p	Pitot tube constant (85.49)
m_a	Mass of residue of acetone wash after evaporation, mg
M_{aw}	Mass of residue of acetone blank, mg
m_f	Particulate matter collected on filter, mg
M_d	Molecular weight of gas (dry), lb/lb · mol
M_s	Molecular weight of gas (wet), lb/lb · mol
m_n	Total particulate matter collected, mg
ρ_a	Density of acetone, mg/mL
P_{bar}	Barometric pressure, in Hg
P_m	Absolute pressure of dry gas meter, in Hg
P_s	Absolute stack gas pressure, in Hg
P_{std}	Barometric pressure; standard conditions, 29.92 in Hg
Q_a	Volumetric flow rate, actual conditions, ACF/min
Q_s	Volumetric flow rate, dry standard conditions, dscf/min
Q_h	Heat input, MMBtu/h
T_m	Absolute average dry gas meter temperature, °R
T_{std}	Absolute temperature at standard conditions, 528°R
t_m	Average dry gas meter temperature, °F
t_s	Average stack gas temperature, °F
θ	Total sampling time, min
V_a	Volume of acetone blank, mL
V_{aw}	Volume of acetone used in wash, mL
V_f	Final volume of impinger contents, mL
V_i	Initial volume of impinger contents, mL
V_{lc}	Total volume collected in impingers and silica gel, mL
V_m	Volume of gas sampled through gas meter, ft^3
V_{ma}	Stack gas volume sampled, ACF
V_{mstd}	Volume of gas sampled through gas meter, dscf
V_{wstd}	Volume of water vapor in gas sampled, standard, ft^3

v_s	Average stack gas velocity, ft/s
W_a	Weight of residue in acetone wash, mg
W_f	Final weight of filter or probe wash beaker, g
W_i	Initial weight of filter or probe wash beaker, g
Y	Dry gas meter calibration factor

Air Techniques, Inc.
Gas Flow Data Sheet

Plant_____   Source_____

Date____/____/____ City_____ State_____ Operators_____/_____

Identification: Pitot # _____ Pitot Cp _____ Digital Therm. #_____ Barometer #_____

Ambient Conditions: Temp. (°F) _____

Stack Data: Length (in.) _____ Width (in.) _____ Equivalent Diameter (in.) _____

Diameter (in.) _____ Port Extension (in.) _____ Area (ft^2) _____

Nearest Disturbances: Upstream, $U_{L/D}$ (L/D) _____ Downstream, $D_{L/D}$ (L/D) _____

Fyrite Analysis: % CO_2 ____/____/____   % O_2 ____/____/____

Number of Points: _____

Point # (pt#)	Diameter Fraction of Traverse Points ($K_{pt#}$)		Distance		Flow Data							
			From Wall	From Wall + Ext.	Test # ____		Test # ____		Test # ____		Test # ____	
	6 Pts.	8 Pts.	inches	inches	Initial	Final	Initial	Final	Initial	Final	Initial	Final
1	0.044	0.032										
2	0.146	0.105										
3	0.296	0.194										
4	0.704	0.323										
5	0.854	0.677										
6	0.956	0.806										
7		0.895										
8		0.968										
Time												
Barometric Pressure (Pb)												
Wet Bulb Temperature (°F)												
Dry Bulb Temperature (°F)												
Static Pressure (inches H2O)												

The following criteria is used to determine the sampling location and the number of points for each traverse:

1. The sampling location must have a $U_{L/D} > 2$ and a $L_{L/D} > 1/2$.
2. If $U_{L/D} > 6$ and $D_{L/D} > 1.5$, then 6 points should be used for each traverse.
3. If $2 < U_{L/D} < 6$ and $0.5 < D_{L/D} < 1.5$, then 8 points should be used for each traverse.

where $U_{L/D}$ is the distance from the sampling location to the nearest upstream disturbance and $L_{L/D}$ is the distance from the sampling location to the nearest downstream disturbance.

The distance from the wall for each point is determined using the following formula:

$$\text{Distance from Wall} = (\text{Stack Diameter}) \times (K_{pt#})$$

where $K_{pt#}$ is the diameter fraction for each pt#

Industrial
Ventilation

Ronald L. Jorgenson

Donaldson Company, Inc.
Minneapolis, MN

4.1 Introduction

With few exceptions, buildings require the use of forced ventilation to assure the comfort and health of the occupants. Consistent air exchange maintains proper levels of oxygen and appropriate levels of relative humidity and rids the building of the chemical off-gassing from building materials, carpets, fabrics, wall coverings, and furnishings. Industrial buildings without ventilation, or with improperly engineered ventilation, cause occupants to suffer the effects of what has come to be termed *sick building syndrome.*

Even if we assume that the air inside the building is adequately clean, the air drawn in from the outside brings with it a wide variety of pollutants. Products of combustion, auto exhaust, carbon, sulfur compounds, lead compounds, dust, plant spores, pollen, bacteria, and viruses are just a few of the potentially harmful contaminants found in outdoor air.

Obviously, a comfort control system must provide clean air at the proper temperature, with a moisture content that provides a comfortable environment. A complete system will heat or cool the airstream as required and will blend a percentage of outside air with the air recirculated within the building. Adding a high percentage of fresh air will increase both heating and air-conditioning costs. Too little fresh air and the occupants will find the air stuffy, high in moisture, and occasionally malodorous.

Inside the industrial factory, contaminants injurious to health must be controlled to levels specified by law. However, even inert dust must be controlled to levels less than 0.003 g/ft^3 [5 mg/m^3]. Outdoor air is quite clean in relationship to this level, typically in the range of 0.000 03 g/ft^3 [0.05 mg/m^3] in rural areas to about twice that level in urban areas.

Even as clean as the average outdoor, or makeup, air is, it still makes good sense to filter it as well as the recirculated air, to remove the airborne solids, since there will be times when the loading will be far above the average. Proper filtration will reduce housekeeping costs by keeping the walls, ceilings, carpets, drapes, and furnishings clean for longer periods.

The typical makeup-air filtration system for an office or light-manufacturing facility will consist of roll media filter or panel filters with an efficiency ranging from 30 to 95 percent. The panel filters are changed periodically when the airflow drops below the acceptable level. Roll filters frequently have an automatic indexing system that advances the filter roll, exposing a clean section of media to the airstream. Indexing is based on a pressure drop or on the time elapsed since the last indexing movement. Some facilities require significantly better filtration than the average, dictating use of the highest efficiency American Society of Heating, Refrigeration, and Air Conditioning Engineers (ASHRAE)-rated filters available. Good examples of this include hospitals, nursing homes, schools, and electronic assembly facilities.

The low levels of contamination in the atmosphere lead to the conclusion that even though it is important to filter all makeup and recycled air entering the facility, the majority of the contaminants in the industrial workplace are generated inside the building. The remainder of this chapter focuses on control of that self-generated contaminant load.

4.2 Control of Fumes and Odors

By far the best method of control for fumes and odors, as well as particulate, is source capture. Source capture requires the immediate isolation of the offending contaminant before it can mix with the general atmosphere. Source capture is the most economical method of control since it minimizes the amount of air that needs to be filtered or discharged. Once the contaminant mixes with the general atmosphere in the plant, large volumes of air need to be filtered, which means money spent for energy as well as to purchase the equipment to do the job.

To be successful, the airflow direction and placement of people in relationship to the fume generation point are critical. The airflow should always carry fumes away from the people affected, rather than carrying them through the breathing zone. A bad example of carrying fumes through the breathing zone is the typical hand-operated welding gun. The workpiece is in front of and below the face of the welder, and the heat generated in the welding process carries the weld plume directly up into the welder's face. A better method would be to use an articulated hood to intercept that plume or a concentric nozzle pickup right at the welding gun tip; either will keep the plume away from the welder. By capturing the plume directly at the source, the amount of air to be filtered is quite small.

Obviously, it is not always possible to capture fumes or odors at their source, especially when the generation area is large. If the fumes or odors are not toxic, it may be possible to provide control of the fumes or odors by general ventilation. *Industrial Ventilation, A Manual of Recommended Practice*[1] is a good source for information on dilution ventilation for control of solvent and other vapors. The recommendations include suggested flow patterns and ventilation rates to protect people from exposure levels exceeding the threshold limit values (TLV) and to avoid reaching the lower explosive limit (LEL) of flammable vapors.

Another excellent method of pollutant collection when the area of generation is large is to use a commercially designed and fabricated control booth or a custom-designed booth that incorporates lighting, noise control when needed, and a filtration system that draws adequate airflow over the worker, past the workpiece, and on to the collection device. This effectively captures the particulate and carries it away from the worker, not through the worker's breathing zone. Abatement efforts aimed at vapors, fumes, and odors usually rely on control systems based on adsorption, absorption, and chemical or thermal destruction.

4.3 Clean Rooms

A clean room is an isolated space inside a building designed to provide a carefully controlled environment, with very little particulate contamination. Clean rooms are used for electronic assembly, pharmaceutical applications, photographic processing, and other applications that cannot tolerate particulate in the airstream. In addition to particulate, clean rooms control the airflow direction and speed, temperature, humidity, and lighting and are typically equipped to provide fire protection.

Federal Standard 209E, Airborne Particulate Cleanliness Classes in Cleanrooms and Clean Zones, sets not-to-exceed limits on particle counts per volume of circulated air, grading performance by classes. Class 100,000 [M 6.5]* allows counts of 100,000/ft^3 [3,530,000/m^3] for 0.5-μm or larger particles, plus 700/ft^3 [24,700/m^3] for 5-μm or larger particles. Class 1 [M 1.5] allows counts of 1/ft^3 [35.3/m^3] for 0.5-μm particles, 3/ft^3 [106/m^3] for 0.3-μm particles, 7.5/ft^3 [265/m^3] for 0.2-μm particles, and 35/ft^3 [1240/m^3] for 0.1-μm particles. There are also intermediate classes of 10,000 [M 5.5], 1000 [M 4.5], and 100 [M 3.5].

Meeting these stringent requirements requires the use of high-efficiency particulate air (HEPA) filters which have a minimum efficiency of 99.97 percent for 0.3-μm particles or ultralow penetration air (ULPA) filters which are rated for a minimum of 99.999 percent for particles in the 0.1- to 0.2-μm particle size range. In order to prolong the life of these rather expensive filters, most systems use less costly prefilters to remove large particulate. Clean rooms also require large airflows. A class 100,000 [M 6.5] system may require 18 to 35 complete air changes per hour, increasing to 540 to 650 for class 10 [M 2.5]. For all classes, all air entering into the clean room must pass through an HEPA or ULPA filter.

4.4 Nuisance and Process Dust Collection Systems

Dust collection systems serve one basic purpose, collecting an airstream that contains solid particulate, separating the solids from the airstream, and containing the solids for disposal or reuse. Nuisance dust systems are further defined as systems that collect dust that is generated by and incidental to your primary process and products. The material collected may or may not have value as scrap or recyclable product. Process dust systems also collect and separate particulate from an airstream, but in this case the particulate is not a nuisance or trash, it is your product. A good example of a process dust would be the collection of flour or starch in a milling facility.

The particulate control device used in a nuisance dust collection system can be any of the devices described in the control technologies chapters of this book, with the exception of those listed for control of gaseous contaminants. Particulate control devices used in process

---

*SI Class designation equivalent to the English Class 100,000. The "6.5" is an exponent indicating a cleanliness level not to exceed $10^{6.5}$ particles/m^3 0.5 μm and larger.

dust collection systems obviously rely on cyclones and media filters in order to preserve the material collected for use. The choice of control technology should be based on the exact process, the amount of particulate, the size and shape of the particulate, and the efficiency required. All the collectors work in the applications that fit their capabilities; however, careful selection will tend to point to one or two possible choices that give the best performance and economy.

Dust collection systems require a number of components in addition to the particle collection device and the blower. Other chapters of this book discuss these major components in detail. The following sections briefly cover the components required to complete the physical dust collection system.

4.5 Ductwork Design

Ductwork is the tubular, square, or rectangular enclosure used to contain, control, and transport air and the contaminants contained therein from the point of entry to and through the particulate control system(s) and system blower and to the point of discharge. The ductwork description usually includes any hoods placed at the inlet(s) to the ductwork.

4.5.1 Hoods

The ductwork starts at the hood. A hood is essentially a funnel designed to provide a smooth transition from air at rest to air moving at the proper velocity to keep the solids in motion while in the ductwork. Depending on the application, the mouth of the funnel may be completely open or a portion may be closed off to provide advantageous distribution of the air. A properly designed hood provides the maximum control of the solids you wish to contain, while using the least amount of air possible and utilizing the least amount of energy to accelerate the airstream. The most effective hoods also take advantage of particulate velocity, when present, rather than working against the particles' momentum. The same is true when the airstream is heated. The most effective hood will take advantage of the natural tendency of heated air to rise, rather than trying to force the heated column of air to reverse its course.

The two most important aspects of the hood design are to keep the hood as close to the generation point as possible and to provide as much containment as possible. The amount of air required to capture a particle increases as the square of the distance between the hood and the particle, meaning if you double the distance between the generation point and the hood, you will need four times the airflow to effect capture. To illustrate the value of containment, consider a grinding wheel enclosed in a hood. If the hood encloses 75 percent of the grinding wheel, the particle collection rate will far exceed that of a similar wheel operating without an enclosure, even at 10 times the airflow.

There are a number of generic hood types with innumerable permutations. The hood can be as simple as the raw, open end of the duct. This type hood is very inefficient since the air entering the duct will be turbulent, and air will be drawn from every direction.

The next step is to provide a flange around the duct end opening, generally at right angles to the wall of the ducting. This flange will prevent airflow from entering the area behind the opening, back around the end of the ducting. This improved design will increase the ability to control the direction of the airflow. It will also provide greater flow at the same energy usage or provide equal flow at lower energy usage.

Transitions in size or cross section from the hood face to the ductwork can be either abrupt or gradual. Smoothly tapered transitions conserve energy, providing the most airflow per horsepower. The last major consideration is that air follows all available paths, inversely proportional to resistance. On large hoods, the airflow velocity will not be uniform across the face of the hood without proper hood opening design. With a large open face, the majority of the airflow will be drawn from the center, with only small amounts drawn from the periphery. Generally, equal airflow velocity can be obtained by utilizing a series of slots to control the open area or a series of vanes that divide the open area into smaller segments, or by using multiple connections to the ductwork.

All the permutations of hoods contain or combine one or more of the general hood design configurations (containment, flanges, shape and/or size transitions, slots, and directional vanes) with a specialized shape to fit the specifics of the job.

Industrial Ventilation, A Manual of Recommended Practice[1] provides illustrations of hood designs for a wide range of equipment for metalworking, material handling, woodworking, open tanks, painting, and many other applications. These specific recommendations also provide guidelines for the proper velocities and/or airflow to use for efficient and safe collection systems.

4.5.2 Ducting

The ducting of a system can be anything from a short length of tubing connected between a single hood and a control device to highly complex systems that tie dozens or even hundreds of hoods to a single filtration or control system and blower. Ducting provides the means to channel the flow of air from the point of contamination to the control device while containing the offending materials.

In the vast majority of ducting systems, the intent is to maintain transport velocity at all places in the ducting. This simply means that the velocity of the airstream is sufficient to keep any solid particulate in the duct moving, not necessarily uniformly mixed but, as a minimum, drifting along the bottom of the ducting.

The consequences of not maintaining transport velocity can be minor or serious depending on the magnitude of the error. If the error is small, solids may fill in portions of the bottom of the duct until the duct open area decreases sufficiently to cause the system to reach transport velocity. In some cases, the ducting may plug completely over time. If the error is too large and the particulate heavy, the ducting may collapse from the weight of the collecting solids before the cross section becomes small enough to create transport velocity.

If the error swings the opposite way, with velocities too high, the price includes increased energy costs due to the greater friction losses and premature component failure due to erosion and abrasion. Table 4.1 below shows typical transport velocities.

TABLE 4.1   Duct Transport Velocities

Particulate class	Transport velocity, ft/s [m/s]	Typical examples
Clean gas steams	17–34 [5–10]	Gas streams with little or no settleable solids, ambient air
Smoke and fumes	33–42 [10–13]	Welding fumes, laser cutting
Dry, powdered products	42–67 [13–20]	Carbon, flour, graphite, leather dust, starch, textile dust, toner dust, pigments, powder coating, pharmaceuticals (dry)
Average industrial dusts	58–67 [18–20]	Abrasive blast, buffing and polishing dust (dry), cement, clay, coal dust, grinding, lime, metallizing, powdered metal, rock dust, tobacco, wood sanding
Heavy particulate and chips	67–75 [20–23]	Cement clinker, coal, composites machining, fiberglass, foundry shakeout and sand handling, lead, metal chips, wood hog waste
Wet, sticky, oily products	75 [23] and up	Buffing lint with rouge, cement dust (moist), milk solids, oily food products

Ducting intended for transporting particulate should be cylindrical. Most systems of this type are under negative pressure, and cylindrical ducting has the hoop strength required to operate at high negative static pressures. The round cross section also yields to a more consistent velocity profile, and fewer solids build up in the ducting.

Ducting on the clean air side of the control device, between the control device and the blower, typically is round since the negative static pressure is at the highest point in the system, and the inlet to the blower is also round. On the downstream side of the blower, the ducting can be of round, rectangular, or square cross section; typically the choice is dictated by the size and shape of the blower discharge. HVAC systems frequently use inexpensive rectangular ducting on the pressure side of the blower since strength is less of an issue and pressures are much lower than in the typical pollution control system.

4.5.3 Balanced-by-design or blast-gate ducting

Since air will follow all available paths, with the volume of air inversely proportional to the resistance, proper duct design must take into account the resistance to flow caused by friction of air movement through the hood and ducting. Directional changes, ducting diameter changes, turbulence, and the energy usage or regain during acceleration or deceleration of the airstream are also critical factors governing airflow.

When two ducts join to make a single duct, the static pressure (suction) will be identical in both ducts at the intersection point. The amount of air flowing through either duct will be based on the same principle of the air following all available paths. The proportioning of the flow between joining ducts can be accurately predicted and controlled by either of two generally accepted methods.

Balanced-by-design systems use a combination of the hood designs, duct sizes, velocities, elbow and fitting radius, and flow adjustment to balance the static pressures at the desired flow rates at duct intersections. This simply means that the system designer starts with the hood and branch duct farthest from the control device and blower and calculates the static pressure required to generate the desired flow of air in that duct segment. At the point the designed duct segment joins with another duct segment, the designer must calculate and match the static pressures of both ducts. If the static pressures calculated for the two ducts are more than a few percent different, the designer will modify one or both ducts to create a balance. This matching process continues at each intersection of ducts until the system main duct reaches the control device.

The designer works within the constraints that transport velocity must be maintained; the airflow should be the lowest that will provide acceptable capture of fumes, vapors, and dust; and the static pressure should be as low as possible. Excesses of airflow and static air both increase capital and operating expense.

Creating balanced-by-design systems requires thorough knowledge of the effects of changing hood parameters, duct sizes, fittings, and airflow. Because of the amount of repetitive calculations, it also fits well with commercial computer programs designed to speed the process and improve on the accuracy of the results.

Balancing the system with the use of blast gates moves the fine flow corrections from the design and calculation stage to the test stage of an installed system. Instead of changing fittings or the hood design to modify the static pressure loss and airflow, the installer adjusts the blast gate to artificially increase the resistance to flow by partially blocking off part of the duct cross section. The system design must still maintain minimum transport velocity in systems transporting solid particulate.

Blast-gate balancing provides flexibility but also builds in the ability for unauthorized changes to the system design. The common perception that more is better leads operators to open blast gates, increasing the airflow at their machine while decreasing the flow in other parts of the system. Installing the blast gates high enough so that any adjustment requires a ladder helps control this problem. Blast gates may also be the source of maintenance problems due to particulate accumulation.

Blast-gate balancing is not recommended when the safety of personnel depends on the proper airflow. This would include applications with toxic or explosive dusts, vapors, and gases.

4.5.4   Extended plenum systems

Extended plenum designs utilize transport velocity only in branch lines. The main plenum allows the airstreams entering from the individual branch lines to fall below transport velocity, dropping the bulk of the solids to the bottom of the plenum. The bottom of the plenum incorporates some method of removing the collected solids. This may be pneumatic conveying, screw conveyor, drag or belt conveyor, or, in cases of very light duty, access doors for manual cleanout.

Applications that require frequent moves of machinery, and consequently moves and changes to the ductwork, use this design to advantage. Changes and additional duct connections have little or no effect on performance of the system, as long as the additions stay within the system flow capacity. This ability to accept change is significantly different from the typical ductwork system that must maintain transport velocity and static pressure balance throughout the full system layout.

The increased flexibility of this system design also increases the initial cost, significantly increases the physical size of the plenum, and adds the cost and potential upkeep expense of locating moving parts in the plenum when using mechanical solids removal devices.

4.6   Feeders and Air Locks

There are several types of air locks and feeders. Air locks permit solids to pass between areas of differing air pressure, while maintaining a barrier to the free exchange of the air or other gases involved. A feeder also might maintain a pressure differential at times but may act only to control a measured flow of a product. The most common types of air locks and feeders utilized in dust control devices are the rotary air lock and the double dump valve.

4.6.1   Rotary air locks and feeders

A rotary air lock is similar, in both appearance and function, to a revolving door in a building. The revolving door allows people to enter or leave the building but maintains a seal to prevent escape of the cooled or heated air to the outdoors and at the same time keeps summer's heat and winter's cold where they belong. In the same manner, a rotary air lock permits the discharge of solids from the hopper of a dust collector, while preventing a loss of vacuum or pressure in the vessel.

A rotary air lock has a series of blades mounted radially on a center shaft. This rotor assembly fits inside of a housing generally cylindrical in shape with openings 180° apart which act as the inlet and outlet for the flow of material through the air lock. The shaft rotates at low speeds, usually less than 50 r/min. The most common drive is a chain-and-sprocket arrangement powered by a gear head motor (Fig. 4.1).

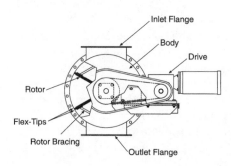

Figure 4.1 Rotary air lock cutaway. (*Courtesy of Torit Products, Donaldson Company, Inc.*)

The most common type of rotary air lock is the straight-through valve which has full openings on top and bottom. Variations of this valve have the opening for the inlet in the area of the rotor which is coming toward the top. In other words, when the rotor is moving in a clockwise direction, the inlet would be between nine o'clock and twelve o'clock. The outlet would be 180° from the inlet (Fig. 4.2). This style valve protects free-flowing frangible products, such as pellets, from degradation due to shearing and pulverization. The product can also be protected by installing a baffle that meters the flow to a fraction of the valve capacity, ensuring each rotor pocket is only partially filled.

An air-swept feeder valve utilizes moderately pressured compressed air, mixing the air with the product as each segment of the rotor drops its load into the mixing area at the bottom of the valve. This aerated product then flows into a pneumatic conveying line.

Rotary valves fall into two categories of construction. These include fabricated valves using rolled sheet stock bodies, inlets, end plates, flanges, and rotors, and cast valves which normally have all body parts machined from castings. Rotors may be either cast or fabricated.

Valves can be manufactured out of almost any material necessary to meet the customer's needs. The most common materials are steel, cast iron, cast and fabricated stainless steel, and cast and fabricated aluminum. Bodies and rotors are frequently coated or plated to modify the characteristics of the base materials. Teflon-coated rotors help prevent material buildup when handling sticky products. Also, hard chrome and electroless nickel plating slow wear.

4.6.1.1 Capacity ratings. The capacity of an air lock expresses its ability to pass solids through the valve. The capacity is normally expressed in cubic feet per revolution, with a recommended fill ratio. With the exception of valves used as metering devices, valves are operated at 40 to 70 percent of volumetric capacity. Models of valves frequently are expressed in sizes based on dimensions of the inlet and outlet flanges, such as 8, 10, or 12 in; however, this has little to do

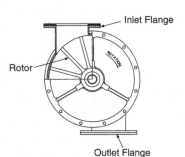

Figure 4.2 Rotary air lock with offset inlet and outlet. (*Courtesy of Torit Products, Donaldson Company, Inc.*)

with the valve's capacity. All valve specifications and purchases should be based on a minimum throughput. At times the size of a valve will be much larger than the solids loading requires. This may be based on the chip size or surge loading, to prevent product bridging over the valve opening, or simply to match a pipe size or the size opening on the dust collector hopper.

4.6.1.2 Design features. Rotary air locks have a flange on both the inlet and the outlet. The majority of the flanges are either round or square. There are specialty valves that have rectangular flanges. The choice of flange shape usually is matched to the connecting flange on the inlet side and/or the outlet side. Valves are available with nonmatching flanges, one round and one square.

The valve throat shape, which includes the area between the flange and the rotor, also varies. Either the valve has a large rotor allowing for a straight throat with no reduction in size, or the rotor is smaller in diameter and/or length necessitating a tapered throat to funnel the cross section from the size of the flange down to the size of the rotor (Fig. 4.3).

Valve rotor designs vary with the application (Fig. 4.4). The most common rotor is called an open rotor. This style has open ends between the rotor blades, and the blades seal or restrict the passage of air and solids on three sides of the blade. Closed rotors have a clo-

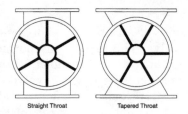

Straight Throat   Tapered Throat

Figure 4.3 Rotary air locks with straight and tapered throats. (*Courtesy of Torit Products, Donaldson Company, Inc.*)

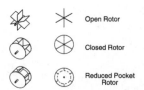

Open Rotor

Closed Rotor

Reduced Pocket Rotor

Figure 4.4 Rotary air lock open, closed, and reduced-pocket rotor designs. (*Courtesy of Torit Products, Donaldson Company, Inc.*)

sure plate on either end of the blades, which creates enclosed prism-shaped pockets, open only on the circumference of the rotor. Rotary valves used for metering product flow may also use a rotor that has reduced pocket volume. This can be envisioned as a cylindrical drum with short radial blades on the exterior.

The rotor blades can also be designed to meet diverse needs. The plain blade is simply machined to closely match the inside diameter of the valve body (Fig. 4.5a). Another style rotor has adjustable tips bolted on. The tips are made from either the same material or a material designed to provide sacrificial wear without damaging the body. Applications handling sugar or similar temperature-sensitive products can suffer from caramelization of the product, which can cause a valve to seize. This application uses a combination of rotor blades, some with clearance, and one or two very close fitting cutter blades that remove caramelized product as it is deposited. On applications that develop high levels of friction caused by product passing between the rotor tips and the body, the rotor blade tips and ends can be relieved (Fig. 4.5b). This means the diameter measured at the leading edge and a flat area just behind the leading edge closely matches the bore, but the remainder of the rotor tip is ground away at a taper to limit friction and stress on the rotor and drive train. If the application involves abrasive product, the rotor tips might be coated with a material such as a hard-facing alloy prior to the final grinding to finished size. If the product contains chips as well as fine granular material, the rotor will frequently utilize a flexible rotor tip fabricated from material such as neoprene rubber, polyurethane, ethylene propylene diene methylene (EPDM), silicone, or Buena-N (Fig. 4.5c). The flex tip usually seals itself on the face of the rotor as well as the ends, and since it can deflect, chips cannot easily jam between the rotor and body.

Chip clearance is application-dependent. If the chips are very fragile, a standard rotor may be acceptable since the rotor would shear the chip as it leaves the open area of the inlet. If the chips will not shear readily, the rotor must be designed to accommodate their passage. Flex-tipped rotors may have a very small amount of flexible

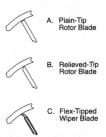

A. Plain-Tip
   Rotor Blade

B. Relieved-Tip
   Rotor Blade

C. Flex-Tipped
   Wiper Blade

Figure 4.5 Rotary air lock blade tip designs: (a) plain tip, (b) relieved tip, (c) flex-tipped wiper blade. (*Courtesy of Torit Products, Donaldson Company, Inc.*)

material extending beyond the rigid backup blade; however, if the application will generate larger chips, the backup blade can be cut back, exposing additional flexible material. The potential pressure differential across the valve limits the amount of exposure of the flexible sealing strip. As the amount of unsupported flexible sealing strip area increases, the ability of the valve to seal against differential pressure decreases.

The exact amount of differential pressure capability depends on the design features. Generally the cast valves will withstand a greater differential than the fabricated styles. Other design features affecting the pressure rating include the number of blades on the rotor, the clearance between the rotor and housing, and rotor strength. As a reference, cast valves with closely machined rotors may carry a differential rating of 10 or 15 lb/in^2 [0.7 to 1.1 kg/cm^2]. A fabricated valve with flexible wipers may be rated in the range of 0.3 to 0.7 lb/in^2 [0.02 to 0.05 kg/cm^2].

Abrasion resistance can be critical on applications handling large volumes of solids or highly abrasive materials. As mentioned earlier, the valve body can be plated to increase the life of the body, or the body can have a replaceable sleeve. The rotor can be fitted with replaceable tips or hard-faced tips, and the profile of the rotor tips can be relieved to minimize the chance of material wedging between the rotor tip and the valve body. A flex-tipped rotor, using a highly abrasion-resistant material such as polyurethane, may also extend the useful life of the valve.

Body, end plates, and rotors can exhibit erosion when the differential pressures are high. The pressure can cause air and the product to pass between the rotor and body or end plate, gradually eroding a channel into the metal. Careful selection of the rotor clearances can minimize this problem.

When rotary valves discharge or meter material into a positive-pressure pneumatic conveying line, each rotor pocket becomes a contained pressurized volume that depressurizes as the pocket reaches the lower-pressure area of the rotary valve inlet. This depressurization and expansion create a burst of air that tends to reentrain the solids, putting the dust back into the collector as a dust cloud instead of allowing the solids to flow into the air lock. The answer is a pressure-relief connection that bleeds off the pressure in each pocket as it passes from the outlet back to the inlet and discharges it harmlessly into the dust collector above the discharge flange.

The valve rotor shaft has a bearing at each end to support the shaft. Valves fabricated from sheet stock frequently use flange-mounted bearings fastened directly to the end plates of the valve (Fig. 4.6). The bearings must be protected from the material inside the valve to prevent premature failure due to contamination. This is espe-

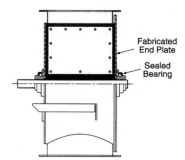

Figure 4.6 Rotary air lock—fabricated construction. (*Courtesy of Torit Products, Donaldson Company, Inc.*)

cially important when the valve is under positive pressure. In addition to the sealed bearing, most designs use a lip seal or packing gland between the valve body and the bearing. When a tough application requires better protection, compressed air injected between the bearing and the seal will force air past the seal into the rotor chamber. This helps keep solids out of the bearing.

Cast valves typically have a lip seal or a packing gland seal at the end plate, with an air gap open to the atmosphere between the seal and the bearing (Fig 4.7). The bearings mount in a boss supported by cast legs. This air gap construction provides protection for the bearing, preventing the pressure inside the valve from forcing solids into the bearing. Another style makes the end plate a double-wall chamber. The wall closest to the rotor has a seal. The outboard wall carries the bearing. This chamber is pressurized with filtered air at a higher pressure than that in the rotor chamber. This arrangement ensures that any leakage past the seal is from the clean side to the dirty side, assuring protection for the bearing.

The drive to rotate the rotor typically uses an electric motor driving through a gear reducer and a chain-and-sprocket arrangement. The gear reducer gives the lower speed needed for the rotor, and the chain and sprocket allows flexibility of speed to match the customer's needs. Small to moderate valves typically have the gear head motor mounted directly to the valve body. The drive can also be floor-mounted when

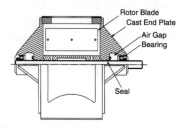

Figure 4.7 Rotary air lock—cast construction. (*Courtesy of Torit Products, Donaldson Company, Inc.*)

desired. This can be desirable when the drive is large or heavy. Rotary valves frequently are powered by common drives operating both a screw conveyor and a valve.

4.6.1.3 Temperature considerations. The temperature of the product and the airstream will affect the design and materials used for the valve. Both the axial and radial clearance of the rotor in relationship to the body must be adjusted to compensate for operating temperature. When operated at high temperatures, the rotor will expand more than the body since the rotor is completely bathed by the airstream and is in contact with the solids on all sides. The body has the ambient air temperature on the outside which will lessen the expansion due to the lower overall temperature. The valve body can be insulated to minimize this problem. High temperatures may also require castings to be stress-relieved and determine the choice of lubricant used in the bearings, the choice of seals or packing materials, the type of elastomer used in flex tips and flange gaskets, and even the choice of paint.

4.6.1.4 Dangers. Rotary valves can be very dangerous when installed or used incorrectly. The inlets, outlets, and drive components must be guarded to prevent accidental contact with rotating parts. Before attempting to service, inspect, or clean a valve, the electric power must be disconnected and locked out. Rotary valves must never be operated without all safety guards and covers in place.

4.6.2 Double dump valves

The valves commonly known as double dump valves are also known as tip valves, flapper valves, flap gates, air lock dust valves, and trickle valves. The double dump valve consists of a pair of chambers, each with a swinging valve plate that seals airtight against a seat (Fig. 4.8). The valve maintains a barrier to uncontrolled airflow by opening the upper valve, allowing any collected material to fall into the chamber below the valve plate. The upper valve then closes, and the lower valve opens to the storage vessel or to the atmosphere. The lower valve closes before the upper valve reopens. In this manner, the solids can flow out of the dust collector, while the valve maintains an airflow barrier.

Double dump valves are available in both sheet-metal–fabricated versions and cast valves. The most common materials of construction are fabricated steel, cast iron, stainless steel, and various alloys.

4.6.2.1 Capacity ratings. The valves are rated in a similar manner as rotary valves. The nominal rated capacity of the valve is stated in

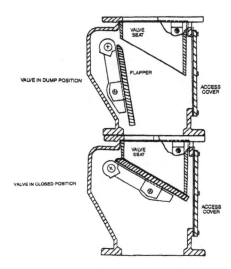

Figure 4.8 Air lock dust valve (double dump valve). (*Courtesy of Plattco Corporation.*)

cubic feet per hour at multiple rates of cycling the valve, typically up to 6 or 10 cycles per minute. The actual useful capacity will depend on the physical characteristics of the solids. The model designations in most cases reflect either the inside dimensions of the flanges or the valve seal opening.

4.6.2.2 Design features. The double dump valves can be purchased in models with square, round, and rectangular flanges. The available models range in size from 4 to 36 in [100 to 900 mm] and possibly larger custom designs. They are available in a variety of pressure ratings from 0.5 or 1.0 lb/in^2 [0.04 or 0.08 kg/cm^2] to 17 or 18 lb/in^2 [1.2 or 1.3 kg/cm^2]. Standard and special configurations of this style valve allow operational temperature from 350 to 1800°F [175 to 1000°C].

The double dump valve can be operated in a variety of ways. When the dust collector operates under very low negative or positive pressure, the valves can be operated completely by gravity. The valve has a counterweight that holds the valve shut until enough solids collect to overcome its weight plus atmospheric pressure if the system is under negative pressure. The weight of the solids swings the valve down and away from the valve seat. Once the solids drop free, the valve again swings shut, powered only by the counterweight.

If the vessel operates under higher levels of vacuum or pressure, a mechanical operating system is required. If the system is under negative pressure, the mechanical opener must overcome the air pressure holding the valve shut in order to open; however, the valve can either be closed mechanically or using only a counterweight. If the system is under positive pressure, the mechanical operator must hold the valve shut against both the weight of the accumulated product and the interior pressure. In this case, the valve can be released to open by gravity, but it must be mechanically forced to close.

The mechanical operating systems available include electric gear head motors operating through chain drives or mechanical linkages, motor-driven cams, air and hydraulic cylinders, and hand-operated levers.

4.6.2.3 Dangers. To a certain extent, the double dump valves are less of a threat to fingers than the rotary valves. They do have pinch points, and the mechanically driven models have great power with the ability to severely injure if not treated with caution and respect.

4.7 Dampers and Control Valves

A damper is a moveable plate, or multiple adjustable vanes, set into an airstream to control the flow volume and, occasionally, the direction of flow. Control valves generally have the same function: control of flow in a duct, pipe, or other enclosed conduit for air or other fluids and gases. Dampers and control valves generally have two purposes in a system: either modulation of the flow volume, or shutting off of the flow in all or part of the system. Sometimes, a damper or valve designed to provide shutoff capability is also used for flow control. However, because of the design differences, flow control dampers cannot provide complete shutoff of flow.

Dampers specifically designed for flow modulation do not have seats; consequently, they cannot completely shut off flow. When system design requires complete stoppage of flow, a valve or damper with a seat should be chosen. Dampers come in every size and shape, round, square, and rectangular. They are available in many styles, each suited for different applications.

The guillotine valve is a plate-type valve designed just like its namesake. The valve blade moves in and out of the airstream from one side. It is well-suited for applications that cannot have obstructions in the airstream. Low pressure drop and little turbulence make the guillotine valve a good choice for highly abrasive applications. This style valve usually has a seat that the blade fits against to minimize leakage. The valve uses a drive system such as a threaded shaft(s) with a gear head drive, chain drive, or rack-and-pinion drive. Other styles include chain and sprockets and pneumatic or hydraulic cylinders.

The slide-gate valve is another version of the guillotine valve. This valve has a sliding plate that completely blocks the opening or moves completely out of the opening to allow unrestricted movement. This style controls flow of solids rather than airflow. A common example is a slide gate mounted on the discharge flange of a hopper. The slide plate may be a simple plate that moves out of the opening. This style allows solids to get into the mounting channels normally occupied by the plate, at times causing jamming of the plate. A slide-plate style known as *hide the penny* has a longer blade. One end is solid to block the flow; the other end has an opening that matches the flange opening. This longer version of the blade ensures the cleanliness of the blade mounting channels and provides a sealing contact surface. Slide gates come in a wide variety of styles, simple models with manual push-pull handles, as well as handwheel drives, pneumatic cylinders, and rack-and-pinion–drive models. Some types completely seal the drive and blade in all positions by totally enclosing the drive mechanism. Nonenclosed models may allow small amounts of solids to leak during the open-close cycle.

A very similar device, called a blast gate, uses a movable blade to adjust the flow in a duct system. This style typically has only a manual adjustment. It has a set screw or other method to lock the blade in the proper adjustment position.

Butterfly valves have a single plate or disk that rotates through 90° to open and close. The plate or disk mounts on a shaft or stem that usually passes through the geometric center of the bore. In a typical butterfly valve, the disk seats against an elastomer radial sealing surface, and the stems are usually fitted with O-ring seals to prevent stem leakage. Those valves designed without seats, able to rotate a full 360°, will usually seal only 95 percent of the flow. Butterfly valves are available with manual handles, handwheel gear operators, chain wheel drives, electric operators, air cylinders, and diaphragm controls. Butterfly valves with radial seals provide leakproof closure when required. They are not likely to be the best choice for throttling flows that contain abrasive material.

Diverter valves do not control the volume of flow but rather change the direction of flow. Typical models for pollution control systems provide two flow directions for an airstream by movement of a single vane or flapper. Control of the valve can be a manual lever, chain wheel, electric motor drive, or pneumatic cylinder.

Back-draft dampers allow airflow in only one direction in a system. Common types include square or rectangular parallel blade dampers with counterweights to allow them to open with almost no pressure loss. A style designed for round ducting has a hinge pin through the geometric center of the duct. The hinge pin carries two valve blades that are slightly elliptical in shape. The blades swing to a position parallel to the duct centerline with proper flow direction. With a back-flow, the blades swing back to a position nearly perpendicular to the duct bore. The slight oval shape prevents the blades from moving through a full 90° or more, positioning the blade for easy reopening without jamming.

The ducting system for a pollution control system may find many uses for dampers and control valves. We will consider their use in control of flow, blowers, fire, temperature, condensation, direction of flow, isolation of components, and service.

4.7.1 Flow control

Filtration and pollution control systems frequently require the use of blast-gate dampers to balance flow in multiple lines within the same system. The damper can introduce an artificial increase in the static pressure (resistance to flow) or provide an adjustable opening for a portion of the system that requires added flow to maintain transport velocity. The blast-gate damper has no sealing provisions and is manually adjusted by sliding it partially into the airstream, and then locking it in place. The slide-gate damper is also frequently used to completely close off a branch of the system when machines are used infrequently.

4.7.2 Blower control

Blower inlet vane dampers have radially mounted vanes that are narrow sectors of a circle. Each vane in the set opens and closes by rotating on a radial axis. The air passing through the partially open vanes takes on a rotational motion in the same direction as the impeller, thus lowering the output of the blower. Since the horsepower required by a blower is proportional to the mass of air that passes through it, the blower operates at a lower horsepower. If the blower has excess capacity for periodic needs, the inlet vane damper is a good choice. It will lower your operating expenses for the periods the blower operates at reduced capacity.

Inlet vane dampers put considerable amounts of hardware into the inlet of the blower. Blower sizes smaller than 18 in [450 mm] make poor candidates for this style damper due to the high percentage of inlet area obstructed even when the vanes are wide open. The inlet vane damper works best with backwardly inclined wheels but has the least control on radial wheels.

Adjusting an inlet vane damper, in effect, generates parallel static pressure flow curves, providing a wide range of adjustments and good control of air volume. An inlet vane damper can provide up to a 30 percent adjustment of the volume without crossing into an unstable portion of the fan curve. This adjustment range can provide the flexibility to meet the requirements of systems with variable inlet conditions; however, inlet vane dampers are noisy. At 30 percent flow reduction settings, this style damper may add 5 dB (A) or more to the noise level.

Blower outlet dampers mount on the discharge flange of the blower and dampen or reduce flow by partially closing off a portion of the outlet area. This reduction artificially increases the total resistance (static losses) in the system, decreasing the airflow. This is commonly referred to as *riding the fan curve*. An outlet damper will provide less adjustment range than an inlet damper. All blowers do not react the same, but an effective range of 10 to 15 percent reduction in flow without crossing to the unstable side of the performance curve would be typical.

The most common types of outlet dampers are multiple-blade-type dampers, with either parallel-blade movement or opposed-blade movement. Parallel-blade dampers rotate all blades in the same direction, with the leading edge of one blade sealing against the tail of the adjacent blade. Opposed-blade dampers rotate adjacent blades in opposite directions such that leading edges seal against each other and trailing edges seal against the adjacent trailing edge. The opposed-blade dampers provide a more linear flow response than parallel-blade models.

Comparing the two styles, inlet vane dampers provide the most horsepower reduction for equal reductions in airflow. The inlet vane damper generates new fan curves. An outlet damper moves the operating point on an existing fan curve. Outlet dampers will operate successfully on a wider range of blower types than will the inlet vane damper, and they add less noise.

Whether you choose an inlet or outlet damper, it is poor practice to use a damper for major adjustments in flow. You run the risk of moving the operating point to the unstable portion of the fan curve. This increases the operating costs, both in terms of energy consumption and in repair of fatigue failure.

Alternatives on large blowers include variable-speed drives utilizing motor controllers, adjustable belt drives, or fluid transmissions. If the issue is present versus future needs, consider two alternatives to major flow reductions with a damper. Use lower speed now and increase the speed with a sheave change when needed, or design the system to use two blowers. The use of two blowers makes great sense if the flow will increase with no change in static pressure requirements.

4.7.3 Fire control

Pollution control systems that contain flammable or explosive materials should have quick-acting isolation dampers in the system. If the control system is outdoors and the air is returned back into the building, both the inlet and outlet ducting should have isolation dampers. They can prevent a fireball from propagating back through the inlet ducting to the operator workstation, or through the clean air ducting back into the workspace. Fire or explosion sensors can trigger fast-acting isolation dampers or valves. They can close in a matter of a few milliseconds, protecting the remainder of the system from the effects of the deflagration.

4.7.4   Temperature control

Systems that handle high-temperature air frequently require control dampers. If the system has a barrier filter, the media used may have a high-temperature limitation. In that case, cool air metered into and mixed with the heated airstream may prevent damage to the filter media. There are other methods of lowering the airstream temperature; however, tempering with cooler air is a common option.

High-temperature systems also require flow control upon cold start-up. Since cold air is significantly more dense than hot air, and blower horsepower is proportional to the weight of the air that passes through the blower, the airflow must be reduced during the start-up period until the airstream reaches normal operating temperature. If the system does not include flow control, the blower will not have adequate horsepower unless it is a nonoverloading blower design.

4.7.5   Condensation control

Frequently, pollution control systems handling hygroscopic products must be located outdoors. The control system may suffer from deposited layers of product on the inside of the housing if exposed to high moisture conditions. This can cause problems ranging from just the nuisance of cleanout to the system being plugged up or sanitation problems for food handling systems. Moist outside air drawn into the system exhaust by the negative pressure inside the building can contribute to this problem. The use of a backdraft damper to prevent air from flowing in the reverse direction in the system can significantly reduce the problem. In addition to using backdraft dampers, have a start-up and shutdown procedure that thoroughly warms the control system prior to feeding any product and a period of time preceding shutdown to allow all the product to be flushed from the system.

4.7.6   Flow direction

Flow diverter valves allow clean, filtered air to be recycled back into the building during winter months and discharged outdoors during those months that do not require heated air. The same valve can allow discharge of the air during an incident of filter leakage or other problem.

4.7.7 Isolation

Complex pollution control systems with multiple compartments that require cleaning while at a zero-flow condition need isolation valves. In systems that use reverse-flow cleaning, the clean air outlet has an isolation valve and a bypass valve that allow ambient air to be drawn into the clean air side of the filter and through the filters in the opposite direction as that filtering flow. The air then exits through what is normally the dirty air inlet and is filtered in other modules that are still active. In systems with mechanical shakers or reverse air blowers providing the filter cleaning action, isolation valves close off the clean air outlet allowing cleaning at zero-flow conditions. After the cleaning down period, the isolation valve reopens and another module goes through the same cycle. Isolation valves are also used when the pollution control system has excess capacity for future needs or has a standby module to ensure there is always enough filter capacity even if a module has a failure.

4.7.8 Service

Control valves properly designed into the system allow service with minimum disruption. As an example, a rotary air lock on the discharge flange of a hopper may fail, allowing collected material to fill the hopper. Without a slide-gate service valve above the rotary valve, repair of the rotary valve becomes very difficult, requiring manual cleanout of the hopper before disassembly. With a slide gate above the rotary valve, service can take place without disruption of flow.

System vessels or enclosures that fall within the definitions set forth in OSHA's Confined Space Standards, 1910.146, require "absolute closure of a pipe, line, or duct by the fastening of a solid plate that completely covers the bore and that is capable of withstanding the maximum pressure of the pipe, line, or duct with no leakage beyond the plate." Proper system design and execution can mean closing valves rather than disassembling the ducting system to allow service.

Reference

1. *Industrial Ventilation, A Manual of Recommended Practice,* 21st ed., American Conference of Governmental Industrial Hygienists, Cincinnati, OH, 1992.

Fans and Blowers

Michael J. Franklin

Twin City Fan and Blower Co.
Minneapolis, MN

Nomenclature

Roman letters

df	Density factor
L_p	Sound pressure level, dB
L_W	Sound power level, dB
ME	Mechanical efficiency
P_{bar}	Barometric pressure, inHg [Pa]
Q_d	Directivity factor
R	Room constant, ft^2 [m^2]
SE	Static efficiency
SP	Static pressure, in w.g. [Pa]
SP_f	Fan static pressure, in w.g. [Pa]
TP	Total pressure, in w.g. [Pa]
TP_f	Fan total pressure, in w.g. [Pa]
VP	Velocity pressure, in w.g. [Pa]

Greek letter

α   Room sound absorption coefficient

5.1 Fan Types and Their Selection

Fans provide the driving force to move air through air pollution control systems. Over the years, different requirements of ventilation and air pollution control systems have resulted in the development of different types of fans. Selecting the right kind of fan is important to ensure that an air pollution control system operates efficiently and reliably.

One way to classify the different types of fans is by the way they generate pressure. Centrifugal fans generate pressure using centrifugal force. Air enters the impeller near its center, then flows through and exits the impeller in a mostly radial direction. Centrifugal fans are sometimes called blowers. Axial fans generate pressure by accelerating the air through the fan in an axial direction. Airflow into, through, and out of axial fans is mostly in an axial direction. Even though they may be used to blow air, axial fans are not usually referred to as blowers. Figures 5.1 and 5.2 show the different parts that make up centrifugal and axial fans, and the nomenclature most commonly used. Table 5.1 presents an overall description of the different fan types.

5.1.1 Centrifugal fans

One way to classify centrifugal fans is by the blade shape at the impeller's outside diameter. This feature affects the ability of a fan to generate pressure, its efficiency, and its strength.

5.1.2 Backward-inclined fans

Backward-inclined fans are the most efficient type of centrifugal fan. The trailing edges of the blades, located at the outside diameter of the impeller, are inclined away from the direction of rotation. This causes the leading edge of the blades, located at the inside diameter of the impeller, to be inclined toward the direction of rotation. With the blades arranged this way, the air enters the impeller smoothly, with very little shock and flow separation. Blades can be either flat plates, curved plates, or airfoil sections, as shown in Figs. 5.3 through 5.5. A spiral- or scroll-shaped housing captures the air that comes out of the impeller and guides it toward the outlet. The mechanical efficiency of backward-inclined fans with airfoil-shaped blades may approach 90 percent.

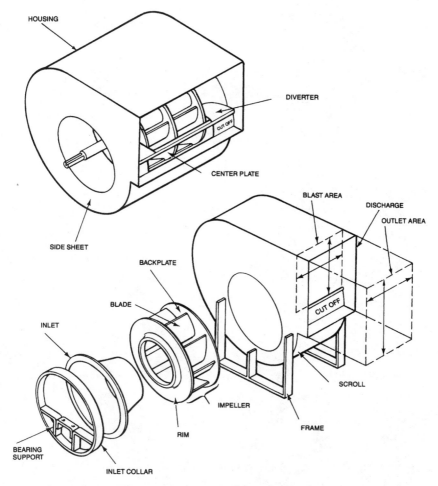

Figure 5.1 Centrifugal fan nomenclature. (*Reprinted from AMCA publication 201-90,* Fans and Systems, *with the express written permission of the Air Movement and Control Association, Inc.*)

The disadvantages of backward-inclined fans are that they must operate at a high speed to generate pressure and can only be used with relatively clean air. Dust tends to accumulate on the inside surface of the blades, which can lead to problems keeping the fan in balance and adversely affect fan performance.

Airfoil-shaped blades are commonly made by die-forming light-gauge material into an airfoil shape. This results in little resistance to even light abrasive dust loading. Sometimes different airfoil manufacturing methods are used to allow airfoil-shaped blades in industrial applications, but these are more costly. Because airfoil blades are

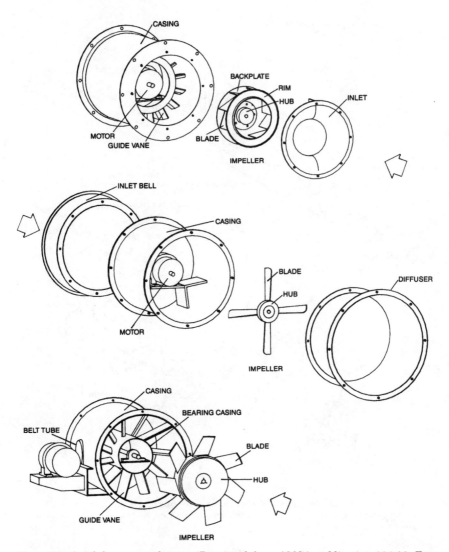

Figure 5.2 Axial fan nomenclature. (*Reprinted from AMCA publication 201-90,* Fans and Systems, *with the express written permission of the Air Movement and Control Association, Inc.*)

hollow, moist or very humid air may cause water to seep or condense on the insides of the blades, causing the fan to go out of balance. Weep holes located near the outside diameter of the impeller are occasionally added to ensure that moisture is drained.

Tubular centrifugal fans have a backward-inclined impeller installed in a tubular housing. Figure 5.6 shows a tubular centrifugal fan. This modification allows for easier installation in round duct-

TABLE 5.1   Types of Fans

	TYPE	IMPELLER DESIGN	HOUSING DESIGN
CENTRIFUGAL FANS	AIRFOIL	• Highest efficiency of all centrifugal fan designs. • Ten to 16 blades of airfoil contour curved away from direction of rotation. Deep blades allow for efficient expansion within blade passages. • Air leaves impeller at velocity less than tip speed. • For given duty, has highest speed of centrifugal fan designs.	• Scroll-type design for efficient conversion of velocity pressure to static pressure. • Maximum efficiency requires close clearance and alignment between wheel and inlet.
	BACKWARD-INCLINED BACKWARD-CURVED	• Efficiency only slightly less than airfoil fan. • Ten to 16 single-thickness blades curved or inclined away from direction of rotation. • Efficient for same reasons as airfoil fan.	• Uses same housing configuration as airfoil design.
	RADIAL	• Higher pressure characteristics than airfoil, backward-curved, and backward-inclined fans. • Curve may have a break to left of peak pressure and fan should not be operated in this area. • Power rises continually to free delivery.	• Scroll. Usually narrowest of all centrifugal designs. • Because wheel design is less efficient, housing dimensions are not as critical as for airfoil and backward-inclined fans.
	FORWARD-CURVED	• Flatter pressure curve and lower efficiency than the airfoil, backward-curved and backward-inclined. • Do not rate fan in the pressure curve dip to the left of peak pressure. • Power rises continually toward free delivery. Motor selection must take this into account.	• Scroll similar to and often identical to other centrifugal fan designs. • Fit between wheel and inlet not as critical as for airfoil and backward-inclined fans.
AXIAL FANS	PROPELLER	• Low efficiency. • Limited to low-pressure applications. • Usually low cost impellers have two or more blades of single thickness attached to relatively small hub. • Primary energy transfer by velocity pressure.	• Simple circular ring, orifice plate, or venturi. • Optimum design is close to blade tips and forms smooth airfoil into wheel.
	TUBEAXIAL	• Somewhat more efficient and capable of developing more useful static pressure than propeller fan. • Usually has 4 to 8 blades with airfoil or single thickness cross section. • Hub usually less than transfer by velocity pressure.	• Cylindrical tube with close clearance to blade tips.
	VANEAXIAL	• Good blade design gives medium- to high-pressure capability at good efficiency. • Most efficient of these fans have airfoil blades. • Blades may have fixed, adjustable, or controllable pitch. • Hub is usually greater than half fan tip diameter.	• Cylindrical tube with close clearance to blade tips. • Guide vanes upstream or downstream from impeller increase pressure capability and efficiency.
SPECIAL DESIGNS	TUBULAR CENTRIFUGAL	• Performance similar to backward-curved fan except capacity and pressure are lower. • Lower efficiency than backward-curved fan. • Performance curve may have a dip to the left of peak pressure.	• Cylindrical tube similar to vaneaxial fan, except clearance to wheel is not as close. • Air discharges radially from wheel and turns 90° to flow through guide vanes.
	POWER ROOF VENTILATORS CENTRIFUGAL	• Low-pressure exhaust systems such as general factory, kitchen, warehouse, and some commercial installations. • Provides positive exhaust ventilation which is an advantage over gravity-type exhaust units. • Centrifugal units are slightly quieter than the axial unit described below.	• Normal housing not used, since air discharges from impeller in full circle. • Usually does not include configuration to recover velocity pressure component.
	POWER ROOF VENTILATORS AXIAL	• Low-pressure exhaust systems such as general factory, kitchen, warehouse, and some commercial installations. • Provides positive exhaust ventilation which is an advantage over gravity-type exhaust units.	• Essentially a propeller fan mounted in a supporting structure. • Hood protects fan from weather and acts as safety guard. • Air discharges from annular space at bottom of weather hood.

(Courtesy of the American Society of Heating, Refrigeration, and Air-Conditioning Engineers, Atlanta, Georgia)

work. They are also used in cases where there is no room for the normal 90° turn of the airflow common with a scroll-type housing.

Backward-inclined fans are used on the clean air side of high-efficiency filters and on supply air systems. On systems that have light dust loadings, these fans are often constructed with the motor, shaft,

Figure 5.3a Backward-inclined fan and impeller with flat blades. (*Courtesy of Twin City Fan and Blower Company, Minneapolis, MN.*)

Figure 5.3b (*Continued*)

impeller, bearings, and bearing supports on a door, which allows the impeller to be swung out of the housing for cleaning, inspection, or maintenance.

5.1.3 Radial-bladed fans

The oldest and most common type of radial-bladed fan uses an open paddle wheel impeller. Figure 5.7 is a photograph of a typical paddle wheel. While they are not as efficient as backward-inclined fans, radial-bladed fans are still widely used because of their ruggedness and

Figure 5.4   Backward-inclined impeller with curved blades. (*Courtesy of Twin City Fan and Blower Company, Minneapolis, MN.*)

Figure 5.5   Backward-inclined impeller with airfoil-shaped blades. (*Courtesy of Twin City Fan and Blower Company, Minneapolis, MN.*)

Figure 5.6   Tubular centrifugal fan. (*Courtesy of Twin City Fan and Blower Company, Minneapolis, MN.*)

Figure 5.7a Industrial fan with paddle wheel style radial-bladed impeller. (*Courtesy of Twin City Fan and Blower Company, Minneapolis, MN.*)

Figure 5.7b (*Continued*)

high-pressure capability. Peak mechanical efficiency of radial-bladed fans is usually in the 65 to 75 percent range. The radial surfaces of their blades cause most materials to slide off due to centrifugal force and resist a buildup of solids if particulate is present. Most radial-bladed fans are best suited for systems where material must pass through the fan.

Many materials that have good abrasion-resisting properties are not easily formed. The flat blades of paddle wheels allow for the use of abrasion-resistant material for the blades or for the addition of wear plates. Adding rings or side plates to the sides of the paddle wheel enables the fan to deal with the impact loads created when chunks of material go through it. These rings, shown in Fig. 5.8, distribute the impact loads to all the blades.

Long, stringy material can get wrapped around the hub of an open paddle wheel, creating an obstruction to airflow and causing the fan to go out of balance. Paddle wheels that have a backplate, like the one shown in Fig. 5.9, solve this problem. This type of impeller is also

Figure 5.8 Radial-bladed impeller with end rings. (*Courtesy of Twin City Fan and Blower Company, Minneapolis, MN.*)

Figure 5.9 Radial-bladed impeller with end backplate. (*Courtesy of Twin City Fan and Blower Company, Minneapolis, MN.*)

used on fans that have the impeller mounted on the motor shaft (arrangement 4). Since the hub is closer to the back of the housing, an extra long motor shaft is not required.

Air exits radial-bladed impeller fans at high velocity. It is important to have a well-designed housing to help convert some of the energy contained in the velocity to pressure. These fans also have relatively small inlets and outlets to maintain sufficient material conveying velocities through the fan. To stand up to the abuse from material going through the fan, and because of the high-pressure capability, housings are usually constructed of heavy-gauge or plate material.

Because of their durability, paddle wheel radial-bladed fans are used on systems that require a fan to be located in a dirty airstream. Examples include pneumatic conveying, the clean side of low-efficiency air cleaning devices, or where the possibility of material getting past the air cleaning device exists. Their high-pressure capacity often makes them a good choice on high-pressure systems.

Radial-tip fans combine the desirable features of paddle wheels and backward-inclined fans. These fans are sometimes referred to as *modified* radial-bladed fans. At the outside diameter, the blade tips are radial. At the inside diameter, the blades are inclined into the direction of rotation, similar to a backward-inclined blade, as shown in Fig. 5.10. The result is an impeller that has some of the material handling and pressure capability of a purely radial-bladed fan with better efficiency. Peak mechanical efficiencies of radial-tip fans are approximately 75 percent. As with a paddle wheel, wear plates added to the backplate and blades will give increased life when dealing with abrasive materials.

Pressure blowers are another variation of radial-bladed fans. As shown in Fig. 5.11, these fans have narrow housings and impellers. Pressure capacities can be in excess of 100 in w.g. [24.8 kilopascals (kPa)] at low volumes. Sometimes pressure blowers are placed in

Figure 5.10a Industrial fan with radial-tip impeller. (*Courtesy of Twin City Fan and Blower Company, Minneapolis, MN.*)

Figure 5.10b (*Continued*)

Figure 5.11a Typical pressure blower and impeller. (*Courtesy of Twin City Fan and Blower Company, Minneapolis, MN.*)

Figure 5.11b (*Continued*)

series, with the air going from the outlet of one impeller to the inlet of another. Pressure blowers with several of these stages can generate pressures as high as 10 psig [280 in w.g., 70 kPa]. The high pressure requires that clearances between the impeller and the housing be kept tight to prevent excessive losses in performance. The high speed requires that the impellers be light to keep start-up times at a minimum and reduce heat buildup in the motor. Because of these factors, most pressure blowers are suitable only for clean air applications.

5.1.4 Forward-curved fans

Forward-curved fans have impellers with many short cupped blades that are curved into the direction of rotation at the tip. Figure 5.12 shows a typical double-width, double-inlet, forward-curved fan. A double-width fan is essentially two single-width impellers placed back to back with a common backplate in a common housing. They have the same pressure capacity but have approximately twice the flow rate and require twice the power of a single-width fan operating at the

Figure 5.12a Double-width, double-inlet, forward-curved fan and impeller. (*Courtesy of Twin City Fan and Blower Company, Minneapolis, MN.*)

Figure 5.12b (*Continued*)

same speed. The squirrel cage fan found in most household furnaces is an example of a forward-curved fan. These types of fans generate the most pressure and volume for a given speed. As with the radial-bladed fans, the air comes out of the impeller at a high velocity, so a well-designed housing is important to get good efficiency.

The closely spaced, cupped-blade shape requires relatively light gauges of construction and limits the use of forward-curved fans to air systems with relatively light dust loading. On dirty systems, dust accumulating on the blades causes a loss in performance and may cause the fan to go out of balance.

These fans are about as efficient as radial-bladed fans (67 to 72 percent), but do not have their pressure capability. Forward-curved fans are typically used on systems that require less than 4 in w.c. [1000 Pa].

The low-pressure capacity, efficiency, and clean air requirement of forward-curved fans limit their application in air pollution control systems. They are used in industrial processes and in builtup units that provide heated or cooled replacement air, although backward-inclined fans are more common. When suitable, these are the smallest centrifugal fans that will provide a required performance.

5.2   Axial Fans

5.2.1   Panel fans

The simplest form of axial fan is the panel fan. A window box fan is a small version of a panel fan. Designed to move a large volume of air at low pressure (less than 1.5 in w.c. [400 Pa]), these fans usually have no more than five long blades. Figure 5.13 shows a typical panel fan. The fan housing consists of a round orifice with a smooth opening. Because no attempt is made to control the flow coming out of the impeller, fan efficiency is not high.

Panel fans are used primarily to supply fresh air to a space or to exhaust contaminated or stale air from a space. Some panel fans have impellers that can move practically the same volume of airflow in either direction, when the direction of rotation is reversed. This enables a fan to supply air in one season and exhaust it in another.

5.2.2   Tubeaxial fans

A tubeaxial fan has an axial flow impeller mounted in a cylindrical or tubular-shaped housing as shown in Fig. 5.14. Often tubeaxial fans use the same impellers as panel fans, although higher-pressure designs are also used. The tubular housing contains the rotational

Figure 5.13 Panel fan. (*Courtesy of TCF Aerovent Company, Minneapolis, MN.*)

Figure 5.14 Tubeaxial fan. (*Courtesy of TCF Axial Division, Twin City Fan and Blower Company, Minneapolis, MN.*)

velocity coming out of the impeller, which allows the conversion of some of this energy to pressure. As a result, tubeaxial fans have more pressure generating capacity and higher efficiency than panel fans. Depending on the fan design, peak efficiency can be anywhere from 65 to 75 percent.

Tubeaxial fans are used primarily for general ventilation. The blades are usually made of cast aluminum or sheet steel, so these fans are more reliable when used on clean air applications. As with panel fans, reversible-flow impellers can be used if a change in airflow direction is needed.

5.2.3 Vaneaxial fans

Vaneaxial fans, as shown in Fig. 5.15, have turning vanes located downstream of the impeller to generate more pressure at higher efficiencies than tubeaxial fans. These vanes catch some of the rotational velocity of the air exiting the impeller and turn it in an axial direction. This rotational energy is lost in fans that do not have turning vanes. This process of straightening the flow converts some of the rotational energy into pressure. Vaneaxial fans can generate pressures as high as 8 in w.g. [2000 Pa], with mechanical efficiencies as high as 85 percent.

Figure 5.15 Vaneaxial fan. (*Courtesy of TCF Axial Division, Twin City Fan and Blower Company, Minneapolis, MN.*)

Vaneaxial fans are mostly used for general ventilation but are occasionally used on the clean air portion of air pollution control systems. They are commonly used in paint spray booth systems. Vaneaxial fans have the smallest physical size of any type of fan for a given flow and pressure. This feature is often useful where there is not much space in which to place a fan.

5.3   Other Types of Fans

5.3.1   Regenerative blowers

Regenerative blowers are similar to centrifugal fans since they generate pressure by centrifugal force, but the general airflow is around the circumference of the impeller. Figures 5.16 and 5.17 show a cutaway view and an overall view of a typical regenerative blower. The impellers in these blowers have a large number of short radial blades. Air enters the blower at one side of the impeller and at the inside diameter of the blades. The air then moves through the impeller in a radial direction. As it exits the impeller, the vanes catch the air and redirect it toward the inside where it reenters the inside of the impeller. The air spirals around the circumference this way, alternating between going through the impeller and the vanes, until it reaches the outlet located next to the inlet. The pressure increases each time the air passes through the impeller.

The pressure-volume curve for regenerative blowers is steeper than fans. This results in a smaller change in flow if something in the ductwork is changed. Regenerative blowers can generate pressures up to 8 lb/in^2 [220 in w.g., 55 kPa] with flow rates up to 1000

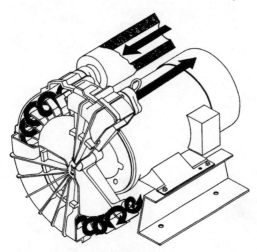

Figure 5.16 Cutaway view of a typical regenerative blower. (*Courtesy of EG&G Rotron Industrial Products, Saugerties, NY.*)

Figure 5.17 Regenerative blower. (*Courtesy of EG&G Rotron Industrial Products, Saugerties, NY.*)

ft^3/min [0.5 m^3/s]. Their efficiency is less than 50 percent, but they are generally used where the flow rates and the power requirement are not very high and where efficiency may not be a major concern. They can be noisy and are often sold with an integral inlet or outlet silencer. Being compact, they are a good choice where space is at a premium. Like pressure blowers, they must be used in the clean air portions of systems.

5.3.2 Positive displacement blowers

As their name implies, air is moved through positive displacement blowers by being pushed by a pair of intermeshed lobed rotors. An illustration showing the cross section of a positive displacement blower is given in Fig. 5.18, and a labeled cutaway view is shown in Fig. 5.19. With each rotation of the rotors, a nearly constant volume of air is displaced. The amount of pressure generated depends on the resistance to flow in the duct system and internal leakage in the blower. The leakage is minimized by keeping small clearances between the rotors and also between the rotors and the blower housing. To maintain these tight clearances, it is essential that the air passing through the blower is clean. Dirty air will cause wear, which increases clearances, which increases leakage and lowers the performance of the blower.

Positive displacement blowers are high-pressure and low-volume devices. Pressures can be as high as 15 lb/in^2 [105 kPa], with flow rates up to 8000 ft^3/min [4 m^3/s]. The power required is proportional to the speed and pressure. Peak efficiencies range from 55 to 83 percent.

Because these blowers are positive displacement devices, changes in air volume can only be obtained by changing the speed. If throttling is attempted by means of a valve or damper, the same volume of air is forced through a smaller area, which increases the pressure rise and power of the blower. The high-pressure air supplied by positive displacement blowers is used for applications such as injection-type pneumatic conveying systems and reverse-flow cleaning of filters.

5.4 Fan Ratings

5.4.1 Air performance ratings

To describe a fan rating, it is necessary to know how much air it is moving, how much pressure it is generating, how much power it

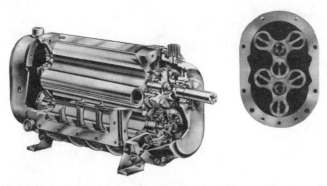

Figure 5.18 Rotary lobe positive displacement blower. (*Courtesy of M-D Pneumatics Division, Tuthill Corporation, Springfield, MO.*)

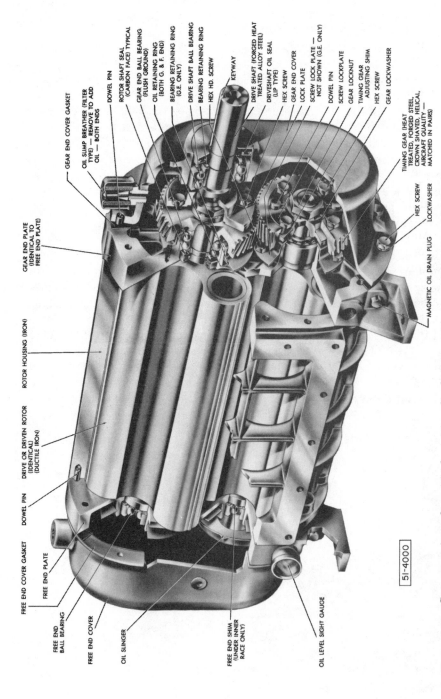

GEAR END COVER GASKET

OIL SUMP BREATHER (FILTER TYPE) — REMOVE TO ADD OIL — BOTH ENDS

DOWEL PIN

ROTOR SHAFT SEAL (CARBON FACE) TYPICAL

GEAR END BALL BEARING (FLUSH GROUND)

OIL RETAINING RING (BOTH G. & F. END)

BEARING RETAINING RING (G.E. ONLY)

DRIVE SHAFT BALL BEARING

BEARING RETAINING RING

HEX HD. SCREW

KEYWAY

DRIVE SHAFT (FORGED HEAT TREATED ALLOY STEEL)

DRIVESHAFT OIL SEAL (LIP TYPE)

HEX SCREW

GEAR END COVER

LOCK PLATE

SCREW LOCK PLATE — NOT SHOWN (G.E. ONLY)

DOWEL PIN

GEAR LOCKNUT

SCREW LOCKPLATE

TIMING GEAR ADJUSTING SHIM

HEX SCREW

GEAR LOCKWASHER

TIMING GEAR (HEAT TREATED, FORGED STEEL, CROWN SHAVED, HELICAL, AIRCRAFT QUALITY — MATCHED IN PAIRS)

HEX SCREW LOCKWASHER

MAGNETIC OIL DRAIN PLUG

GEAR END PLATE (IDENTICAL TO FREE END PLATE)

ROTOR HOUSING (IRON)

DRIVE OR DRIVEN ROTOR (IDENTICAL) (DUCTILE IRON)

DOWEL PIN

FREE END COVER GASKET

FREE END PLATE

FREE END BALL BEARING

FREE END COVER

OIL SLINGER

FREE END SHIM (UNDER INNER RACE ONLY)

OIL LEVEL SIGHT GAUGE

5I-4000

Figure 5.19 Cutaway view of a typical positive displacement blower. (*Courtesy of M-D Pneumatics Division, Tuthill Corporation, Springfield, MO.*)

takes, and the speed at which it is running. The most common method of specifying air quantity is the volumetric flow rate, expressed in cubic feet per minute in the English system of units or cubic meters per second in the SI system of units. Volumetric flow rate can be determined from the mass flow rate by dividing by the density. Pressure in the English system of units is usually expressed in terms of inches water column rise with respect to ambient pressure (in w.g.). In the SI system of units, the common unit of measurement is pascals. Some pressure blowers are rated in ounces per square inch, and some others in inches or millimeters of mercury column rise. Power is expressed in horsepower (hp) or kilowatts [kW], and speed in revolutions per minute.

Fan ratings and the methods for obtaining them are defined by the Air Movement and Control Association (AMCA), which is an association of manufacturers. They define the fan total pressure to be the total pressure rise across the fan inlet and outlet:

$$TP_f = TP_2 - TP_1 = (SP_2 + VP_2) - (SP_1 + VP_1) \qquad (5.1)$$

where TP_f = total pressure generated by fan, in w.g. [Pa]
  TP_2 = total pressure at fan outlet, in w.g. [Pa]
  TP_1 = total pressure at fan inlet, in w.g. [Pa]
  SP_2 = static pressure at fan outlet, in w.g. [Pa]
  SP_1 = static pressure at fan inlet, in w.g. [Pa]
  VP_2 = velocity pressure at fan outlet, in w.g. [Pa]
  VP_1 = velocity pressure at fan inlet, in w.g. [Pa]

Example 1   The following readings are the results of a fan test. What is the fan total pressure?

$$SP_2 = 1.35 \text{ in w.g. } [335 \text{ Pa}]$$

$$SP_1 = -3.50 \text{ in w.g. } [-869 \text{ Pa}]$$

$$VP_2 = 0.25 \text{ in w.g. } [62 \text{ Pa}]$$

$$VP_1 = 0.25 \text{ in w.g. } [62 \text{ Pa}]$$

solution   Using Eq. (5.1),

$$TP_f = (1.35 + 0.25)-(-3.50 + 0.25) = 4.85 \text{ in w.g.}$$

$$= [(335 + 62)-(-869 + 62) = 1204 \text{ Pa}]$$

The velocity pressure of a fan is defined as the velocity pressure at the fan outlet. Since the static pressure is equal to total pressure minus velocity pressure, the equation for fan static pressure can be derived as follows.

$$SP_f = TP_f - VP_2 = (SP_2 + VP_2) - (SP_1 + VP_1) - VP_2 \qquad (5.2)$$

$$= SP_2 - SP_1 - VP_1$$

Example 2   What is the fan static pressure of the fan tested in Example 1?
solution   Using Eq. (5.2),

$$SP_f = 1.35 - (-3.50) - 0.25 = 4.60 \text{ in w.g.}$$

$$= [335 - (-869) - 62 = 1142 \text{ Pa}]$$

Be careful to use Eq. (5.2) correctly. There is a tendency to take the difference between the static pressure at the inlet and the outlet and call it static pressure rise. This is only correct when the velocity pressure of the inlet is zero, such as when fans have open, nonducted inlets. The inlet and outlet areas of most fans are about the same, and as a result so are the velocity pressures. By looking at Eq. (5.1) for a fan with a ducted inlet ($VP_o \cong VP_i$), it can be seen that the difference between the static pressure at the outlet and inlet is actually the fan total pressure. What causes even more confusion is the term *total static pressure,* taken to be the static pressure rise between the inlet and the outlet. This term has no meaning or definition by the AMCA or any other industry standard and should not be used.

The method of testing fans to determine their ratings is specified by a joint standard of the AMCA and the American Society of Heating, Refrigeration, and Air Conditioning Engineers (ASHRAE). This test code can be specified as AMCA standard 210-85 or ANSI/ASHRAE standard 51-1985, Laboratory Methods of Testing Fans for Rating. The test code specifies the test setup, the instrumentation, the procedure, and the calculations required to measure a fan's rating.

Figure 5.20 shows a typical backward-inclined–fan curve with test points of cubic feet per minute, static pressure, and power plotted on it. This plot shows only six test points; actual tests usually have a minimum of ten. A similar plot could be made using total pressure. The bottom horizontal scale is used for cubic feet per minute, the lower vertical scale on the left is for static pressure, and the upper vertical scale is used to plot power. Point A_1 is on a system that has no resistance to flow. As a result, the flow is the highest, measured at 50,000 ft³/min [23.6 m³/s], and the static pressure is zero. The power is measured to be 26 hp [19.4 kW] and is plotted at point A_2. The term *brake horsepower* shown on the upper vertical scale means that this is the power measured at the fan shaft and does not include any losses due to belts, couplings, or bearings. Point A_1 is sometimes referred to as *wide open volume* (WOV) or *free delivery*.

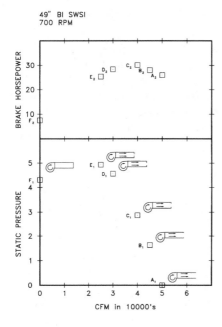

Figure 5.20   Fan testing points.

Other points on the curve, representing other duct systems, can be generated by adding resistance to the flow, usually by closing a damper or adding an orifice plate that has a smaller opening. At point B_1, the added resistance reduced the flow to 45,000 ft³/min [21.2 m³/s], increased the static pressure to 1.63 in w.g.[405 Pa], and changed the power to 28 hp [20.9 kW], as plotted at point B_2. Points C, D, and E are measured the same way by adding increasing resistance to flow. At Point F_1, the resistance has been increased to the point where the outlet duct has been completely closed off preventing any airflow (ft³/min = m³/s = 0). The static pressure at this point is 4.3 in w.g. [1068 Pa], and the power requirement is 7.5 hp [5.6 kW]. This point on the curve is sometimes referred to as *shutoff no delivery* (SND) or *no flow static pressure*. Smooth lines are drawn through the static pressure and power points to develop a fan curve.

The brake horsepower curve has its highest level at point C_2. This is typical for most backward-inclined fans and many axial fans. Power curves that have this feature are called *nonoverloading*, since a fan that has its motor selection based on this power can operate anywhere else on the fan curve without overloading the motor. Radial-bladed and forward-curved fans do not have this feature. Their power is highest at free delivery and lowest at shutoff no delivery. Some axial fans will have the highest power at the SND point.

The static pressure peaks at point E_1. As the resistance is increased passed this point, the fan blades stall and the flow can become unstable. Operation to the left of the peak static pressure should be avoided. It can cause noisy and unstable airflow and may cause structural damage to some fans and their attached ductwork.

The fan operates with the highest efficiency at point D. This region of the fan curve, a little bit below the peak static pressure point, has the highest efficiency for practically every fan type.

This demonstration of how fans are tested shows that there is an intimate relationship between the amount of air a fan moves and the duct system attached to the fan. Frequently the fan is blamed for not delivering a sufficient amount of air when, in fact, the problem is a duct system that has resistance to a flow that is higher than anticipated.

Figure 5.21 shows the same curve as Fig. 5.20, with the exception that the test points are removed and efficiency curves are added. Fan efficiency can be stated in terms of total efficiency or static efficiency. Total or mechanical efficiency, as given in Eq. (5.3), is the ratio of the power added to the airstream to the power put into the fan at the fan shaft.

$$\text{ME} = \frac{\text{TP} \times Q}{6362 \times W} = \left[\frac{\text{TP} \times Q}{1000 \times W} \right] \tag{5.3}$$

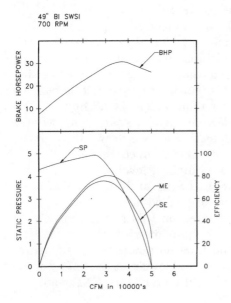

Figure 5.21 Fan performance curve with efficiency.

where ME = mechanical efficiency
       TP = fan total pressure, in w.g. [Pa]
        Q = volumetric flow rate, ft³/min [m³/s]
        W = power input at fan shaft, hp [kW]

Example 3 The test data taken on the fan in Example 1 also indicates that the airflow rate is 12,000 ft³/min [5.66 m³/s] and the power consumption at the fan shaft is 11.3 hp [8.43 kW]. What is the mechanical efficiency of this fan ?

solution Using Eq. (5.3),

$$\text{ME} = \frac{4.85 \times 12,000}{6362 \times 11.3} = 0.81 = 81\%$$

$$= \left[\frac{1204 \times 5.66}{1000 \times 8.43} = 0.81 = 81\% \right]$$

Some fans, such as panel fans, do not make use of the air velocity. The velocity and the energy associated with it are dumped into the space downstream of the fan. In this case, it is appropriate to consider the static efficiency, as calculated in Eq. (5.4), as the efficiency of the fan.

$$\text{SE} = \frac{\text{SP} \times Q}{6362 \times W} = \left[\frac{\text{SP} \times Q}{1000 \times W} \right] \qquad (5.4)$$

where SE = static efficiency and SP = static pressure.

Example 4 What is the static efficiency of the fan used in the previous examples?

solution Using Eq. (5.4),

$$\text{SE} = \frac{4.60 \times 12,000}{6362 \times 11.3} = 0.77 = 77\%$$

$$= \left[\frac{1142 \times 5.66}{1000 \times 8.43} = 0.77 = 77\% \right]$$

Figures 5.20 and 5.21 are examples of fan curves at a single speed. Figure 5.22 is a plot of a family of fan curves showing what happens when the speed of a fan is changed. These additional curves could be generated by testing the fan at each speed, but a much easier way to achieve this is to use a set of relationships known as the fan laws. The first fan law states that the change in airflow is proportional to the change in speed. This can be written as

$$\frac{Q_N}{Q_O} = \frac{\text{RPM}_N}{\text{RPM}_O} \qquad (5.5)$$

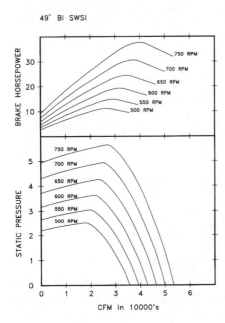

49" BI SWSI

Figure 5.22 Multispeed fan performance curve.

where Q = volumetric flow rate, ft³/min [m³/s]
         N = new
         O = old
        RPM = fan speed in revolutions per minute.

The second fan law states that the change in pressure is proportional to the square of the change in speed:

$$\frac{SP_N}{SP_O} = \frac{TP_N}{TP_O} = \left(\frac{RPM_N}{RPM_O}\right)^2 \tag{5.6}$$

where SP = fan static pressure, in w.g. [Pa], and TP = fan total pressure, in w.g. [Pa]. The change in power is proportional to the cube of the change in speed, as is shown in the equation for the third fan law:

$$\frac{W_N}{W_O} = \left(\frac{RPM_N}{RPM_O}\right)^3 \tag{5.7}$$

where W = fan power requirement, hp [kW]. These fan laws are valid only if the duct system remains constant (e.g., no dampers changing position or no filters loading up) and only the fan speed is varied.

Example 5   We have a fan that is operating at 10,000 ft³/min [4.72 m³/s], 4 in w.g. [993 Pa] static pressure, 1000 r/min, and is drawing 9 hp [6.7 kW]. We would like to increase the speed of the fan to get 11,000 ft³/min [5.19 m³/s].

solution   We solve the problem by using the fan laws. Rearranging Eq. (5.5) we get

$$\text{RPM}_N = \text{RPM}_O \times \frac{Q_N}{Q_O} = 1000 \times \frac{11,000}{10,000} = 1100$$

$$= \left[1000 \times \frac{5.19}{4.72} = 1100 \right]$$

Rearranging Eq. (5.6) we get

$$\text{SP}_N = \text{SP}_O \times \left(\frac{\text{RPM}_N}{\text{RPM}_O} \right)^2 = 4 \times \left(\frac{1100}{1000} \right)^2 = 4.84 \text{ in w.g.}$$

$$= \left[993 \times \left(\frac{1100}{1000} \right)^2 = 1202 \text{ Pa} \right]$$

Rearranging Eq. (5.7) we get

$$W_N = W_O \times \left(\frac{\text{RPM}_N}{\text{RPM}_O} \right)^3 = 9 \times \left(\frac{1100}{1000} \right)^2 = 12 \text{ hp}$$

$$= \left[6.7 \times \left(\frac{1100}{1000} \right)^2 = 8.9 \text{ kW} \right]$$

In other words, a 10 percent increase in speed results in a 10 percent increase in flow, a 21 percent increase in pressure, and a 33 percent increase in power.

The power relationship can get you in trouble if you size a motor very close to the power required for a rating and you are not sure of the actual duct system pressure requirement. If the fan speed is selected based on an underestimated pressure, it will be necessary to speed up the fan to get the required flow. The consequence of the third fan law is that you cannot increase the speed of a fan very much without also increasing the motor size. Not only will this involve the cost of a new motor, mounting, and drive, but it can result in unexpected costs due to replacing the motor lead wires and motor switch gear. This problem can be avoided by estimating the system pressure requirements as accurately as possible and sizing the motor with some reserve capacity. Normal methods of estimating system pressure requirements give a result that is within approximately ± 25 percent of the actual static pressure required for a given flow rate. Errors of 50 percent or more are not uncommon. In Example 5, we would have been in trouble had we selected a 10-hp [7.5-kW] motor but in good shape if we had selected a 15-hp [11.2-kW] motor. If the system were to be installed as designed, and the power requirement remained 9 hp [6.7 kW], the energy consumption with the 15-hp [11.2-kW] motor would not be much greater than with the 10-hp [7.5-kW] motor since motor efficiency does not fall off much until the load is less than 50 percent of the motor's nameplate rating. Also keep in mind that the power ratings are based on the power required at the fan shaft and do not include belt drive losses.

Figure 5.23 shows a typical fan performance table with lines of constant revolutions per minute overlaid on top, demonstrating the relationship between performance curves and performance tables. An advantage to showing performance in tables is that information on a large range of fan speeds can be shown in a small space. Let's say we want to find the speed and power required to run this fan at 15,300 ft³/min [7.22 m³/s] and 1 in w.g.[248 Pa]. By finding 15,300 ft³/min in the ft³/min column, and reading across horizontally to the two columns below the 1″ SP heading, the table tells us that this fan needs to run at 629 r/min and requires 4.48 hp [3.34 kW] of power.

5.4.2   Nonstandard air ratings

Most published fan data, whether it be in tables or as curves, is shown at standard air density. In the fan industry, 0.075 lb/ft³ [1.2 kg/m³] is considered standard air and corresponds to air at 68°F [21°C] and at sea level (barometric pressure = 29.92 inHg [101.3 kPa]), with 50 percent relative humidity. If a fan handles air that is greater than 120°F [50°C] , less than 20°F [−10°C], is at an elevation greater than 3000 ft [1000 m] above sea level, or has a relative humidity greater than 50 percent, corrections for nonstandard density should be taken. Table 5.2 gives values for the ratio of the density at various temperatures and altitudes to standard density, or the following equation can be used:

$$df = \frac{\rho_G}{0.075} = \frac{SP + 13.6 \times P_{bar}}{407} \times \frac{530}{T + 460}$$

$$= \left[\frac{\rho_G}{1.2} = \frac{SP + P_{bar}}{101.1 \times 10^3} \times \frac{293}{T + 273} \right] \tag{5.8}$$

where   df = ratio of actual density to standard density
ρ_G = actual density, lbm/ft³ [kg/m³]
SP = static pressure, in w.g. [Pa]
P_{bar} = barometric pressure, inHg [Pa]
T = dry bulb temperature, °F [°C]

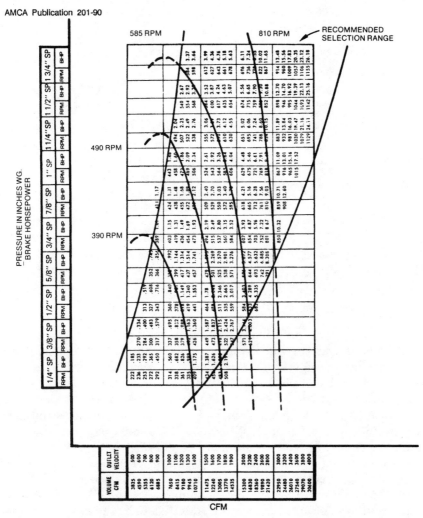

Figure 5.23 Typical fan performance table showing relationship to a family of constant-speed performance curves. (*Reprinted from AMCA publication 201-90,* Fans and Systems, *with the express written permission of the Air Movement and Control Association, Inc.*)

This equation will only work for air with low humidity. A psychrometric chart should be used to determine the density of air with high humidity.

Example 6   A fan is to operate on the clean side of a baghouse where the airstream temperature is to be 250°F [121°C], and the static pressure at the fan inlet is to be −20 in w.g. [−4967 Pa]. The elevation at the location of the fan is 2000 ft [610 m] above sea level. What is the density ratio and actual air density at the fan inlet?

solution   Using Table 5.2, the barometric pressure at 2000 ft [610 m] above sea level is found to be 27.82 inHg [94.2 kPa]. Calculate the density factor using Eq. (5.8).

$$df = \frac{-20 + 13.6 \times 27.82}{407} \times \frac{530}{250 + 460} = 0.66$$

$$= \left[\frac{-4967 + 94.2 \times 10^3}{101.1 \times 10^3} \times \frac{293}{121 + 273} = 0.66 \right]$$

The inlet air density is

$$\rho_1 = 0.075 \times 0.66 = 0.050 \text{ lb/ft}^3$$
$$= [1.2 \times 0.66 = 0.79 \text{ kg/m}^3]$$

Fans are considered to be constant-volume devices. This means that if the density of the air flowing through a constant-speed fan is changed, the actual volumetric flow rate will remain constant. The static pressure and power will vary in proportion to the change in density.

In most systems, air density and volumetric flow rate are not constant throughout the duct system. It is important to remember to select the fan based on the actual flow rate calculated for the density at the fan, as opposed to the flow rate at standard conditions or at some other point in the system. In the English system of units, the flow rate at the actual air density is called acfm, which stands for actual cubic feet per minute; scfm or *standard* cubic feet per minute is the volumetric flow rate at standard conditions that gives the same mass flow rate (lbm/min). In the SI system of units, make sure to specify what density the volumetric flow rate (m³/s) corresponds to. If the pressure rise across the fan is significant, specify the flow rate based on the inlet air density, since AMCA defines the fan air density as the density of the air entering the fan inlet.

If all other considerations are equal (abrasion, corrosion, design temperature, etc.), locate the fan in the portion of the duct system that has the highest density. Since as the density increases, the acfm decreases, this portion of the duct system has the lowest acfm. The pressure requirement does not change with location in the system, so a smaller fan can be selected, and, because of thermodynamic effects, there is an efficiency advantage also.

Example 7   A fan is to be selected to generate 6 in w.g. [1490 Pa] of static pressure for a system handling 25,000 acfm [11.8 m³/s] of dry air at 3000 ft [914 m] above sea level and 400°F [204°C]. In order to look at performance tables that contain data at standard air density, this performance needs to

TABLE 5.2 Air Density Factors for Various Temperatures and Altitudes

	Altitude in Feet Above Sea Level												
	0	1000	2000	3000	4000	5000	6000	7000	8000	9000	10000	15000	20000
Air temp °F	Barometric Pressure in Inches of Mercury												
	29.92	28.86	27.82	26.82	25.84	24.90	23.98	23.09	22.22	21.39	20.58	16.89	13.75
70	1.000	.964	.930	.896	.864	.832	.801	.772	.743	.714	.688	.564	.460
100	.946	.912	.880	.848	.818	.787	.758	.730	.703	.676	.651	.534	.435
150	.869	.838	.808	.770	.751	.723	.696	.671	.646	.620	.598	.490	.400
200	.803	.774	.747	.720	.694	.668	.643	.620	.596	.573	.552	.453	.369
250	.747	.720	.694	.669	.645	.622	.598	.576	.555	.533	.514	.421	.344
300	.697	.672	.648	.624	.604	.580	.558	.538	.518	.498	.480	.393	.321
350	.654	.631	.608	.586	.565	.544	.524	.505	.486	.467	.450	.369	.301
400	.616	.594	.573	.552	.532	.513	.493	.476	.458	.440	.424	.347	.283
450	.582	.561	.542	.522	.503	.484	.466	.449	.433	.416	.401	.328	.268
500	.552	.532	.513	.495	.477	.459	.442	.426	.410	.394	.380	.311	.254
550	.525	.506	.488	.470	.454	.437	.421	.405	.390	.375	.361	.296	.242
600	.500	.482	.469	.448	.432	.416	.400	.386	.372	.352	.344	.282	.230
650	.477	.460	.444	.427	.412	.397	.382	.368	.354	.341	.328	.269	.219
700	.457	.441	.425	.410	.395	.380	.366	.353	.340	.326	.315	.258	.210
800	.420	.404	.389	.375	.362	.350	.336	.323	.311	.300	.290	.237	.193

Unity basis = standard air density of 0.075 lbs/ft³. At sea level (29.92 inHg barometric pressure) this is equivalent to dry air at 70°F. When desired performance is at other than standard conditions (0.075 lbs/ft³), it must be converted to equivalent standard condition before entering the tables.

Source: of Twin City Fan and Blower Company, Minneapolis, Mn.

be corrected to standard conditions. The density factor for 400°F [204°C] and 3000 ft [914 m] above sea level is found in Table 5.2 to be 0.55. At standard conditions, the static pressure is then 6/0.55 = 10.9 in w.g. [2707 Pa]. This value is used to look up the revolutions per minute and power required in the performance table. If the power in the table turns out to be 60 hp, then the power at design conditions would be 60 × 0.55 = 33 hp [24.6 kW]. Lets say that the air temperature is 70°F [20°C] at start-up. We have to make sure that the motor is sized so that there is enough power to operate under this condition. Again looking in Table 5.2, this time for 70F [20°C] and 3000 ft [914 m] above sea level, a density factor of 0.9 is found. So the power at start-up or the *cold* power is 60 × 0.9 = 54 hp [40.3 kW].

5.4.3 Sound ratings

Sound is our ears' response to pressure fluctuations that travel through air. Microphones also respond to these pressure waves and translate their response into electric signals. In this way different sounds can be described by their amplitude or how *loud* they are and by their frequencies or *pitch*.

Compared to the normal barometric pressure of 29.92 inHg [407 in w.g., 101,100 Pa], these fluctuations can be very small. The human ear can sense fluctuations with amplitudes as small as 0.000 000 1 in w.g. [0.000 024 8 Pa] and as high as 407 in w.g. [101,100 Pa]. Normally in acoustics, microbars are used as the unit for pressure. One microbar is approximately one millionth of the normal barometric pressure. To work with a property that has such a large variation in amplitude, the logarithmic relationship decibel is used. For sound pressure the following equation is used:

$$L_p = 20 \times \log \frac{P}{0.0002} \tag{5.9}$$

where L_p = sound pressure, dB
    P = sound pressure, μbar
    0.0002 = reference sound pressure which is approximately at the lower limit of human perception, μbar

Example 8   A sound source produces a sound pressure level of 9 μbar [0.9 Pa]. What is the sound pressure level in decibels?

solution   From Eq. (5.9),

$$L_p = 20 \times \log \frac{P}{0.0002} = 20 \times \log \frac{p}{0.0002} = 93 \text{ dB}$$

The problem with using sound pressure to rate the sound levels produced by a fan is that these levels can change depending on how and where the fan is installed. Sound power, which is used to rate fan sound levels, is the acoustic power produced by a fan. Sound power is independent of installation. A common analogy used to describe the relationship between sound power and sound pressure is the wattage and illumination of a light bulb. A 100-W light bulb mounted on the ceiling of a small room with glossy white walls will provide better illumination than it will in a large room with dark walls. Likewise a fan with a given sound power rating will produce higher sound pressure levels in a small room with hard cinder block walls than it will in a large auditorium with carpeting and curtains. Similar to sound pressure, the equation for sound power is

$$L_w = 10 \times \log \frac{W}{10^{-12}} \tag{5.10}$$

where L_w = sound power, dB
        W = sound power, W
        10^{-12} = reference sound power, W

Example 9   A sound source produces 0.002 W of sound power. What is the sound power level in decibels?

solution   From Eq. (5.10),

$$L_w = 10 \times \log \frac{W}{10^{-12}} = 10 \times \log \frac{9}{0.0002} = 93 \text{ dB}$$

Because decibels are logarithmic values, adding sound levels is more complicated than just adding the decibel values. The chart shown in Fig. 5.24 is used to combine decibel levels.

Example 10   One fan produces 85 dB of sound pressure when operating by itself, and the one next to it produces 90 dB when operating by itself. What is the sound level when both are turned on?

solution   The expected sound level can be found by first noting that the difference between the two levels is 5 dB. From Fig. 5.24, we see that we need to add slightly more than 1 dB to the higher level, so the combined sound level would be 91 dB.

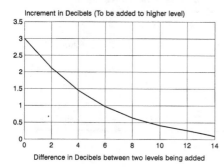

Increment in Decibels (To be added to higher level)

Difference in Decibels between two levels being added

Figure 5.24 Chart for combining decibels.

The second characteristic used to describe sound is frequency. Frequency is the number of pressure peaks per second expressed as hertz [Hz] that the sound pressure fluctuations exhibit. The human ear is sensitive to frequencies between 20 and 20,000 Hz. To reduce the quantity of numbers required to describe a sound's frequency spectrum, electric filters were developed that divided the frequency range into octave bands. The octave bands, as shown in Table 5.3, are set up so that the center frequency of each band is twice that of the band below it. The electric filters average the sound level over the frequency range of each octave band. The octave bands with center frequencies of 63, 125, 250, 500, 1000, 2000, 4000, and 8000 Hz are the ones that are of interest when analyzing fan sound characteristics.

Sound power values are obtained for fans by testing them using the procedures outlined in AMCA standard 300-85, Reverberant Room Method for Sound Testing Fans. Like the air performance tests, test points are taken from wide open volume to shutoff no delivery. Care should be taken to make sure how the test data is shown. Some manufacturers present the sound radiating from the inlet and the outlet separately. Others present the total sound level, which would be the case if both the inlet and outlet were left open. The most common method is to show a single number that can be used for either the inlet or outlet sound. The last two methods are the most practical considering sound test accuracy. While the test code is the current state of the art for sound testing, it should be noted that measuring sound is not as precise as one would think. In AMCA publication 303, "Application of Sound Power Level Ratings," there is the following statement:

> Within the present state of the art, differences in sound power levels of 2 dB or less are not considered significant. In comparing products of different manufacturers, it is good practice to disregard differences of less than 4 dB. This is particularly true in the first octave band where differences of 6 dB or less should be disregarded.

TABLE 5.3   Octave Band Frequencies and A-Weighted Scale Corrections

Octave band	1	2	3	4	5	6	7	8
Frequency range, Hz	45 to 90	90 to 180	180 to 355	355 to 710	710 to 1400	1400 to 2800	2800 to 5600	5600 to 11200
Center frequency, Hz	63	125	250	500	1000	2000	4000	8000
A-weighting correction, dB	−25	−15	−8	−3	0	+1	+1	−1

Source: From ANSI S1.4-1971, ANSI S1.6-1984.

Sound power ratings need to be converted to sound pressure values to be of practical use. The following equation is used to estimate sound pressure from sound power:

$$L_w - L_p = 10 \times \log \frac{1}{Q_d/(4 \times \pi \times r^2) + (4/R)} - 10.5$$

$$= [10 \times \log \frac{1}{Q_d/(4 \times \pi \times r^2) + (4/R)} - 0.3] \quad (5.11)$$

where Q_d = directivity factor
r = distance from sound source, ft [m]
R = room constant, ft² [m²]

The directivity factor is a function of where the fan is installed. If the fan could be installed so that the sound radiates in a spherical pattern, the directivity factor is 1. A fan mounted on the floor, so that the sound radiates in a hemispherical pattern, has a directivity factor of 2. If it is mounted close to a wall, the factor increases to 4, and if it is mounted in a corner, the directivity is 8.

The 4/R factor takes into account room effects, with the room constant R calculated using the following equation:

$$R = \frac{A \times \alpha}{1 - \alpha} \quad (5.12)$$

where A = total surface area of the room, ft² [m²] and α = room absorption coefficient (see Table 5.4).

Equation (5.12) assumes that all the surfaces in the room have the same absorption coefficient. More often, this is not the case, and it is necessary to calculate the average absorption coefficient. Calculate the average using the following equation:

TABLE 5.4 Sound Absorbtion Coefficients

Material	\multicolumn{6}{c}{Octave band center frequencies, Hz}					
	125	250	500	1000	2000	4000
Concrete Block, Coarse	0.36	0.44	0.31	0.29	0.39	0.25
Concrete Block, Painted	0.10	0.05	0.06	0.07	0.09	0.08
Brick	0.03	0.03	0.03	0.04	0.05	0.07
Plaster	0.14	0.10	0.06	0.05	0.04	0.03
Glass, Heavy Plate	0.18	0.06	0.04	0.03	0.02	0.02
Glass, Window Grade	0.35	0.25	0.18	0.12	0.07	0.04
Wood	0.15	0.11	0.10	0.07	0.06	0.07
Sound Absorption Panel*	0.89	1.20	1.16	1.09	1.01	1.03

*Noishield-4-in-thick perforated metal clad sound absorption material.
Source: of M. Hirschorn, Industrial Accoustics Company, Inc., Bronx, NY.

$$\bar{\alpha} = \frac{A_1 \times \alpha_1 + A_2 \times \alpha_2 + \cdots + A_n \times \alpha_n}{A_1 + A_2 + \cdots + A_n} \tag{5.13}$$

where n = number of different types of surfaces ($i = 1$ to n)
  A_i = surface area for surface i, ft^2 [m^2]
  α_i = absorption coefficient for surface i

Of course, if the fan is installed outdoors, with no walls nearby, then there are no room effects and the $4/R$ term in Eq. (5.11) is zero.

OSHA regulations often specify a dBA requirement for sound levels. This is an *A-weighted* single sound pressure value. The human ear does not respond to all frequencies the same way, and the *A*-weighting compensates for this. The procedure for calculating dBA is to calculate the sound pressure at the specified distance using Eq. (5.11), add the *A*-weighting correction factors shown in Table 5.3, and then combine the eight octave bands into one using Fig. 5.24.

Example 11 Calculate the dBA levels at 5 ft for a fan installed with its outlet connected to a room 10 ft high, 20 ft long, and 15 ft high. The walls of the room are painted concrete block and the floor and ceiling are wood.

solution The outlet is located in the center of one of the 15- by 20-ft walls, so use a directivity factor of 2. The total wall area is

$$A_w = (10 + 10 + 20 + 20) \times 15 = 900 \text{ ft}^2$$

The area of the floor and ceiling is

$$A_{fc} = 2 \times (10 \times 20) = 400 \text{ ft}^2$$

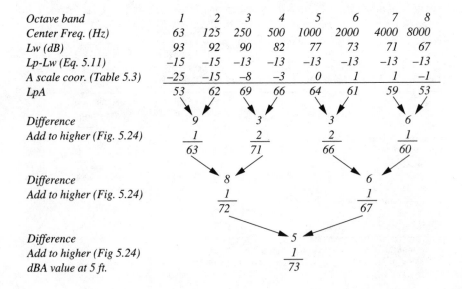

Octave band	1	2	3	4	5	6	7	8
Center Freq. (Hz)	63	125	250	500	1000	2000	4000	8000
Lw (dB)	93	92	90	82	77	73	71	67
Lp-Lw (Eq. 5.11)	-15	-15	-13	-13	-13	-13	-13	-13
A scale coor. (Table 5.3)	-25	-15	-8	-3	0	1	1	-1
LpA	53	62	69	66	64	61	59	53

Difference   9   3   3   6
Add to higher (Fig. 5.24)   1   2   2   1
                               63   71   66   60

Difference   8   6
Add to higher (Fig. 5.24)   1   1
                                    72   67

Difference   5
Add to higher (Fig 5.24)   1
dBA value at 5 ft.   73

The absorption coefficient for the room surfaces is found in Table 5.4. As an approximation, use the value given in the table for the 125-Hz octave band, and the value for the 4000 Hz for the 8000 Hz octave band. Using Eq. (5.13) to calculate the average absorption coefficient and Eq. (5.12) to calculate the room constant, we get the following:

Octave band	1	2	3	4	5	6	7	8
Center freq., Hz	63	125	250	500	1000	2000	4000	8000
For the walls, α_w	0.10	0.10	0.05	0.06	0.07	0.09	0.08	0.08
Floors and ceiling, α_{fc}	0.15	0.15	0.11	0.10	0.07	0.06	0.07	0.07
α	0.12	0.12	0.07	0.07	0.07	0.08	0.08	0.08
R	177.3	177.3	97.8	97.8	97.8	113.0	113.0	113.0

Using the outlet sound power values for this fan shown below, the difference between sound power and sound pressure using Eq. (5.11) ($Q_d = 2$, $r = 5$, R from above), the A-weighting factors from Table 5.3, and the adding decibel chart in Fig. 5.24, we can calculate the dBA at 5 ft.

It is important to note that the distance is specified with A-weighted decibels. Because A-weighted decibel values are based on sound pressure, distance has an effect on these measurements. If the sound levels obtained are too high, silencers are available. They mount onto the fan flange and reduce the sound levels coming out.

If a fan has ductwork connected to both the inlet and the outlet, and the ductwork openings are a long distance away, casing radiated sound can be the primary source of sound coming from the fan. The AMCA test code does not address casing radiated sound, so an industry standard does not exist. Casing radiated sound levels can be estimated based on the sound transmission loss of the casing material, which will vary with the fan type and size and from one manufacturer to another. Contact the manufacturer when casing radiated sound levels are required.

Techniques exist to calculate the sound absorption or sound generation in ductwork. These techniques, which are outside the scope of this work, can be used to estimate the sound levels coming out of open ducts.

5.5   Fan and Duct System Interaction

5.5.1   System lines

Because of the first fan law [Eq. (5.5)], the flow rate can be directly substituted for revolutions per minute in the second fan law resulting in the following:

$$\frac{P_N}{P_O} = \left(\frac{Q_N}{Q_O}\right)^2 \qquad (5.14)$$

where P = static or total pressure, in w.g. [Pa], and Q = volumetric flow rate, ft³/min [m³/s]. This is the equation for a system line and describes the relationship between flow and pressure for a given duct system. As long as there are no changes in the ductwork, blast gate, or damper positions or changes in filter pressure drop, this equation can be used to predict the pressure requirements for a desired flow once the pressure at a given flow is known.

A fan can only operate at the intersection of its fan curve and the system line determined by the duct system that the fan is connected to. This intersection point defines the *point of operation*. Figure 5.25 shows the same fan curve as Fig. 5.20, only with system lines added. Each test point represents the point of operation for a duct system that has a different resistance to flow and has its own unique system line. Note that as the system resistance increases, the slope of the system lines increases and the flow decreases.

Example 12   Point D_1 in Fig. 5.25 shows a point of operation of 30,200 ft³/min [14.25 m³/s] and 4.60 in w.g. [1142 Pa]. What pressure is required if the system is to deliver 20,000 ft³/min [9.44 m³/s]? If the new flow rate is to be achieved by slowing the fan down, what is the new speed?

solution   Rearranging Eq. (5.14), we get

$$SP_N = SP_O \times \left(\frac{Q_N}{Q_O}\right)^2 = 4.60 \times \left(\frac{20,000}{30,200}\right)^2 = 2.02 \text{ in w.g.}$$

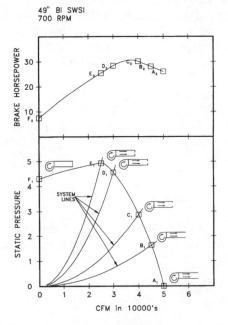

Figure 5.25   Fan performance curve with system lines shown.

$$= [1142 \times \left(\frac{9.44}{14.25} \right)^2 = 501 \text{ Pa}]$$

On Fig. 5.25, notice how this point also lies on the system line that passes through point D_1. Other points along the system line can be calculated using the equation above. To calculate the new speed, use Eq. (5.5).

$$\text{RPM}_N = \text{RPM}_O \times \frac{Q_N}{Q_O} = 700 \times \frac{20,000}{30,200} = 464$$

$$= [700 \times \frac{9.44}{14.25} = 464]$$

Figure 5.26 shows the results of an underestimated duct system. A system is designed for 15,000 ft³/min [7.1 m³/s], the fan static pressure required is calculated to be 4 in w.g. [993 Pa], and a fan is selected to operate at 886 r/min and 12.4 hp [9.2 kW]. This point of operation is shown as point 1. When the system was installed, several elbows had to be added to circumvent an obstruction, which increased the actual resistance to flow to 5 in w.g. [1242 Pa] at 15,000 ft³/min [7.1 m³/s]. This is plotted as point 2. Since an actual duct system can only operate on a system line passing through point 2, and the fan can only operate on its fan curve, the actual point of operation is point 3, which is read to be 13,700 ft³/min [6.5 m³/s], at 4.2 in w.g. [1043 Pa] of static pressure and 11.8 hp [8.8 kW]. To reach the design flow, the

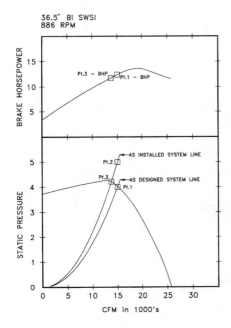

Figure 5.26 Effect of underestimating system resistance.

fan speed needs to be increased. By using the first and third fan laws, the new speed is determined to be 969 r/min and the required power is determined to be 15.4 hp [11.5 kW].

A similar example is shown in Fig. 5.27. This time a system is designed for 25,000 ft³/min [11.8 m³/s] and 7 in w.g. [1739 Pa] of static pressure, and a fan is selected to operate at 1072 r/min and 36.32 hp [27.1 kW]. Included in the static pressure is 4 in w.g. [993 Pa] for the pressure drop across a dust collector's filters, but when the system is first started up and the filters are clean, the pressure drop is only 1 in w.g. [248 Pa]. So at start-up, the duct system requirement is only 4 in w.g. [993 Pa] as plotted at point 2. The system line passing through point 2 intersects the fan curve at 29,877 ft³/min [14.1 m³/s] and 5.7 in w.g. [1416 Pa] and requires 38.5 hp [28.7 kW]. The increased flow will allow the hoods in the duct system to do a better job in capturing contaminant, and as the filters become dirty, the point of operation moves toward the design point. Sometimes the increase in airflow causes problems with the dust collector if the new flow is higher than what the dust collector is designed for. The resulting higher velocity through the filters may cause them to become plugged or blinded before they are properly seasoned, so that the pressure drop will keep on increasing with time and never stabilize at the pressure of 4 in w.g. [993 Pa]. As in the previous example, the speed of the fan could be changed, and this would be advantageous from an energy consumption standpoint because of the third fan law, but the speed would have to be increased

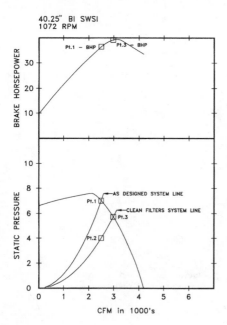

Figure 5.27 Effect of overestimating system resistance.

as the filters become seasoned. This could be accomplished with a variable-frequency drive. Another approach is to add a control damper to the system. When the new system is started up, the damper is closed down to a 3-in-w.g. [745-Pa] pressure drop to compensate for the initial low filter pressure drop. As the filters become loaded, the damper is opened to maintain the pressure drop across the filters, damper, and duct system at the 7-in-w.g. [1739-Pa] design static pressure.

5.5.2 System effects

To provide their customers with fans that have the best performance possible, manufacturers test them with ideal duct connections. Most of the time, fans are tested with open inlets, and designs that do not have an integral inlet bell will be tested with one to simulate inlet ductwork. Figure 5.28 shows the streamlines the air follows as it enters a fan with a well-designed inlet bell. The airflow is smooth and the velocity at the entrance is uniform.

The velocity profile coming out of even the most efficient fans is not uniform. Especially on centrifugal fans there are regions of high-velocity air coming out of the impeller. Figure 5.29 shows typical outlet velocity profiles for centrifugal and axial fans. A length of straight duct is required at the outlet to allow the high velocity (kinetic energy) to be converted to static pressure (potential energy). To get full pressure conversion, a duct at least as long as the 100 percent effective duct length is required. The AMCA defines the 100 percent effective duct length to be a minimum of two and one-half equivalent duct diameters for outlet velocities up to 2500 ft/min [12.7 m/s], with one duct diameter added for each additional 1000 ft/min [5.1 m/s]. Also it is important that the outlet connection be smooth. In the test code,

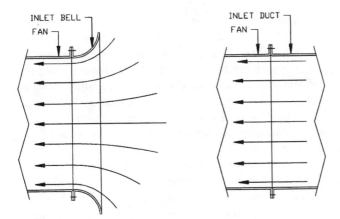

Figure 5.28 Ideal inlet flow conditions.

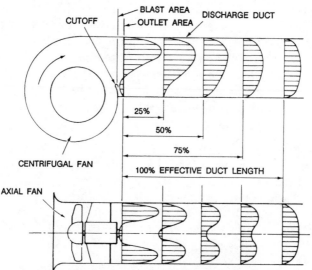

TO CALCULATE 100% EFFECTIVE DUCT LENGTH, ASSUME A MINIMUM OF 2-1/2 DUCT DIAMETERS FOR 2500 FPM OR LESS. ADD 1 DUCT DIAMETER FOR EACH ADDITIONAL 1000 FPM.

EXAMPLE: 5000 FPM = 5 EQUIVALENT DUCT DIAMETERS. IF THE DUCT IS RECTANGULAR WITH SIDE DIMENSIONS a AND b, THE EQUIVALENT DUCT DIAMETER IS EQUAL TO $(4ab/\pi)^{0.5}$

CONTROLLED DIFFUSION AND ESTABLISHMENT OF A UNIFORM VELOCITY
PROFILE IN A STRAIGHT LENGTH OF OUTLET DUCT

Figure 5.29 Fan outlet velocity profiles. (*Reprinted from AMCA publication 201-90,* Fans and Systems, *with the express written permission of the Air Movement and Control Association, Inc.*)

the AMCA specifies that the area of the outlet duct be between 95 and 105 percent of the fan outlet area and that any transitions used have sides that are sloped no more than 15° on converging sections and 7° on diverging sections.

Duct fittings or fan installations that do not provide smooth, uniform flow at the fan inlet or complete pressure recovery at the fan outlet cause the fan to perform at a level below its catalog rating. Elbows cause nonuniform airflow at the fan inlet. More air enters the fan at the outside of the elbow than at the inside, as illustrated in Fig. 5.33a. Figure 5.30 shows what happens to a centrifugal fan's performance curve, first, when an elbow is located at its inlet and, second, when an elbow is located at the inlet and there is no outlet duct. This loss in performance is known as *system effect.*

Duct configurations that cause system effects should be avoided whenever possible. To help in those situations when it is unavoidable, the AMCA has published a method of estimating the system effects for some of the most common situations. A series of tests were run in the AMCA's laboratory. Although variations due to factors such as fan

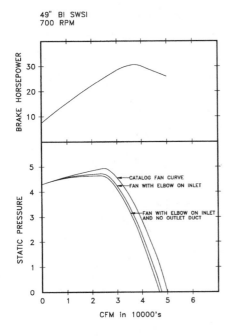

49" BI SWSI
700 RPM

Figure 5.30 Fan performance curve with system effects.

type and design were observed, figures showing system effect factors were put together to allow fan users to estimate system effects. These have been reproduced in Figs. 5.32 through 5.44. The figures give a system curve letter for the various configurations, which is used with the appropriate velocity (inlet or outlet) to determine the system effect loss factor on Fig. 5.31.

The system effect curves are plotted at standard density. If the system effect at a nonstandard density is needed, multiply the standard density system effect factor by the density factor. The system effect factor must be added to the system design pressure to select the fan and determine the required speed and power. The system effect will cause the fan to generate only the system design pressure, not the pressure that the fan was selected for.

5.5.3 Inlet system effects

Elbows mounted near the inlet of a fan cause the most common inlet system effect. Elbows should be installed at least three equivalent duct diameters upstream of a fan inlet. Not only do elbows mounted closer cause a system effect, but they can also cause pressure fluctuations, increased sound levels, and structural damage to the fan.

System effect curves for mitered and four-piece elbows near the inlets of axial fans are given in Fig. 5.32. Determining which curve to use is a function of whether the fan is tubeaxial or vaneaxial, the dis-

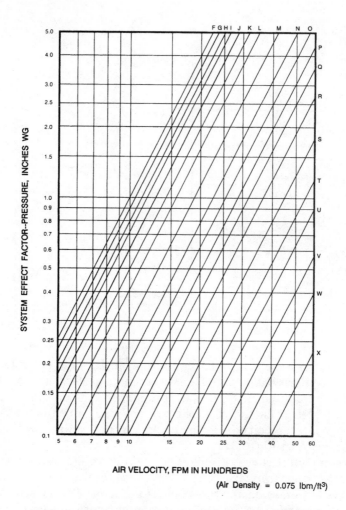

AIR VELOCITY, FPM IN HUNDREDS

(Air Density = 0.075 lbm/ft³)

Enter the chart at the appropriate air velocity (on the abcissa) read up to the applicable curve, then read across from the curve (to the ordinate) to find the SEF at standard air density.

Figure 5.31 System effect curves. (*Reprinted from AMCA publication 201-90,* Fans and Systems, *with the express written permission of the Air Movement and Control Association, Inc.*)

tance to the fan, whether the elbow is mitered, and the hub-to-tip ratio. The hub-to-tip ratio is the ratio of the diameter of the hub (see Fig. 5.2) to the diameter of the impeller at the tips.

As depicted in the previous example, elbows at a centrifugal fan's inlet cause a system effect. System effect curves for round, mitered ducts are shown in Fig. 5.34, and curves for square elbows are shown in Fig. 5.35. Often the elbow at the inlet is not a true elbow but a drop box as illustrated in Fig. 5.33b. The system effect for this type of inlet elbow is nearly impossible to predict as it is a function of fan type,

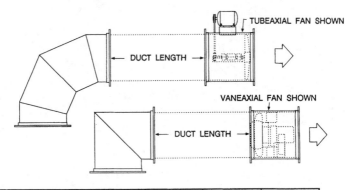

	H/T	90° Elbow	No Duct [1][2]	0.5 D [1][2]	1.0 D [1][2]	3.0 D
Tubeaxial Fan	.25	2 piece	U	V	W	---
Tubeaxial Fan	.25	4 piece	X	---	---	---
Tubeaxial Fan	.35	2 piece	V	W	X	---
Vaneaxial Fan	.61	2 piece	Q-R	Q-R	S-T	T-U
Vaneaxial Fan	.61	4 piece	W	W-X	---	---

Figure 5.32 System effect curves for inlet duct elbows—axial fans (determine the system effect factor by using Fig. 5.31). (*Reprinted from AMCA publication 201-90,* Fans and Systems, *with the express written permission of the Air Movement and Control Association, Inc.*)

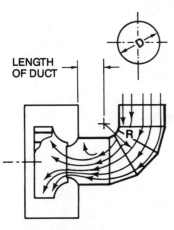

Figure 5.33a Flow conditions for various inlet elbow configurations. (*a*) Nonuniform flow into a fan inlet induced by a 90°, three-piece section elbow—no turning vanes. (*b*) Nonuniform flow induced into a fan inlet by a rectangular inlet duct. (*c*) Improved inlet flow conditions with a specially designed inlet box. (*Reprinted from AMCA publication 201-90,* Fans and Systems, *with the express written permission of the Air Movement and Control Association, Inc.*)

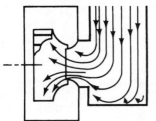

Figure 5.33b (*Continued*)

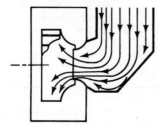

Figure 5.33c (*Continued*)

R/D	SYSTEM EFFECT CURVES		
	NO DUCT	2D DUCT	5D DUCT
–	N	P	R-S

TWO PIECE MITERED 90° ROUND SECTION ELBOW--NOT VANED

Figure 5.34a System effect curves for various mitered elbows without turning vanes (determine the system effect factor by using Fig. 5.31). (*Reprinted from AMCA publication 201-90, Fans and Systems, with the express written permission of the Air Movement and Control Association, Inc.*)

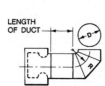

R/D	SYSTEM EFFECT CURVES		
	NO DUCT	2D DUCT	5D DUCT
0.5	O	Q	S
0.75	Q	R-S	T-U
1.0	R	S-T	U-V
2.0	R-S	T	U-V
3.0	S	T-U	V

THREE PIECE MITERED 90° ROUND SECTION ELBOW--NOT VANED

Figure 5.34b (*Continued*)

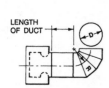

R/D	SYSTEM EFFECT CURVES		
	NO DUCT	2D DUCT	5D DUCT
0.5	P-Q	R-S	T
0.75	Q-R	S	U
1.0	R	S-T	U-V
2.0	R-S	T	U-V
3.0	S-T	U	V-W

FOUR OR MORE PIECE MITERED 90° ROUND SECTION ELBOW--NOT VANED

Figure 5.34c (*Continued*)

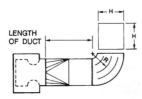

SYSTEM EFFECT CURVES

R/D	NO DUCT	2D DUCT	5D DUCT
0.5	O	Q	S
0.75	P	R	S-T
1.0	R	S-T	U-V
1.0	S	T-U	V

SQUARE ELBOW WITH INLET TRANSITION--NO TURNING VANES

Figure 5.35a System effect curves for various square duct elbows (determine the system effect factor by using Fig. 5.31). D = diameter of the inlet collar. The inside area of the square duct (H × H) should be equal to the inside area of the fan inlet collar. The maximum permissible angle of any converging element of the transition is 15°, and for a diverging element, 7°. (*Reprinted from AMCA publication 201-90,* Fans and Systems, *with the express written permission of the Air Movement and Control Association, Inc.*)

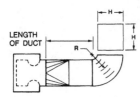

SYSTEM EFFECT CURVES

R/D	NO DUCT	2D DUCT	5D DUCT
0.5	S	T-U	V
1.0	T	U-V	W
2.0	V	V-W	W-X

SQUARE ELBOW WITH INLET TRANSITION--3 LONG TURNING VANES

Figure 5.35b (*Continued*)

SYSTEM EFFECT CURVES

R/D	NO DUCT	2D DUCT	5D DUCT
0.5	S	T-U	V
1.0	T	U-V	W
2.0	V	V-W	W-X

SQUARE ELBOW WITH INLET TRANSITION--SHORT TURNING VANES

Figure 5.35c (*Continued*)

inlet size, and box width and depth. Losses in flow rate as high as 45 percent have been observed with this type of inlet. Most fans can be purchased with an optional inlet box, as shown in Fig. 5.33c. These boxes are designed to have minimal and predictable system effects.

Often the duct configuration can cause an airflow pattern that is detrimental to fan performance. Figure 5.36 shows one example of a duct configuration that causes the air to spin in the direction opposite the impeller rotation (counterrotating swirl). When this happens,

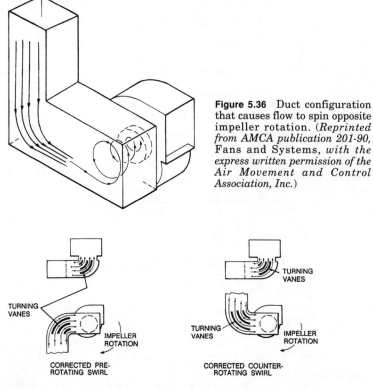

Figure 5.36 Duct configuration that causes flow to spin opposite impeller rotation. (*Reprinted from AMCA publication 201-90,* Fans and Systems, *with the express written permission of the Air Movement and Control Association, Inc.*)

Figure 5.37 Corrections for inlet spin. (*Reprinted from AMCA publication 201-90,* Fans and Systems, *with the express written permission of the Air Movement and Control Association, Inc.*)

there may be a slight increase in pressure produced by the fan, but with a significant increase in power.

Similar duct configurations can cause the air to spin in the direction of impeller rotation (prerotating swirl). This reduces the pressure produced and the power consumed by the fan. Inlet vanes produce this effect to control flow with power savings.

Like drop boxes, the severity of the system effect caused by counter-rotating and prerotating spin is impossible to predict. Avoid duct configurations that have the potential to cause spin. Figure 5.37 shows how turning vanes placed in the elbows can minimize system effect. If there is room, straighteners can be installed into the duct just upstream of the fan inlet to break up spinning air. Figure 5.38 shows one type of straightener that can be used.

Fan inlets located too close to a wall cause nonuniform flow and a resulting system effect. Figure 5.39 gives the system effect curves to correct for this situation.

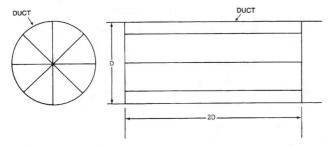

Figure 5.38 Airflow straightener. (*Reprinted from AMCA publication 201-90,* Fans and Systems, *with the express written permission of the Air Movement and Control Association, Inc.*)

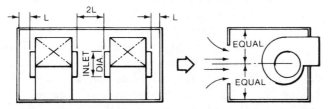

Figure 5.39a System effect curves for fans located in plenums and cabinet enclosures and for various wall to inlet dimensions. (*a*) Fans and plenum. (*b*) Axial fan near wall. (Determine the system effect factor by calculating inlet velocity and using Fig. 5.31). (*Reprinted from AMCA publication 201-90,* Fans and Systems, *with the express written permission of the Air Movement and Control Association, Inc.*)

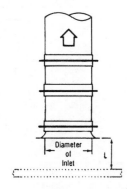

Figure 5.39b   (*Continued*)

L DISTANCE INLET TO WALL	SYSTEM EFFECT CURVES
0.75 X DIA OF INLET	V-W
0.5 X DIA OF INLET	U
0.4 X DIA OF INLET	T
0.3 X DIA OF INLET	S

Figure 5.39c   (*Continued*)

Duct sized smaller than the fan inlet and butterfly valves and belt guards placed too close to the fan inlet are two examples of obstructions that cause system effects. To estimate the system effect factor, project the area of the obstruction onto the inlet area. Calculate the unobstructed area, the percentage of unobstructed area, and the inlet velocity based on the obstructed inlet. Use the percentage of unobstructed area and Fig. 5.40 to determine which system effect curve to use, and then the obstructed inlet velocity and Fig. 5.31 to determine the system effect factor.

5.5.4 Outlet system effects

The most common cause of system effect on fan outlets is an insufficient length of straight duct that prevents the full conversion of velocity to static pressure. As defined previously, the length of straight duct required is the 100 percent effective duct length. The system effect curves for no duct or a short piece of duct on a fan outlet can be determined from Fig. 5.41 for axial fans and Fig. 5.42 for centrifugal fans. To use Fig. 5.42, the blast-area–to–outlet-area ratio is required. The blast area is the outlet area minus the projected area of the cutout (see Fig. 5.1). The blast area is not normally included in the catalogs but is available by contacting the manufacturer. Note that tubeaxial fans do not have a system effect curve for no outlet duct. Tests performed by the AMCA indicated very little loss in performance with tubeaxial fans without outlet ducts and with

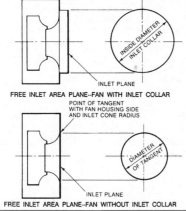

Figure 5.40 System effect curves for inlet obstructions (determine the system effect factor by calculating inlet velocity and using Fig. 5.31). (*Reprinted from AMCA publication 201-90,* Fans and Systems, *with the express written permission of the Air Movement and Control Association, Inc.*)

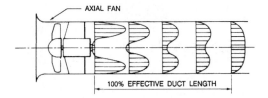

TO CALCULATE 100% EFFECTIVE DUCT LENGTH, ASSUME A MINIMUM OF 2-1/2 DUCT
DIAMETERS FOR 2500 FPM OR LESS.   ADD 1 DUCT DIAMETER FOR EACH ADDITIONAL
1000 FPM.

EXAMPLE: 5000 FPM = 5 EQUIVALENT DUCT DIAMETERS

	No Duct	12% Effective Duct	25% Effective Duct	50% Effective Duct	100% Effective Duct
Tubeaxial Fan	---	---	---	---	---
Vaneaxial Fan	U	V	W	---	---

Figure 5.41 System effect curves for outlet ducts—axial fans (determine the system effect factor by using Fig. 5.31). (*Reprinted from AMCA publication 201-90,* Fans and Systems, *with the express written permission of the Air Movement and Control Association, Inc.*)

vaneaxial fans with outlet ducts as short as 50 percent effective duct length.

Elbows are frequently mounted close to fan outlets. This placement will disrupt the energy conversion from velocity pressure to static pressure. Estimates of the system effect factor can be obtained using Fig. 5.43 for axial fans and Fig. 5.44 for centrifugal fans. As with short ducts, tests indicate that the factors are negligible with tubeaxial fans and greater with centrifugal fans than with vaneaxial fans. The position of the elbow influences how great the system effect is with centrifugal fans. When the elbow follows the curvature of the scroll, as in position A in Fig. 5.44, the system effect is minimized, but when the elbow goes against the curvature of the scroll, as in position C, the effect is worsened.

Example 13 A fan is installed on a system that requires 14,000 ft³/min [6.6 m³/s] at 7 in w.g. [1739 Pa]. The fan has a four-piece mitered 90° round elbow $(R/D = 1)$ mounted directly on its inlet and no outlet duct. The elbow has a cross-sectional area of 3.14 ft² [0.29 m²], but the fan has an inlet area of 3.69 ft² [0.34 m²]. The outlet area of the fan is the same as the inlet area, and the blast area is 70 percent of the outlet area. What flow rate and static pressure should the fan be selected for to account for these system effects?

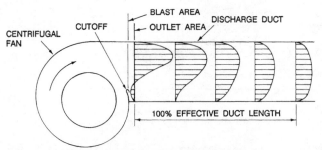

TO CALCULATE 100% EFFECTIVE DUCT LENGTH, ASSUME A MINIMUM OF 2-1/2 DUCT DIAMETERS FOR 2500 FPM OR LESS. ADD 1 DUCT DIAMETER FOR EACH ADDITIONAL 1000 FPM.

EXAMPLE: 5000 FPM = 5 EQUIVALENT DUCT DIAMETERS.   IF THE DUCT IS RECTANGULAR WITH SIDE DIMENSIONS a AND b, THE EQUIVALENT DUCT DIAMETER IS EQUAL TO $(4ab/\pi)^{0.5}$

	No Duct	12% Effective Duct	25% Effective Duct	50% Effective Duct	100% Effective Duct
Pressure Recovery	0%	50%	80%	90%	100%
Blast Area Outlet Area	System Effect Curve				
0.4	P	R-S	U	W	—
0.5	P	R-S	U	W	—
0.6	R-S	S-T	U-V	W-X	—
0.7	S	U	W-X	—	—
0.8	T-U	V-W	X	—	—
0.9	V-W	W-X	—	—	—
1.0	—	—	—	—	—

Figure 5.42 System effect curves for outlet ducts—centrifugal fans (determine the system effect factor by using Fig. 5.31). (*Reprinted from AMCA publication 201-90,* Fans and Systems, *with the express written permission of the Air Movement and Control Association, Inc.*)

solution   There are three system effects to consider: the elbow on the inlet, the inlet duct being smaller than the fan inlet, and the absence of an outlet duct. In order to determine the system effect, we need to know what the inlet and outlet velocities are. Since this fan has equal inlet and outlet areas, the velocity at both locations is

$$V = \frac{14,000}{3.69} = 3794 \text{ ft/min [19.3 m/s]}$$

Figure 5.34 shows that for a four-piece 90° elbow, $R/D = 1$, with no duct between the elbow and the fan, we should use system effect curve R. From Fig. 5.31, with a velocity of 3794 ft/min [19.3 m/s] and system effect curve R, we find that the system effect factor for the elbow is 1 in w.g. [248 Pa]. Using Fig. 5.40, with 85 percent (3.14/3.69) unobstructed inlet area, we find that we should use system effect curve T. Figure 5.31 shows that with system effect curve T and 3794 ft/min [19.3 m/s] the system effect factor for the obstructed inlet is 0.45 in w.g. [112 Pa]. Figure 5.42 tells us that for a blast area 70 percent of the outlet area and no outlet duct we should use system effect curve S. Again going to Fig. 5.31, with 3794 ft/min [19.3 m/s] and system effect curve S, we find that the system effect factor for the lack of ductwork on the outlet is 0.70 in w.g. [174 Pa]. The total system effect factor is 2.15 in w.g. (1.00 + 0.45 + 0.70) [534 Pa]. The fan should be selected for 14,000 ft³/min [6.6 m³/s] at 9.15 in w.g. [2272 Pa],

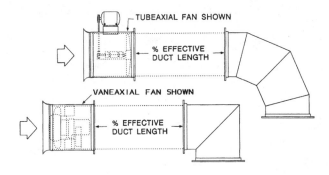

	90# Elbow	No Duct	12% Effective Duct	25% Effective Duct	50% Effective Duct	100% Effective Duct
Tubeaxial Fan	2&4 Pc	---	---	---	---	---
Vaneaxial Fan	2 Pc	U	U-V	V	W	---
Vaneaxial Fan	4 Pc	W	---	---	---	---

Figure 5.43 System effect curves for outlet duct elbows—axial fans (determine the system effect factor by using Figs. 5.29 and 5.31). (*Reprinted from AMCA publication 201-90,* Fans and Systems, *with the express written permission of the Air Movement and Control Association, Inc.*)

but, because of the system effects, it will only deliver 14,000 ft³/min [6.6 m³/s] at 7 in w.g. [1739 Pa].

Not all system effects are bad. Frequently cones or rectangular transitions called *evases* are added to fan outlets to increase the amount of static pressure regain over what is possible with a straight duct section that matches the outlet area. Figure 5.45 is a photograph of a radial-tip fan with an evase. The increase in pressure can be estimated using the static pressure regain for expansion as calculated for any piece of ductwork or piping, or corrected performance data can be obtained from the manufacturer.

5.6 Fan Construction

5.6.1 Standard discharges

Figure 5.46 is a reproduction of AMCA standard 99-2406-83 which specifies the standard designations for rotation and discharge for centrifugal fans. Often, the selection of the proper discharge eliminates duct configurations that cause system effects. Note that the direction of fan rotation is determined by looking at the side that is normally the drive side, i.e., the side of the fan that is opposite the inlet. Most manufacturers will construct fans with discharges at nonstandard angles, and this figure can be used to designate the angle required.

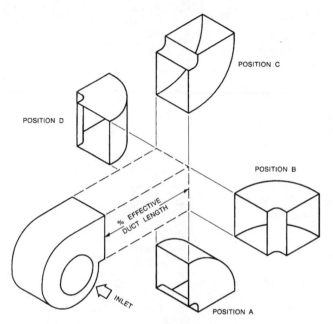

POSITION C

POSITION D

POSITION B

% EFFECTIVE
DUCT LENGTH

INLET

POSITION A

Figure 5.44a System effect curves for outlet duct elbows—centrifugal fans (determine the system effect factor by using Figs. 5.29 and 5.31). (*Reprinted from AMCA publication 201-90,* Fans and Systems, *with the express written permission of the Air Movement and Control Association, Inc.*)

5.6.2 Fan arrangements

Fan arrangement refers to the relative location of the fan impeller, the fan bearings, and the motor. Standard drive arrangements for centrifugal and axial fans are specified in Figs. 5.47 and 5.48. Each arrangement has is own advantages and disadvantages and does not apply to all fan types. The difference between some of the arrangements is that some are intended to be direct-driven at the motor speed while some are belt-driven. Inlet box arrangement positions for centrifugal fans are shown in Fig. 5.49.

Belt drives for fans are most common up to 200 hp; above this, fans are normally direct-driven. Belt drives offer the advantage that they can be selected so that the fan operates at just the right speed to produce a given performance. Motor location with respect to the drive end of the shaft should be called out per AMCA standard 99-2407-66, which is reproduced in Fig. 5.50. Avoid belt-drive combinations that have the fan running at more than 150 percent of the motor speed. At drive ratios higher than this, drive maintenance becomes more critical and is required more frequently. Also avoid adjustable pitch sheaves for fans having 15 hp [11.2 kW] of power or greater. While

Blast Area Outlet Area	Outlet Elbow Position	No Outlet Duct	12% Effective Duct	25% Effective Duct	50% Effective Duct	100% Effective Duct
0.4	A	N	O	P-Q	S	
	B	M-N	N	O-P	R-S	
	C	L-M	M	N	Q	
	D	L-M	M	N	Q	
0.5	A	O-P	P-Q	R	T	
	B	N-O	O-P	Q	S-T	
	C	M-N	N	O-P	R-S	
	D	M-N	N	O-P	R-S	
0.6	A	Q	Q-R	S	U	
	B	P	Q	R	T	
	C	N-O	O	Q	S	
	D	N-O	O	Q	S	
0.7	A	R-S	S	T	V	
	B	Q-R	R-S	S-T	U-V	
	C	P	Q	R-S	T	
	D	P	Q	R-S	T	
0.8	A	S	S-T	T-U	W	
	B	R-S	S	T	V	
	C	Q-R	R	S	U-V	
	D	Q-R	R	S	U-V	
0.9	A	T	T-U	U-V	W	
	B	S	S-T	T-U	W	
	C	R	S	S-T	V	
	D	R	S	S-T	V	
1.0	A	T	T-U	U-V	W	
	B	S-T	T	U	W	
	C	R-S	S	T	V	
	D	R-S	S	T	V	

NO SYSTEM EFFECT FACTOR

SYSTEM EFFECT CURVES FOR SWSI FANS

For **DWDI** fans determine SEF using the curve for SWSI fans. Then apply the appropriate multiplier from the tabulation below

MULTIPLIERS FOR DWDI FANS
ELBOW POSITION A = ΔP X 1.00
ELBOW POSITION B = ΔP X 1.25
ELBOW POSITION C = ΔP X 1.00
ELBOW POSITION D = ΔP X 0.85

Figure 5.44b (*Continued*)

they provide an easy way to change fan performance by changing speed, they require more maintenance, do not run as true, are more prone to have vibration problems, and are more expensive than fixed pitch drives.

Direct-drive fans require less maintenance and provide a more efficient package since there are no belt-drive losses. Since the fan must run at the motor speed, nonstandard impeller widths or diameters may be required to provide a given performance. Because the fan speed cannot be changed unless an expensive variable-frequency drive is used, it is very difficult to compensate for unexpected changes in the duct system such as underestimating or overestimating system pressure requirements.

- Arrangement 1 is the most popular fan arrangement and can be used on a wide range of fan sizes and motor powers. Normally

Figure 5.45 Radial-tip fans with evase outlets. (*Courtesy of Twin City Fan and Blower Company, Minneapolis, MN.*)

arrangement-1 fans are belt-driven. Accessories such as external inlet vanes and inlet boxes can be easily added by bolting them to the inlet flange.

- Arrangement-2 fans are relatively uncommon. They afford some space saving but do not have as strong a bearing pedestal as arrangement 1 and as a result are only practical with fans whose diameters are less than 36 in [0.93 m] and whose motors have less than 30 hp [22.4 kW] of power.

- Arrangement-3 single-width, single-outlet (SWSI) fans are used where there is no room for the length of an arrangement-1-style pedestal. The disadvantage with this type of arrangement is that one bearing is in the airstream unless an inlet box is used. It is undesirable to have the bearing in the airstream on systems handling corrosive, flammable, or high-temperature airstreams. Also, properly maintaining the inlet bearing on a ducted inlet application is difficult. If an inlet box is used, a longer shaft is required and the inlet box must be an integral part of the fan as opposed to a bolt-on accessory. Independent bearing pedestals are used on large, high-horsepower fans that require strong, rigid support for the bearings. Experience has shown that with large fans, center hung arrangement 3 works better than overhung impeller arrangements in the following situations:

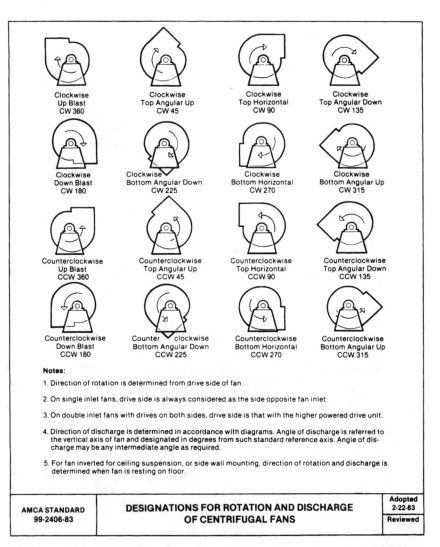

Notes:

1. Direction of rotation is determined from drive side of fan.

2. On single inlet fans, drive side is always considered as the side opposite fan inlet.

3. On double inlet fans with drives on both sides, drive side is that with the higher powered drive unit.

4. Direction of discharge is determined in accordance with diagrams. Angle of discharge is referred to the vertical axis of fan and designated in degrees from such standard reference axis. Angle of discharge may be any intermediate angle as required.

5. For fan inverted for ceiling suspension, or side wall mounting, direction of rotation and discharge is determined when fan is resting on floor.

| AMCA STANDARD 99-2406-83 | DESIGNATIONS FOR ROTATION AND DISCHARGE OF CENTRIFUGAL FANS | Adopted 2-22-83 |
| | | Reviewed |

Figure 5.46 Designations for rotation and discharge for centrifugal fans. (*Reprinted from AMCA publication 99-86*, Standards Handbook, *with the express written permission of the Air Movement and Control Association, Inc.*)

Speed, r/min	Impeller diameter, in [m]
1800	>66 [1.7]
1200	>96 [2.4]
900	>106 [2.7]

- Arrangement-3 double-width, double-inlet (DWDI) fans can be thought of as fans with two single-width impellers mounted back to

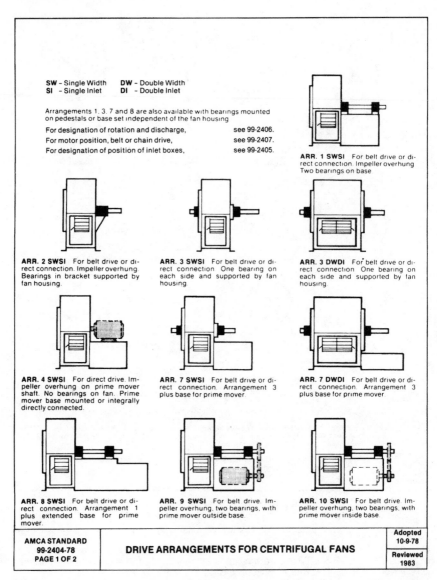

SW – Single Width   **DW** – Double Width
SI – Single Inlet   **DI** – Double Inlet

Arrangements 1, 3, 7 and 8 are also available with bearings mounted on pedestals or base set independent of the fan housing

For designation of rotation and discharge,   see 99-2406.
For motor position, belt or chain drive,   see 99-2407.
For designation of position of inlet boxes,   see 99-2405.

ARR. 1 SWSI For belt drive or direct connection. Impeller overhung Two bearings on base

ARR. 2 SWSI For belt drive or direct connection. Impeller overhung. Bearings in bracket supported by fan housing.

ARR. 3 SWSI For belt drive or direct connection. One bearing on each side and supported by fan housing.

ARR. 3 DWDI For belt drive or direct connection. One bearing on each side and supported by fan housing.

ARR. 4 SWSI For direct drive. Impeller overhung on prime mover shaft. No bearings on fan. Prime mover base mounted or integrally directly connected.

ARR. 7 SWSI For belt drive or direct connection. Arrangement 3 plus base for prime mover.

ARR. 7 DWDI For belt drive or direct connection. Arrangement 3 plus base for prime mover.

ARR. 8 SWSI For belt drive or direct connection. Arrangement 1 plus extended base for prime mover.

ARR. 9 SWSI For belt drive. Impeller overhung, two bearings, with prime mover outside base.

ARR. 10 SWSI For belt drive. Impeller overhung, two bearings, with prime mover inside base.

| AMCA STANDARD 99-2404-78 PAGE 1 OF 2 | **DRIVE ARRANGEMENTS FOR CENTRIFUGAL FANS** | Adopted 10-9-78 Reviewed 1983 |

Figure 5.47a Drive arrangements for centrifugal fans. (*Reprinted from AMCA publication 99-86,* Standards Handbook, *with the express written permission of the Air Movement and Control Association, Inc.*)

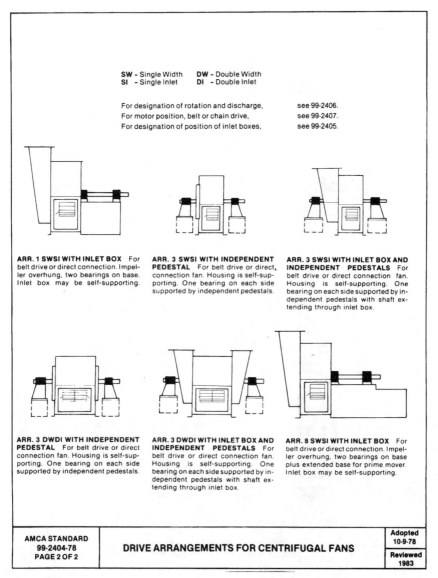

SW - Single Width   DW - Double Width
SI - Single Inlet   DI - Double Inlet

For designation of rotation and discharge,   see 99-2406.
For motor position, belt or chain drive,   see 99-2407.
For designation of position of inlet boxes,   see 99-2405.

ARR. 1 SWSI WITH INLET BOX For belt drive or direct connection. Impeller overhung, two bearings on base. Inlet box may be self-supporting.

ARR. 3 SWSI WITH INDEPENDENT PEDESTAL For belt drive or direct, connection fan. Housing is self-supporting. One bearing on each side supported by independent pedestals.

ARR. 3 SWSI WITH INLET BOX AND INDEPENDENT PEDESTALS For belt drive or direct connection fan. Housing is self-supporting. One bearing on each side supported by independent pedestals with shaft extending through inlet box.

ARR. 3 DWDI WITH INDEPENDENT PEDESTAL For belt drive or direct connection fan. Housing is self-supporting. One bearing on each side supported by independent pedestals.

ARR. 3 DWDI WITH INLET BOX AND INDEPENDENT PEDESTALS For belt drive or direct connection fan. Housing is self-supporting. One bearing on each side supported by independent pedestals with shaft extending through inlet box.

ARR. 8 SWSI WITH INLET BOX For belt drive or direct connection. Impeller overhung, two bearings on base plus extended base for prime mover. Inlet box may be self-supporting.

| AMCA STANDARD 99-2404-78 PAGE 2 OF 2 | **DRIVE ARRANGEMENTS FOR CENTRIFUGAL FANS** | Adopted 10-9-78 |
| | | Reviewed 1983 |

Figure 5.47b (*Continued*)

back on a common shaft and with a common housing. A DWDI fan provides a shorter but wider profile than an SWSI fan for a given performance. The inlet duct connection is more difficult than a single-inlet fan and is usually done with an inlet box on each inlet and a "pair of pants" duct fitting. This arrangement is frequently used on large fans that are handling dirty or corrosive air. The inlet box

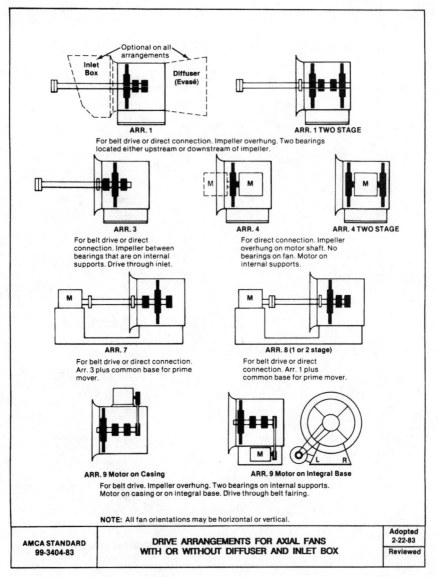

Figure 5.48 Drive arrangements for axial fans with or without diffuser and inlet box. (*Reprinted from AMCA publication 99-86,* Standards Handbook, *with the express written permission of the Air Movement and Control Association, Inc.*)

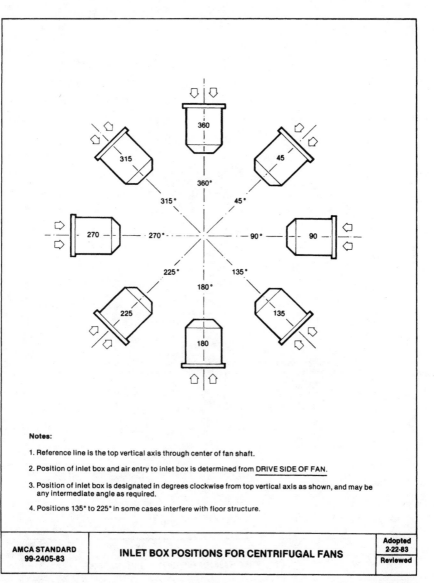

Notes:

1. Reference line is the top vertical axis through center of fan shaft.

2. Position of inlet box and air entry to inlet box is determined from <u>DRIVE SIDE OF FAN.</u>

3. Position of inlet box is designated in degrees clockwise from top vertical axis as shown, and may be any intermediate angle as required.

4. Positions 135° to 225° in some cases interfere with floor structure.

AMCA STANDARD 99-2405-83	INLET BOX POSITIONS FOR CENTRIFUGAL FANS	Adopted 2-22-83
		Reviewed

Figure 5.49 Inlet box positions for centrifugal fans. (*Reprinted from AMCA publication 99-86,* Standards Handbook, *with the express written permission of the Air Movement and Control Association, Inc.*)

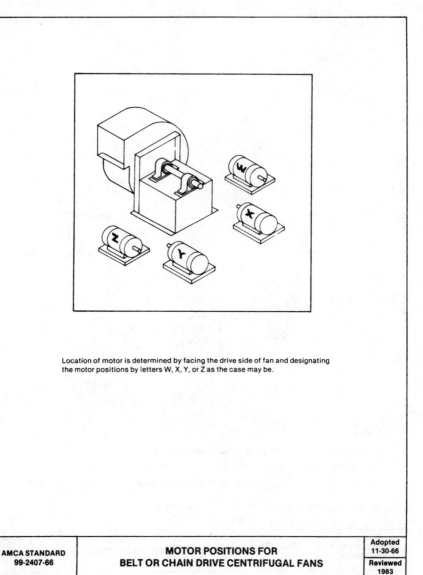

Location of motor is determined by facing the drive side of fan and designating the motor positions by letters W, X, Y, or Z as the case may be.

AMCA STANDARD 99-2407-66	MOTOR POSITIONS FOR BELT OR CHAIN DRIVE CENTRIFUGAL FANS	Adopted 11-30-66 Reviewed 1983

Figure 5.50 Motor positions for belt-drive centrifugal fans. (*Reprinted from AMCA publication 99-86,* Standards Handbook, *with the express written permission of the Air Movement and Control Association, Inc.*)

position should be specified using the standard designations shown in Fig. 5.49.

- Arrangement-4 direct-driven fans have the impeller mounted on the motor shaft. Standard motor bearings can usually handle the weight and thrust loads for impellers with diameters up to about 40 in [1 m].

- Arrangement-7 SWSI and DWDI fans are similar to arrangement-3 fans, except that a motor pedestal is added to support a motor at a height allowing attachment to the fan shaft with a flexible coupling.

- Arrangement-8 fans are similar to arrangement-1 fans, except that they are direct-driven. A motor pedestal is added to support a motor at a height so that it can be attached to the fan shaft with a flexible coupling.

- Arrangement-9 fans are belt-driven and are similar to arrangement 1 except that the motor is attached to the side of the bearing pedestal. This provides a compact package, and often the motor and drives are provided premounted from the manufacturer. The size of the motor is limited by what will practically fit on the pedestal.

- Arrangement-10 fans also provide a compact package. The motor is mounted inside the bearing pedestal. The motor and drives are often provided premounted. Since the motor is mounted under the pedestal, a weather cover to protect the motor and bearings from the elements is easily added, so arrangement-10 fans are frequently used for fans that are installed outdoors. Like arrangement-9 fans, the size of the motor is limited by what will fit inside the pedestal.

5.6.3 Special construction—spark resistance

Frequently fans are used to move air that has explosive or flammable characteristics. The AMCA has developed three standard classifications for spark-resistant construction that are shown in Fig. 5.51. Type A provides the greatest resistance to sparking and specifies that all parts of the fan that are in contact with the airstream must be made of nonferrous material and that the bearings, shaft, and impeller must be restrained to prevent movement. Usually aluminum or bronze is used as the nonferrous alloy, although nickel-based alloys are occasionally used for high-temperature applications. Type B calls for a nonferrous impeller and for nonferrous rubbing rings that prevent ferrous parts from rubbing against each other if the shaft or bearings shift. Type C is the least stringent, requiring that if a shift of

Fan applications may involve the handling of potentially explosive or flammable particles, fumes or vapors. Such applications require careful consideration of *all* system components to insure the safe handling of such gas streams. This AMCA Standard deals only with the fan unit installed in that system. The Standard contains guidelines which are to be used by both the manufacturer and user as a means of establishing *general* methods of construction. The exact method of construction and choice of alloys is the responsibility of the manufacturer; however, the customer must accept both the type and design with full recognition of the potential hazard and the degree of protection required.

TYPE   CONSTRUCTION

A   All parts of the fan in contact with the air or gas being handled shall be made of nonferrous material. Steps must also be taken to assure that the impeller, bearings, and shaft are adequately attached and/or restrained to prevent a lateral or axial shift in these components.

B   The fan shall have a nonferrous impeller and nonferrous ring about the opening through which the shaft passes. Ferrous hubs, shafts, and hardware are allowed provided construction is such that a shift of impeller or shaft will not permit two ferrous parts of the fan to rub or strike. Steps must also be taken to assure that the impeller, bearings, and shaft are adequately attached and/or restrained to prevent a lateral or axial shift in these components.

C   The fan shall be so constructed that a shift of the impeller or shaft will not permit two ferrous parts of the fan to rub or strike.

Notes

1. No bearings, drive components or electrical devices shall be placed in the air or gas stream unless they are constructed or enclosed in such a manner that failure of that component cannot ignite the surrouding gas stream.

2. The user shall electrically ground all fan parts.

3. For this Standard, nonferrous material shall be any material with less than 5% iron or any other material with demonstrated ability to be spark resistant.

4. The use of aluminum or aluminum alloys in the presence of steel which has been allowed to rust requires special consideration. Research by the U. S. Bureau of Mines and others has shown that aluminum impellers rubbing on rusty steel may cause high intensity sparking.

The use of the above Standard in no way implies a guarantee of safety for any level of spark resistance. "Spark resistant construction also does not protect against ignition of explosive gases caused by catastrophic failure or from any airstream material that may be present in a system."

This Standard applies to:

Centrifugal Fans
Axial and Propeller Fans
Power Roof Ventilators

This Standard applies to ferrous and nonferrous metals. The potential questions which may be associated with fans constructed of FRP, PVC, or any other plastic compound were not addressed.

| AMCA STANDARD 99-0401-86 | CLASSIFICATIONS FOR SPARK RESISTANT CONSTRUCTION | Adopted 2-12-86 |
| | | Reviewed |

Figure 5.51 Classifications for spark-resistant construction. (*Reprinted from AMCA publication 99-86,* Standards Handbook, *with the express written permission of the Air Movement and Control Association, Inc.*)

the impeller or shaft occurs, ferrous parts of the fan may not run into or strike each other.

5.6.4 Special construction—high temperature

High-temperature fans require special construction. What is considered high temperature varies with different manufacturers, but most often fans handling airstreams with temperatures above 300°F [150°C] are considered high-temperature fans. To maintain acceptable bearing life, it is essential that the bearings are kept cool. This is usually accomplished by using a shaft seal and a shaft cooling wheel as shown in Fig. 5.52. The shaft seal prevents hot air from blowing out the shaft hole, and the shaft cooler acts as a heat sink, preventing heat from being conducted down the shaft and into the bearing. Sometimes even this is not sufficient to keep the bearings cool, and external cooling using forced air, water, or circulating oil must be used. Fan housings can be insulated, and bearing pedestals are often modified to minimize the heat conduction into the bearing. Because of the loss of strength at elevated temperatures, the structural design of the fan is often modified. If the temperature rises very quickly, at rates higher than 15°F per minute, special design modifications may be required to ensure that the impeller remains secured to the shaft. Very rapid temperature changes also may cause excessive thermal stresses as some parts heat up quicker than others.

5.6.5 Special construction—corrosion resistance

Special modifications are often required for corrosion protection. Sometimes a special material such as stainless steel, aluminum, or fiberglass-reinforced plastic must be used. A special coating such as epoxy, vinyl, elastomer, or zinc can be applied to steel to give adequate protection. Selection of the proper material is beyond the scope of this section. Many factors such as the type of corrosive material, fume concentration, airstream temperature, and moisture content must be taken into account when selecting the proper material of construction or coating. Bearings must be kept out of corrosive airsteams. Shaft seals are usually used to keep corrosive fumes away from bearings outside of corrosive airstreams.

5.6.6 Special construction—abrasion resistance

Abrasive materials also require special construction. As with high-temperature fans, shaft seals are used, but in this case to prevent the

Figure 5.52 Gasket-type shaft seal and shaft cooler. (*Courtesy of Twin City Fan and Blower Company, Minneapolis, MN.*)

abrasive material from getting into the bearings. If the environment on the outside of the fan contains abrasive dust, taconite-grade bearing seals may be needed. The components of the fan that are in contact with the abrasive material may be made of an abrasion-resistant steel or be coated with an abrasion-resisting material such as rubber, ceramics, or a welded-on chromium carbide layer. On low-speed fans, wear plates made of, or coated with, abrasion-resisting material can be attached to the fan blades and scroll. On high-speed fans, where it is important to keep the blade stresses down, abrasion-resistant patches are welded to just the critical areas where abrasion occurs. Frequently, spare impellers and shafts are maintained for fans that are critical to a process since it takes less time to replace the impeller than to repair it in place.

5.7   Fan Accessories

5.7.1   Shaft seals

Shaft seals are used for a variety of reasons, such as for sealing in corrosive fumes, high-temperature air, and abrasive material. Gasket shaft seals, like the one shown in Fig. 5.52, are the simplest and most common type. Gasket material can be selected to operate at temperatures over 1000°F [540°C]. This type of seal is not completely airtight and should not be used on very high pressure applications where a tight seal is important.

Stuffing box shaft seals, like the one shown in Fig. 5.53, provide a tighter seal than gasket shaft seals. The packing material can be selected to meet the requirements of corrosive or high-temperature applications. Because the packing rubs against the shaft, the resulting friction increases the power slightly, especially when new. Allow for a little extra power when selecting motors for fans with stuffing box shaft seals. The heat caused by the friction prevents the use of stuffing box shaft seals on very high speed shafts.

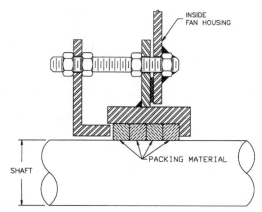

Figure 5.53 Cross section of typical stuffing box shaft seal. (*Courtesy of Twin City Fan and Blower Company, Minneapolis, MN.*)

Special mechanical seals, which provide very low leakage, can also be used. As with the other types of shaft seals, the seal material can be selected to stand up to corrosive environments. Often circulation oil or water is used to cool the seal on high-speed or high-temperature fans.

The previous two shaft seals may require special maintenance and care during installation. Because of the tight clearances involved, care should be taken to properly align the shaft and the surface that the seal is mounted to. Also ample access should be provided around the seal to allow for servicing.

5.7.2 Shaft coolers

Shaft coolers are used with shaft seals on high-temperature fans. They act as a heat sink that prevents heat from being conducted down the shaft and into the bearing. Fins on the cooler increase convective heat transfer on the cooler and cause air movement around the bearing that increases convective heat transfer on the bearing housing. If insulation is added to the fan housing, clearance should be left around the shaft cooler to allow for this air movement. A cooler guard should be added if there is the possibility that people will be near the fan.

5.7.3 Shaft and belt guards

If a fan is to be operated in an area where people are present, shaft and belt guards are an essential accessory to provide protection from moving parts. Because bearings and belt drives need occasional servicing, they should be designed for ease of removal or opening.

Tachometer holes should be located in belt guards so that the fan speed can be checked without removing the guard. Excessive heat shortens bearing and drive life, so guards should be provided with plenty of ventilation openings, especially on high-temperature fans.

5.7.4 Inlet and outlet screens

Screens are essential for fans that have open inlets or outlets whenever people have access to an operating fan. Inlet screens can also be used to keep large foreign objects from being drawn into the fan.

5.7.5 Inlet bells

All fans require an inlet duct or an inlet bell to deliver their rated flow. Some fans, such as panel fans and most backward-inclined fans, have an inlet bell as an integral part of their design. Fans that do not have this feature require inlet bells to achieve their catalog ratings.

5.7.6 Access doors

Access doors are used to inspect the condition of the inside of a fan, to gain access for cleaning, and for field balancing of the impeller. Several different styles are often available, as shown in Fig. 5.54, depending on the purpose of the door.

Bolted access doors are the simplest and least expensive. Since much hardware has to be removed, they should be used when infrequent access to the fan interior is required. Quick-release access doors should be used when frequent cleaning or inspection is required. These doors are designed to be quickly opened without the use of tools. Raised access doors are used on fans that have insulated housings. The door's mounting flange attaches to a flange that is raised several inches out from the fan housing.

To prevent the buildup of material around the inside of the door, access doors are often constructed with a double skin, so that the

Figure 5.54 Access panels. (*Courtesy of Twin City Fan and Blower Company, Minneapolis, MN.*)

inside of the door is flush with the inside of the housing. These doors are also used in the impeller area of axial fans to minimize turbulence in this critical area.

5.7.7 Drains

Housing drains are used on centrifugal fans with scroll housings. They should be used on all fans that handle moist or humid air and on fans that are located at the bases of stacks. If the fan is located outdoors, the drain can simply be a hole drilled in the bottom of the housing, or if the fan is located indoors, a pipe fitting can be attached to the outside of the housing to pipe the drained material away.

5.7.8 Outlet dampers

Outlet dampers are adjustable dampers that are installed on the outlets of fans to control the amount of airflow. Figure 5.55 shows fan curves for several damper blade positions with the damper considered to be part of the fan. If the damper blade position is the only thing in the duct system that changes, the point of operation will remain on a system line. Figure 5.55 shows a fan and system operating at points A_1 and A_2, 15,000 ft³/min [7.1 m³/s] at 4 in w.g. [993 Pa] static pressure and 13.6 hp [10.1 kW], respectively, with the damper blades 100 percent open. As the damper is closed to 75 percent open, the point of

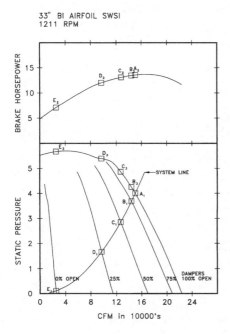

Figure 5.55 Fan performance curve with outlet damper—damper considered as a part of the fan. (*Courtesy of Twin City Fan and Blower Company, Minneapolis, MN.*)

operation moves to the intersection of the system line and the 75 percent open pressure-volume curve at point B_1, a flow rate of 14,400 ft³/min [6.8 m³/s] at 3.7 in w.g. [919 Pa] static pressure. The power stays on the horsepower curve and moves to point B_2, 13.5 hp [10.1 kW]. Likewise, as the damper is closed further to 50 percent open, the point of operation and power move to points C_1 and C_2. As the damper is closed further, the process continues until the point of operation moves to the 0 percent open curve at point E_1 and the power moves to point E_2. The flow rate at point E_1 will depend on how much leakage there is in the damper with the blades fully closed. Dampers can be modified with blade seals to move point E_1 as close to no flow as possible.

If the damper is considered to be a part of the duct system, the point of operation moves from point A_1 to point B_3 as the damper is closed to 75 percent open. The difference between points B_2 and B_3 is the pressure drop across the damper. Looking at the system this way, the point of operation moves to points C_3, D_3, and E_3 as the damper is closed to 50, 25, and 0 percent open, respectively. Points D_3 and E_3 are located to the left of the peak static pressure and may not be stable.

Outlet damper blades can be arranged so that the blades remain parallel as they close, as depicted in Fig. 5.56a, or they can oppose each other, as shown in Fig. 5.56b. Parallel blades tend to deflect the air to one side of the duct, which can cause problems if elbows or branch takeoffs are located just downstream of the damper. Opposed-blade dampers leave a more uniform velocity profile downstream of them and should be used if duct fittings are located in this area. They also have a more linear response of flow with respect to damper position.

5.7.9 Variable inlet vanes

Variable inlet vanes are also used to control the air volume flowing through fans. The blades on these dampers are arranged so that they

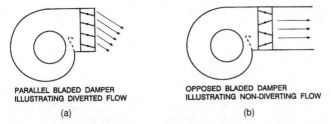

PARALLEL BLADED DAMPER
ILLUSTRATING DIVERTED FLOW

OPPOSED BLADED DAMPER
ILLUSTRATING NON-DIVERTING FLOW

(a)   (b)

Figure 5.56 Parallel- and opposed-blade dampers. (*Reprinted from AMCA publication 99-86,* Standards Handbook, *with the express written permission of the Air Movement and Control Association, Inc.*)

cause the air entering the fan to spin in the direction of impeller rotation as they are closed. With the spin in this direction the fan power is reduced. Figure 5.57 shows a fan and system operating at points A_1 and A_2, a flow rate of 15,000 ft^3/min [7.1 m^3/s] at 4 in w.g. [993 Pa] static pressure and 13.6 hp [10.1 kW], respectively, with the inlet vane blades 100 percent open. As the blades close to 75 percent open, the point of operation moves to the intersection of the system line and the 75 percent open pressure-volume curve at point B_1, a flow rate of 13,800 ft^3/min [6.5 m^3/s] at 3.4 in w.g. [844 Pa] static pressure. The power moves to the intersection of this flow rate and the 75 percent open power curve, which is plotted as point B_2, 11.4 hp [8.5 kW]. Likewise, as the vanes are closed further to 50 percent open, the point of operation and power move to points C_1 and C_2. As the vanes are closed further, the process continues until the point of operation moves to the 0 percent open curve at point E_1 and the power moves to point E_2. Because inlet vane blades do not seal as well as outlet damper blades, the leakage is greater. When the power curves on Figs. 5.55 and 5.57 are compared, the energy savings that are obtained with inlet vanes become obvious. The spin produced by inlet vanes stabilizes the flow at reduced vane settings, so that greater

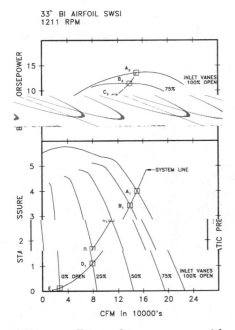

Figure 5.57 Fan performance curve with variable inlet vanes. (*Courtesy of Twin City Fan and Blower Company, Minneapolis, MN.*)

reductions in flow rates are possible with inlet vanes than with outlet dampers.

Variable inlet vanes can be manufactured in their own cylindrical housing which is then bolted to the fan inlet flange, or if the fan has an integral inlet bell, the vanes can be nested into the bell. Figure 5.58 shows the two most common types of inlet vanes. Nested vanes should not be used on systems that handle corrosive materials.

5.7.10   Inlet boxes

Inlet boxes are specially designed elbows that are mounted on fan inlets. A well-designed inlet box minimizes the system effect due to elbow placement at the fan inlet. Most fan manufacturers can provide performance data with an inlet box correction. If this data is not available, the system effect factor curves S or T on Fig. 5.31 can be used to estimate the correction required. Inlet boxes can be either a bolt-on accessory, or they can be made to be an integral part of the fan housing. The information in Fig. 5.49 should be used to specify the inlet box orientation.

The tendency of a parallel-bladed damper to turn the air provides an energy advantage when the parallel-bladed damper is mounted on the inlet of inlet boxes. If the blades are arranged so that they turn the air in the direction of wheel rotation, power savings similar to those obtained with inlet vanes are obtained as the damper is closed.

Figure 5.58 Nested and external variable inlet vanes. (*Courtesy of Twin City Fan and Blower Company, Minneapolis, MN.*)

5.7.11 Split housings

Fan housings can be split to allow the removal of the fan impeller, shaft, and bearings as a unit without the removal of any ductwork. The split can be made along the fan centerline or as a pie-shaped segment. Split housings are a useful feature in areas where ductwork removal would be difficult or where there is no room to remove the impeller through the fan inlet.

5.7.12 Swing-out construction

Fans that handle dirty air and have ducted inlets and outlets frequently are constructed so that the impeller, shaft, bearings, and motor are mounted on a door. This allows the impeller to swing out of the housing for easy cleaning without ductwork having to be removed. Figure 5.59 is a photograph of a backward-inclined fan with swing-out construction. Typically, axial, backward-inclined, and tubular centrifugal fans can be built with this feature.

5.8 Fan Installation and Maintenance

5.8.1 Fan mounting

The best location for mounting a fan is on a concrete pad that has a plan area no more than twice that of the fan's and weighs at least three times as much as the fan. The small plan area ensures that the pad is rigid, and the weight provides inertia to keep vibration levels low. Foundations with larger areas should have a correspondingly larger mass. Secure the fan using L- or T-shaped anchor bolts placed in a pipe sleeve that allows for adjustment. If there are spaces between the fan and the foundation near the mounting holes, fill the

Figure 5.59 Swing-out construction. (*Courtesy of Twin City Fan and Blower Company, Minneapolis, MN.*)

space with shim material. Do not draw the fan down to the foundation as this will distort the fan structure and lead to problems such as vibration or the impeller or shaft rubbing against the housing. Often grout is used to distribute the load and fill any spaces between the fan and the foundation. In many cases, it is not possible or practical to mount the fan on the ground, which requires supporting the fan by structural steel. In these instances it is important to design the structure for rotating equipment. Static design is often not adequate. The structure must be designed so that any frequencies that cause structural resonance are less than 80 percent or greater than 120 percent of the fan operating speed.

Determining all the resonant frequencies of a structure usually is not practical. In these cases, it may be necessary to mount the fan on vibration isolators to isolate the fan vibration from the structure and reduce the likelihood of damaging resonance. The best method for isolating the fan vibration is by placing the fan and motor on an inertia base. An inertia base is a steel-reinforced concrete pad that floats on spring steel vibration isolators. Inertia bases are recommended for large fans with motor powers of 75 hp [56 kW] or higher. Large direct-driven fans may require inertia bases to maintain proper coupling alignment.

The next best method of fan mounting is on a structural steel vibration base. Vibration bases are designed to avoid resonance with the fan and motor and yet be strong enough to support the weight and drive loads. Small, light fans sometimes may be mounted directly on spring or rubber vibration isolators. If there is any doubt whether this is acceptable, check with the manufacturer to see if the fan structure is stiff enough for this type of mounting.

5.8.2   Duct connections

To avoid fan vibrations from causing resonance in the ductwork, use flexible duct connections at the fan inlet and outlet whenever possible. On high-temperature systems, these connections need enough flexibility to allow for thermal expansion of the ductwork. Most fans are not designed to carry duct loads. All ductwork connected to a fan should be independently supported. Avoid duct configurations that cause a system effect on performance whenever possible. If it is unavoidable, use the procedures outlined earlier to account for the changes in fan performance.

5.8.3   Pre-start-up checklist

Before starting any fan for the first time, perform all the procedures in the manufacturer's installation. If a pre-start-up checklist is not available, use the following to make sure a fan is ready to start.

- [ ] All nuts, bolts, and set screws are tight.
- [ ] All system connections are properly made and tightened.
- [ ] All access doors are secure.
- [ ] All bearings are properly lubricated. All set-screw–mounted bearings have tight set screws.
- [ ] The fan impeller, fan housing, and ductwork are clean and free of debris.
- [ ] All rotating components turn freely.
- [ ] The impeller is located properly in the fan housing.
- [ ] If used, the belt drive is properly aligned and tightened.
- [ ] If used, the coupling is properly aligned.
- [ ] All safety guards are in place and securely mounted.
- [ ] The motor is properly wired and grounded, and all lead wires are properly insulated.
- [ ] A trial *bump* is performed, i.e., the power is turned on just long enough to start the assembly rotating.
- [ ] The impeller is rotating in the proper direction.
- [ ] There are no unusual noises such as an impeller rubbing against the housing.
- [ ] The unit runs up to speed.
- [ ] There is no excessive vibration.
- [ ] Bearing temperatures are monitored during initial operation.
- [ ] After a week of operation, all nuts, bolts, and set screws are checked and tightened if necessary. The belt tension is checked and, if necessary, adjusted.

Antifriction bearings are often packed full of grease at the factory to provide protection from corrosion during shipment and storage. After initial start-up, the bearings will purge the excess grease through their seals and may run hot until the excess grease is purged. If split pillow block bearings are used, the excess grease can be removed by opening the bearing caps and removing grease until the housing is one-third full.

5.8.4 Maintenance

To ensure troublefree operation and long life, regularly check fan components. Fan bearings, in particular, require regular attention.

Belt tension, drive alignment, and coupling alignment may need occasional adjustment. Periodically inspect the shaft and impeller for signs of dirt buildup, corrosion, wear, or fatigue. Clean and repaint components when appropriate.

The relubrication schedule should be initially performed as recommended by the manufacturer. The lubrication frequency may need to be adjusted due to factors such as the cleanliness of the fan area, temperature, and humidity. With grease lubrication, the best indication whether the lubrication interval is adequate is to check the condition of the grease purged from the bearings when adding new grease. If the grease is darkened and oxidized, increase the relubrication frequency. When adding grease, rotate the shaft to avoid damaging the seals. Avoid mixing lubricants made from different bases, and use lubricants that are recommended by the fan or bearing manufacturer.

Excessive vibration will damage the bearings and fan structure. Regularly clean fans that handle dirty air to prevent the buildup of material that may cause unbalance. Use the chart in Fig. 5.60 to evaluate the severity of vibrations. Correct vibration levels that exceed the slightly rough region by having the impeller balanced.

GENERAL MACHINERY
VIBRATION SEVERITY CHART

For use as a GUIDE in judging vibration as a warning of impending trouble.

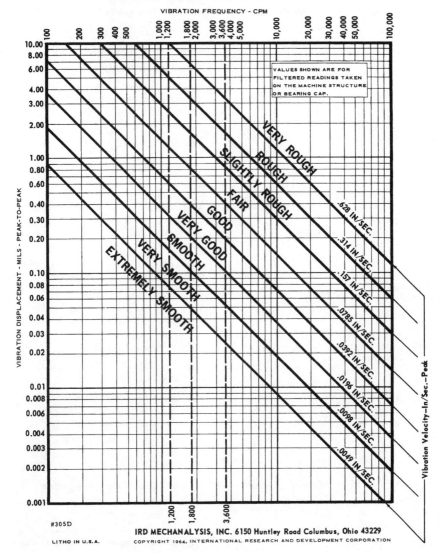

Figure 5.60 Vibration severity chart. (*Reprinted with permission of IRD Mechanalysis, Inc., 6150 Huntley Road, Columbus, OH.*)

Fan and Blower Troubleshooting Guide

Problem	Cause	Possible solution
Airflow rate low, pressure high	System resistance to flow too high	Check to make sure all dampers or inlet vanes in the system are open. Check for foreign material in the ductwork, hoods, or screens. Is the duct transport velocity high enough?
		Check for plugged filters.
		Check to make sure the system was installed as designed. Were extra nonessential duct fittings or ductwork added?
		Increase the fan speed. Use the fan laws to calculate the new speed and power requirements [Eqs. (5.5) to (5.7)].
Airflow rate low and fluctuating, pressure high or low and fluctuating	System resistance to flow too high—fan in stall	Check for the cause of the high system resistance per the above (changing fan speed usually will not solve stall problems).
		It may be necessary to use a differently designed impeller—check with the fan manufacturer.
		Do not operate axial fans under these conditions for extended periods.
Airflow rate low, pressure low	Fan speed too low	Increase the fan speed.
	System effects	Redesign the ductwork to eliminate system effects.
		Add flow straighteners, turning vanes, or straight sections of duct to minimize the system effect.
		Increase the fan speed. Use the fan laws to calculate the new speed and power requirement [Eqs. (5.5) to (5.7)].
	Centrifugal fan impeller rotating in the wrong direction	Change the rotation to the proper direction.
	Axial fan blade angle too low	If the impeller is an adjustable type, adjust the blades to a higher angle of attack. Keep all the blades set to the same angle.
		If the angle is not adjustable, replace the impeller with one with a higher angle.
		Increase the fan speed.

Fan and Blower Troubleshooting Guide (*Continued*)

Problem	Cause	Possible solution
Airflow rate low, pressure low	Impeller improperly located in the fan housing.	Check with the fan manufacturer for the proper impeller location. Adjust as required.
	Actual air density lower than design air density	Select a new fan speed based on the actual air density.
Airflow rate high, pressure low	System resistance to flow too low	Check for access or cleanout doors that are left open.
		Check for missing or broken filters.
		If the filters are new, the flow may be correct once the filters become seasoned, but the filters may become blinded if the flow rate is excessive.
		Check for improperly adjusted control dampers or inlet vanes.
		Reduce the fan speed. Use the fan laws to calculate the new speed and power requirements [Eqs. (5.5) to (5.7)].
		Repair leaks in the duct system.
Airflow rate high, pressure high	Fan speed too high	Reduce the fan speed.
	Actual air density higher than design air density	Select a new fan speed based on the actual air density.
Airflow in the wrong direction	Axial fan impeller rotating in the wrong direction	Change the rotation to the proper direction.
Power consumption high	Fan speed too high	Reduce the fan speed.
	System resistance too low—centrifugal fan impeller with radial or forward-curved blades	Increase the resistance to flow or reduce the fan speed.
	Inlet duct configuration causing the air entering the fan to spin in the direction opposite of impeller rotation	Redesign the ductwork to eliminate system effects.
		Add flow straighteners, turning vanes, or straight sections of duct to minimize the system effect.
	Actual air density higher than design air density	Select a new fan speed based on the actual air density.

Fan and Blower Troubleshooting Guide (*Continued*)

Problem	Cause	Possible solution
Motor overload protection trips before the fan reaches full speed	Inertia (WR^2) of the impeller at or above the motor's inertia load capacity.	Use time delay or slow trip motor protection. It is not uncommon for fans with across the line starts to take 20 s to get to full speed.
		Use a soft starting method (autotransformer, wye-delta, etc.) to bring the motor up to speed more slowly.
		If a variable-frequency drive is being used, adjust the drive to a longer acceleration time.
		Use a motor with more horsepower (higher WR^2 capacity).
Excessive vibration levels	Fan not adequately supported	Make sure all bearing bolts, motor bolts, and fan mounting bolts are tight.
		Make sure the structure supporting the fan is rigid.
		Stiffen the support structure if necessary.
		Check for cracked welds on the fan and its support.
	Belt drive worn or improperly mounted	Check the belts and sheaves for wear. Replace them if necessary.
		Make sure the sheave mounting bolts are tight.
		Check the alignment (angular and parallel) of the two sheaves.
		Tension the belts properly.
	Impeller out of balance	Clean off the buildup of foreign materials.
		Check for cracked welds. Repair the welds or replace the impeller.
		Check for excessive wear or other damage on the impeller. Replace the impeller if necessary.
		Balance the impeller.
	Out of balance sheaves	Balance the sheaves.
	Bent fan or motor shaft	Straighten or replace the shaft.
	Worn or misaligned coupling	Realign the motor and fan shafts.
		Replace the coupling.

Fan and Blower Troubleshooting Guide (*Continued*)

Problem	Cause	Possible solution
Excessive vibration levels	Worn or failed bearings	Replace the bearings.
Excessive noise	Impeller or shaft rubbing against the fan housing	Adjust the position of the impeller in the housing by moving or shimming the bearings.
		Adjust the position of the inlet.
	Defective bearings	Replace the bearings.
	Squealing belts	Tension the belts properly.
		Replace the belts if they are worn or glazed.
		Check the drive alignment.
		Adjust multigroove sheave so that all the grooves have the same pitch diameter.
	Turbulent or nonuniform airflow entering the fan	Redesign the ductwork to eliminate system effects.
		Add flow straighteners, turning vanes, or straight sections of duct to minimize the system effect.
		Add a silencer to the duct system.

Bibliography

Aberbach, R. J., "Fans—A Special Report," *Power,* March 1968, p. 2.

AMCA publication 99-86, *Standards Handbook,* Air Movement and Control Association, Inc., 30 West University Drive, Arlington Heights, IL 60004-1893, Phone: 708-394-0150, Fax: 708-253-0088, Publications: 708-394-0404, 1986.

AMCA publication 200, *Fans and Systems,* Air Movement and Control Association, Inc., 30 West University Drive, Arlington Heights, IL 60004-1893, Phone: 708-394-0150, Fax: 708-253-0088, Publications: 708-394-0404, 1987.

AMCA publication 201-90, *Fans and Systems,* Air Movement and Control Association, Inc., 30 West University Drive, Arlington Heights, IL 60004-1893, Phone: 708-394-0150, Fax: 708-253-0088, Publications: 708-394-0404, 1990.

AMCA standard 210-85, ANSI/ASHRAE Standard 51-1985, Laboratory Methods of Testing Fans for Rating, Air Movement and Control Association, Inc., 30 West University Drive, Arlington Heights, IL 60004-1893, Phone: 708-394-0150, Fax: 708-253-0088, Publications: 708-394-0404, 1985.

AMCA standard 300-85, Reveberant Room Method for Sound Testing Fans, Air Movement and Control Association, Inc., 30 West University Drive, Arlington Heights, IL 60004-1893, Phone: 708-394-0150, Fax: 708-253-0088, Publications: 708-394-0404, 1985.

AMCA publication 303-79, *Application of Sound Power Level Ratings,* Air Movement and Control Association, Inc., 30 West University Drive, Arlington Heights, IL 60004-1893, Phone: 708-394-0150, Fax: 708-253-0088, Publications: 708-394-0404, 1979.

American Society of Heating, Refrigerating, and Air-Conditioning Engineers, *ASHRAE Handbook—Systems and Equipment,* Chap. 18, "Fans," Atlanta, GA, 1992.

American-Standard Power and Controls Group, *Installation, Operation and Maintenance Instructions: American Blower—Heavy Duty Centrifugal Fans,* Dearborn, MI.

Champion Blower and Forge, Inc., *Installation, Operation, and Maintenance Instructions,* Roselle, IL.

Hirschorn, M., "Compendium of Noise Control Engineering—Part 2," *Sound and Vibration,* February 1988, p. 16.

Jorgensen, R. (ed.), *Fan Engineering,* 8th ed., Buffalo Forge Company, Buffalo, NY, 1983.

M-D Pneumatics Division, Tuthill Corporation, *Maintenance Manual—Models 4000 and 5500,* Springfield, MO, 1993.

Meyers, R. C. "Industrial Fans—Guidelines for a Successful Installation," *Iron and Steel Engineer,* October 1976, p. 38.

Peterson, A. P. G., and E. E. Gross, Jr., *Handbook of Noise Measurement,* 7th ed., General Radio Company, Concord, MA, 1972.

Rogers, A. N., "Proper Location of Fans," *Air Conditioning and Heating,* August 1955, p. 83.

Roppenecker, W. F., "New Options in Choosing Blowers," *Machine Design,* March 23, 1978, p. 56.

Twin City Fan and Blower Company, *Installation, Operation and Maintenance Manual,* Minneapolis, MN, 1993.

6

Construction and Standards

Ronald L. Jorgenson

Donaldson Company, Inc.
Minneapolis, MN

6.1 Industry Standards

From the choice of raw materials; through design, fabrication, and installation; to the final inspection at the installation job site, every choice and every decision carries the burden of complying with a wide array of codes, suggested practices, and laws. The list below represents the authorities most commonly cited for compliance.

ACI	American Concrete Institute
AFBMA	Anti-Friction Bearing Manufacturers Association
AGMA	American Gear Manufacturers Association
AISC	American Institute of Steel Construction
AISI	American Iron and Steel Institute
AMCA	Air Moving and Conditioning Association
ANSI	American National Standards Institute
ASHRAE	American Society of Heating, Refrigeration, and Air Conditioning Engineers
ASME	American Society of Mechanical Engineers

ASTM	American Society of Testing and Materials
AWS	American Welding Society
AWWA	American Water Works Association
CEMA	Conveyor Equipment Manufacturer's Association
CRSI	Concrete Reinforcing Steel Institute
FMEC	Factory Mutual Engineering Corporation
IEEE	Institute of Electrical and Electronics Engineers
IGCI	Industrial Gas Cleaning Institute
IPCEA	Insulated Power Cable Engineers Association
ISA	Instrument Society of America
JIC	Joint Industrial Conference
MPTA	Mechanical Power Transmission Association
MSS	Manufacturers Standardization Society
NAANN	National Association of Architectural Metal Manufacturers
NACE	National Association of Corrosion Engineers
NCMA	National Concrete Masonry Association
NEC	National Electrical Code
NEMA	National Electrical Manufacturer's Association
NFPA	National Fire Protection Association
NPTA	National Power Transmission Association
OSHA	Occupational Safety and Health Administration
RMA	Rubber Manufacturers Association
SMACNA	Sheet Metal and Air Conditioning Contractors National Association
UBC	Uniform Building Code
UL	Underwriters Laboratories

In addition to the list above, there are numerous state and local building codes and requests to certify compliance with regulations concerning hiring practices, nonuse of restricted materials, and manufacturing techniques.

6.2   Materials of Construction

Carbon steel is by far the most common material used in the construction of pollution control system devices. Hoods; ducting; blowers; and control devices such as dust collectors, inertial separators, and electrostatic precipitators all routinely utilize carbon steel. The relatively low material costs, ease of fabrication, physical strength, and availability of numerous structural member shapes all contribute to the popularity of this material. To a lesser extent, galvanized steels, stainless steel, fiber-reinforced plastics, and aluminum are part of the family of materials used in control systems.

6.3   Hoods, Ducting, and Control Devices

The most common industrial pollution control system fabrication material is hot rolled steel. This material will handle system temperatures up to 650°F [343°C]. All carbon steels have poor corrosion resistance, especially in the presence of salts and acids. They resist corrosion due to alkali atmospheres at moderate temperatures.

Galvanized steel and the most common aluminum ducting both have temperature limitations of 400°F [204°C]. Galvanized construction generally provides protection from normal atmospheric corrosion; however, it will not provide adequate protection for corrosive airstreams. Aluminum provides corrosion protection from most organic acids and many salts but has no protection from hydrochloric or sulfuric acids and heavy metal salts. Corrosion characteristics are somewhat alloy-specific.

Stainless-steel construction provides corrosion resistance and extended life for high-temperature systems, although at a significantly higher cost than carbon steel. Polyvinyl chloride and fiber-reinforced plastics provide protection from the effects of corrosive airstreams. However, plastic ducting creates fire and explosion risks in dry collection systems due to the static electricity generated by the airstream and particulate. These materials also fare poorly when exposed to abrasive applications.

6.4   Nonstandard Materials of Construction

6.4.1   Corrosion

A corrosive atmosphere or product requires the use of nonstandard materials of construction and, at times, unusual designs and fabrication techniques. To ensure success, the buyer and the manufacturer of the pollution control system must cooperate, each supplying the other with the accumulated knowledge and historical information detailing what has succeeded and failed in the past.

For example, sodium azide, a propellant for automotive air bags, reacts with a wide variety of materials, including copper, bronze, brass, mercury, lead, silver, gold, and some organic materials. The chemical reaction creates explosive compounds. In this case, the customer's knowledge of the product collected in the system will be far greater than the level of knowledge expected of the manufacturer of the control system. This intimate knowledge must be combined with the manufacturer's knowledge of the materials of construction. In this case, the manufacturer must certify that the components will not contain any materials that would create a hazard for the user.

The complex matrix of conditions, including temperature, humidity, airstream gaseous components, solids in the airstream, ambient conditions, and other unknowns, make it impossible to suggest specific materials of construction within this text. Instead, the intent is to provide general information to begin your search.

In general, austenitic stainless steels such as 304 and 316 are the most common stainless-steel alloys used in fabrication of dry dust collectors. They provide high resistance to atmospheric corrosion. The low-carbon grades, 304L and 316L, have less tendency to corrode in the weld zones. There are many specialty stainless steels with corrosion resistance to chlorides, acids, salt, marine atmospheres, and very high temperatures.

The aluminum alloys provide good corrosion resistance to most organic acids and many salts. Many alloys resist corrosion resulting from industrial and marine atmospheres.

Fiber-reinforced plastics (FRPs) can provide corrosion resistance in pollution control systems. Wet scrubbers and occasionally a baghouse may utilize this material. Polyester resins provide good resistance to both dilute and strong acids and weak alkalis, with less resistance to strong alkalis. Epoxies provide good resistance to alkalis and somewhat less resistance to acids.

Titanium provides very good resistance to most acids except hydrofluoric and nitric. It also exhibits good resistance to alkalis and organic salts.

The nickel-iron-chromium alloys are good choices for systems exposed to most acids (not hydrochloric and hydrofluoric), alkalis, and salts and provide protection from chloride ion stress cracking. Other nickel alloys allow high-temperature operations in both oxidizing and reducing atmospheres.

The costs of the exotic alloys make fabrication of complete systems or even portions of the system expensive. When the cost cannot be justified, coatings or lining materials may be the answer.

The choices of coatings can include elastomeric coatings such as polyurethane, fluorocarbon, isobutylene isoprene, polysulfide, nitrile, silicone, chlorosulfonated polyethylene, and coatings such as latex rubber and epoxy. Fused glass, ceramics, and acid or fire brick linings can also be utilized. In addition, sheet rubber can be cemented or vulcanized to metal substrates.

If you expect a corrosion problem and do not have a successful application on which to base your buying decisions, obtain professional advice. The cost of a guess can be short life, extensive and expensive repairs, and unwanted downtime. If you cannot obtain definitive advice, test samples of the suggested material under conditions as near to the intended operating conditions as possible.

6.4.2 Abrasion

Abrasion-resistant materials include steels specifically formulated to give added life in abrasive situations. Coatings and linings provide abrasion protection for substrates. This list includes weld-deposited hard-facing alloys, refractories, ceramic coatings, carbides, and a wide variety of flame spray coatings.

On the opposite end of the hardness scale are resilient coatings that also provide excellent protection from abrasion. The most common of these is polyurethane. Others include rubber, chloroprene, chlorosulfonated polyethylene, and some epoxies. Sometimes combinations of materials give the desired results. An example would be cast polyurethane loaded with ceramic beads to prevent cutting of the soft polyurethane.

6.4.3 Sanitation

Stainless steel is the most commonly used material in systems that require sanitary practices. It is pleasing in appearance, resistant to most cleaning chemicals, and does not lose the desired characteristics over time. Other choices include chromium plating, electroless nickel plating, and epoxy coatings. A sanitary system may require special gaskets and sealant materials such as white neoprene or silicone.

6.5 Fabrication Techniques

6.5.1 Ductwork

SMACNA provides excellent design information for ductwork. Their guide, *Round Industrial Duct Construction Standards,*[1] provides information on material gauge specification for four classes of service relating to the volume and abrasiveness of the product transported. The guide range is from diameters of 4 to 60 in [100 to 1500 mm], for both steel and aluminum, with static pressure ratings from -0.07 to 1.1 lb/in^2 [0.005 to 0.08 kg/cm^2]. Fabricators produce several types of ducting and a number of different joining systems, each with differing installation methods.

Simple rolled and seam-welded tubing as well as spiral-wound strip, seam welded into a cylindrical duct, are both commonly used in industrial pollution control ducting systems. Rectangular ducting requires angle or bar reinforcement to take the place of the natural hoop strength of round ducting.

Fabrication of the ductwork system revolves in great part around the joining method chosen. The choices include butt-welded joints; lap joints; flanged joints; sleeve joints; and O-ring–sealed, quick-disconnect–style joints.

Ducting that is simply butt welded has a lower purchase price but higher labor costs for installation. This style of ducting has great flexibility; the parts can be easily cut to the proper length, fitted in place, and welded. Since duct gauges are typically light, welding the duct while it is suspended in place can be challenging.

A second common method of duct installation uses angle iron or bar flanges, gaskets, and bolts to join sections of ducting. The flanges are either welded to the duct, or the end of the duct has a rolled or formed lip that falls between the two mating flanges. Flanged ducting costs more than plain duct and must be fabricated to a set of scaled plans but requires little skill to assemble.

Lap joints, the stovepipe joints, have one end crimped to allow it to enter into the noncrimped end of the next length. Manufacturers provide lap joint ducting in sizes up to at least 15 in [350 mm]. The joint requires sheet-metal screws or rivets for mechanical strength and duct tape or caulking to seal the joints. This ducting style appears most frequently in systems installed in small shops, do-it-yourself installations, or temporary installations, handling small quantities of materials with low abrasion characteristics. A common application is small woodworking equipment in small- to moderate-size shops. In applications where equipment moves frequently, this ducting easily comes apart and can be reconfigured and reused in the next setup or installation.

The male end of the duct sections and fittings must always face downstream, away from the hood or pickup point. Installing the joints backwards causes disruption of flow, greater pressure drop, and provides a place for solids to collect. This can cause recurring plugging problems when handling chips, lint, and fluffy or stringy solids. Sleeve joints, also called draw band joints, use a nonwelded butt joint, fastened with a bolted metal wrapper, designed to tightly encircle and clamp onto the ducting.

A quick-disconnect–style duct provides the user with the flexibility to install, remove, and reuse ducting components. One version of the quick-disconnect ducting has a rolled bead at each end of the duct sections. A circular band with a "U" cross section wide enough to span both beads, lined with a foam tape sealant and equipped with an over-center latch, holds the two beads tightly together. The foam tape seals the circumference when the clamp tightens down on the beads. The initial cost is higher than that for plain duct; however, installation is very quick and the duct can be disassembled and kept for reuse or reconfigured and reassembled to meet changing needs.

Poor construction techniques include lap joints that have the raw edge of the ducting, fitting, or crimped end facing into the airflow. This causes airflow disturbance and can cause plugging when handling fibrous materials. The use of excess amounts of sealant, sheet-metal screws, and rivets also provides projections into the airstream. They can be the nucleus for material buildup and eventual plugging problems when handling fibrous or stringy materials. Excess turns and small-radius fittings use up static pressure, lowering the airflow in the system and increasing operating costs. Do not ignore system effects. Use the proper inlet design for the system blower, avoid elbows close to the blower inlet, use an outlet duct, and use the recommended no-loss blower discharge design.

6.5.2 Control device fabrication

The general fabrication techniques used in the construction of control devices include all-welded construction, partial assembly, panel construction, and site built. All-welded construction means the control device primarily relies on welded joints and the device arrives at the job site structurally complete including supporting legs when required, ready to set in place on its foundation. Once set in place, completion of the installation requires only attachment of inlet and outlet ducting, ladders, platforms, railings, and other accessories.

Partial assembly construction typically means the control device arrives at the job site in major subassemblies. The size of the subassemblies may either be dictated by shipping considerations or to provide assembly joints at logical points in the structure. Examples would be assembly joints between the body and hopper sections or shipping the leg assembly separately from the body section. This is probably the most common method of fabrication. The customer must do some assembly, but this is offset by lower shipping costs due to better use of truck bed space and by less likelihood of transportation permit expense due to oversize loads.

Although not very common, some control systems are manufactured such that all assembly relies on bolted joints. This is rarely used except when intended for international shipment, conserving shipping space and lowering the shipping expense. This style typically requires some preassembly by the vendor or technical support at the job site to ensure proper assembly and function.

Site-built construction is usually reserved for only the largest control devices. These are the baghouses, scrubbers, and electrostatic precipitators used to control emissions from power plants and similar very large airflow systems. Devices in this group are so large that they cannot be shipped by truck or rail. Even in this product group, critical subassemblies arrive prefabricated or in modular form when possible.

6.6   Explosive and Flammable Application Design

6.6.1   Introduction

Improper design or poor installation of equipment intended to handle highly flammable materials in hazardous environments can have grave consequences resulting in fire or deflagrations. This is an area that demands the expertise of designers, engineers, and manufacturers with the experience and knowledge to provide a safe, effective system.

The ideal design eliminates all ignition sources; operates the process either well below or above the flammable limits; or provides an inert atmosphere of nitrogen, carbon dioxide, steam, or possibly cleaned flue gas. However, you must assume the failure of some component or process will put the system at risk. The prudent designer protects against that failure with detection devices and extinguishing or suppression systems.

Prior to discussions of equipment design and application in hazardous atmospheres, it is important to learn the language, definitions, and standards that apply. Two excellent sources of information are the NFPA and the FMEC.

The NFPA publishes a wide variety of codes, standards, and training and educational materials pertaining to fires, explosions, hazardous materials, and the National Electrical Code.[2] NFPA 70, the National Electrical Code, article 500, Hazardous (Classified) Locations, provides definitions of three classes of atmospheres. Class I pertains to ignitable gases or vapors, class II covers combustible dusts, and class III pertains to ignitable fibers. Article 500 also subdivides the classes into divisions based on the probability of a flammable or explosive mixture being present. Division-1 applications assume that the flammable or explosive material is present. Division 2 deals with applications in which the hazard exists only when a fault condition occurs. NFPA 497M, Classification of Gases, Vapors, and Dusts for Electrical Equipment in Hazardous Locations[3] provides additional classification information.

There are so many NFPA codes specific to products, contaminants, and classes of industry that they are simply too numerous to list here. However, there are several with information pertinent to many pollution control applications.

NFPA 68: Guide for Venting of Deflagrations[4]

NFPA 69: Explosion Prevention Systems[5]

NFPA 77: Static Electricity[6]

NFPA 91: Exhaust Systems for Air Conveying of Materials[7]

NFPA 650: Pneumatic Conveying Systems for Handling Combustible Materials[8]

6.6.2 Component selection

The safety of the facility and personnel requires the use of care and knowledge in the choice of equipment intended for collection of explosive or flammable products. Electrical control and electrically operated components utilized in the system must meet the requirements of the National Electrical Code,[2] articles 500 and 501, 502, or 503. This reference applies to wiring, wireways, fittings, boxes and enclosures, switches, circuit breakers, motor starters, fuses, motors, and all other electrically powered devices or those devices used to distribute or control electric power. Be sure you properly identify the application as to the appropriate class and division. Improper identification may result in a catastrophic loss of property, equipment, and/or life.

Enclosures, motors, and other equipment intended for use in hazardous locations will cost significantly more than similar devices rated for nonhazardous service. Good economic design includes locating the system in a nonhazardous area when possible. Even locating the electrical control panel in a remote, low-hazard location would likely significantly reduce the capital expense.

The collector or control device you choose should be based primarily on meeting the filtration requirements of the system. Usually, efficiency and loading ranges overlap sufficiently to allow you to consider more than one type of system. Your choice will then reflect the compromise of weighing the costs against the varying risk potentials.

Cyclones present a very low risk of fire or deflagration. They have no moving parts and typically do not utilize combustible materials in their construction. Materials collected in storage containers can burn, ignited by burning materials carried into the system or by spontaneous combustion. If the system collects flammable materials, provide deflagration venting and either locate the equipment outdoors or provide properly designed ducting between the vent and the outdoors.

Wet scrubbers also present little risk when operated properly. The water level must cover any collected sludge, and the sludge should be removed frequently. Reactive metals may generate hydrogen when finely divided and immersed in water. The enclosure should provide for dissipation of this highly flammable gas, and work procedures should require the removal of all collected sludge on a daily basis. If the scrubber has a plastic liner or contains packing materials that might burn, you should provide appropriate fire protection for the system.

Electrostatic precipitators can collect flammable solid materials, although they are not recommended if the concentration would approach or reach the lower explosive limit. The risk of an electric arc precludes their use in applications with explosive or flammable gases or vapors because of the very low energy levels required to ignite flammable gases. Wet precipitators lower the fire risk since the collected material will be wetted prior to discharge to the hopper. Precipitators can be fitted with sprinkler systems if properly interlocked to disconnect the electrical supply if the sprinkler system activates.

Media collectors pose a greater fire and explosion risk since they collect the flammable dust and store a portion on the filters. The periodic cleaning of the filters, by pulsing or other method, may generate a cloud of dust that exceeds the lower explosive or flammable limit during the cleaning cycle. The filter media is frequently combustible as well. However, when collecting ignitable dusts, even nonflammable medias will rarely survive a fire that ignites the dust collected on the filter material.

Although the media choice will not likely affect the bag survival chance in a fire, choosing a media that dissipates static electricity may prevent a fire. Most of the synthetic fibers are not conductive and may store sufficient static charge to cause arcing. Medias containing approximately 5 percent stainless-steel fibers provide continuous dissipation of the charge.

Three ingredients are required for most explosions or deflagrations to occur. These are

1. Fuel

2. Oxygen

3. Ignition source

Usually, if any one of these items is not present, an explosion cannot occur. Depending on the severity of an explosion, should it occur, the system designer may wish to eliminate the possibility of occurrence for any one or all of these items in addition to designing the system to withstand an explosion.

6.6.3 Deflagration vents

Deflagrations, more commonly called explosions, generate large volumes of high-temperature gases. In a confined space, such as a dust collector body, these products of combustion generate a very rapid pressure rise. Explosion vents are mechanical devices designed to release the pressure, preventing the catastrophic failure of the enclosure. NFPA 68, Guide for Venting of Deflagrations,[4] provides very detailed information on choosing and sizing explosion vents.

Keep in mind that explosion vents do not prevent explosions; they only relieve the pressure from the enclosed vessel. The vent prevents fragmentation or rupture of the vessel but otherwise does little to protect the facility or personnel. The escaping gases typically will take the form of a fireball, expelling significant amounts of burning materials and possibly shrapnel. If this fireball is not discharged to the outdoors, in a safe direction, a secondary fire or explosion may result. If the explosion vent discharges inside the building, the shock wave may produce a dust cloud from materials that collect on ledges, lighting fixtures, and other surfaces. If that dust cloud ignites, the building may not survive the overpressure and shock wave.

All dust collectors or other pollution control devices that handle flammable dusts and/or gases must be protected by an appropriately chosen sprinkler system, fire- or explosion-suppression system, explosion venting, or a combination of these. These protective devices and systems must be routinely inspected, tested, and maintained. Too often after a few years, the explosion vent leaks and is replaced with a piece of steel plate or plywood, the sprinkler system freezes and the valves are closed and the pipes drained, and the annual inspection and recharging of the suppression system costs too much and so it is not done. The most common reason is that operator attitudes become lax if they have not witnessed an explosion in their operation. In many operations, the probability of an explosion is very low, but the consequence if an explosion occurs may be very high especially if proper precautions have not been taken. Another common problem is an explosion vent that was never ducted to the outside of the building, although that was the original intention.

Since the explosion vent on the control device only protects that device, not people, additional protection is usually required. Far too often, installations lack provisions to prevent the explosion or fire from propagating back into the workplace through the ductwork and ignore the venting requirements for the ductwork volume.

After providing the appropriately sized explosion venting for the ducting, the choices to protect the remainder of the system include fast-acting fire or explosion dampers or valves that physically close off the ducting, suppression systems designed to quench the fireball, and diverter or abort dampers that route the return air in a safe direction rather than returning it to the building. Activation of these devices can be triggered by spark detection, infrared flame detectors, temperature detectors, pressure sensors, or combinations of these devices.

6.6.4 Static electricity

Static electricity can add an additional risk of fire and explosion when collecting dry dusts. The greatest construction design, or material-of-construction–related risk, appears to come from the use of plastic pipe or other nonconductive construction materials for ducting or control devices. High-velocity airstreams passing through plastic pipe can generate dangerously high voltages. This energy level can ignite dust clouds and flammable gas or vapor mixtures. NFPA 77, Static Electricity,[6] provides very explicit information on generation, accumulation, and dissipation of static electricity.

In addition to nonconductive ducting, there are other design and construction issues to address. Collection system components that come in contact with particulate should be fabricated from conductive materials and not coated with high-dielectric-strength surface finishes on the inside. Avoid the high-speed impact of nonconductive dusts with nonconductive surfaces. Material handling containers used for collection of low-conductivity solids should either be conductive or, if you are using flexible bulk containers, a fabric with a low dielectric strength should be chosen.

6.6.5 Grounding requirements

Ducting, control devices, blowers, rotary air locks, double dump valves, screw conveyors, pneumatic conveying lines, and similar devices used in any system that handles flammable materials should be positively grounded. Bonding or grounding of all components will not prevent the generation of static electricity; it only provides a safe path to allow the charge to leak or bleed away.

The National Electrical Code requires that all metallic non–current-carrying parts of an electrical system and all metal items attached to the electrical system must be grounded. However, do not confuse the electrical grounding system with equipment grounding. Use grounding lugs and suitable ground wires across flex joints, gaskets, or other nonconductive materials; from body section to body section; from body to support structure; and from support structure to earth. The solids collection container(s) should also have a positive ground to the collector if separated by a nonconductive flexible connector section. Ground the system in multiple places to properly installed earthing rods, underground metal pipe, or the structure of a metal building. Do not rely on conduit, flexible conduit, and similar materials to maintain the ground; use copper wire.

Media filtration systems provide ground wires sewn into the bag seam and grounded to the metal shell of the collector. The manufacturers of needle-punched felts containing 3 to 5 percent stainless-steel fiber claim much more effective static bleed-off than bags with only a ground wire. The wire cages that support the filter tubes should be solidly grounded to the housing and should not have high-dielectric-strength coatings. Pleated media filter cartridges with antistatic-treated media, conductive potting compounds, and special grounding methods provide similar protection.

6.6.6 Vessel and duct strength

A dust collector used for testing fire detection, where deflagration detection and suppression systems are also used to demonstrate the difference between suppressed and nonsuppressed deflagrations, has withstood more than 100 deliberate explosions. This collector did not have any extraordinary reinforcing, just the normal commercial-grade construction with explosion venting based on NFPA 68 guidelines.[4] The sheet metal between reinforcing ribs resembles a series of pillows rather than the crisp, straight lines that left the factory. However, in all 100+ deliberate explosions, the housing withstood the internal overpressure developed as a result of the deflagration.

Ignoring all issues except vessel strength, this example of inelastic deformation without rupture, fragmentation, or other catastrophic failure appears to provide the best economic balance. Designs that prevent any deformation either through massive reinforcement or through very large vent areas in proportion to the enclosed volume will significantly add to the unit cost.

The decision must be based not only on the initial cost but also on the costs associated with any repairs, interruption of production, or other added downtime associated with less than absolute protection. In the majority of deflagrations, the fire that follows will typically necessitate repairs even if the enclosure perfectly withstood the forces generated by the deflagration. The lowest threshold of adequate protection is absolute assurance that the system remains intact, with no fragmentation or ruptures that endanger lives. Equipment can be repaired or replaced, and vessels that bulge rarely sue for their injuries; however, personnel injuries and building destruction caused by secondary explosions can bankrupt or destroy a business.

NFPA 68 recommends that the vessel or enclosure design allows at least a 33 percent margin between the anticipated stress and the ultimate strength of the weakest point in the design.[4] If the installation requires a duct between an explosion vent and the outside of the building, ensure that the duct is capable of withstanding the same stress loads as the system enclosure and take into account the added pressures generated in the enclosure due to the vent ducting.

6.6.7 Vent sizing

NFPA 68 provides a great deal of information on the process of sizing vents to protect buildings and equipment from the forces generated in a deflagration.[4] The process is a well-documented mathematical process once you gather all the required information. The information required includes

- *Material identification.* You must properly identify all materials, including gases, liquids, and solids, that could be oxidized in a deflagration. Also, you need to determine the median particle size of solid materials and the concentrations of all the ignitable materials. After proper identification of the fuel, determine from published values the minimum explosive concentration and the maximum rate of pressure rise. Many products have values generated by experimental testing. If your product lacks published data, there are laboratories available to provide custom testing services. Be very cautious of mixtures in which both the gases and solids are flammable. Mixtures may create pressures and rate–of–pressure-rise values that are higher than for either product alone.

- *The ultimate strength of the enclosure.*

- *The volume of the enclosure,* and, for some calculations, *the surface area of the interior of the vessel.*

- *The pressure required to release the vent.*

Lower-magnitude factors that may affect the sizing of explosion vents include the turbulence of the flow, moisture content of solids, mixtures that contain inert materials in addition to flammable materials, and the strength of the ignition source. NFPA 68 contains nomographs and the appropriate formulas to enable determination of the appropriate ratio of vent area to the vessel enclosed volume.[4] Manufacturers of pollution control equipment and deflagration vents will typically provide technical support for choosing the proper vent ratios.

6.6.8 Spark detection and extinguishment

Applications that have the potential to generate sparks may benefit from the use of a spark detection system used in conjunction with an extinguishing system. Mechanical sources such as grinding, sanding, size reduction, or other applications that might generate sparks by impact, metal-to-metal contact, tramp metal ingestion, bearing failure, or other friction sources are all candidates for spark detection and extinguishing systems.

Infrared detectors located in ductwork between the source and the inlet to a pollution control device will protect the downstream equipment. Infrared sensor lenses mounted flush with the duct wall will remain relatively free from deposits that might otherwise blind the sensor. A typical installation includes at least two sensors, mounted in a staggered pattern, on opposite sides of the duct to prevent blind spots.

The sensors detect sparks by the infrared radiation given off by the particle. A control panel monitors the sensors, and when a sensor detects a spark, it triggers a short-duration burst of water spray from opposed nozzles. The distance between the nozzles and the sensors is based on the duct velocity and the response time of the system, ensuring the nozzles are functioning when the spark reaches that point in the ducting. The control panel should monitor the frequency of detection, and exceeding a preset lower limit should trigger an alarm or shut down the source of the sparks.

6.6.9 Fire detection and extinguishment

In addition to the mechanical sources for ignition mentioned above, the electrical sources for ignition include both electrostatic discharges and electrical faults. Thermal ignition sources can be from external sources such as welding or cutting torches and even spontaneous combustion or other exothermic reactions.

Detecting fires in the earliest stage in a pollution control device requires special attention because of the high rate of airflow through the enclosure. With the rapid turnover of air, the heat and products of combustion will be carried away and dissipated in the early stages. Also, in dust collectors, the solids in the air may mask the fire from radiant energy detectors.

One method of detection, measurement of any rise in temperature across the pollution control device, can detect fires fairly rapidly, often within 40 to 60 s. Since the trigger is differential temperature, this system functions with little regard for the operating temperature or even changes in the incoming airstream temperature.

Fixed temperature detectors typically are set to detect temperatures that exceed the highest expected ambient temperature by 100°F [38°C]. This detection method has less flexibility and requires relatively stable operating conditions. Collectors that are outside in intemperate climates may experience temperature variations of 140°F [60°C] from winter to summer. This means at times the fixed detector may require more than twice the desired temperature rise before actuating.

A combination of these two temperature detectors allows the differential detector to shut down the blower; the internal temperature then rises rapidly, setting off the extinguishing system once the fixed temperature detector reacts.

Fire detectors can be based on properties other than heat. Other types include detection of the gases or particulate generated in a fire and detection of radiant energy, such as infrared or ultraviolet rays. Each type of sensor has limitations based on the types of collection devices and the airstream contaminants present during normal operation of the pollution control system.

The choices of extinguishing systems also vary with the type of pollution control device and the products in the airstream. Water delivery systems operated by sprinkler or deluge valves are relatively inexpensive to install. Potential weak points include problems with freezing, cleanup expense, and the damage to equipment from the weight of the water. Hoppers and support structures are often not designed for the loads imposed by being full of water and should have automatic water drain valves included in the system design.

High-pressure, fixed-volume fire-extinguishing or fire-suppression delivery systems usually have a pressure vessel with a scored plate designed to be ruptured by an explosive charge. The extinguishing materials contained within the vessel are typically pressurized to 300 to 750 lb/in^2 [21 to 53 kg/cm^2] using a gas such as nitrogen. When the system actuates, the charge ruptures the plate and the high-pressure gas expands to violently discharge the extinguishing material, minimizing the time between detection and delivery.

The delivery system might contain water; dry chemicals such as sodium bicarbonate, potassium bicarbonate, potassium chloride, or monoammonium phosphate; carbon dioxide; or one of the halogenated extinguishing agents. Use of carbon dioxide or nitrogen requires duct isolation or flow stoppage to provide adequate dilution of the oxygen content to a point low enough to extinguish a fire.

6.6.10 Deflagration detection and suppression

Deflagration detection and response must be almost instantaneous to be of any value, usually within the first 100 to 150 ms. Obviously, the speed of light is the ultimate, so ultraviolet radiation is the first choice. Unfortunately, this method will not work reliably on dust collection systems due to the danger of the airborne dust cloud obscuring the sensor and flame front pathway.

For those applications that cannot reliably use ultraviolet sensors, the second choice is to sense the pressure wave generated by the deflagration, which moves at the speed of sound. Pressure sensors can detect a pressure wave of 0.54-lb/in^2 [0.04-kg/cm^2] magnitude above the ambient pressure, well within the suggested practice of detection prior to reaching a 3-lb/in^2 [0.21-kg/cm^2] pressure rise.

The high-pressure, fixed-volume delivery systems described above suppress deflagrations by interfering with the oxidization process, absorption of the thermal energy needed to sustain the flame front, or displacement or exclusion of oxygen. The key is rapid response in order to deliver the suppressant within the initial flame stage. The suppressant delivery systems need to be close, typically within a range of about 10 ft [3 m] of any point that requires protection. Assuming a discharge rate of 200 ft/s [60 m/s], the reaction time plus just a 10-ft [3 m] delivery uses 50 ms of the 100- to 150-ms allowable response time.

6.7. Product Storage

6.7.1 Introduction

Pollution control devices separate the solids from the airstream, thus concentrating any combustible dust as *fuel*. Any ignition source, such as a spark drawn into a hood, will also be filtered out of the airstream and will lodge in the control device. Add the forced draft produced by the blower, and all the elements for a fire are present. Thus, pollution control devices are frequently damaged in fires and deflagration and are often incorrectly identified as the source of the fire. Minimizing the volume of the fuel is one way to limit the extent of a fire or deflagration. Do not use hoppers for product storage. Product stored in hoppers just adds an additional fuel source. If the collection device has storage containers such as drums under the hopper, using rotary air locks will separate the storage from the collector, also reducing the amount of fuel available. The air lock should be interlocked with the fire-deflagration detection system to ensure it stops if a fire or deflagration is sensed.

Collection containers located inside the housing (dust pans) or under the collection device should be emptied as frequently as practical, with a minimum of once a day. Again, the issue is to minimize the fuel supply. In some cases, this applies even to a product stored wet. Reactive metals will liberate hydrogen when finely divided and stored under water.

Product storage away from the collection device decreases the immediate fuel supply but may require additional protective systems, including explosion venting, if there exists any chance burning materials could be transported from the collection system into the storage container. Screw conveyors and pneumatic conveying systems can easily transport burning material into a storage vessel. Frequently the volume of the storage container will be ignored, providing no protection in case of a deflagration. In the case of silos, the distances are great, making successful detection and timely response difficult or impossible. In some cases, an inert atmosphere or adequate explosion vent area may be the only possible answer.

The delivery of a product into the storage vessel should minimize the dust cloud and, as much as possible, prevent classification of the material. Allowing the course material to fall freely can create a dust cloud of the finest fraction and with the greatest deflagration risk. Avoid high-velocity jets of air aimed directly at the product surface, which increase the dust loading in the container freeboard space.

6.7.2 Corrosion

Corrosion cannot be ignored in many pollution control systems. Even if the economic considerations of repair, premature replacement, and the cost of downtime were not as significant, the safety of the plant and occupants still requires close attention by the purchaser. Designing a pollution control system to withstand corrosive airstreams or corrosive products entails knowing every detail concerning the products, the application, materials of construction, design, fabrication techniques, operating temperatures, and any cleaning products and procedures.

The conditions that contribute to corrosion are very situation specific. Even knowing the materials in the airstream and the materials used in construction of the system components will often still not be enough. Temperature, humidity, concentration of the contaminants, dew points, and even the effects of abrasion or erosion can significantly affect the results of corrosion on the life or premature failure of components.

In general, construction of equipment for corrosive applications requires expert advice in the choice of construction materials, design, and fabrication technique. The design and fabrication techniques are often given less attention than the material selection but are no less critical to the success of the project.

Crevices are the cause of most corrosion failures, resulting from improper design and/or fabrication. In wet systems such as scrubbers, crevices may create galvanic cells. In dry systems, crevices create areas that may trap corrosive solids and moisture. Skip welds, spot welds, riveted joints, weld slag inclusion, welds with incomplete penetration and pinholes, and similar quality problems all provide opportunities for corrosion to gain a foothold.

Abrasion or erosion can accelerate the corrosion process, affecting either wet or dry systems. Some materials, such as aluminum, have a corrosion process that can passivate or seal the base metal away from further damage. Abrasion or erosion of this protective layer exposes the base material to further corrosion, significantly accelerating any corrosive effects.

Dry systems without insulation and with significant temperature differences between the airstream and atmospheric temperature may have condensation problems. Condensation can cause acceleration of corrosion plus other problems in a dry system such as buildup of product, unwanted chemical reactions, and even system plugging. Even insulated systems may have condensation problems related to voids in the insulation, reinforcement, or structure or to leaks at access doors and penetrations. Insulation can also trap moisture between the insulating material and the duct or control system skin, providing an external site for corrosion to occur.

Dry systems handling warm, moist air may benefit from start-up and shutdown procedures. The procedure would allow heated air to warm the system prior to introduction of moist air. At the end of the cycle, allow sufficient time for the system to completely dry prior to shutdown.

Although filter media does not truly corrode, choosing the correct filter media requires the same care as making the proper choices of the other materials of construction. That choice requires knowledge of the chemical composition and moisture content of the product and the airstream and knowledge of the airstream temperature. Reputable vendors will be able to provide information on the correct filter media choice to withstand the operating conditions of the proposed system.

Scrubbers frequently operate at elevated temperatures, often increasing the severity of corrosion problems. Liquid scrubbers may absorb gases from the airstream that combine with the scrubbing liquid to create corrosive liquids. The scrubbing liquid may require pH control or continuous dilution by adding fresh fluid and overflowing contaminated fluids to a waste treatment system.

Scrubber design and fabrication must reflect several areas of concern. Perforations of the body such as access openings, shaft penetrations and seals, gasket materials, support or reinforcing structure attachment, expansion joints, linings or laminated construction, interior baffles, and abrasive wear points must all be carefully engineered to prevent premature failure due to corrosion. Flat-bottom tanks installed on wood or concrete floors may trap liquids that can create corrosive conditions and short life.

Electrostatic precipitators, like scrubbers, frequently handle high-temperature airstreams. Warm, moist air may cause condensation and corrosion of the enclosure skin and roof. The hopper will be the coolest part of the system and can be a problem area for condensation. The condensed moisture combined with collected solids can trap moisture allowing corrosion. As in other system types, air leaks at access points or penetrations of the enclosure can create conditions for condensation. Sharp-edged collector plates or misaligned electrical system components may allow electrical erosion. Dissimilar materials, especially at electrical connections, can mean rapid galvanic erosion.

6.7.3   Abrasion

The effects of abrasion show up in areas of high velocity, such as in ducting and in areas of directional change, mechanical motion, and sliding of bulk solids. Motion must be combined with abrasive particulate to cause a problem. Usually, ducting systems require high velocities to transport the solids from the hood or pickup point to the control device. Transport velocities vary from 25 ft/s [7.6 m/s] for fume control to 75 ft/s [23 m/s] for chips and heavy solids. Even mildly abrasive materials moving at these speeds can cause erosion of the ductwork.

Elbows and other devices that change the direction of movement of the airstream require special attention. As an abrasive-laden airstream enters an elbow, momentum and centrifugal force will carry much of the solids to the outside of the elbow. The increased concentration adversely affects the life span of the fitting. Several special designs address this problem by providing replaceable plates on the outside circumference of the elbow.

Other devices that force directional change of the airstream use wear plates or shallow wells that trap a layer of the product. Trapped material allows the impact to be against a layer of a product rather than against part of the enclosure. This approach conserves the hardware and minimizes contamination of the product stream.

Mechanical motion of parts that must come in contact with abrasive material will accelerate wear. In pollution control systems, examples of some common mechanical wear points include screw conveyers, feeders and rotary air locks, material handling blowers, control valves, slide gates and dampers, and slurry pumps. All these components require special attention during the process of specification to ensure satisfactory performance.

6.7.4   Abrasion control

Careful choice of the hood velocity and position can be one of the most effective methods of controlling abrasion and often will be completely ignored. As an example, dust control systems for abrasive blasting operations frequently ingest large quantities of the abrasive as well as the dust. A customer called complaining of premature filter media wear. A field inspection revealed excessive velocity at the hood, coupled with a hood much closer to the source than required. Adjustments of flow and hood position ended the premature filter failures and conserved useable abrasive.

Duct velocities should be no greater than required to maintain transport velocity. Higher velocities will not only increase wear but also increase costs due to higher energy consumption. There is also a danger in choosing too low a velocity. In extreme cases, ducting systems have torn loose from their supports and collapsed due to the weight of accumulated solids built up over time due to inadequate velocity.

Ducting systems that must transport abrasive material require careful balancing of the individual duct legs. Systems handling nonabrasive materials may permit balancing at flow junctions utilizing higher than normal velocity as a means to match static pressures. However, systems transporting abrasive material may exhibit premature failure due to abrasion and erosion if higher velocities are utilized. Sometimes, the only recourse is to choose heavy-duty materials of construction. As an example, a customer wished to transport dry cast-iron chips in a dust collection system. In this case, schedule 40 black iron pipe with sweep elbows made effective, although heavy and expensive, duct material.

Some styles of control devices may benefit from abrasion-resistant inlet designs. In brief, an abrasion-resistant inlet provides a method to slow the airstream from the duct transport velocity, absorb inertial energy of the particulate, and deliver both to the control device at a velocity that will minimize the potential for abrasion of internal components and filter media. This usually relies on a combination of a duct size transition to a larger diameter in order to lower velocity and forcing the air and particulate to undergo a directional change to encourage absorption of the energy stored in particulate velocity. The design must not permit solids buildup that would cause support failure.

The control device should incorporate a proper flow distribution design to ensure that airflows are relatively uniform throughout. Zones of high velocity may mean a nonuniform distribution of loading and less than optimal filtration efficiency, as well as increasing the possibility of premature wear and failure.

When the system will handle highly abrasive particulate, the solids handling portion of the system must take into account the expected severity of service. Rotary valves, slurry pumps, and recirculation pumps should be heavy-duty and designed with wear-resistant materials. Slide gates must take into consideration the probability that abrasive material will get between the slide and the sealing mechanism. There are few things more frustrating than a slide gate that will not slide when you really need to close it! Screw conveyors should not have internal hangers and bearings if at all possible and should operate at the lowest practical speed.

CEMA publishes an excellent guide to the characteristics of a wide variety of materials. CEMA standard no. 550, Classification and Definitions of Bulk Materials,[9] is a source of information concerning the abrasive nature of products as well as many other physical characteristics.

References

1. Round Industrial Duct Construction Standards, Sheet Metal and Air Conditioning Contractors' National Association, Inc., Vienna, VA, 1977.
2. National Electrical Code, NFPA 70, National Fire Protection Association, Quincy, MA, 1996.
3. Classification of Gases, Vapors, and Dusts for Electrical Equipment in Hazardous Locations, NFPA 497M, National Fire Protection Association, Quincy, MA, 1992.
4. Guide for Venting of Deflagrations, NFPA 68, National Fire Protection Association, Quincy, MA, 1994.
5. Explosion Prevention Systems, NFPA 69, National Fire Protection Association, Quincy, MA, 1992.
6. Static Electricity, NFPA 77, National Fire Protection Association, Quincy, MA, 1993.
7. Exhaust Systems for Air Conveying of Materials, NFPA 91, National Fire Protection Association, Quincy, MA, 1995.
8. Pneumatic Conveying Systems for Handling Combustible Materials, NFPA 650, National Fire Protection Association, Quincy, MA, 1990.
9. Classification and Definitions of Bulk Materials, CEMA standard no. 550, Conveyer Equipment Manufacturers Association, Manassas, VA, 1970.

7

The Economics
of Air Pollution
Control

David L. Amrein and William L. Heumann

Fisher-Klosterman, Inc.
Louisville, KY

7.1 Introduction

Economics is the primary force driving the implementation and selection of industrial air pollution control (APC) technologies. The level to which pollution must be abated and the technologies selected to achieve these goals are driven by economics. Public or community economics is considered in setting air pollution standards, while plant and process economics primarily affect the specific equipment and technologies selected.

In the cases utilizing APC technologies for product recovery, the specification and selection of equipment are purely plant and process economics. In cases where APC technologies are to be implemented to prevent or reduce the emission of pollutants to the atmosphere, the economics will be largely social. In most cases, social economic issues are addressed in the form of regulations and the agencies that enforce them. Most nations and states have agencies devoted to the setting, monitoring, and enforcement of environmental regulations (e.g., U.S. EPA).

The social economic issues will be addressed by an evaluation of the *net social cost* (NSC). The NSC may be expressed as[1]

$$NSC = CD_1 - CD_2 + EC \qquad (7.1)$$

where CD_1 = annual damage cost without control
CD_2 = annual damage cost reduction with controls
EC = annual cost of control equipment

The usage of Eq. (7.1) is extremely difficult to set for those attempting to make public policy since there is no simple method available to determine or predict CD_1 and CD_2 accurately. The factors that must be considered in evaluating CD_1 and CD_2 include expected mortality rates, expected illness and/or injury rates, aesthetic losses, and property damage. Subsequently a cost or value must be placed on each of these factors. As a result of the difficulty in arriving at meaningful values for CD_1 and CD_2, much public policy on environmental issues is, to some degree, arbitrary. Since many of the economics issues that must be considered are catastrophic in nature, it is common and logical to err on the side of enforcement and implementation of regulations controlling pollution than to await further data and risk serious consequences.

Equation (7.1) may be used to evaluate implementation of control technologies on a plant level in some cases. For example, assume an operating industrial plant has emissions of a pollutant at a level that meets emission standards but results in an annual cost of $68,000 from the repainting of neighbors' houses that the company feels is necessary for public relations. These emissions may be reduced by 90 percent by installing a system that has a total annualized cost of $32,000. What is the NSC after implementation?

$$NSC = CD_1 - CD_2 + EC$$
$$CD_1 = \$68,000$$
$$CD_2 = 0.9 \cdot \$68,000 = \$61,200$$
$$EC = \$32,000$$
$$NSC = 68,000 - 61,200 + \$32,000 = \$38,800$$

The NSC of $38,800 compares favorably with the original NSC before implementation of $68,000.

In many industrial APC applications, various technologies are under consideration. Although these various technologies may differ in almost every imaginable way, cost and the evaluation thereof is relevant to all the technologies. Furthermore, if care is taken during the evaluation to ensure consistency, the lowest overall cost solution may be selected from very different offerings.[2]

7.2 Specification of an APC System

The solution to any dust collection, product recovery, or pollution control problem depends on a combination of understanding the problem and utilization of the right tools. Understanding the problem is the foremost ingredient. If the problem is not properly understood, the results will invariably turn out to be less than desired.

To successfully solve any pollution control problem, one must have as much knowledge or expertise of the system in question as is reasonably possible. In many instances a lack of proper understanding results in an incomplete investigation of all the options available. All too often, the solution selected is not the best option available.

Within this book are discussions of various collection technologies available. Within each technology, multiple options exist, based on the final results being sought. Collection equipment should always be evaluated and purchased based upon an accurate assessment of all variables including performance, quality, features, initial cost, operating cost, and equipment life.

Most engineers involved in industrial pollution control and dust collection projects have a basic knowledge of the physical aspects of the collection problem in question. Sometimes, however, the problem of choosing the right course of action is complicated by inadequate budget or other restrictions that may be affecting the decision-making process. A complete assessment of the problem involves an evaluation of all reasonable options available. The following considerations should be included in a complete evaluation:

- *Emission reduction or product recovery efficiency.* The total collection efficiency.

- *Operating cost.* The cost of energy consumed, chemicals, operators, etc.

- *Capital cost.* The installation cost including permits, plant preparation, equipment, and installation services.

- *Package size.* The length, width, and height of the space utilized by the installed equipment.

- *Installation logistics.* The assessment of any obstructions that may interfere with the placement of equipment in its final location; for example, you may need to remove a section of roof to gain access to the installation site. In addition, loss of production due to the installation and commissioning process should be considered.

- *Accessories.* May include transitions, ductwork, hangers, support steel, air locks (although always considered as necessary), access ways, equipment access features (doors, pokeholes, break apart construction, etc.), and explosion vents or fans.

- *Operating life.* Factors affecting operating life include abrasion, corrosion, and external factors.

- *Maintenance and service.* Preventive measures and continuous upkeep required to keep the equipment performing satisfactorily.

- *Reliability.* Since many industries run at or near capacity for 24 h/day, it is essential for the equipment in service to operate with minimal interruptions.

- *Installation costs.* Includes preparation equipment, erection, and start-up services.

- *Materials of construction.* Must consider abrasion, corrosion, stresses, and the type of service intended for the equipment.

- *Risk analysis.* Determine the operating variables the system is likely to encounter.

The specification probably represents the most significant step in the selection and acquisition of a system or equipment. The specification should not only convey a complete description of the APC problem but also the desired outcome and priorities the purchaser will choose in evaluating proposals. Commonly the specification itself will limit the range of technologies offered to prevent the selection of those determined to be unacceptable by the user or engineer.

From a contractual perspective the specification will serve as one of the main documents describing the equipment purchased, as well as the performance and operating characteristics thereof. A specification that is unclear or vague may not only result in excessive cost but possibly the complete failure of the selected APC system to serve its essential purpose, that is, the reliable control of an air pollution problem.

Although the importance of the specification is obvious and logical, incorrect or incomplete specifications are responsible for most cases of vendor-purchaser disputes involving performance problems with APC systems. It is all too common within industry for an APC system to not operate as expected due to incorrect initial design resulting from an inaccurate or incomplete specification.

The APC equipment or system received will rarely exceed the expectations requested in the inquiry specification. When specifications lack proper definition, vendor proposals may reflect large variances in cost because of undefined considerations. Some of the specification items that frequently produce significant differences between competing quotes if left unclear are

Inlet gas and particulate definition

Quality of materials of construction

Minimum thickness or weights of the materials of construction

Construction quality

Surface finish of plates and/or welds (interior and exterior)

Accessories

Paint specifications

Allowable pressure drop or energy consumption

Required collection efficiency

Abrasion considerations

Corrosion considerations

Erection considerations

Access considerations

Instrumentation considerations

Load considerations

Testing requirements

Structural requirements

Vendors of APC systems will typically quote and provide the minimum solution to meet the specification since the marketplace is competitive and most frequently the lowest bidder who meets the specification is awarded the contract. Great care must therefore be given to ensure that the minimum requirements of the specification are adequate and acceptable.

The outline of a typical specification for APC equipment is

I. General
   1.1 Scope
   1.2 Exclusions from scope
   1.3 Drawings, data sheets
   1.4 Requirements of regulatory agencies
   1.5 Duty (operation parameters)

II. Technical
   2.1 Design data
   2.2 Performance requirements
   2.3 Design considerations (including access, features, and erection)

2.4 Construction requirements (including materials, welding, gaskets, hardware, and erection features)

2.5 Finish requirements (internal and external)

2.6 Assembly requirements (including accessories)

2.7 Installation requirements

2.8 Start-up and commissioning requirements

2.9 Testing requirements (including construction testing and operating tests)

III. Execution

3.1 Schedule for design construction, erection, testing, etc.

3.2 Technical information required with bid

3.3 Technical information required with contract

3.4 Warranties and/or guarantees

IV. Miscellaneous

4.1 Contact personnel (including names, addresses, phones, faxes, etc.)

4.2 Attachments

A complete specification should convey all the requirements and expectations of the purchaser as previously discussed. In some cases specifications are written with design data that is intentionally overstated in an effort to be conservative. This can prove to be disastrous unless the writer is familiar enough with the characteristics of the technologies in question to ensure that safe design values are given. As an example, consider a dust collection system that will handle 21,300 acfm [36,194 m³/h]. To allow for a margin of error, the engineer may specify an allowable pressure drop of 6 in w.c. [152 mmH$_2$O] at 24,000 acfm (40,782 m³/h) while achieving a 99 percent collection efficiency. This overstating of the gas flow rate may result in a safe baghouse selection but could be catastrophic if a cyclone or fixed-throat venturi scrubber is selected. In these latter cases, the collection efficiency may be lower than the specified level of 99 percent at the actual operating conditions. In the following examples, the desired result is to achieve 98 percent total collection efficiency at less than a 7-in-w.g. pressure drop utilizing a cyclone. Table 7.1 gives what may appear to be two similar particle size distributions for use in this example. An airflow variance based upon a selected cyclone example could cause variations in pressure drop as shown in Table 7.2.

A particle size variance can cause a dramatic difference in collection efficiency as well, without any alteration in pressure drop. To illustrate this, let us examine the effect of the two different particle size distributions as shown in Table 7.1. The effect of the change of

TABLE 7.1   Example Particle Size Distributions

Stokes equivalent diameters, μm	Percentage Finer than specified size (by weight)	
	Example A	Example B
1.0	0.07	0.10
1.5	0.33	
2.0	0.60	
3.0	1.87	
4.0	3.50	
5.0	4.63	0.35
6.0	5.17	
8.0	8.17	
10.0	11.2	1.70
15.0	20.40	
20.0	28.43	
30.0	42.73	11.90
40.0	56.27	
50.0	65.70	29.20
60.0	78.13	
80.0	87.48	71.60
100.0	99.99	99.99

TABLE 7.2   Example Performance Summary

Airflow, acfm	Pressure drop,* (inches w.g.)	Example A		Example B	
		Total collection efficiency, %	Outlet emission, lb/h	Total collection efficiency, %	Outlet emission, lb/h
4000	2.73	97.16	7.09	99.58	1.05
6000	6.74	98.09	4.77	99.76	0.60
8000	12.77	98.59	3.52	99.86	0.34

*Based on an inlet loading of 250 lb/h for these examples.

particle size alone is illustrated in Table 7.2 utilizing a cyclone collector by comparing example A versus example B at any given flow rate.

Particle size data is vital for proper selection of many types of collection devices. The results from a particle size distribution may vary greatly depending on the test method of particle sizing selected. Therefore, it is important to provide the test method by which particle sizes were determined with the specification. Further discussion on particle sizing methods is covered in Chaps. 3 and 8 of this text.

It becomes quite apparent that proper input data is a prerequisite to proper equipment selection and sizing prior to addressing all the other considerations. As shown in Table 7.2, outlet emissions prove to be significantly different for the same cyclone based upon the different particle size distributions and/or gas flow rate and pressure drop.

7.3 The Costs of Air Pollution Control

Air pollution control generally includes the basic collection equipment application groups as described in Chaps. 8 to 14. Each application group and combination of groups should be evaluated as possible choices, for any potential collection need, based upon the type of pollutant(s), performance requirements, and other regulatory requirements. In many instances several technologies could be applied as possible selections, and the engineer must select the most economical overall solution that will meet the desired objectives.

Many equipment selections are based upon intangible considerations such as a company's preference for a previously used technology. In some cases these preferences are unfounded prejudices while in others their basis is in an intuitive interpretation of real economic factors. In all cases the engineer should endeavor to quantify the basis for prior equipment or technology preferences in the body of the specification and the subsequent economic analysis. This will allow equipment suppliers to use their expertise in proposing the best possible equipment selection to meet the process requirements and preferences of the end user.

Although there is a great deal of general published data available addressing the average costs of various control technologies, there is also enough sufficient diversity within the range of plant operations to make an accurate cost analysis for a specific application based upon that data unlikely. In addition to variations within plant operations, new technologies, materials, and processes are constantly changing the historical cost data. Unless the APC application is too small to justify the expense of analysis, extensive analysis has been performed on a very similar application, or the regulations dictate the use of a particular technology; the engineer should approach each application with an open mind and evaluate proposed technologies with a detailed cost analysis. Within a specific industrial application, the cost of installation, operation, and maintenance for an APC system can be very dependent on characteristics that are unique to that plant or operation. For instance, the cost of operation for a wet scrubber can be significantly affected by the disposal costs of the liquid slurry. If the plant already has an existing settling pond with excess capacity, the operating cost will be dramatically lower than if one must be built new for this application.

Although generalized cost data may be very misleading for individual industrial applications, it can be very useful for macroeconomical decision-making purposes. Since these data represent average costs, they are appropriate if a large number of cases are under consideration.

There are numerous methods of comparing the costs of APC equipment and systems. The available methods of cost comparison are the same as those available for cost comparison of general capital equipment and systems. Usually total-annualized-cost (TAC) methods are selected for APC systems although other methods may be more desirable if the APC system will serve more as an integral part of a manufacturing process than as an add-on control system.

Several equations are common for calculating the TAC for an APC system. One such equation is[3,4]

$$TAC = \frac{CI}{LI} + MOC + CC - REC \tag{7.2}$$

where   TAC = total annual costs, dollars/year
         CI = total installed capital investment, dollars
         LI = expected system life, years
        MOC = maintenance and operating costs, dollars/year
         CC = capital costs, dollars/year
        REC = recovery credits from pollution control credits or product recovery, dollars/year

Costs from many of the categories used in Eq. (7.2) can be in any one of three forms. These are fixed, variable, and semivariable. Fixed costs are those that accrue regardless of the rate of production or, for that matter, whether the plant is operating. An example is the cost of capital used to purchase and install the APC system. Variable costs are those that will change with system operation and the level thereof. Energy costs are usually variable. Semivariable costs are those that vary with system operation but also have a fixed component. Maintenance costs are usually semivariable since they vary with the level of operation, but some maintenance is required even if the system is not operated at all.

The level to which costs must be analyzed at various levels of operation and how detailed the engineer must describe variable and semivariable cost components will be different for each application. If plant operation is reasonably constant, it may be possible to calculate the variable costs at the expected operating condition and treat them as fixed costs from that point forward. If the level of operation is uncertain and/or changes are expected over time, it is important to prepare total annual cost estimates that reflect the range of conditions expected.

Total installed capital investment (CI) as used in Eq. (7.2) should include all costs that may be capitalized as part of the installed system. Some of the costs that may be included are

Equipment
    Primary equipment cost
    Auxiliary equipment cost
    Instruments and controls
    Taxes
    Freight
Installation
    Foundations
    Support steel
    Truck unloading
    Erection
    Electrical
    Plumbing
    Insulation
    Painting
    Site preparation
Other
    Real estate
    Preliminary studies
    Project engineering and supervision
    Lost production or downtime for installation

The system life (LI) used in Eq. (7.2) represents the useful system life in industrial operation. This equation utilizes a straight-line depreciation of the capital costs over the system life. Guidelines for equipment and parts lives are shown in Table 7.3.

Annual maintenance and operating costs (MOC) as used in Eq. (7.2) may include

1. Operating labor cost including overhead (see Table 7.4)

2. Supervisory labor costs including overhead

3. Maintenance labor cost including overhead (see Table 7.4)

4. Maintenance materials such as oil, nuts and bolts, welding supplies, cleaning supplies, solder, and duct tape

TABLE 7.3   Equipment and Parts Life, Years[5]

| Item | Types of service | | |
	Light	Average	Severe
Equipment			
Cyclones	50	35	5
Baghouses	40	20	5
Cartridge collectors	40	20	5
Particulate scrubbers	20	10	3
Electrostatic precipitators	40	20	5
Absorbers	20	10	5
Adsorbers	20	15	5
Parts			
Filter bags	5	1.5	0.3
Cartridges (cellulose)	3	1	0.2
Cartridges (fabric)	5	1.5	0.3
Adsorbents	8	5	2
Refractories	10	5	1
Ceramics	40	10	5

TABLE 7.4   Typical Labor Hours per 24-Hour Day[6]

| Equipment | Operation | | Maintenance | |
	High	Low	High	Low
Cyclone	11	0.25	4	0.25
Fabric filters	12	6	6	3
Particulate scrubbers	24	6	6	3
Electrostatic precipitators	6	1.5	3	1.5
Absorbers	6	1.5	3	1.5
Adsorbers	6	1.5	3	1.5
Thermal oxidizers	6	1.5	3	1.5

5. Replacement materials and spare parts

6. Utilities which may include electricity, natural gas, propane, oil, water, compressed air, and steam

7. Cost of waste treatment which may include on-site treatment and/or landfill and the use of special haulers

8. Supervision and engineering not included elsewhere

9. Property taxes

10. Testing and monitoring

11. Insurance

The capital costs (CC) used in Eq. (7.2) represent the cost of money utilized to implement an APC system. This may include the interest or administrative costs on commercial loans, corporate bonds, and industrial revenue bonds. Capital cost should represent the opportunity cost of the capital since presumably the capital represents a resource that could have been used elsewhere.

The recovery credits (REC) used in Eq. (7.2) are only appropriate in systems that

1. Receive credits under APC regulations for the removal of a pollutant
2. Collect usable product from the process

To illustrate the use of Eq. (7.2), let us look at an example cost comparison of two alternative technologies in an industrial process. The design conditions are given as follows:

1.0 Conveying gas definition

Define gas being delivered to collector inlet _____ Air _____

Gas analysis (by weight or by volume) _____ 100% Air _____

Gas volume (acfm) _36,000 acfm_

Gas temperature _____ 70° _____

Gas pressure _____ ≈ Standard atmospheric _____

2.0 Particulate definition

Particulate description _____ Particleboard fines _____

Particle size distribution (as determined by liquid sedimentation @ 5.6 = 1.1)

Stokes equivalent diameter, μm	Percentage finer (by weight)
20	0.001
30	0.002
40	0.007
50	0.014

Dust/particulate load at collector inlet 8,000 lb/hr (97.2 grains/acf)

Particle density (specific gravity) _____ 1.1 _____

Bulk density for air lock sizing _____ 12 lb/ft^3 _____

3.0 Performance requirements

Required efficiency __ 99.98% removal (by weight), minimum, __ _ recovered material value $0.0l/lb _

Maximum allowable pressure drop across collector _ 6 in w.g. _

Operating pressure/vacuum of collector _____ −15 in w.g. _____

An example summary of total installed capital investment for the proposed system (CI) is shown in the table on the next page. Another

EXAMPLE TOTAL INSTALLED CAPITAL INVESTMENT

EQUIPMENT	HIGH EFFICIENCY CYCLONE	PULSE JET BAGHOUSE
PRIMARY EQUIPMENT COST:	$ 54,000.00	$ 36,000.00
AUXILIARY EQUIPMENT COST:	$ 12,200.00	$ 12,200.00
DUCTWORK AND HOODS:	$ 8,000.00	$ 9,000.00
INSTRUMENTS AND CONTROLS:	$ 3,600.00	$ 3,600.00
TAXES(4%):	$ 3,112.00	$ 2,432.00
FREIGHT:	$ 1,600.00	$ 1,000.00
SUBTOTAL:	$ 82,512.00	$ 64,232.00
INSTALLATION		
FOUNDATIONS:	$ 2,400.00	$ 3,400.00
SUPPORT STEEL:	$ 3,800.00	$ 3,200.00
TRUCK UNLOADING:	$ 400.00	$ 600.00
ERECTION:	$ 12,000.00	$ 18,500.00
ELECTRICAL:	$ 2,000.00	$ 3,000.00
PLUMBING:	$ 100.00	$ 1,200.00
INSULATION:	$ 8,000.00	$ 4,000.00
PAINTING:	$ 1,000.00	$ 1,000.00
SITE PREPARATION:	$ 2,000.00	$ 2,000.00
SUBTOTAL:	$ 31,700.00	$ 36,900.00
OTHER		
REAL ESTATE:	$ 10,000.00	$ 12,000.00
PRELIMINARY STUDIES:	$ 10,000.00	$ 5,000.00
PROJECT ENGINEERING:	$ 2,000.00	$ 2,000.00
LOST PRODUCTION:	$ 22,500.00	$ 22,500.00
SUBTOTAL:	$ 44,500.00	$ 41,500.00
TOTAL INSTALLED CAPITAL INVESTMENT:	$ 158,712.00	$ 142,632.00

EXAMPLE ANNUAL MAINTENANCE AND OPERATING COSTS

PLANT OPERATION:	320	DAYS/YEAR		24	HOURS/DAY

	HIGH EFFICIENCY CYCLONE				PULSE JET BAGHOUSE			
	QUA	UNITS	PRICE	SUM	QUA	UNITS	PRICE	SUM
OPERATOR LABOR:	160	HOURS	14	$ 2,240.00	2560	HOURS	14	$ 35,840.00
MAINTENANCE LABOR:	320	HOURS	14	$ 4,480.00	1280	HOURS	14	$ 17,920.00
SUPERVISION:	15	%	OF LABOR	$ 1,008.00	15	%	OF LABOR	$ 8,064.00
MAINTENANCE SUPPLIES:	100	%	OF MAINT.	$ 4,480.00	100	%	OF MAINT.	$ 17,920.00
SPARE PARTS:				$ 300.00				$ 3,650.00
UTILITIES:				$ 23,040.00				$ 26,880.00
DISPOSAL COSTS:				$ -				$ -
MISC. SUPER. AND ENG.:				$ 2,400.00				$ 2,400.00
PROPERTY TAXES:				$ 1,580.00				$ 1,420.00
TESTING AND MONITORING:				$ 8,000.00				$ 8,000.00
INSURANCE:				$ 1,467.00				$ 3,892.00
TOTAL MOC:				$ 48,995.00				$ 125,986.00

table shows an example summary of maintenance and operating costs (MOC). We may then calculate the TAC as follows:

$$\text{TAC} = \frac{\text{CI}}{\text{LI}} + \text{MOC} + \text{CC} - \text{REC}$$

$$\text{CI} = \begin{cases} \$158,712 & \text{for the high-efficiency cyclone} \quad \text{(see Table 7.5)} \\ \$142,632 & \text{for the pulse jet baghouse} \end{cases}$$

$$\text{LI} = \begin{cases} 35 \text{ years} & \text{for the high-efficiency cyclone} \quad \text{(see Table 7.3)} \\ 20 \text{ years} & \text{for the pulse jet baghouse} \end{cases}$$

$$\text{MOC} = \begin{cases} \$48{,}995 & \text{for the high-efficiency cyclone} \quad \text{(see Table 7.6)} \\ \$125{,}986 & \text{for the pulse jet baghouse} \end{cases}$$

$$\text{CC} = 10\% \text{ of CI} = \begin{cases} \$15{,}871 & \text{for the high-efficiency cyclone} \\ \$14{,}263 & \text{for the pulse jet baghouse} \end{cases}$$

$$\text{REC} = 2 \cdot E_F \cdot 320 \text{ days/year} \cdot 24 \text{ h/day} \cdot \text{material value}$$

Assume

$$E_T = \begin{cases} 99.985 & \text{for the high-efficiency cyclone} \\ 99.999 & \text{for the pulse jet baghouse} \end{cases} \quad \text{and} \quad E_F = \frac{E_T}{100}$$

Then

$$\text{REC} = \begin{cases} 8000 \cdot 0.99985 \cdot 320 \cdot 24 \cdot 0.01 = \$614{,}308 & \text{for high-} \\ \text{efficiency cyclone} \\ 8000 \cdot 0.99999 \cdot 320 \cdot 24 \cdot 0.01 = \$614{,}394 & \text{for pulse jet} \\ \text{baghouse} \end{cases}$$

Then

$$\text{TAC} = \begin{cases} \dfrac{\$158{,}712}{35} + \$48{,}995 + \$15{,}871 - \$614{,}308 = -\$544{,}907.37 \\ \text{for high-efficiency cyclone} \\ \dfrac{\$142{,}632}{20} + \$125{,}986 + \$14{,}263 - \$614{,}394 = -\$467{,}013.40 \\ \text{for pulse jet baghouse} \end{cases}$$

   This example highlights a case where an APC system is part of a production process and some form of collector would be used for product recovery even in the absence of environmental regulations. The negative sign in the answer represents that the system makes money for the user, hence the negative total annualized cost. The cyclone would represent the obvious choice in this example since it represents a TAC that is over $77,000 lower than that for the pulse jet baghouse.

7.4   Vendor Selection

Once cost analyses have been performed on various offerings, a vendor or supplier of that technology must be selected. In addition to direct cost differences between vendors, there are numerous other factors that should be considered. The equipment vendor selected should meet the following criteria:

1. Technical expertise to address all concerns that are involved
2. Ability to manufacture the desired equipment to the required standards
3. Capable to assist the customer in understanding all the options available
4. Can meet the promised schedules for manufacture and shipment
5. Can offer expertise and service in the following areas as needed:
   a. Manufacturing
   b. Operation
   c. Installation and erection services
   d. Start-up assistance if desired
   e. Troubleshooting
   f. Consulting

Far too often a purchaser is attracted to price difference without an understanding of how that may relate to the underlying services available. It is not uncommon in the APC industry for proposals from two vendors to have significant cost differences. Some of the factors that can justify a difference may be

1. Differences in the type of equipment being offered between vendors

2. Materials of construction (type and thickness)

3. Lightweight construction versus heavy duty

4. Weld class or quality

5. Features and accessories included

6. Physical size and weight

A common problem with vendor evaluation is that when numerous vendors are requested to provide a quotation for a piece of equipment, it is assumed that they are furnishing the same thing. This may be an incorrect assumption. APC equipment items are not standard. Each vendor has engineered its own lines of equipment with various features and benefits. One vendor's selection may physically be one-half the height of another's recommendation. Another may have a net weight that could be several times another. Therefore, it becomes the responsibility of the purchaser or user to determine what factors or priorities are the most important. In the case of APC equipment, variances may go far beyond size, weight, and cost. Other factors that may enter into the evaluation are

Efficiency

Equipment life expectancy

Power consumption

Logistics with other associated equipment

Vendor experience, customer service, and reliabilities

7.5 APC System Warranties

There are two general types of warranties that are given from vendors of APC equipment and systems to the purchaser. These are material warranties and performance warranties. Material warranties protect the purchaser against substandard quality materials, poor quality, mechanical and electrical defects, etc. Performance warranties are intended to protect the purchaser against system operation in a manner that is less than promised by the vendor. Warranties of either type provide the following:

1. A clear contractual listing of the contract requirements
2. Legal methods and remedies in cases of breaches of warranties

Although much emphasis is placed upon the legal aspects of warranties, including remedies in the event of breach, there are few cases in which performance warranties have been enforced through litigation within the APC industry. There are several likely explanations for this, despite numerous situations in which APC equipment and systems have failed to meet the warranted performance. The primary difficulty in enforcement of performance warranties is cost. For a warranty to be enforced, it is usually required that not only is it shown that the APC system is performing inadequately but also that the design conditions of the equipment are those that were originally specified. This may be very expensive and in some cases impossible to show. In virtually all industrial applications, the actual operating conditions are different to some degree from those specified. Although these variances may not be significant, it is often difficult to argue this point to those who are less technical and/or familiar with the particular technical issues.

For these reasons, the best warranty for both purchaser and vendor is one that is not used after the system is operational. The best function for both material and performance warranties is to clearly convey the project requirements and serve as part of the contractual enforcement of the project quality assurance. Whether the quality assurance aspect relates to material, work quality, engineering, etc., is irrelevant. Both purchaser and vendor should understand all aspects of the scope of supply and feel confident they will be met. If the purchaser only has confidence in the vendor's offering as a result of the warranty given, trouble is likely to occur.

Enforcement of warranties between disputing parties is likely to be a lose-lose situation. Both parties to the conflict will suffer economically and emotionally. Further, litigation is a great distraction away from the main business of either party.

Performance warranties are a risk assumed by the equipment supplier based on input provided as the basis for the selection. A performance warranty can be a controversial issue when it is found that the operating parameters are different from the specified parameters. The degree of risk can be based on the confidence level that comes with the design data furnished. If the design data provided has been inaccurately or incompletely specified, the potential for a conflict at the warranty stage increases. An APC equipment manufacturer may have perfect knowledge of how the selected equipment will function given certain inlet conditions, but what contingencies or safety factors provide a safe design if there is uncertainty as to the operating conditions?

Performance warranties may be a result of an air pollution code regulated level of emissions as listed on the purchaser's permit to operate. (In the United States, Title V of the Clean Air Act of 1990 ordered all states to develop programs to establish permit levels acceptable from emissions sources.) Even though a lack of uniformity still exists within the United States, most states require a permit for any equipment that emits an air pollutant. Since nearly all manufacturing operations produce some form of waste, it is a matter of time before most manufacturing operations will be forced to control their air pollutants or cease operations.

Because of the increasing level of regulation and severity thereof, many companies overstate current-day regulations so that they are protected from any anticipated future changes in the regulations. For example, a forest product operation may require that their collector not emit more than 5 lb/h of sawdust even though they have a current allowable limit of 15 lb/h from a given operation.

References

1. W. Strauss, *Industrial Gas Cleaning,* Pergamon Press, Oxford, 1975, p. 515.
2. W. M. Vatouk, chap. 14 in S. Calvert and H. M. England (eds.), *Handbook of Air Pollution Technology,* Wiley Interscience, NY, 1984, p. 331.
3. The Clean Air Act as amended August 1977 (serial no. 95-11), Washington, D.C., U.S. Government Printing Office, 1977.
4. Gad Hetroni, *Handbook of Multiphase Systems,* Hemisphere Publishing, Washington, D.C., 1982, pp. 9–15.
5. Vatouk, op. cit., p. 342.
6. Vatouk, op. cit., p. 339.

Control Technologies: Particulate

8

Cyclones

William L. Heumann

Fisher-Klosterman, Inc.
Louisville, KY

8.1 Introduction

Cyclone is a generic name given to several different devices that have the common attribute of utilizing centrifugal force to separate particulate from a gas or liquid flow stream. As such, they represent the most efficient and generally the most economical form of an inertial separator for an industrial pollution control problem. Cyclones are the most common dust removal devices used within industry.[1]

Despite their longevity within industry, cyclones remain some of the most highly versatile and misunderstood devices for separation of particulate from gas. They have an inherent simplicity of design that usually encompasses no moving parts, with the ability to collect fine particulate. Current designs of industrial high-efficiency cyclones can provide collection efficiencies of 90 percent and higher for particles with diameters as small as 2 μm (particulate specific gravity = 1). The centrifugal force placed on a particle in a cyclone is much greater than the force of gravity or inertia that can be placed on it in a gravity settling chamber or inertial separator, respectively. The centrifugal force applied to particles is usually on the order of several hundred times greater than the gravitation acceleration.[2] Therefore, cyclones currently represent one of the most efficient methods of particle separation for particles with diameters of 2 μm and larger.

Cyclones may be readily produced for service in the most severe conditions experienced in industrial processes, and they frequently require little or no maintenance. Because of their simple design with corresponding low initial capital costs, flexibility in operation, low maintenance costs, and low operating costs, cyclones should be the first devices considered for usage in the separation of particulate from gas. The use of cyclones, however, may be ruled out because of one of the following reasons:

1. A cyclone cannot reasonably achieve the required collection efficiencies.

2. It would be too costly (relative to the magnitude of other solutions) to ascertain the design data that can allow for an answer to reason 1 above.

8.2 How Cyclones Work

There are many different styles and designs of cyclones, but each operates on the same basic principles. This section deals with operational characteristics that are consistent among all cyclonic devices. If you are unfamiliar with the operation of cyclones in a general manner, then read Sec. 8.3 first. Section 8.3 describes many of the various devices which are considered cyclones and gives simple descriptions of the flow patterns and operation of each. In all cyclones particulate is separated from the gas stream by centrifugal force. Particulate is thrown toward the outside of a spinning column of gas, while the relatively clean gas exhausts from the center of the spinning vortex. The two main factors affecting the performance of any cyclone include (1) the velocity at which a given particle is moving toward the wall of the cyclone (where it is theoretically collected) and (2) the length of time available to move the particle into a region where it will be collected before the gas exits the device. This is called the cyclone *residence time*. There are two primary measures of cyclone performance: (1) the energy consumption in moving the gas particulate through the cyclone and (2) the fractional or size efficiency curve for the cyclone.

8.2.1 Pressure drop

The energy consumption is most frequently expressed as the *pressure drop* across the cyclone. Pressure drop is the difference between the gas static pressures measured at the inlet versus the outlet of the cyclone. Gases always flow from an area of high pressure toward that of low pressure, so the inlet static pressure is always higher than the outlet. The cyclone pressure drop is then expressed as

$$\Delta P = P_1 - P_2 \qquad (8.1)$$

where ΔP = pressure drop
      P_1 = absolute pressure at cyclone inlet
      P_2 = absolute pressure at cyclone outlet

Pressure Drop Example 1 A cyclone inlet static pressure is measured at + 2 in w.c. [+51 mmH$_2$O], and the outlet static pressure is measured as −6 in w.c. [−152 mmH$_2$O]. The atmospheric pressure is 14.7 lb/in^2 [1.034 kg/cm^2]. What is the cyclone pressure drop?

solution

$$\Delta P = P_1 - P_2$$
$$P_1 = 2 + 14.7\,(27.67) = 408.7 \text{ in w.c.}$$
$$= [51 + 1.034\,(10.340) = 10{,}743 \text{ mmH}_2\text{O}]$$
$$P_2 = -6 + 14.7\,(27.67) = 400.7 \text{ in w.c.}$$
$$= [-152 + 1.034\,(10{,}340) = 10{,}540 \text{ mmH}_2\text{O}]$$
$$\Delta P = 408.7 - 400.7 = 8 \text{ in w.c.}$$
$$= [10{,}743 - 10{,}540 = 203 \text{ mmH}_2\text{O}]$$

The units are the same as the individual pressure measurements. When attempting to accurately measure cyclone pressure drop, it is very important to ensure that the measurement truly represents static pressures only and does not include some component of velocity pressure. This is frequently difficult on the cyclone outlet where the gas is rapidly spinning causing a velocity pressure not only in the direction of the gas flow but also normal to that of the gas flow. This normal or radially directed pressure is caused by the centrifugal force on the spinning gases. It is important to measure static pressures as positive or negative before adding to the atmospheric pressure to get the absolute pressures. Pressure drop is most frequently expressed in inches of water column in the United States and millimeters of water column elsewhere in the world. This should by no means preclude the use of another measure of pressure if it is more appropriate for a given application. In process applications where the pressure drop is high, it is common to express pressure drop in inches of mercury, millimeters of mercury, atmospheres, bars, pounds per square inch, etc. The main factors affecting pressure drop in a cyclone are the geometry of the cyclone itself, the gas flow rate, gas density, and dust loading.

8.2.1.1 The geometry of the cyclone itself. The most pronounced effects are usually caused by the inlet and outlet sizes and geometry, although it is virtually impossible to evaluate these independently of the total cyclone geometry.

8.2.1.2 The gas flow rate. With all other conditions remaining constant, pressure drop will increase exponentially with the gas flow rate. Pressure drop as a function of changes in flow rate is expressed as

$$\Delta P_2 = \Delta P_1 \cdot \left(\frac{Q_2}{Q_1}\right)^{C_1} \tag{8.2}$$

where ΔP_2 = new cyclone pressure drop
ΔP_1 = original cyclone pressure drop
Q_2 = new cyclone flow rate
Q_1 = original cyclone flow rate

Pressure Drop Example 2   A cyclone has a pressure drop of 8 in w.c. [203 mmH$_2$O] at a gas flow rate of 10,000 acfm [16,992 m^3/h]. Assuming no change in gas density and C_1 = 2.1, what is the pressure drop at 16,000 acfm [27,188 m^3/h]?

solution

$$\Delta P_2 = \Delta P_1 \cdot \left(\frac{Q_2}{Q_1}\right)^{C_1}$$

ΔP_1 = 8 in w.c. = [203 mmH$_2$O]

Q_2 = 16,000 acfm = [27,188 m^3/h]

Q_1 = 10,000 acfm = [16,992 m^3/h]

C_1 = 2.1

$$\Delta P_2 = 8 \cdot \left(\frac{16,000}{10,000}\right)^{2.1} = 22 \text{ in w.c.} = \left[203 \cdot \left(\frac{27,188}{16,992}\right)^{2.1} = 545 \text{ mmH}_2\text{O}\right]$$

For most cyclones the exponent C_1 ranges between 1.9 and 2.3 and is best determined by actual measurements on a cyclone from a given family of cyclones. A family of cyclones is defined by each member of that family being geometrically proportional in all internal dimensions to all other members of that family. In other words, two cyclones are of the same family if they represent scaled versions of each other. The exponent then is the mathematical embodiment of the entire cyclone geometry as described above.

8.2.1.3 Gas density. Pressure drop is directly proportional to gas density. In a given cyclone at some operating condition, pressure drop will double with a doubling of gas density if all else remains the same. The pressure drop equation then becomes

$$\Delta P_2 = \Delta P_1 \cdot \left(\frac{Q_2}{Q_1}\right)^{C_1} \cdot \frac{\rho_{G2}}{\rho_{G1}} \tag{8.3}$$

where ΔP_2 = new cyclone pressure drop
ΔP_1 = original cyclone pressure drop
Q_2 = new cyclone flow rate
Q_1 = original cyclone flow rate
ρ_{G2} = new gas density
ρ_{G1} = original gas density

Pressure Drop Example 3   A cyclone has the same characteristics as described in Example 2 except the original gas density is 0.075 lb/ft³ [1.202 kg/m³] and the new gas density is 0.046 lb/ft³ [0.737 kg/m³]. What is the new cyclone pressure drop?

solution

$$\Delta P_2 = \Delta P_1 \cdot \left(\frac{Q_2}{Q_1}\right)^{C_1} \cdot \frac{\rho_2}{\rho_1}$$

$$\Delta P_1 = 8 \text{ in w.c.} = [203 \text{ mmH}_2\text{O}]$$

$$Q_2 = 16,000 \text{ acfm} = [27,188 \text{ m}^3/\text{h}]$$

$$Q_1 = 10,000 \text{ acfm} = [16,992 \text{ m}^3/\text{h}]$$

$$C_1 = 2.1$$

$$\rho_{G2} = 0.046 \text{ lb/ft}^3 = [0.737 \text{ kg/m}^3]$$

$$\rho_{G1} = 0.075 \text{ lb/ft}^3 = [1.202 \text{ kg/m}^3]$$

$$\Delta P_2 = 8 \cdot \left(\frac{16,000}{10,000}\right)^{2.1} \cdot \frac{0.046}{0.075} = 13 \text{ in w.c.}$$

$$= \left[203 \cdot \left(\frac{27,188}{16,992}\right)^{2.1} \cdot \frac{0.737}{1.202} = 334 \text{ mmH}_2\text{O}\right]$$

8.2.1.4 Dust loading.   Cyclone pressure drop decreases with increased dust loading, and vice versa. Although several theories exist to explain this phenomenon, the mechanism has not been empirically verified. Numerous empirical measurements have shown the loading effect to be

$$\Delta P_L = \Delta P_0 C_2$$

where for $0.433 < W < 5340$ [for $1.249 < W < 12,220.056$],

$$C_2 = 0.96 - 0.04722 \cdot \ln W = \left[0.96 - 0.4772 \cdot \ln \frac{W}{2.288}\right] \tag{8.4}$$

$$C_2 = \begin{cases} 1 & \text{if } W < 0.433 \ [W < 1.249] \\ 0.55 & \text{if } W > 5340 \ [W > 12{,}220.056] \end{cases}$$

where ΔP_0 = pressure drop at $W = 0$
$\quad\quad\ W$ = dust loading, grains/acf (gr/acf) [g/m^3]
$\quad\ \Delta P_L$ = pressure drop at specified dust loads

The general equation for cyclone pressure drop incorporating all the above becomes

$$\Delta P_2 = C_2 \cdot \Delta P_1 \cdot \left(\frac{Q_2}{Q_1}\right)^{C_1} \cdot \frac{\rho_{G2}}{\rho_{G1}} \tag{8.5}$$

where ΔP_1 is original pressure drop at $W = 0$.

8.2.2 Cyclone fractional efficiency

The true measure of a cyclone's performance as it relates to particulate removal is the fractional efficiency curve. Cyclones, as well as any other inertial devices, collect aerodynamically larger particles (i.e., those with greater terminal velocities) more readily than smaller ones. The continuous curve that shows the removal efficiency of some device versus particle size is the fractional efficiency curve. If the particulate entering a cyclone is ball bearings, we would expect a very high collection efficiency. Conversely, if the particulate was coal fly ash, we would expect a much lower rate of collection because the particles we are trying to collect are of such obvious difference in size. This property is true of all inertial separators and is the reason that their ability to collect particulate is expressed as size efficiency or fractional efficiency. This denotes that the ability of an inertial device to collect particulate is dependent on the aerodynamic particle size of the particles entering it. A fractional efficiency curve for a cyclone operating under certain conditions may be as shown in the table below.

Particle size (Stokes equivalent diameters), μm	Percentage Collection (by weight)
1	10.1
3	15.6
5	34.8
7	76.9
10	88.1
15	94.7
20	96.1
30	98.3
40	99.2
50	99.6

Graphical and tabular representations of actual cyclone fraction efficiency curves are shown in Figs. 8.1 to 8.3. Fractional efficiency curves will generally appear as S-shaped curves when plotted on normal scales and will become asymptotic toward 0 and 100 percent efficiencies and never actually reach either value. It is proper to represent fractional efficiency curves in either a graphical or tabular form as shown.

In a normal industrial application the user has no real concern for the fractional efficiency of the cyclone but rather for the total collection efficiency of that particular application. If we use a cyclone with any of the fractional efficiency curves shown in Figs. 8.1 to 8.3 for a given application, it would have a total collection efficiency that is dependent on the particle size distribution of the particulate entering the cyclone. On coal fly ash we might achieve only 80 percent total collection, while on wood sawdust we realize + 99 percent collection using the same device with all conditions being identical except the particle size distribution. In both cases, the true measures of cyclone performance, the fractional efficiency curve and the pressure drop, remained constant. For this reason it is vitally important prior to selecting a cyclone for a given application where total collection efficiency is important that extreme care be taken in accurately measuring the characteristics of the particulate that is to be collected, especially the particle size distribution.

If we have determined the fractional efficiency curve for a given cyclone, then the new fractional efficiency curve for any member of that family of cyclones may be determined by use of the following equation:

$$d_n = d_o \cdot K_1 \tag{8.6}$$

where K_1 = efficiency curve shift constant
d_n = new particle size
d_o = original particle size

$$K_1 = K_2 \cdot K_3 \cdot K_4 \cdot K_5 \tag{8.7}$$

Constant K_2 is determined by

$$K_2 = \left(\frac{\rho_{po}}{\rho_{pn}} \right)^{0.5} \tag{8.8}$$

where ρ_{po} is original particle density and ρ_{pn} is new particle density. (Efficiency goes up with increased particle density and vice versa.)

Constant K_3 is determined by

CYCLONE DATA

CYCLONE TYPE:	HIGH CAPACITY			
L/D RATIO:[1]	4			
		ENGLISH		SI
BODY DIAMETER:	2.2	FT	0.7	M
INLET VELOCITY:	59.1	FT/S	18.0	M/S
OUTLET VELOCITY:[2]	38.2	FT/S	11.6	M/S

PROCESS DATA

GAS FLOW RATE:	5000.0	FT³/MIN.	2942.5	M³/HR.
PRESSURE DROP:[3]	8.1	" W.C.	205.74	mm H_2O
GAS DENSITY:	0.075	LB/FT³	1.201	KG/M³
GAS VISCOSITY:	1.20E-05	LB/FT-S	0.0179	Cp
PARTICLE DENSITY:	62.4	LB/FT³	999.2	KG/M³

FRACTIONAL EFFICIENCY CURVE

PARTICLE SIZE STOKES EQUIVALENT DIAMETERS (MICRONS)	% COLLECTION (BY WEIGHT)[3]
1.0	2.73
2.5	15.81
5.0	37.95
7.5	53.99
10.0	65.13
15.0	78.69
20.0	86.25
30.0	93.46
40.0	96.47
50.0	97.93
75.0	99.30
100.0	99.71

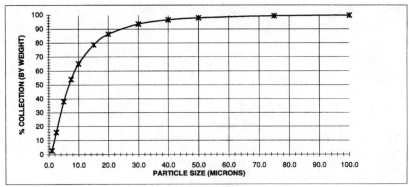

(1) L/D RATIO DOES NOT INCLUDE DUST RECEIVER WHICH IS INCLUDED IN THE PERFORMANCE CALCULATION.
(2) AVERAGE VELOCITY AS MEASURED AT THE BOTTOM OR ENTRANCE TO THE OUTLET PIPE.
(3) PERFORMANCE DATA SHOWN IS AT PARTICULATE LOADINGS BELOW .3 GR./FT³ [.69 G/M³].

Figure 8.1 High-capacity cyclone performance data. (*Courtesy of Fisher-Klosterman, Inc.*)

CYCLONE DATA

CYCLONE TYPE:	HIGH EFFICIENCY		
L/D RATIO:[1]	4		

		ENGLISH		SI
BODY DIAMETER:	3.1	FT	0.9	M
INLET VELOCITY:	59.1	FT/S	18.0	M/S
OUTLET VELOCITY:[2]	48.6	FT/S	14.8	M/S

PROCESS DATA

GAS FLOW RATE:-	5000.0	FT³/MIN.	2942.5	M³/HR.
PRESSURE DROP:[3]	8.0	" W.C.	203.2	mm H₂O
GAS DENSITY:	0.075	LB/FT³	1.201	KG/M³
GAS VISCOSITY:	1.20E-05	LB/FT-S	0.0179	Cp
PARTICLE DENSITY:	62.4	LB/FT³	999.2	KG/M³

FRACTIONAL EFFICIENCY CURVE

PARTICLE SIZE STOKES EQUIVALENT DIAMETERS (MICRONS)	% COLLECTION (BY WEIGHT)[3]
1.0	0.53
2.5	18.18
5.0	63.14
7.5	85.38
10.0	91.09
15.0	96.08
20.0	98.00
30.0	99.32
40.0	99.71
50.0	99.86
75.0	99.97
100.0	99.99

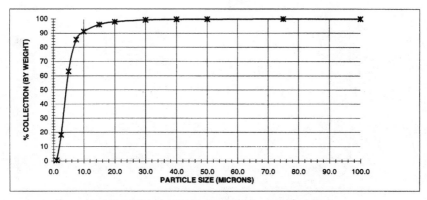

(1) L/D RATIO DOES NOT INCLUDE DUST RECEIVER WHICH IS INCLUDED IN THE PERFORMANCE CALCULATION.

(2) AVERAGE VELOCITY AS MEASURED AT THE BOTTOM OR ENTRANCE TO THE OUTLET PIPE.

(3) PERFORMANCE DATA SHOWN IS AT PARTICULATE LOADINGS BELOW .3 GR./FT³ [.69 G/M³].

Figure 8.2 High-efficiency cyclone performance data. (*Courtesy of Fisher-Klosterman, Inc.*)

CYCLONE DATA

	ENGLISH		SI	
CYCLONE TYPE:	ULTRA HIGH EFFICIENCY			
L/D RATIO:[1]	4			
BODY DIAMETER:	4.4	FT	1.3	M
INLET VELOCITY:	59.1	FT/S	18.0	M/S
OUTLET VELOCITY:[2]	52.7	FT/S	16.1	M/S

PROCESS DATA

GAS FLOW RATE:	5000.0	FT^3/MIN.	2942.5	M^3/HR.
PRESSURE DROP:[3]	7.8	" W.C.	198.12	mm H$_2$O
GAS DENSITY:	0.075	LB/FT^3	1.201	KG/M^3
GAS VISCOSITY:	1.20E-05	LB/FT-S	0.0179	Cp
PARTICLE DENSITY:	62.4	LB/FT^3	999.2	KG/M^3

FRACTIONAL EFFICIENCY CURVE

PARTICLE SIZE STOKES EQUIVALENT DIAMETERS (MICRONS)	% COLLECTION (BY WEIGHT)[3]
1.0	0.02
2.5	22.16
5.0	89.11
7.5	95.12
10.0	97.49
15.0	99.14
20.0	99.64
30.0	99.91
40.0	99.97
50.0	99.99
75.0	99.99
100.0	99.99

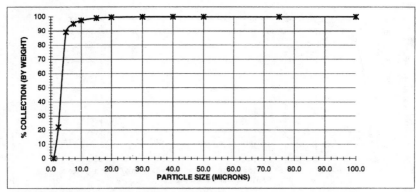

(1) L/D RATIO DOES NOT INCLUDE DUST RECEIVER WHICH IS INCLUDED IN THE
PERFORMANCE CALCULATION.
(2) AVERAGE VELOCITY AS MEASURED AT THE BOTTOM OR ENTRANCE TO
THE OUTLET PIPE.
(3) PERFORMANCE DATA SHOWN IS AT PARTICULATE LOADINGS BELOW
.3 GR./FT^3 [.69 G/M^3].

Figure 8.3 Ultrahigh efficiency cyclone performance data. (*Courtesy of Fisher-Klosterman, Inc.*)

$$K_3 = \left(\frac{\eta_n}{\eta_o}\right)^{0.5} \tag{8.9}$$

where η_o is original inlet gas viscosity and η_n is new inlet gas viscosity. (Efficiency goes down with increased gas viscosity and vice versa.)
    Constant K_4 is determined by

$$K_4 = \left(\frac{U_o}{U_n}\right)^{0.5} \tag{8.10}$$

where U_n is new gas velocity and U_o is original gas velocity.[3]
(Efficiency goes up with increased inlet velocity and vice versa.)
    Constant K_5 is determined by

$$K_5 = \left(\frac{D_n}{D_o}\right)^{1.5} \tag{8.11}$$

where D_n is new cyclone diameter and D_o is original cyclone diameter.
(Efficiency goes down with increased cyclone body diameter and vice versa.)
    The generalized equation combining all the above may be written as Eq. (8.7) or as follows:

$$K_1 = \left(\frac{\rho_{po} \cdot \eta_n \cdot U_o}{\rho_{pn} \cdot \eta_o \cdot U_n}\right)^{0.5} \cdot \left(\frac{D_n}{D_o}\right)^{1.5} \tag{8.12}$$

Cyclone Efficiency Curve Shifting Example   A high-capacity test cyclone of given geometry and performance is described as

Body diameter

$$D_0 = 2.2\,ft = [0.67\ \text{m}]$$

Inlet velocity

$$U_0 = 60\ \text{ft/s} = [18.29\ \text{m/s}]$$

Inlet gas viscosity

$$\eta_0 = 1.2029 \times 10^{-5}\ \text{lbm/ft} \cdot \text{s} = [0.0179\ \text{cP}]$$

Particle specific gravity

$$\rho_{po} = 1$$

A new cyclone of the same family will operate at the following parameters:

Body diameter

$$D_n = 8\ \text{ft} = [2.44\ \text{m}]$$

Test Cyclone Fraction Efficiencies

Particle size, μm	% Collection weight)
1	2.73
3	20.81
5	37.95
7	50.42
10	61.10
15	78.69
20	86.25
30	93.46
40	96.46
50	97.93
75	99.30
100	99.71

Inlet velocity

$$U_n = 65 \text{ ft/s} = [19.81 \text{ m/s}]$$

Inlet gas viscosity

$$\eta_n = 2.01 \times 10^{-5} \text{ lbm/ft} \cdot \text{s} = [0.0299 \text{ cP}]$$

Particle specific gravity

$$\rho_{pn} = 4$$

Determine the new fractional efficiency curve.

solution

$$K_1 = \left(\frac{\rho_{po} \cdot \eta_o \cdot U_n}{\rho_{pn} \cdot \eta_n \cdot U_0} \right)^{0.5} \cdot \left(\frac{D_n}{D_0} \right)^{1.5}$$

$$= \left[\frac{1 \cdot (1.2029 \times 10^{-5}) \cdot 65}{4 \cdot (2.01 \times 10^{-5}) \cdot 60} \right]^{0.5} \cdot \left(\frac{8}{2.2} \right)^{1.5} = 2.79$$

$$= \left[\left\{ \frac{1 \cdot 0.0179 \cdot 19.81}{4 \cdot 0.0299 \cdot 18.29} \right\}^{0.5} \cdot \left(\frac{2.44}{0.671} \right)^{1.5} = 2.79 \right]$$

Test cyclone		K_1	=	Particle size, μm New cyclone	% Collection (by weight)
1	·	2.79	=	2.79	2.73
3	·	2.79	=	8.37	20.81
5	·	2.79	=	13.95	37.95
7	·	2.79	=	19.53	50.42
10	·	2.79	=	27.90	61.10
15	·	2.79	=	41.85	78.69
20	·	2.79	=	55.80	86.25
30	·	2.79	=	83.70	93.46
40	·	2.79	=	111.60	96.46
50	·	2.79	=	139.50	97.93
75	·	2.79	=	209.25	99.30
100	·	2.79	=	279.00	99.71

As dust loading increases, so does cyclone collection efficiency. There are numerous proposed mechanisms to describe why efficiency increases with particle loading. The most compelling is that particles form agglomerates which behave as larger particles. The likelihood of particles forming agglomerates is related to the concentration of particles, the particle size distribution, and the physical characteristics of the particulate.[4,5] Although the effect of dust loading has been well-observed and quantified for specific conditions, a general description from the basic physical parameters has not yet been developed.

It appears likely that research in the area of particle agglomeration will provide the greatest advances in the understanding of the operation of many systems of solids and gases over the next decade. One prime example is the observed effect that sometimes collection efficiency in a cyclone will decrease with increased inlet velocity. This effect, although not consistently observed for all conditions, has been described as a function of particle reentrainment after collection. The most significant discussion of this principle describes the loss of collection efficiency as a function of saltation velocity of the particulate.[6] The logical difficulty with this description of the phenomenon is that not only does the ability of the cyclone to reentrain collected particulate increase with increased inlet velocity but so does the cyclone's ability to collect particulate and recollect entrained particulate. It is probable that the effect of decreased collection with increased inlet velocities for some conditions will be linked to the disruption of particle agglomerates, thus creating a much smaller apparent particle size distribution.

An empirical adjustment factor Z for loading should be applied to the cyclone fractional efficiency curve after the K coefficient has been used to shift the fractional efficiency curve for all conditions except dust loading. The generalized equation for applying Z is

$$E_{dn} = 100 - [Z \cdot (100 - E_{do})] \tag{8.13}$$

where E_{dn} = new efficiency at particle size d
$\quad\;\; E_{do}$ = original efficiency of particle size d
$\quad\;\; W$ = dust loading, gr/acf [g/m^3]
$\quad\;\; Z = 2.095 \cdot W^{-0.0197} - 1.09$

$$= \left[2.095 \cdot \left(\frac{W}{2.288} \right)^{0.0197} - 1.09 \right] \tag{8.14}$$

Dust Loading Fraction Efficiency Curve Shift Example   In the prior example, assume a dust loading of 110 gr/acf [251.68 g/m^3]. What is the new fractional efficiency curve?

solution

$Z = 2.095 \cdot W^{0.0197} - 1.090 = 0.8197$

$\quad = \left[2.095 \cdot \left(\dfrac{W}{2.288} \right)^{-0.0197} - 1.090 = 0.8197 \right]$

Particle size, μm	E_{do} [% collection (by weight)]	$E_{dn} = 100 - [Z \cdot (100-E_{do})]$ [% collection (by weight)]
2.79	2.73	20.27
8.37	20.81	35.09
13.95	37.95	49.14
19.53	50.42	59.36
27.90	61.10	68.11
41.85	78.69	82.53
55.80	86.25	88.73
83.70	93.46	94.64
111.60	96.46	97.10
139.50	97.93	98.30
209.25	99.30	99.43
279.00	99.71	99.76

It should be noted that the correction of E_{do} to E_{dn} using the constant Z is based upon observations and measurements in actual operating equipment at various conditions with numerous types of particulate. This correction for dust loading does not take into account any variance in particle characteristics except the loading itself and is generally used in the prediction of cyclone efficiency for selection of an emission control device or for product recovery applications. For these reasons, Z is generally conservative. Actual observed dust load-ing effects are generally greater than those predicted using this method. Use of this method will provide fractional efficiencies and therefore total collection efficiency calculations that are lower than actually observed. Although this is generally desirable, those who are selecting cyclones for use as particle classifiers should realize that in the prediction of particle separation for these purposes, the most accurate fractional efficiency curve possible is desirable. It is advis-able to perform pilot testing on the actual material at conditions that are as close to those of operating as possible when classification is desired.

A more liberal dust loading correction is presented by the American Petroleum Institute (API) and has been used heavily in the prediction of cyclone performance in conjunction with fluidized bed systems. Further discussion of dust loading effects follows in Sec. 8.2.8.[2]

The cyclone laws listed above may be used with a high degree of confidence and accuracy assuming the original fractional efficiency curve was determined accurately. Unfortunately, determining a cyclone fractional efficiency curve is a difficult and expensive process if done properly, and most engineers and manufacturers of cyclones have opted for the use of theoretically based mathematical models for the generation of fractional efficiency curves instead of the much more accurate empirical method.[7]

There is a vast array of literature citing methods of predicting cyclone fractional efficiencies utilizing generalized equations. One of the best theoretical approaches is presented by Leith and Licht as well as comparisons with prior methods of calculating fractional effi-ciencies.[8] It is strongly suggested that the designer work with empiri-cally based fractional efficiency curves whenever possible as it will provide the most accurate predictions of cyclone efficiencies. For this reason, a broad array of empirically based fractional efficiency curves are included as references herein. By selection of a fractional efficien-cy curve from those available and utilizing the cyclone laws listed previously, the engineer will be able to develop an efficiency curve that is more accurate than that predicted by any of the current theo-retical models.

The resulting variation between predicted and actual cyclone per-formance has led many to believe that the process is more art than science. The difficulty in accurately predicting cyclone fractional effi-ciency curves from theoretical models is caused by the extremely com-plicated flow patterns created within a cyclone (see Figs. 8.4 to 8.6). These flow patterns should be considered a general representation of the movement of gases within the device and not an exact description. Measurements of the direction and velocity of gas flow within

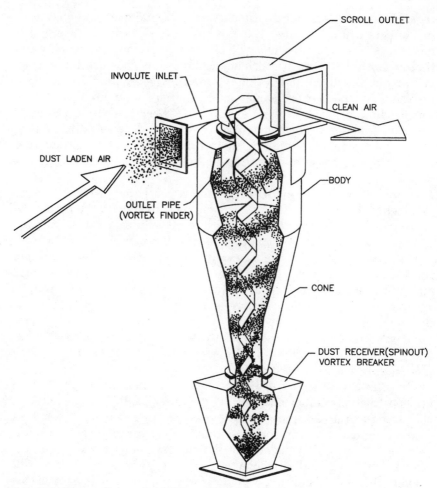

Figure 8.4 Reverse-flow cyclone. (*Illustration by Alex Donenberg. Courtesy of Fisher-Klosterman, Inc.*)

cyclones are extremely difficult due to the complexity of the flow patterns and the effect of the measuring device on the flow patterns themselves. Once these flow patterns have been measured for a given cyclone, it remains to be determined how they will be affected by an infinite array of possible variations in cyclone geometry, gas conditions, and particulate influences.

Much research into cyclones has been based upon the concept that there is some optimum or ideal geometry for a cyclone.[9] Within the broad context of industrial applications, there is no such ideal geometry for all applications. The design of any cyclone must balance variables of cyclone capacity (gas flow rate at a given pressure drop), cyclone collection efficiency, physical size, and cost. In most applica-

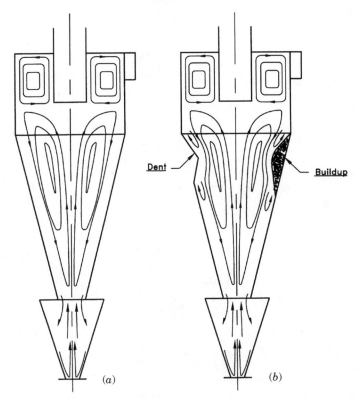

Figure 8.5 Cyclone secondary flow patterns. (*a*) Normal. (*b*) Effects of dents and buildups. (*Illustration by Alex Donenberg. Courtesy of Fisher-Klosterman, Inc.*)

tions, all these variables are of some importance. In general, the least expensive cyclone that will meet a performance requirement at the lowest energy consumption will be the ideal selection. A very high efficiency, high residence time cyclone would not be the ideal selection for an application where high collection efficiency is not required.

Further, given a known gas flow pattern, the ability to predict the removal of particulate from that flow pattern is as complicated as the determination of the original flow pattern and is also affected by many variables. Conversely, many have sought to describe cyclone gas flow patterns by observation of the visible bands of dust within a cyclone or the number of *turns* within the cyclone. It is inaccurate to assume that this is a reasonable description of the gas flow patterns. For these reasons, it is strongly recommended that only empirically based models of cyclone performance be used when cyclone performance is a critical aspect in the selection of equipment and will be vital to the process in

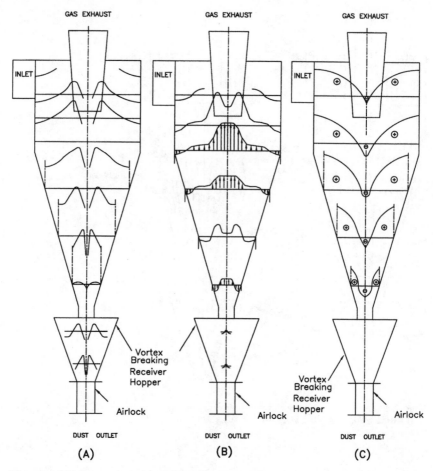

Figure 8.6   Relative variation of tangential, radial, and vertical components of velocity. (*Illustration by Alex Donenberg. Courtesy of Fisher-Klosterman, Inc.*)

which it is used. General calculations of cyclone pressure drop or fractional efficiencies may not be accurate unless they have been verified empirically for the family of cyclones under consideration.[10]

It is most important to understand the uses and misuses of the cyclone laws listed previously. The use of these laws for comparison of cyclones of a given family provides the most accurate tool currently available for prediction of cyclone efficiency. Unfortunately they are all too frequently used incorrectly in cyclone analysis. A prime example of this is the misunderstanding concerning large-diameter cyclones versus small-diameter cyclones. As described in Eq. (8.11), for a given family of cyclones with all other factors being equal, a small-diameter cyclone is more efficient than a large one.

By definition since both cyclones are of the same family and all other factors are held constant, including inlet velocity, the small-diameter cyclone would have a lower gas flow rate. If the smaller-diameter cyclone had the same gas flow rate, it would also benefit from a higher inlet velocity and, as a consequence, would have a higher pressure drop.

The correct usage of Eq. (8.11) then leads us to conclude that for a given set of operating conditions and a given cyclone family it is more efficient to use multiple smaller-diameter cyclones in parallel as opposed to fewer larger-diameter cyclones. This cyclone law is one of the most significant tools available to the cyclone designer for achieving improved collection efficiencies and has been well-proven over decades of cyclone usage and design. The incorrect usage of the law leads the designer to conclude, in general, that large-diameter cyclones are less efficient than smaller-diameter cyclones, even when comparing cyclones of different families.

It is impossible to perform a simple comparison of cyclone efficiency when the cyclones are of different families. It is not only possible, but common, for a large-diameter cyclone of a given family to be more efficient than a smaller-diameter cyclone of a different family. It is quite difficult to compare cyclones of different families without empirically based operating data for each individual family. In the prediction of performance for cyclone families where empirically developed operating data is not available from a member of that family, theoretical models are often used. Unfortunately, theoretically based models will rarely provide an accurate prediction of a cyclone's fractional efficiency curve or pressure drop to be used for selection or final design of a cyclone.

8.2.3   Cyclone total collection efficiency

The calculation of total collection efficiency in a cyclone once we have a known fractional efficiency curve and particle size distribution is a relatively simple mathematical operation. If we describe the fractional efficiency curve as a mathematical function of particle size $f(d)$ and the particle size distribution also as a function of particle size $g(d)$, then the total collection efficiency E_T is described as

$$E_T = \int_{d=0}^{d=\infty} \frac{\partial g(d)}{\partial y} \cdot f(d)\partial d \cdot 100 \qquad (8.15)$$

In practice it is common to perform the above operation as a series calculation, since the integral is not apparent or readily known in many cases for the product of $f(d) \cdot \partial g(d)/\partial y$. In this case the generalized form of the calculation of total collection efficiency is

$$E_T = \sum_{d=0}^{d=\infty} \left\{ [g(d)_{N+1} - g(d)_N] \cdot f \left[\frac{(d)_N + (d)_{N+1}}{2} \right] \right\} \cdot 100 \quad (8.16)$$

Total Collection Efficiency Calculation Example Using the fractional efficiency curve developed in the prior example and the following particle size distribution determine the total collection efficiency.

Particle size (Stokes equivalent diameters), μm	% Less than (by weight)
1	0.2
5	2.32
10	6.55
20	17.81
50	54.94
100	90
200	99.87

solution The solution of this problem using Eq. (8.16) is best shown utilizing a tabular format (see Table 8.1). To arrive at the values for fractional efficiencies and particle size percentages at particle sizes not given specifically in the prior tabular representation of the data, one may utilize one of the following methods:

1. Develop a mathematical regression model of the data that will allow for accurate prediction of points between existing data points. In general, mathematical functions that describe probability density functions will best fit the data that is typical for particle size distributions and fractional efficiency curves. Some of the common functions used are normal, log normal, Weibull, gamma, and beta distributions. Many times it is desirable for accuracy to use several regression segments, each with its own parameters, to describe the data accurately. In other words, a linear regression may be made up of several straight-line segments, each with its own intercept and slope, to accurately describe observed data.

2. Use graphs of data to read the desired data points. Understand that use of a graphical method also fits the observed data to some trend line. The quality of the regression methods, whether by eyeballing or least-squares regression, is critical to the accuracy of the calculation.

Table 8.1 is a graphical representation of the calculation of total collection efficiency E_T as described in Eq. (8.16). As a logical by-product of this calculation, one may also solve for total emissions E_E and/or the particle size distribution of collected or emitted particulate. Frequently, the particle size distribution of the effluent as well as the mass emissions is required to properly design downstream air pollution control equipment.

In Table 8.1, columns 1 and 2 divide the particle size distribution into 1-μm increments. Column 3 shows the average particle size for that incre-

TABLE 8.1 Total Collection Efficiency Calculation*

1	2	3	4	5	6	7
Particle size range, μm						
Low	High	Mid	Mass fraction	$E_{MID}/100$	Catch	Effluent
0.0	1.0	0.5	0.002 000 00	0.079 128 03	0.000 158 26	0.001 841 74
1.0	2.0	1.5	0.003 766 32	0.145 526 76	0.000 548 10	0.003 218 22
2.0	3.0	2.5	0.004 931 76	0.191 332 94	0.000 943 61	0.003 988 16
3.0	4.0	3.5	0.005 868 02	0.228 043 46	0.001 338 16	0.004 529 85
4.0	5.0	4.5	0.006 664 34	0.259 214 16	0.001 727 49	0.004 936 85
5.0	6.0	5.5	0.007 360 96	0.286 533 11	0.002 109 16	0.005 251 80
6.0	7.0	6.5	0.007 980 28	0.310 966 33	0.002 481 60	0.005 498 68
7.0	8.0	7.5	0.008 536 41	0.333 130 45	0.002 843 74	0.005 692 67
8.0	9.0	8.5	0.009 038 93	0.354 702 13	0.003 206 13	0.005 832 80
9.0	10.0	9.5	0.009 494 76	0.382 993 23	0.003 636 43	0.005 858 33
10.0	11.0	10.5	0.009 909 14	0.409 700 41	0.004 059 78	0.005 849 36
11.0	12.0	11.5	0.010 286 14	0.434 957 78	0.004 474 04	0.005 812 10
12.0	13.0	12.5	0.010 629 05	0.458 880 53	0.004 877 47	0.005 751 59
13.0	14.0	13.5	0.010 940 60	0.481 568 97	0.005 268 65	0.005 671 94
14.0	15.0	14.5	0.011 223 06	0.502 782 26	0.005 642 76	0.005 580 30
15.0	16.0	15.5	0.011 478 41	0.522 675 02	0.005 999 48	0.005 478 93
16.0	17.0	16.5	0.011 708 35	0.541 593 42	0.006 341 17	0.005 367 19
17.0	18.0	17.5	0.011 914 42	0.559 602 42	0.006 667 34	0.005 247 08
18.0	19.0	18.5	0.012 097 95	0.576 760 59	0.006 977 62	0.005 120 33
19.0	20.0	19.5	0.012 260 20	0.593 121 00	0.007 271 78	0.004 988 42
20.0	21.0	20.5	0.012 402 28	0.605 475 41	0.007 509 27	0.004 893 00
21.0	22.0	21.5	0.12 525 22	0.617 170 05	0.007 730 19	0.004 795 03
22.0	23.0	22.5	0.012 630 00	0.628 346 91	0.007 936 02	0.004 693 98
23.0	24.0	23.5	0.012 717 52	0.639 041 02	0.008 127 01	0.004 590 50

⋅ ⋅ ⋅

TABLE 8.1 Total Collection Efficiency Calculation*

1		2	3	4	5	6	7
	Particle size range, μm						
Low	High	Mid	Mass fraction	$E_{MID}/100$	Catch	Effluent
216.0	217.0	216.5	0.000 029 09	0.994 819 50	0.000 028 94	0.000 000 15
217.0	218.0	217.5	0.000 027 65	0.994 886 74	0.000 027 51	0.000 000 14
218.0	219.0	218.5	0.000 026 28	0.994 952 97	0.000 026 15	0.000 000 13
219.0	220.0	219.5	0.000 024 98	0.995 018 21	0.000 024 85	0.000 000 12
220.0	221.0	220.5	0.000 023 73	0.995 082 46	0.000 023 62	0.000 000 12
221.0	222.0	221.5	0.000 022 55	0.995 145 76	0.000 022 44	0.000 000 11
222.0	223.0	222.5	0.000 021 42	0.995 208 11	0.000 021 31	0.000 000 11
223.0	224.0	223.5	0.000 020 34	0.995 269 53	0.000 020 24	0.000 000 10
224.0	225.0	224.5	0.000 019 32	0.995 330 04	0.000 019 23	0.000 000 09
225.0	226.0	225.5	0.000 018 34	0.995 389 66	0.000 018 26	0.000 000 08
226.0	227.0	226.5	0.000 017 41	0.995 448 39	0.000 017 33	0.000 000 08
227.0	228.0	227.5	0.000 016 53	0.995 506 26	0.000 016 45	0.000 000 07
+		250.0	0.000 295 12	0.995 563 28	0.000 293 81	0.000 001 31
			1.000 000 00		0.774 604 45	0.225 395 55

*Note: $E_T = 77.46\%$ and
$E_E = 22.54\%$

ment. In column 4, the mass fraction of particulate within the size range listed in columns 1 and 2 is given. Column 5 shows the cyclone fractional efficiency at the average particle size (for the size range under consideration) divided by 100. Column 6 calculates the fraction of particulate collected for each size range by multiplying the value of column 4 by column 5. Column 7 calculates the fraction of particulate emitted for each size range either by column 4−column 6 or by column 4 × (1−column 5). The total collection efficiency E_T is then the sum of column 6 times 100, and the total emissions E_E is the sum of column 7 times 100.

In the above example, the linear regression of the Weibull function was used to determine the data point values between or beyond those specified. It is important to note that when using Eq. (8.16) the greater the number and smaller the size of the increments used, the greater the accuracy of the calculation. It is impossible to give an absolute rule for the number of increments required as it will vary depending on the specific cyclone fractional efficiency curve, particle size distribution, and the accuracy desired.

When performing this calculation manually, many will opt for steps of 10 μm and greater to keep the calculation to approximately 8 to 12 increments for most applications. No fewer than 100 increments are desirable if an accurate prediction of total cyclone collection efficiency is required. The continuation of the calculation for particle sizes greater than where cyclone fractional efficiencies have reached virtually 100 percent and/or the particle size distributions are 100 percent less than the particle diameter $d(N)$ is superfluous. Likewise, below a certain particle size the fractional efficiencies and particle size distribution both approach 0 percent and the product becomes zero and the calculation is superfluous. In general, using a computer, the number of increments can be set at a great enough value so that the calculation is accurate to several decimal places. To properly apply and use cyclones, it is important to have a basic understanding of the operating physics that are expressed in the cyclone laws.

8.2.4 Inlet velocity

In most cyclonic devices, the source of the circular flow of the gas and particulate is the conversion of a linear flow in a pipe or duct to a rotational flow by virtue of the manner in which the flow stream is introduced into the cyclone. The conversion of linear to circular flow takes place in or immediately behind the cyclone inlet. There are numerous different designs of the cyclone inlet (see Fig. 8.7). This is one of the most important points within the cyclone as the efficiency of a cyclone is directly related to the magnitude of centrifugal force

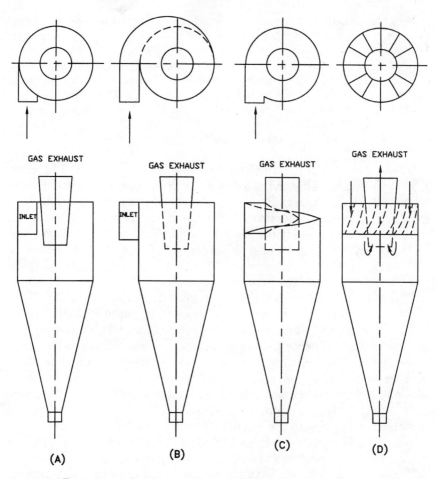

Figure 8.7 Basic types of cyclone entry. (*a*) Tangential entry. (*b*) Involute entry. (*c*) Curved entry. (*d*) Axial entry. (*Illustration by Alex Donenberg. Courtesy of Fisher-Klosterman, Inc.*)

placed on any given particle.* In general, the centrifugal force on an object that is traveling in a circular path is described by

---

*Many scientists and physicists deplore the use of the term *centrifugal force*. As a particle travels in a circular path, there are generally two forces that are acting on the particle. These are (1) the force acting in the direction of travel of the particle which is tangent to the circular path at any instant in time and (2) a force pulling toward the center of the circle which causes the particle to travel in a circular path. The force that is pulling toward the center of the circle is centripetal force. Without it, the particle would fly off in the direction of travel. We define centrifugal force as that force that is in the opposite direction and equal in magnitude to the centripetal force. Because of the ease of comprehension and widespread common usage of the term centrifugal force, we use it herein.

$$F_c = M \cdot \frac{U_T^2}{r} \qquad (8.17)$$

where F_c = centrifugal force
M = particle mass
U_T = tangential velocity
r = radius of circular path of travel

As you can see, the force on the object is dependent on and increases with an increase in radial (tangential) velocity U_T. The tangential velocity is directly derived from the linear velocity through the inlet of the cyclone. In general, the higher the inlet velocity within a given cyclone, the greater the centrifugal force on the particulate, and subsequently the particle in question will move toward the wall of the cyclone (where it is theoretically collected) at a greater velocity. One may consider the operation of the cyclone as a race where we try to move the particles to the wall of the cyclone before they reach the point of the cyclone where the gases exit. Therefore, we have better probabilities of winning the race with higher inlet velocities. The real effect on a given cyclone's collection efficiency is described by the shift in the fractional efficiency curve as expressed in Eq. (8.10). Unfortunately, there are significant costs associated with performance gained in this manner. Pressure drop (energy consumption) exponentially increases with increased inlet velocity as shown in Eqs. (8.2) and (8.4). In addition, when collecting abrasive particulate, the rate of erosion will usually increase with increased inlet velocity. A frequently used rule of thumb is that the rate of erosion will increase as the cube of the inlet velocity (i.e., if the inlet velocity doubles, the rate of erosion will increase by a factor of 8). In some cases, with friable particles (those that fracture or break) or agglomerates of particles, it is possible to generate fines at a rate that more adversely affects the collection efficiency than the gains offered by increased inlet velocity. As a final warning it is important to note that although the physics of increased collection efficiency with increased inlet velocity is valid for all velocities that may be considered, the requirements for an acceptable installation configuration may become more severe with increased inlet velocity. Because of the increase in severity of the vortex, collected fines may be reentrained. In general, care should be taken when inlet velocities exceed 100 to 120 ft/s [30 to 37 m/s] to ensure the desired performance levels will be met.

8.2.5 Body diameter

The cyclone body forms a physical barrier that will bound the outer edge of the rotating gases within the cyclone. If all other factors are the same, there will be less centrifugal force on particles traveling around a large-radius circle than a small-radius one. Body diameter then has an effect similar to velocity. Changes in diameter increase or decrease the centrifugal force on a particle which causes an increase or decrease in the velocity with which a given particle moves toward the wall of the cyclone where it is theoretically collected. This effect is generally described by $F_c = m \cdot U_T^2/r$, and the specific effect on a given cyclone's fractional efficiency curve is given by the cyclone laws in Eqs. (8.10) and (8.11).

Cyclone body diameter is one of the most powerful tools in cyclone design as well as one of the most misunderstood and misused. The use of the cyclone laws leads us to the correct conclusion that within a given family of cyclones it is possible to achieve higher particulate collection efficiencies with multiple cyclones in parallel as opposed to with a single larger cyclone (assuming proper installation of the parallel arrangement). Misuse of the cyclone laws has led many engineers and scientists to the incorrect conclusion that in general, small-diameter cyclones are more efficient than large-diameter cyclones. If we are comparing cyclones of different geometry (different families) at identical operating conditions, it is impossible to easily predict which cyclone will have the highest particulate collection efficiency. Very frequently, the larger-diameter cyclone will be more efficient than the smaller when the cyclones are of different families due to the complex interplay between the great number of factors which affect the cyclone fractional efficiency curve.

In practical usage, the cyclone laws relating to body diameter provide a reliable and useful tool for shifting the fractional efficiency curve from a known efficiency curve to that of smaller or larger members of the same family. It cannot be easily used for comparing cyclones of different families.

8.2.6   Gas viscosity

The viscosity of the gas from which we desire to remove particulate is of great significance in the cyclone selection and final operation. Viscosity is the measure of the resistance of a gas or liquid to movement. Since gas molecules must move out of the way to allow the passage of an object, viscosity relates directly to the drag forces placed upon the object and the subsequent speed that the object will reach in moving through that gas while under some driving force. This effect is best described by Stokes' law, which is discussed in Chap. 2. As can be readily seen from examination of Stokes' law, the terminal velocity of a particle is inversely proportional to the gas viscosity.

Although it is not practical in industrial applications to control the gas viscosity to achieve our ends, it is vital to understand its effect on cyclone operation. For practical consideration we assume that gas viscosity remains unchanged with pressure. As the gas temperature increases, so does the viscosity, with a resulting shift of the cyclone's fractional efficiency curve toward lower collection efficiencies. As gas temperature decreases, so does the viscosity, which results in an increase in collection if all other factors remain constant. An understanding of the gas viscosity effect on cyclone performance is important since cyclones are frequently selected for service at high temperatures and/or with gases other than air since their relatively simple construction often favors their use in severe processes. Gas viscosity is a measured property and cannot be accurately predicted from other known physical properties of that gas. The engineer should understand that viscosity is an entirely different and independent property from gas density with which it is often confused. With all other factors held constant as temperature increases, gas density decreases, and vice versa. In other words, the gas becomes less *thick* as temperature rises. Conceptually, we then might incorrectly imagine that it would be easier for a particle to move through a gas. Although the gas density has a significant effect upon the energy consumption required to move a body of gas, it has little effect upon the ease with which an object can move through the gas in most industrial conditions.

8.2.7   Particle density

The density of the particulate directly affects the speed with which any given particle will travel through a gas while acted upon by some force. This effect is expressed by Stokes' law. As the particle density increases, so does the velocity at which it travels through the gas, and there is a subsequent increase in cyclone collection efficiency. As particulate density decreases, so does the cyclone collection efficiency if all other factors remain constant. As with gas viscosity, particle density is not a factor that the designer can control, practically, but it is vital to have an understanding of its effect. Many industrial applications may have ranges in particle specific gravities (particle density/water density) from 0.3 or lower to more than 7.

8.2.8   Dust loading

It has been well known for many years that cyclone collection goes up with increased dust loading. Unfortunately, this is an area that has not been fully researched at this writing and methods of predicting the effect vary greatly. Furthermore, the methodology used in measuring the effect in most instances raises a question as to its reliability.

It appears likely that one of the most significant mechanisms causing increased fractional efficiencies with increased dust loading is that of agglomeration. As the dust loading increases, the probability of collision between particles and subsequent formation of agglomerates increases.[4,5] The correction to the fractional efficiency curve of a given cyclone is discussed in Sec. 8.2.2. An adjustment from E_{do} to E_{dn} that is used heavily within the petroleum industry is shown in Fig. 8.8a and b. These graphs show the new collection efficiency E_{dn} for a given particle size whose original efficiency E_{do} is known based upon various dust loading rates.[2] Since the reliability of this calculation is uncertain, it is recommended that the more conservative values shown in Fig. 8.8c and d be used. These values are those expressed previously by Eq. (8.13).

8.2.9 Cyclone geometry

Cyclone geometry is the greatest single variable that designers have at their disposal in affecting the final efficiency of the device. The effect of cyclone body diameter within a given family of cyclones has been discussed previously and leads to the common usage of cyclones in parallel arrangement for high-efficiency applications. This effect is valid while maintaining the geometric ratios or proportions of the cyclone family. There are an infinite number of geometries available

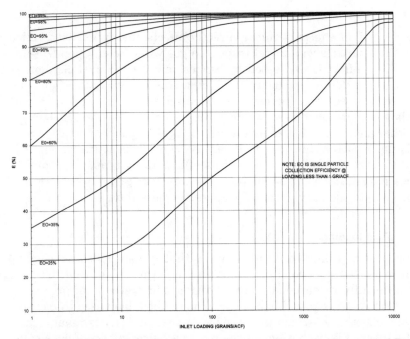

Figure 8.8a Effect of inlet loading on collection efficiency. (*Courtesy of Fisher-Klosterman, Inc.*)

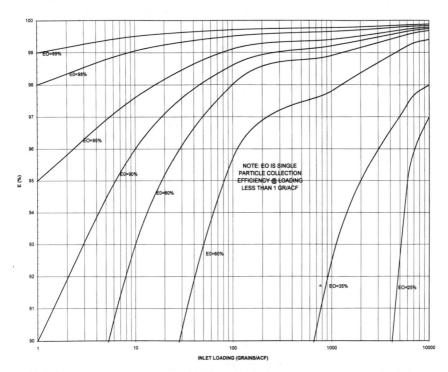

Figure 8.8b Effect of inlet loading on collection efficiency. (*Courtesy of Fisher-Klosterman, Inc.*)

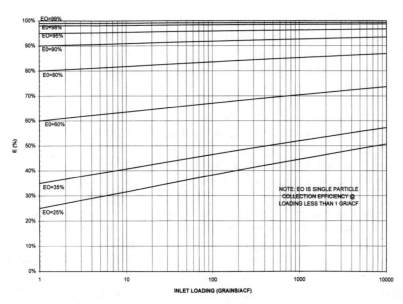

Figure 8.8c Effect of inlet loading on collection efficiency. (*Courtesy of Fisher-Klosterman, Inc.*)

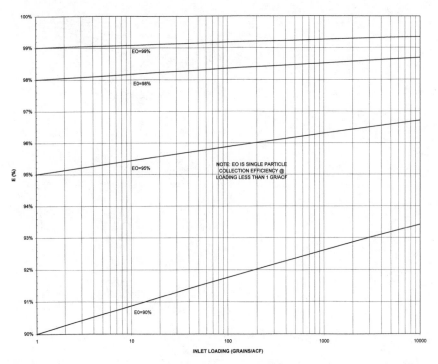

Figure 8.8d Effect of inlet loading on collection efficiency. (*Courtesy of Fisher-Klosterman, Inc.*)

within each of several basic cyclone styles. Some of the different styles of cyclones are described later in Sec. 8.3, but for the most part industrial usage and research has centered around *reverse-flow cyclones,* and this is what is usually envisioned when the term cyclone is used. Much research has been performed in an effort to describe accurately the flow of particulate and gas through the geometries imposed by the construction of cyclones. In general, the research has greatly increased the understanding and appreciation of the complexity of accurate cyclone design. Unfortunately this has done little to create an accurate enough mathematical description of the workings of cyclones to provide the designer with more than a gross approximation of cyclone performance. In fact, in much of the research, there is a lack of proper adherence to scientific methods due to the complexity, difficulty, and cost of research involving microparticulate. Also, many of the published theories and descriptions are inaccurate.

Nonetheless, and despite humankind's inability to accurately describe the mathematics of cyclone physics, variations in cyclone geometry can provide for significant improvements in collection efficiency at equal or lower energy consumption. In fact, since there are an infinite number of cyclone geometries that can operate at a given

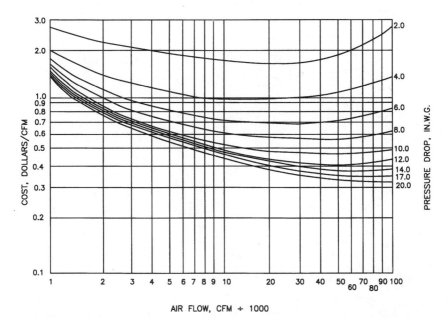

Figure 8.9 Cyclone costs as a function of airflow at a constant collection efficiency of 90 percent at particle diameter of 5 microns and a density of 168.5 lb/ft³. It is assumed that the gas is air at 70°F and 14.7 psia. Dust loading is 10 gr/acf. Prices are for single 10-gauge carbon steel cyclones without accessories or ductwork in 1994 U.S. dollars. (*Illustration by Alex Donenberg. Courtesy of Fisher-Klosterman, Inc.*)

pressure drop at a given set of operating conditions, it is always possible to design a better cyclone than the last to meet a given set of operating conditions.

The limitations on cyclone usage for the removal of fine particulate in most industrial applications are those of economics and not limitations within the physical properties that affect cyclone design. Cyclones are commonly used in the determination of in situ particle size distributions utilizing several small cyclones in series, some of which have cut sizes of less than 1 μm.

Unfortunately, these cyclones are quite small and handle a relatively small gas flow rate by industrial standards. Arranging several thousand of these into a parallel arrangement that maintains the performance achieved with one poses economic, not physical, barriers to common usage. Cyclone costs as functions of air at constant efficiency and various pressure drops are shown in Fig. 8.9. Efficiency in all cases is 90 percent in collecting 5-μm-diameter particles with air at standard temperature and pressure. Dust density in the air passing through the cyclone is 168.5 lb/ft³ [2698 kg/m³]. Figure 8.10 illustrates costs as functions of airflows at constant pressure drop, and various efficiencies are shown by these curves. Pressure drop in all

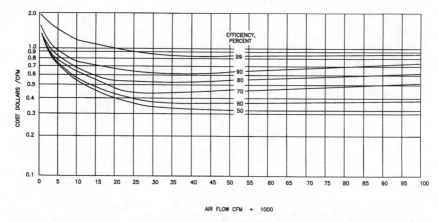

Figure 8.10 Cyclone costs as a function of airflow at a constant pressure drop of 6 in
H$_2$O with air at 70°F and 14.7 psia. Collection efficiencies are for 5-μm particles having
a density of 168.5 lb/ft^3. Dust loading is 10 gr/acf. Prices are for single 10-gauge carbon
steel cyclones without accessories or ductwork in 1994 U.S. dollars. (*Illustration by
Alex Donenberg. Courtesy of Fisher-Klosterman, Inc.*)

cases is 6.0 in w.g. [152 mmH$_2$O] with air at standard temperature
and pressure. Collection efficiencies listed for the cyclones are for 5-
μm-diameter particles having a density of 168.5 lb/ft^3 [2698 kg/m^3].

In recent years we have found that the efficiency of a cyclone is
more dependent on its overall geometry than any other factor, save
proper installation. In general, it can be concluded that the following
guidelines apply to cyclone design.

1. *Body diameter.* If L/D ratios (overall cyclone height/cyclone
body diameter) and all inlet conditions remain constant, then efficiency will be greater for large-diameter cyclones by virtue of the
increased residence time if all other dimensions remain constant. As
discussed previously, one effect of increased diameter is reduced efficiency as a result of reduced centrifugal force. If the above conditions
are maintained, however, the increase in residence time will provide a
net enhancement to the collection efficiency for most industrial particulate despite the reduced centrifugal force. In fact, this change of
body diameter and the net height of the cyclone while holding all
other dimensions constant generates a new family of cyclones. The
only absolute statement one can make about the performance of
cyclones of different families is that they will have different fractional
efficiency curves. Our research has shown that frequently the fractional efficiency curves will intersect and that the high-residence-time
cyclone will be more efficient on larger-diameter particles (above 1
μm), while low-residence-time (high-capacity) cyclones are often more
efficient on extremely fine particulate with diameters below 1 μm.

Since cyclones are most frequently used for particulate that is mostly above the particle sizes at which the curves cross (0.5 to 2 μm), high-residence-time cyclones are usually more efficient. Graphically, the reader may see this effect in Figs. 8.1 to 8.3 by comparing cyclones of similar flow rates and pressure drops.

2. *L/D ratio.*   The length-to-diameter ratio (or height-to-diameter ratio) *L/D* is the total of the cyclone body height plus the cone height divided by the inside diameter of the cyclone body or barrel. Generally, with all other factors remaining constant, cyclone performance improves with increased *L/D* ratios. Normally this ratio is between 3 and 6 for high-efficiency cyclones with 4 being the most common value. *L/D* values should rarely be below 2 if there is any concern with cyclone performance at all. Research has shown minimal value in using *L/D* ratios greater than 6.

3. *Outlet pipe diameter.*   The smaller the bottom end of the outlet pipe, the higher the cyclone pressure drop and collection efficiency if all other factors are held constant. The effects of reducing the bottom diameter of the outlet pipe are multiple. As the gas enters the outlet pipe it is still spinning with the same rotation induced by the inlet flow pattern. As the diameter of the spinning vortex of exiting gases decreases, the centrifugal force on a particle in this flow stream increases and the probability that the particle will be thrown out of this gas stream before it exits the cyclone increases. The smaller-diameter outlet pipe also has the effect of causing a particle that is escaping the cyclone to cross a greater distance to become entrained in the exhaust gas flow. Reducing the diameter of the outlet pipe also reduces the axial (vertical) velocity components within the cyclone, increasing the residence time. Pressure drop will increase with increases in outlet velocity (decreases in outlet pipe diameter).

4. *Outlet pipe length.*   One of the most abused areas and least understood areas of cyclone design is that of outlet pipe length. For any cyclone there is an optimum outlet pipe length. If the outlet pipe is lengthened or shortened from the optimum length, there will be an increased pressure drop and reduced collection efficiency.[11] Although some theoretical calculations show optimum outlet pipe length for some cyclones to be such that the outlet pipe does not extend below the floor of the inlet, we have not found this to be true for any cyclone tested. In general, the outlet pipe should extend below the bottom of the inlet by 10 to 60 percent of the inlet height. Significant losses of efficiency on both small and large particles may occur by short circuiting of the gas flow if the outlet pipe does not extend below the inlet floor. Beyond the optimum length, cyclone collection will decrease while pressure drop increases. The decrease in performance will become dramatic as the bottom of the outlet pipe approaches the cyclone cone.

5. *Inlet design.*   There are several different styles of inlet design that are in common usage in industrial cyclones. The most common styles are the involute or scroll inlet, the tangential inlet, and the axial inlet (see Fig. 8.7). Each style of inlet has advantages and disadvantages as one would expect. The involute inlet is generally considered to be the most efficient design and is more expensive to fabricate than a tangential inlet. The involute inlet, by providing a relatively gentle introduction of the gas flow into the cyclone body, has the lowest energy consumption in achieving a certain inlet velocity. The cyclone efficiency is enhanced while momentum is maintained as the flow stream radius decreases from the inlet into the body and subsequently toward the axis of the cyclone (see Fig. 8.6a). If compared with an identical cyclone with the inlet arranged as a tangential inlet, the tangential inlet cyclone will have a higher pressure drop and slightly higher fractional efficiencies in many cases. This has led many to conclude that tangential inlets are more efficient than involute. If we then use the energy savings in the involute inlet cyclone to increase the collection efficiency by reducing the inlet area (increasing inlet velocity) to the point at which the involute inlet cyclone and tangential inlet cyclone have the same pressure drop, the involute inlet cyclone will have higher fractional efficiencies.

Tangential inlet cyclones provide the lowest cost in most cases, especially if the cyclone is to be designed for high-pressure or vacuum applications. In cyclone designs where a large body diameter and relatively small diameter outlet pipe are used, the increased pressure drop of the tangential inlet versus the involute inlet may be very small. Extremely high pressure drops and erosion with abrasive particulate may occur if a tangential inlet is used on a cyclone where the inside edge of the inlet is inside the point at which it would intersect the wall of the outlet pipe.

The axial inlet is the most expensive design for fabrications from sheet and plate and has no advantage from a performance point of view. This design can be economically advantageous in a multicyclone design (see Fig. 8.11) where the individual cyclones are of cast construction and the top axial inlet simplifies the plenum design.

6. *Outlet pipe shape.*   In high-residence-time cyclones it is possible to utilize a conical-shaped outlet pipe. The cyclone will have the performance of a cyclone with a cylindrical outlet pipe equal to the small diameter at the bottom, but because of pressure regain in the concentric expansion of the outlet pipe, the pressure drop will be lower.

7. *Cone shape.*   The cyclone cone serves the purpose of reducing the strength of the cyclone vortex at the material discharge making it possible to disengage the particulate to an area where it will not be substantially reentrained. For the most part, only reverse-flow

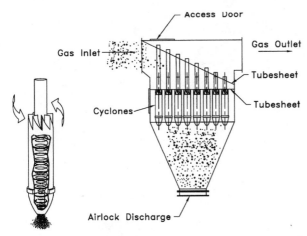

Figure 8.11 Multiple conventional cyclone design. Axial entry type with preliminary settling chamber. (*Illustration by Alex Donenberg. Courtesy of Fisher-Klosterman, Inc.*)

cyclones have conical lower sections as described here. The amount of disengaging area required below a cyclone is dependent on the inlet velocity, particle size, the general geometry of the cyclone, and the diameter at the bottom of the cone. If all other factors are equal, the smaller the cyclone discharge, the lower the turbulence beneath the cyclone and subsequent reentrainment, and vice versa.

If the cone discharge diameter is too small, however, a loss of efficiency will occur due to the movement of the particulate into the axis of returning gases in the center of the cyclone. Further, it is important to note several practical operational problems associated with an undersized discharge diameter. A significant loss of efficiency will occur if the discharge diameter is too small to pass the amount of material being collected. As a minimum, the discharge diameter of the cyclone dust and/or dipleg shall be sized as[2]

$$D = \left(153 \cdot \frac{M}{\rho_B}\right)^{0.4} = \left[3.5 \cdot \left(2450 \cdot \frac{M}{\rho_B}\right)^{0.4}\right] \qquad (8.18)$$

where D = discharge pipe diameter, in [cm]
   M = solids mass flow rate, lb/s [kg/s]
   ρ_B = solid bulk density, lb/ft^3 [kg/m^3]

Discharge Sizing Example   A cyclone has a mass loading of 80 lb/s [36.3 kg/s] of ammonium phosphate dust. What is the minimum cyclone discharge diameter?

solution

$$D = (153 \cdot \frac{M}{\rho_B})^{0.4} = [3.5 \cdot (2450 \cdot \frac{M}{\rho_B})^{0.4}]$$

$$M = 80 \; lb/s = [36.3 \; kg/s]$$

$$\rho_B = 21 \; lb/ft^3 = [336.3 \; kg/m^3] \qquad \text{(See App. 1G in Chap. 1.)}$$

$$D = \left(153 \cdot \frac{80}{21}\right)^{0.4} = 12.8 \text{ in}$$

$$= \left[3.5 \cdot \left(2450 \cdot \frac{36.3}{336.3}\right)^{0.4} = 32.6 \text{ cm}\right]$$

Other critical points for solids and gas passage may also be examined using this operation. In the case of high-capacity cyclones at high dust loadings, it is wise to compare the area of the annular space between the body of the cyclone and the outlet pipe to ensure the cross-sectional area exceeds the area of the diameter calculated in Eq. (8.16). If it is not, abrasion in this area may be severe and plugging due to bridging may be frequent.

8.2.10   Particle sizing for cyclone calculations

Great care must be taken in data analysis for any prediction of total cyclone efficiency to be accurate. Not only must we have an accurate, if not conservative, fractional efficiency curve, but we must also have an accurate and complete aerodynamic particle size distribution. The total collection efficiency calculation will only be as good as these two input parameters.

A very common error in the design and specification of a cyclone system results from selection of an inappropriate level of precision in measurement of vital data or the specification thereof. For example, assume we have an application where 99.9 percent overall collection efficiency is required and that the smallest point measured from the particle size distribution is 0.3 percent less than 10 μm. It is impossible to accurately select a cyclone that will emit 0.1 percent or less of the particulate when 0.3 percent of the sample is undefined. We have no real knowledge about the fraction of the particulate that is below 10 μm. It may or may not be accurately described by a linear regression of the smallest data points. It is possible that not only is there 0.3 percent less than 10 μm but also that there is 0.3 percent less than 0.1 μm. It is also possible that none of the particulate (0 percent) is less than 9 μm.

For this reason, the purchaser of a cyclone should specify the fine end of the particle size distribution using sound engineering judgment if appropriately precise and accurate data is not available. If this information is not specified, the cyclone vendor will make this assumption for the purchaser and may make it impossible for a purchaser to compare cyclone selections from different vendors.

Although testing and data acquisition is covered more fully in Chap. 3, the requirements for accurate particle sizing for the design of inertial separators (including cyclones) are somewhat unique and require special mention. Particle size analysis of a sample of particulate that is to be collected in a cyclone should measure those properties of the sample that are identical to those that will affect the performance and operation of the cyclone. That is, the particle size distribution should be determined by an aerodynamic method. Further, development of cyclone fractional efficiency curves requires extensive measurement of particle size distributions which should also be by aerodynamic methods.

It is possible to measure the size of particulates through a variety of means. The simplest and most direct is to physically measure a particle. This may be done with a ruler or micrometer if the particles are large or a microscope and scale if the particles are small. In industry and science, a number of methods are used which measure the physical size of particles in a sample. The most common is the sieve which is a mesh or screenlike surface with tightly controlled sized openings. If a particle passes through the sieve, it is obviously smaller than the sieve opening size in at least two of its three dimensions. The particle remains on top of the sieve or is considered larger than the sieve opening size if any two dimensions of the particle are greater than the opening. There are numerous methods of particle sizing that measure the same parameters more easily and accurately than sieving as the particle sizes become smaller. These devices generally measure physical size in only two dimensions and make no measurement of particle density. In some of these devices, the known specific gravity or particle density may be input to the device and the device will calculate an *aerodynamic particle size* based upon measured or assumed particle characteristics.

These physical measurements do not measure the aerodynamics of a particle, and subsequently any predictions of aerodynamic behavior from this calculation may be erroneous. Usually, if the measured particles are generally spherical, nonporous, and of homogeneous consistency, physical sizing methods will more closely match aerodynamic sizes than if the particles are of nonspherical shape or nonhomogeneous consistency. Even in the situation where the particles are uniformly spherical and homogeneous, there are significant variations in measured particle sizes between methods, particularly between physical sizing methods versus aerodynamic methods. How a particle acts in a gas stream, or its aerodynamics, is all important in the collection of that particle in an inertial separator like a cyclone or venturi scrubber. In fact, the physical size of a particle, in itself, may be meaningless in describing how well we can remove these particles from a gas stream. In some situations, physical sizing methods coincidentally agree with the aerodynamic description of the particles since the third dimension of the particle is similar to the first two (i.e., a spherical particle) and the particle is of homogeneous consistency and density.

The method used to determine the aerodynamics of a particle is to measure the velocity at which a particle falls through a gas or liquid while subject to a force (e.g., gravity). Of course, it takes some time for the particle to reach a steady-state velocity where the acceleration forces are counterbalanced by the drag forces on the particle. As can be seen in Chap. 2, this time is very short for small particles. If a sky diver jumped from a plane and did not encounter air resistance, he or she would continue to accelerate at 32.2 ft/s^2 [9.8 m/s^2] indefinitely. In actuality, a free-falling sky diver will reach a constant velocity at around 100 mi/h (depending on body position). This steady-state constant velocity is called terminal velocity.

Two particles of equal terminal velocity would have the same probability of being collected in an inertial device regardless of how different any of their other physical parameters may be. For instance, a 10-μm particle of coal fly ash may have the same terminal velocity as the brown skin from a peanut. The physical sizes, the densities, and the chemical compositions are all radically different, but the result is nevertheless an equal collection efficiency in an inertial device such as a cyclone.

When we measure the aerodynamics of a dust sample, we actually determine a terminal velocity distribution. In other words, for a given sample, we may find that 5 percent of the sample has less than a 1-ft/s terminal velocity, 10 percent of the sample has less than a 5-ft/s terminal velocity, etc.

Since it is confusing to describe dusts by their terminal velocity distribution, we equate particle sizes with terminal velocities using Stokes' law (see Chap. 2). Stokes' law lets us calculate a homogeneous spherical particle of a given density that has a diameter which will give us a specific terminal velocity. This is Stokes' equivalent diameter.

Just as with physical sizing methods, there are numerous devices for the determination of aerodynamic particle size distributions. Their results are similar, and they usually correspond with reasonable correlation from one method to another. The four most common methods of aerodynamic particle size analysis are

Bahco Micro Particle Classifier (ASME Power Test Code 28). This is the only method that is covered by an ASME code. This helps control the reproducibility of tests performed by this method. The Bahco is simply a laboratory-quality rotary classifier that makes several cuts between 1 and 40 μm. The method is somewhat difficult to perform and requires experience to get accurate results.

Liquid sedimentation. This is a general method that describes a number of specific devices and test procedures. In liquid sedimentation, a sample of particulate is dispersed in a liquid (frequently water) and its rate of settling is measured.

Micromerograph. This device uses the same principle as liquid sedimentation only the samples are dispersed in air instead of a liquid.

Cascade impactor. The cascade impactor is frequently used in conjunction with stack tests and has the advantage of providing in situ results. The impactor is a series of small inertial collection devices each with progressively smaller cut sizes. By weighing the collection from each stage, we are able to determine the aerodynamic distribution.

8.2.11 Cyclone flow patterns

The internal flow patterns and characteristics thereof have been studied extensively since the 1940s. Unfortunately, a great number of the studies did not follow adequate scientific procedures to definitively support the conclusions drawn. For this reason, we will try to address the subject of cyclone flow patterns with a presentation of what is known.

It has been well-proven by measurement and observation that the primary flow pattern within a cyclone can be described as a double rotating helix pattern, as shown in Fig. 8.4. The gas enters toward the top of the cyclone tangentially. As it spins, it travels downward along the wall of the cyclone. The gas, while continuing to spin in the same direction of rotation, reverses direction and flows upward along the axis of the cyclone to the outlet pipe. The tangential velocity (Fig. 8.6a) is primarily the resultant of the inlet velocity. Numerous studies have shown the tangential velocity at most locations within the cyclone where the general direction of flow is downward (excluding the central return vortex) to be[8,9,11]

$$U_T \cdot r^l = \text{constant}$$

where U_T = tangential velocity measured at any point along r
r = radius of flow path at point velocity is measured
l = cyclone exponent ranging between 0.5 and 0.9

The practical impact of this effect is that U_T will increase with reductions in r. In other words, as the gas moves toward the central core of the cyclone, not only does the angular velocity increase but so does the tangential velocity. The result is that the tangential velocities and subsequent centrifugal force achieved within the cyclone body will be several times higher than that achieved in the inlet of the cyclone itself. A typical vertical or axial velocity profile is shown in Fig. 8.6b.

Recent research we have conducted indicates that the primary determinant of the downward velocity component along the wall of the cyclone is the nozzle effect caused by the annular space between the wall of the cyclone and outlet pipe. An approximation of an average velocity may be arrived at by simply dividing the gas flow rate by the open cross-sectional area at the bottom of the outlet pipe. This downward axial flow component pneumatically transports the particulate to the discharge point of the cyclone. Gravity plays a relatively insignificant part in the discharge of particulate from a cyclone.

The upward axial flow velocity is less easily determined since there is no clear prediction of the size (diameter) of the inner vortex at any given elevation within the cyclone except immediately prior to entry into the outlet pipe. It is known that the volumetric gas flow rate of the inner and outer vortex decreases the further below the cyclone inlet a measurement is taken.

In practical applications, there is significant deviation between the theories of flow and particle collection within cyclones. Much evidence exists to support the conclusion that much of the particulate that escapes a cyclone was theoretically collected. This means that the emitted particulate was moved to within close proximity of the cyclone wall, then moved to the cyclone discharge, and subsequently reentrained and lost from the cyclone. Much research has been carried out on the subject of reentrainment, and, although not quantitative, the qualitative conclusions are

1. Longer cyclones have lower rates of reentrainment, while shorter cyclones have greater rates with all other conditions and geometry held constant.

2. Dust receivers beneath cyclones result in lower rates of reentrainment than in installations where an air lock is placed directly at the cyclone discharge with all other conditions and geometry held constant.

3. Reentrainment from dust receivers is directly related to the size of the cyclone discharge diameter into the dust receiver with all other conditions and geometry held constant.

4. Much of the particulate that is reentrained from the lower portions of the cyclone or dust receiver is subsequently recollected in the

cyclone proper. Therefore, levels of turbulence or fluidization within the lower portions of the device do not necessarily correlate directly to measured mass emission levels or nominal efficiency levels.

Although a discussion of various theories of cyclone flow is not intended, there have been many misconceptions drawn from research in this area that have resulted in cases of less-than-optimum and poor cyclone design. For this reason, a few of the difficulties surrounding research in these areas will be pointed out.

Visual observations of the flow within cyclones is preferred by most research engineers and many scientists, if for no other reason than that it is fascinating to watch. Observations of the return vortex quality by Stairmand;[9] vortex activity by Bryant, Silverman, and Zenz;[12] and natural vortex length by Alexander[13] and others have led to some positive engineering guidelines whether or not the conclusions and mechanisms cited are fully proven. The reported observations from these researchers may provide a basis for understanding and/or visualizing what is occurring within a cyclone, but the reader should be aware of the following problems that may be associated with many of the measurements.

1. Observations of internal flow patterns usually involve observations of entrained particulate or liquid as the gas is usually clear. Visible bands and concentrations of a particulate, which is often behaving as an agglomerate, do not necessarily describe the gas flow patterns accurately and fully.

2. Internal obstructions such as measuring devices and pistons will affect the flow patterns they are being using to measure or highlight.

3. Observed physical phenomena such as poor vortex shape or solids entrainment in the inner vortex are not quantitative. Frequently, the conclusion from such an observation of poor inner vortex quality is that this is obviously bad. In many cases, this effect was not shown quantitatively or in a scientific manner and the "obvious" conclusion has not been properly proven or quantified.

Many aspects of presumed cyclone flow patterns are derived from observations of material flow within cyclones. Although the conclusions drawn are not necessarily accurate in describing gas flow patterns, the observations themselves provide valuable insight into cyclone operation. Visible bands of concentrated particulate are often formed that somewhat resemble that shown in Fig. 8.4. Although this is not an exact representation for all types and loadings of dust particles and all cyclones, the general depiction is reasonable. Erosion of the cyclone surfaces, in the case where the particulate is abrasive, may be concentrated in areas where these bands of particulate continually impact or slide along the cyclone shell.

As the bands of particulate spiral downward into the conical section of a conventional cyclone, these spirals will flatten out and become more horizontal. In some cases, a continuous band of material may be held in suspension within the cone. The band of material is forced to discharge material as new particulate is introduced at the top. This phenomenon is predictable from primary physics as shown in Fig. 8.12. In this case, a given particle will be held in suspension at a given elevation within the cone of a cyclone when the resulting forces on a particle are balanced as expressed by Eq. (8.18).

$$F_{CU} = F_g + F_G \tag{8.18}$$

where F_{CU} = upward (and parallel to cone) force resulting from centrifugal force on particle

F_g = gravitational force (downward parallel to cone)

F_G = drag force of gas on the particle (downward parallel to cone)

$$F_{CU} = \frac{M \cdot U_T^{\ 2}}{r} \cdot \sin \theta \tag{8.19}$$

$$F_g = M \cdot g \cdot \cos \theta \tag{8.20}$$

$$F_G = 3 \cdot \pi \cdot \eta_G \, U_F \cdot d_p \tag{8.21}$$

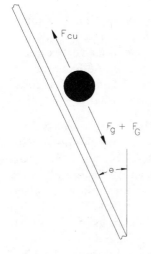

Figure 8.12 Forces on an individual particle in a cyclone core. (*Courtesy of Fisher-Klosterman, Inc.*)

where M = particle mass, g
  U_T = tangential gas velocity at radius r, cm/s
  θ = cone angle from vertical or half the apex angle, degrees
  g = gravitational acceleration, 980 cm/s^2
  η_G = gas viscosity, g/cm · s or P
  U_F = axial gas velocity, parallel to the cyclone wall, cm/s
  d_p = particle diameter, μm
  r = radius of cyclone at elevation under consideration, cm

A sample calculation is shown in Table 8.2. This calculation is based upon the following assumptions:

1. Friction losses are negligible.

2. U_F may be calculated directly from the gas flow rate divided by an open cross-sectional area between the cyclone wall and outlet pipe.

3. The particles are spherical and homogeneous.

4. Particle agglomeration is ignored.

This calculation indicates that in this particular case, if the cyclone has a discharge diameter that is 20 in (6508 mm) or less, particles below 120 μm in diameter will be discharged. Particles that are about 150 μm in diameter and larger will be suspended above the cyclone discharge until forced downward by the additional accumulation of particulate. Although this calculation is based upon a great number of gross assumptions, it does predict and illustrate this commonly observed phenomenon.

Pressure profiles for various cyclones under various operating conditions have been measured and studied extensively. Measuring static pressures at various points within an operating cyclone is somewhat more simple than measurement of velocity pressure, and therefore more empirical data for static pressure exists. Despite the availability of data, the reader should be aware that pressure profiles both in absolute and relative values may vary greatly with cyclone geometry and operating conditions. The simplest description of pressure profiles within a cyclone is that the static pressure will be the highest along the walls of the cyclone while the central core will have the most negative pressures. A typical pressure profile is shown in Fig. 8.6c. It should be understood that since the central vortex of the cyclone has a relatively negative pressure compared to the inlet pressure, the discharge of an operating cyclone may have a negative pressure when compared to the ambient pressure. This may be true even in cases where the cyclone inlet pressure is positive when compared to the ambient pressure. In other words, a cyclone operating at + 8 in w.c. [203 mmH$_2$O] static pressure at the inlet may be negative at the mate-

TABLE 8.2

Cyclone Input Data

	English units		SI units	
Gas flow rate	20,000.00	acfm	94,401,963.56	cm³/s
Inlet CL radius	36.50	in	92.71	cm
Inlet velocity	60.00	ft/s	1828.80	cm/s
Body diameter	53.00	in	134.62	cm
Outlet diameter	40.00	in	101.60	cm
Coefficient, n	0.70			
Gas viscosity	0.00	lb/ft · s	1.79E-04	P
Particle density	85.00	lb/ft³	1.36	g/cm³
r	10.00	in	25.40	cm
Cone angle	7.10	degrees	0.12	rad

Particle Cone Interactions

U_T	64.12 ft/s		1.95E + 03 cm/s					
U_F	50.58 ft/s		1.54E + 03 cm/s					

d, μm	Volume, cm³	Mass, g	F_g	F_G	F_{CU}	$F_g + F_G - F_{CU}$	Direction	Units
10	5.23E-10	7.13E-10	6.93E-07	2.60E-03	1.32E-05	2.59E-03	Down	g · cm/s²
30	1.41E-08	1.92E-08	1.87E-05	7.80E-03	3.57E-04	7.46E-03	Down	g · cm/s²
60	1.13E-07	1.54E-07	1.50E-04	1.56E-02	2.86E-03	1.29E-02	Down	g · cm/s²
90	3.81E-07	5.20E-07	5.05E-04	2.34E-02	9.65E-03	1.42E-02	Down	g · cm/s²
120	9.04E-07	1.23E-06	1.20E-03	3.12E-02	2.29E-02	9.51E-03	Down	g · cm/s²
150	1.77E-06	2.41E-06	2.34E-03	3.90E-02	4.47E-02	-3.36E-03	Up	g · cm/s²
180	3.05E-06	4.16E-06	4.04E-03	4.68E-02	7.72E-02	-2.64E-02	Up	g · cm/s²
210	4.85E-06	6.60E-06	6.42E-03	5.46E-02	1.23E-01	-6.16E-02	Up	g · cm/s²
240	7.23E-06	9.85E-06	9.58E-03	6.24E-02	1.83E-01	-1.11E-01	Up	g · cm/s²

rial discharge. If this cyclone is not properly air-locked, it would suck air in the material discharge, destroying the collection efficiency.

8.3 Different Types of Cyclones

There are a great number of devices which are properly referred to as cyclones although in many cases they are not what we have come to call a cyclone in a particular industrial environment. Since the operation of any cyclone is fully dependent on the flow patterns within the device, it was appropriate to leave the individualized description of how they work for this section. It is important to clarify the terminology *cyclone* at the onset of any project because of the wide array of cyclonic devices that are in industrial usage. The two major types of cyclones are

1. *Reverse-flow cyclone* (Fig. 8.4). The reverse-flow cyclone is the generalized description given to the type of cyclone most commonly envisioned when cyclones are mentioned. In a reverse-flow cyclone the dirty gas enters the cyclone at one end where it is caused to spin by the inlet design (see Sec. 8.2 above). As the gas spins it travels down the length of the cyclone body along the wall of the device. Toward the bottom of the cyclone the gas reverses its direction, while maintaining the same spinning motion, and exits the outlet tube located at the same end of the device as the inlet. Common designs utilize involute, tangential, and axial inlets.

2. *Straight-flow cyclone* (Fig. 8.13). The straight-flow cyclone operates using the same principles as the reverse-flow cyclone only the gas maintains the same direction of flow while spinning without reversing direction. Common designs include involute, tangential, axial with fixed vanes, and axial with rotating vanes. Also see Figs. 8.11 and 8.14 to 8.17.

8.4 Basic Selection Criteria

The most basic description of the data required for proper cyclone selection must include all pertinent information about the gas and particulate that are entering the cyclone. It is evident from the cyclone laws that the designer will need to know the inlet gas flow rate, density, gas viscosity, and particle density prior to determining the cyclone fractional efficiency curve. If the gas density and viscosity are unknown, the designer can determine them from inlet gas composition, pressure, and temperature. If particle density or specific gravity is not known, it may be determined by direct measurement using a pycnometer. In some cases a specific gravity for process products may be known or possibly determined from their chemical composition.

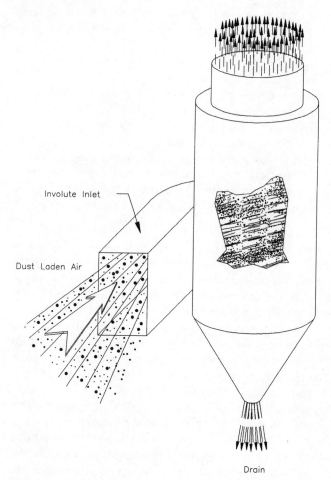

Figure 8.13 Straight-flow cyclone with involute inlet.
(*Illustration by Alex Donenberg. Courtesy of Fisher-Klosterman, Inc.*)

It is important to remember that if a total cyclone collection efficiency will be a requirement of the cyclone design, an inlet particle size distribution will be required. Many times the methods of determining aerodynamic particle size distributions utilize a calibrated or unit particle specific gravity in their representations of particle sizes. This is due to the fact that most methods actually measure terminal velocity distributions of particulate samples and then convert this information to a size distribution using Stokes' law and a specific gravity that is input into the calculation. In any case, it is vital if a particle size distribution is used to specify at what specific gravity the distribution is based. If the distribution is not based upon some parti-

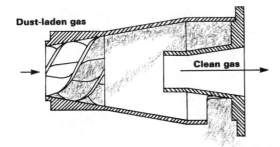

Figure 8.14 Straight-flow cyclone with axial inlet.
(*Courtesy of Flosep SA, Ltd.*)

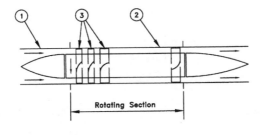

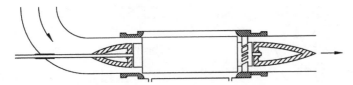

Figure 8.15 Straight-flow cyclone with axial turbine. Item 1 is a
fixed cylindrical inlet duct, item 2 is the rotating inner sleeve of the
centrifuge, and item 3 is compressor turbine blades. Clearances
between the outer casing of the centrifuge and the fixed inlet and
outlet ducts permit free rotation of the centrifuge. (*Illustration by
Alex Donenberg. Courtesy of Fisher-Klosterman, Inc.*)

cle density or specific gravity, then it is not an aerodynamic particle
size distribution and should not be used for cyclone design.

The following list gives the minimum information required for
cyclone selection.

1. Gas flow rate (minimum, operating, and maximum)

2. Gas density of each flow condition

3. Gas viscosity of each flow condition

4. Particulate specific gravity or particle density

5. Particulate loading

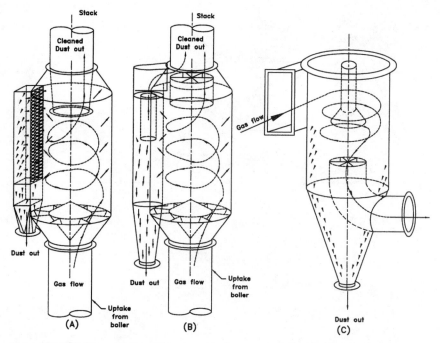

Figure 8.16 Up flow and down flow centrifugal collectors. (*a*) Upflow collector with settling chamber separator. (*b*) Upflow collector with cyclone separator. (*c*) Vortex downflow separator. (*Illustration by Alex Donenberg. Courtesy of Fisher-Klosterman, Inc.*)

6. Aerodynamic particle size distribution method*

7. Aerodynamic particle size distribution*

In addition to the above listed items, there are many attributes of any process that will affect the construction and design of the cyclone selected and to the greatest extent possible should be available to the cyclone designer prior to beginning the cyclone selection. Some of these are

1. Gas temperature (minimum, operating, and maximum)
2. Gas pressure (minimum, operating, and maximum)
3. Gas composition
4. Particulate composition
5. Known particulate characteristics [abrasiveness, friability, explosion characteristics, hygroscopicity (propensity for absorbing water), stickiness, etc.]

---

*Only required if a total collection efficiency is to be determined.

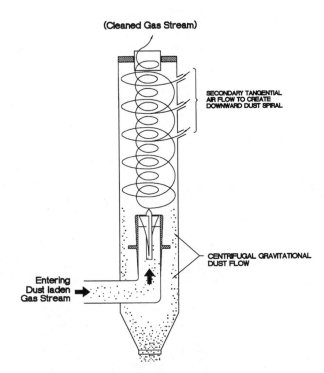

(Cleaned Gas Stream)

SECONDARY TANGENTIAL
AIR FLOW TO CREATE
DOWNWARD DUST SPIRAL

CENTRIFUGAL GRAVITATIONAL
DUST FLOW

Entering
Dust laden
Gas Stream

Figure 8.17 Straight-flow cyclone with reverse dust flow.
(*Illustration by Alex Donenberg. Courtesy of Fisher-Klosterman, Inc.*)

6. Corrosion characteristics

The remaining criteria required for cyclone selection and design include a complete description of what is required of the cyclone. It is vital to completely describe

1. The desired collection efficiency (or fractional efficiency)

2. The maximum allowable pressure drop

Aside from the inlet conditions, which are presumably a function of the process and are not controllable, collection efficiency and allowable pressure drop represent the two most significant design variables outside of cyclone geometry and arrangement. Great care must be given to balancing the desire for particle collection and that of energy consumption (pressure drop) so that the most economical final solution is obtained. Some of the pros and cons of utilizing higher cyclone pressure drops are

Pros

1. Higher collection efficiency
2. Smaller cyclone size
3. Lower cyclone cost

Cons

1. Greater energy consumption
2. Larger and more expensive fans and motors
3. A greater rate of cyclone erosion if the particulate is abrasive

There are additional criteria that may be important for selection of a cyclone that are frequently overlooked until after the cyclone has been selected, many times requiring the redesign of the cyclone. It is impossible to list all the requirements that may be associated with all applications, but some of the more common are

1. Height and size limitations
2. Heat loss minimization
3. Service life from abrasion and/or corrosion
4. Design codes and standards
5. Nozzle connection sizes
6. Desired method of discharging material from cyclone
7. Service requirements for cleaning and/or clearing plugs, solids buildup, etc.

8.5   Applications

There are four general categories of applications for cyclones within industry that the reader should be familiar with. Understanding the difference between these categories and recognizing which is the most appropriate for any given application is important in making the proper cyclone selection. The four main categories of applications are air pollution control, process, precleaner, and liquid entrainment removal.

8.5.1   Air pollution control applications

Although environmental regulations have become more strict over time, a corresponding improvement in cyclone design and prediction capabilities has maintained the applicability of cyclones as the chief emission control equipment for many applications. In many cases, the ease of operation and reliability of a high-efficiency cyclone are much better than those for higher-efficiency types of collectors (scrubber, baghouse, or electrostatic precipitator). Not only does this make the cyclone a more attractive selection from an operating perspective, but it may also be superior from an environmental perspective. A cyclone that emits 4 lb/h of particulate PM10 (-10 μm in diameter) is better from an environmental perspective than the baghouse that emits 0.5 lb/h of PM10 particulate if that collector also has an explosion once every 6 months that emits 16,000 lb of particulate and other pollutants to the atmosphere. This does not take into account what may be the most important issue—safety.

Cyclones are used in several areas of industry as the primary emission control device. Within the woodworking and wood-processing industries, cyclones are commonly used as the primary air pollution control device. There are also numerous applications in metal grinding and cutting, and plastics manufacturing that utilize cyclones for this purpose. There are numerous applications utilizing cyclones as the primary particulate control device for petroleum refining by the fluid catalytic cracking (FCC) process.

Using a cyclone for particulate emission control is obviously attractive whenever possible from an operational point of view. Difficulties in using a cyclone as the primary air pollution control device may arise for the following reasons:

1. It is more difficult to predict cyclone performance to a level that will allow compliance with many environmental regulations.

2. To accurately predict the emissions from a cyclone requires an extremely accurate description of the inlet conditions to the proposed cyclone including dust loading and/or aerodynamic particle size distribution.

3. A cyclone design may not be available that economically meets the required performance levels once good-quality design data has been collected.

The significance of these criteria is that unless the application is large and/or has severe operating conditions, it may not be economical to look at using a cyclone for a given application. An example might be a simple wood-sanding application for 1000 ft^3/min of air (1699 m^3/h). It may be more economical to simply purchase a baghouse for $7000 than to pay $8000 for stack testing to determine if a $6000 cyclone will work. It is therefore most logical to examine the use of cyclones as a particulate emission control device in applications where the benefits of using a cyclone exceed the upfront cost of determining if the cyclone will be suitable. Further allowances must be made for the possibility that use of a cyclone may not be economical despite the proper collection of data.

8.5.2   Process cyclone applications

Cyclone usage is very extensive throughout industry for applications where the cyclone is an integral part of the process and involved in the production of a substance. Although the application is not identical to air pollution control, many of the same general characteristics for design of the cyclone apply. For process cyclone applications, the characteristics of design may include

1. A cyclone of some design will be used at that point of the system, and no other technology is reasonably acceptable.

2. Cyclone performance is a key criterion for selection but so are a number of items such as cost, physical size, quality of construction, and life expectancy. The engineer must select the best balance between all the important design criteria.

3. Some estimate of operating conditions at the cyclone inlet will be used since either the process has not yet been built and/or the process conditions are too severe to allow for reasonable measurement of the inlet conditions to the cyclone.

The general conclusion from these criteria is that the cyclone performance will not usually represent an absolute value that must be attained for successful operation at the plant. In other words, there will be a relative value attached with different levels of cyclone performance. The economic benefits of improved cyclone performance must be weighed against the costs required to achieve that benefit. Accurate and reasonable predictions of cyclone performance are necessary to make that evaluation. In contrast to air pollution control applications, however, a small error is usually not catastrophic. If a process cyclone application is based upon a predicted cyclone emission of 50 lb/day [23 kg/day], it would probably not be catastrophic if the actual emissions were 55 lb/day [25 kg/day]. However, this level of variance could result in the shutdown of the entire plant if this were an air pollution control application.

One of the most common applications for cyclones is in conjunction with fluid beds. Fluidized beds are used in many combustion processes, chemical reactors, petroleum refining, and product cooling and drying. In these processes, a bed material is fluidized with air or some other gas. The bed material may be any number of granulated solids depending on the process. Combustion processes will often use limestone, sand, and/or the fuel itself (i.e., coal) as the bed material, while chemical and petroleum refining reactors will utilize a granulated catalyst. As the fluidization gas leaves the bed, solids are entrained within this gas flow. Cyclones are used to recover and return the entrained solids to the bed. Since the bed is fluidized and behaves as a liquid, the cyclone may be discharged directly to the bed by submerging a drainpipe or dipleg below the bed level. Entrainment levels from fluid beds vary greatly depending on the type of fluid bed, bed material, operating conditions, etc., but levels of 0.2 lb (solids)/ft^3 (gas) [3.2 kg/m^3] are not uncommon.

It is important to retain most of this material in the bed for any number of reasons including

1. Cost of replacement bed material
2. Prevention of heat loss with bed material loss
3. Maintaining fines so that fluidization characteristics are optimized
4. Protection of downstream equipment

There are numerous other process cyclone applications. Cyclones are frequently used as process equipment in conjunction with other drying, cooling, and milling systems. Another common process application for cyclones are as product receivers in pneumatic conveying systems (Fig. 8.18).

In many process systems, the aspect of product recovery may be more significant than all other aspects of the cyclone selection (Fig. 8.19). Consider the case of the manufacture of pharmaceutical prod-

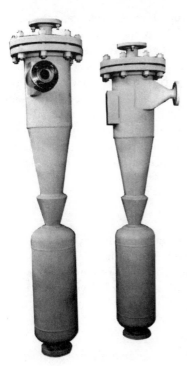

Figure 8.18 ASME code stamped process cyclones. (*Courtesy of Fisher-Klosterman, Inc.*)

Figure 8.19 Stainless-steel polished cyclone and receiver for pharmaceutical process. (*Courtesy of Fisher-Klosterman, Inc.*)

uct that has a $1000/lb [$2205/kg] value. For process reasons, the product is not recoverable if collected in other types of devices. Achieving the highest possible collection efficiency in the cyclone then takes on a new meaning. Because of this cost justification, cyclones for this type of service will frequently represent state-of-the-art designs which will have performance levels that would not have been economical for emission control applications. In other words, to achieve this level of performance for air pollution control, alternate technologies would probably be more economical. Like air pollution control applications, product recovery applications usually require very accurate performance predictions.

8.5.3 Cyclones as precleaners

One of the most common general industrial uses for cyclones is as precleaners for other air pollution control equipment. The reasons vary between applications, but the most common reasons for using cyclone precleaners ahead of other particulate control devices are

Precleaner for media filtration devices

1. Reduced loading results in increased filter life
2. Reduced loading results in reduced frequency of filter cleaning

3. May reduce the required size of the final collector

4. Reduces the fire and explosion hazard due to spark carryover from the process

Precleaner for scrubbers

1. Reduces the abrasion caused by particle loading in the contactor

2. Reduces the amount of liquid bleed-off from the liquid recycle system

3. Reduces the frequency of pipe and nozzle pluggage on recycle systems

4. Reduces recycle pump erosion on recycle loops

5. Reduces the amount of liquid supply required to the scrubber

Precleaner for electrostatic precipitators (ESPs)

1. Removes large particulate that may foul ESP internals and plenums

2. Reduces erosion on expensive ESP components

3. Reduces the size of the collector in some cases

4. Reduces the fire hazard due to spark carryover from the process

In general, the performance requirements for cyclones used as precleaners are less stringent than the categories listed previously. There are even cases where cyclones have been "de-rated," or their efficiency has been reduced, to improve the handling characteristics of particulate that escapes to the final control device by adding some coarse particles to that dust stream. Precleaner cyclones are usually selected on the basis of price, physical size, energy consumption, and construction. Many applications involve cyclones for product recovery (process cyclone) and as a precleaner for the final air pollution control device. A good example of this type of usage is powder coating applications (Fig. 8.20).

8.5.4 Liquid entrainment separators

Cyclones are used very heavily within industry to remove entrained liquid droplets from a gas stream. The most common application of this type is the use of a cyclonic separator (see Chap. 10) for entrainment separation after a scrubbing device. Usually these devices are straight-flow cyclones instead of reverse-flow cyclones. Liquid droplets have some unique characteristics that affect their collection in centrifugal separators. These characteristics are as follows:

1. They will readily adhere to a surface after contact where a dust particle may bounce off. Liquid droplets and films will creep along

Figure 8.20 Powder coating cyclone assembly. Cyclones for product recovery and precleaner for downstream filter. (*Courtesy of Fisher-Klosterman, Inc.*)

solid surfaces. In many cases, the direction of travel will be in the opposite direction as the gas flow.

2. Liquid droplets more readily coalesce into a larger mass.

3. Liquid droplets and/or coalesced and liquid masses can be easily broken back into new individual droplets by gas flow shear braces. Although agglomerates of solid particles may be easily broken, it is usually quite difficult to break apart individual particles.

Numerous studies have been performed in using reverse-flow cyclones for liquid entrainment separation. In general, these studies have shown that reverse-flow cyclones are less effective at separation of liquid and gas than straight-flow cyclones. If a reverse-flow cyclone is required, it should be designed to operate at an inlet velocity below 65 ft/s [20 m/s]. Further, the utilization of a skirt or deflector on the outlet pipe to prevent liquid droplets creeping down along the pipe is recommended. Deflectors or shields in the lower area to prevent the shearing off of droplets and subsequent reentrainment in the axis may also be useful.[14]

8.6 Installation and Maintenance

Cyclones operate by a combination of simple and complicated physical properties that when used properly can often provide substantial collection of particles with diameters as small as 2 μm and smaller. The success, or failure, of a cyclone to achieve the desired results is closely related to the quality of the installation. The quality of the installation may have a great effect on cyclone performance. It is often determined that a given cyclone is not adequate for an application when in fact the problem lies with an installation error.

There are several installation criteria that must be met, or the result will be a significant loss of collection efficiency and/or aggravated occurrences of operational problems such as plugging and severe erosion. Because of the importance of these criteria and their frequent misuse, we list some of the key installation criteria for cyclones separately for emphasis and quick reference as follows:

1. The cyclone discharge must be gas-tight so that no gas is allowed to flow into the material discharge. A small amount of leakage into the cyclone at the material discharge can destroy the cyclone collection efficiency. In addition, any air leakage will suspend solids in the cyclone cone resulting in increased erosion losses and an increased probability of plugging. A common misconception is that cyclones operating under positive pressure need not be air-locked. This is true if

   a. There is a net outflow of gases from the cyclone material discharge.

   b. Adequate provisions have been made to accommodate the secondary collection problem caused by the exiting gases. Quite often, cyclones operating under positive pressure will have a negative pressure at the material discharge due to the very negative relative pressure in the center of the cyclone vortex. The all too common result is that positive-pressure cyclones are installed without an air lock and there is a corresponding loss of efficiency.

2. An adequate disengaging chamber or dust receiver must be used beneath the cyclone to prevent the reentrainment of previously collected particulate. Common descriptions of cyclone operation describe some point within the cyclone at which the downward vortex has reversed direction and for all practical purposes is dead. This representation is far from the truth. The active exchange of gases in cyclonic flow patterns will extend well below the discharge of the cyclone. Laboratory tests frequently show the severity of the vortices to be sufficient to completely fluidize a particulate sample 12 discharge diameters below the cyclone at relatively normal operating conditions. If we assume that the application is one requiring high collection efficiencies on particles with diameters smaller than 10 μm, then the particles that must be reentrained within this turbulent flow for the cyclone to "fail" its purpose are submicrometer particles with terminal velocities of less

than 0.5 in/s. For all high-efficiency cyclone applications, significant portions of the effluent particulate have been *collected* (thrown to the wall of the device and pneumatically conveyed to close proximity of the discharge point) and subsequently reentrained. To minimize the amount of dust that is reentrained, we recommend the use of a dust receiver beneath the cyclone that is not used for material storage and is at least two discharge diameters wide by three discharge diameters tall.

Another method of minimizing reentrainment is the use of *vortex breakers*. Vortex breaker is a generic name for a variety of devices used to block the axis of the cyclone and prevent extension of the vortex below this point. Usually, a vortex breaker is mounted near the cyclone discharge and is nothing more than a disk of a smaller diameter than the cyclone so that there is a gap between the vortex breaker and cyclone where collected material may pass. By killing the vortex, reentrainment from below this point is eliminated. Careful design of the vortex breaker is required so that

a. The location is not so high within the cyclone as to effectively shorten the cyclone and cause a loss of efficiency.

b. If the particulate is abrasive, the vortex breaker supports will not erode away and cause the device to fall down or become misaligned.

c. The addition of the vortex breaker does not cause pluggage of the particulate flow out of the cyclone discharge.

3. No internal obstructions or other disturbances to gas flow should be added to the cyclone aside from those that are a natural part of the cyclone design. The possible sources of added obstructions range from instruments to poorly designed access doors and uneven refractory installation. A small disturbance to the natural gas flow patterns may result in a surprisingly major increase in emissions. The best rule is to make as few breaks in the normal cyclone construction as possible and, where they are required, to take every precaution to ensure they provide the least disturbance to the gas flow as possible. Access doors and instruments should be designed so that they are flush with the surrounding cyclone wall.

4. Inlet ductwork should be designed so that it does not interfere with the flow of the cyclone. Some common considerations for inlet duct design are

a. Horizontal elbows that are within four duct diameters of the cyclone should be of the same rotation as the cyclone.

b. Downward vertical elbows may be close-coupled to the cyclone inlet, but care must be exercised with upward vertical elbows. If the design of the elbow is such that it allows an accumulation of solids to settle in the cyclone inlet, a loss of efficiency will occur. Some precautions that will prevent problems from upward vertical elbows include moving the elbow as far upstream of the cyclone as possible, using as long a radius elbow as is reasonably possible, and using a flat bottom inlet transition as described in consideration c below.

c. Inlet transitions to the cyclone should have angles of less than 30°, with 15 to 20° or less a desirable range. If solids accumulate along the floor of the inlet of the cyclone, a significant loss of efficiency may occur due to the erratic turbulence resulting from material buildup. Many times, an increased frequency of plugging in the cyclone will also be associated with low inlet velocities or other inlet duct arrangements that allow solids to accumulate on the floor of the inlet. For these reasons, we recommend the use of flat bottom inlet transitions when possible. In this transition, the inlet duct floor is level with the cyclone inlet floor and all the transition is upward and to the sides. This moves the eddy currents caused by the transition into the upper area where solids cannot settle and tends to keep the floor of the cyclone inlet swept clean even at relatively low inlet velocities.

d. Size the ductwork and cyclone inlet so that adequate conveying velocities are maintained at the minimum flow conditions. The reasons for this are the same as expressed in considerations b and c above. If there is a large range between the minimum and maximum flow rates, this may result in too high a pressure drop and/or inlet velocities at the maximum flow rate and it may necessitate using parallel cyclones that can be taken off-line as the flow rate decreases.

The above items are a few of the most salient and most frequently abused aspects of cyclone installation. This by no means indicates that there are not other critical requirements of a cyclone installation that must be observed to achieve the desired performance. In many cases, installation and operation requirements are specific to the application and must be addressed on a case-by-case basis.

8.7 Troubleshooting Guide

As with troubleshooting other mechanical devices, the best aid to troubleshooting once a problem has occurred is a good working knowledge of the device and common sense. With these tools and the following guide, most cyclone problems can be discovered and resolved. See Table 8.3.

TABLE 8.3

Symptom	Possible problem	Solution
1. Pressure drop is too high	Too high a gas flow rate resulting from incorrect initial design of the duct-work or fan	Leave alone unless it is causing process problems. If so, change fan operation or add additional flow restrictions in the system to reduce flow rate and cyclone ΔP
	Leakage into the system ahead of the cyclone	Repair ductwork or hood leaks
	Cyclone has an internal obstruction	Clear internal obstruction
	Incorrect cyclone design	Redesign or replace cyclone
2. Pressure drop is too low	Too low a gas flow rate resulting from incorrect initial design of the duct-work or fan	Change fan operation or replace with larger fan. Redesign components with high ΔP to reduce pressure drop. See Chap. 5 for possible fan problems
	Air leakage into the cyclone assembly	Repair
	Air leakage into downstream system components	Repair
	Incorrect initial design of the cyclone	If loss in collection efficiency is not an issue, leave it alone. If collection efficiency is too low, see symptom 3 below
3. Low collection efficiency	Incorrect initial design	If small performance improvements and/or higher ΔP are acceptable, redesign of the existing cyclone may be possible. If higher ΔP is not acceptable and/or major improvements in collection efficiency are required, cyclone replacement will be required
	Leakage into cyclone	Repair leaks and ensure air locks are working properly and are reasonably gas-tight
	Internal obstructions or plugging	Remove obstruction(s). If persistent plugging occurs, consider alternative construction and look for determining and solving root causes such as condensation and too small a discharge diameter
	Poor inlet ductwork design	Redesign and replace

TABLE 8.3 (*Continued*)

Symptom	Possible problem	Solution
4. Plugging	Cyclone discharge is too small for actual loading	Redesign cyclone with larger discharge diameter
	Material may be accumulating in dead space if cyclone has dished head	Replace dished head with flat roof, false roof, or refractory-lined flat roof
	Material may be naturally sticky	Improve internal surface finishes; PTFE coating, electropolish, etc.
		Use vibrators
		Provide easy access for cleaning
	Condensation	Insulate and/or heat trace.
5. Erosion	Too high an inlet velocity	Reduce flow rate
		Redesign inlet for lower velocities
	Naturally erosive particulate	Minimize inlet velocity
		Abrasion-resistant construction
		Ensure proper cyclone geometry
		Design for easy repairs and/or replacement

References

1. W. Strauss, *Industrial Gas Cleaning,* 2d ed., Pergamon Press, Oxford, 1975, p. 216.
2. *Manual on Disposal of Refinery Wastes,* Volume on Atmospheric Emissions, Chap. 11, "Cyclone Separators," API Publication 931, American Petroleum Institute, Washington, D.C., 1975, p. 11-1.
3. Strauss, op. cit., pp. 246–248.
4. J. Abrahamson, "Collison Rates of Small Particles in a Vigorously Turbulent Fluid," *Chemical Engineering Science,* vol. 30, no. 11, Pergamon Press, 1975, pp. 1371–1379.
5. J. Abrahamson, C. G. Martin, and K. K. Wong, "The Physical Mechanisms of Dust Collection in Cyclones," *Transactions of the Institution of Chemical Engineers,* vol. 56, 1978, pp. 168–177.
6. B. Kalen and F. A. Zenz, "Theoretical-Empirical Approach to Saltation Velocity in Cyclone Design," A.I.C.H.E. Symposium Series, vol. 70, no. 137, 1974, pp. 388–396.
7. Gad Hetsroni, *Handbook of Multiphase Systems,* McGraw-Hill, New York, 1982, pp. 9-94.
8. David Leith and William Licht, "The Collection Efficiency of Cyclone Type Particle Collectors—A New Theoretical Approach," A.I.C.H.E. Symposium Series, vol. 68, no. 126, 1972, p. 196.

9. B. Stairmand, "The Design and Performance of Cyclone Separators," *Transactions of the Institution of Chemical Engineers*, vol. 29, 1951, pp. 356–383.

10. W. Heumann, "Cyclone Separators: A Family Affair," *Chemical Engineering*, vol. 98, no. 6, June 1991, pp. 118–123.

11. D. Leith and D. Mehta, "Cyclone Performance and Design," *Atmospheric Environment—An International Journal*, vol. 7, 1973, pp. 527–549.

12. H. Bryant, R. Silverman, and F. Zenz, "How Dust in Gas Affects Cyclone Pressure Drop," *Hydrocarbon Processing*, June 1983, pp. 87–90.

13. R. Alexander, *Proceedings of the Australian Institute of Mineral and Meturallogical Engineers*, New Series, no. 152–153, 1949, pp. 203–208.

14. Strauss, op. cit., pp. 273–274.

Media Filtration

Ronald L. Jorgenson

Donaldson Company, Inc.
Minneapolis, MN

There are a wide variety of styles and types of industrial media filtration equipment. Three general groups of applications include ambient air filtration, nuisance dust, and process dust collection.

9.1 Ambient Air

Ambient air filtration involves taking air from the general atmosphere or from inside a building and removing the contaminants from this air in order to protect people, equipment, or processes. Examples of this category include the heating, ventilation, and air-conditioning (HVAC) systems for buildings; filters on the intake of internal combustion engines; air used in switch gear rooms; and pressurization systems for clean rooms.

The HVAC system in nearly every industrial building utilizes some type of media filtration, generally static filters rather than self-cleaning filters. Static filters build a dust cake until the pressure drop gets so high that sufficient airflow cannot be maintained. At that point, the filter elements are thrown away and replaced with a clean set. These HVAC systems mix outside air with recycled air from inside the building. Both airstreams will contain dust, dirt, lint, insects, pollen, spores, and a wide array of by-products from the manufacturing process. The most common filter types include roll media, a loose non-woven media frequently treated with an oil or surface tackifier to increase the efficiency; bag filters, most commonly fiberglass-multi-compartment filters with efficiency ratings from 30 to 95 percent; and pleated filters, available with a wide array of sizes (2 to 12 in [50 to 300 mm] depth of pleat) and efficiency (25 to 95 percent). The most efficient of this group of static filters, typically used for clean rooms, are the HEPA and ULPA filters, which reach efficiencies of 99.97 percent at 0.3 μm (HEPA) and 99.9999 percent at 0.12 μm (ULPA).

9.2   Nuisance Dust Collection

Nuisance dust collection is a highly diverse application category. This area of pollution control involves the protection of the environment and protection of the workers and machinery in the industrial workplace. This category of media filtration includes static filters of all types, cabinet envelope filters, baghouses, cartridge filters, and multistage systems.

9.3   Process Dust Collection

The third application category, process dust collection, is similar to the nuisance dust application except in this case the dust is not a nuisance; it is a product of the system process intended for sale. Flour and starch milling are good examples of processes that require process dust collection. This group of applications is also called product recovery. Most of the collectors used in nuisance dust applications are also used in process dust collection with heavy usage of baghouses, cartridge filters, and multistage systems.

9.4   Definitions and Examples

A media filter in its most simple form provides a porous barrier to sieve solid particulate from an air or liquid stream. Media filtration is very old and is present in our everyday lives in many ways. A common example is the window screen: a very coarse filter designed to keep bugs, leaves, birds, and other airborne material from entering our homes. The window screen has a mesh fine enough to keep the bugs out, but air, water, and dust will all pass through the openings. Since the screen is a single layer, it is a good example of a surface loading filter. Surface loading means that the filtered material collects on the surface of the filter media.

Other examples of fibrous media filtration would be a coffee filter and the oil filter and carburetor air filter on your automobile. These filters have a much different appearance than the screen or sieve. Close examination under magnification would show a random pattern of fibers instead of the parallel arrangement of woven wire or filament used in a screen. The fibers are typically held in place by saturating a mat of fibers with resinous binder and then curing the media. This filter media also has depth, meaning the air or liquid flowing through the media has to pass through many layers of intersecting fibers rather than one layer as found in a screen. The filtration mechanisms become more complicated, adding impaction, diffusion, tortuous path flow (interception), and particle attractive forces to the sieving functions of a simple screen. This more complex filter mechanism involves both surface loading and depth loading, meaning some of the particles remain on the immediate surface and some of the particles make their way into the depth of the media before being captured and retained.

The filters used in home heating systems also are depth filters, although they are highly porous and very inefficient. Looking through a filter of this type shows pores or openings many times the size of the particulate collected, raising questions about what makes them work.

All media or barrier filters rely on combinations of functions and forces to achieve results. The most obvious mechanism, sieving, requires that the openings in the media be smaller than the particulate you wish to retain. If the particulate is smaller than the openings, the sieving efficiency of the filter obviously will be poor or nonexistent. However, if any particles are larger than the screen openings, they collect on the surface of the screen and become part of the filtration process. Since particles do not normally have uniform size and shape, they nest together forming complex layers with numerous channels for fluid flow, all highly irregular in shape, size, and direction. As smaller particles enter channels, they become trapped by the sieving process or by attractive forces, creating ever smaller openings for airflow. A typical clean filter has both relatively low efficiency and little resistance to flow. As the filter traps particles (the buildup is usually referred to as a filter cake), the efficiency and the resistance to flow both increase. Eventually the resistance to flow becomes so great that the flow effectively ceases and the filter becomes just a barrier.

The sieving mechanism is simple. In industrial filtration, it plays a major role primarily in filtration of dielectric and viscous fluids. Even though the surface of filter media may appear to the eye as a solid surface, the pore space between fibers is often very large. Most of the particulate that poses problems for air filtration is smaller than the media's average pore size.

The practical nature of this book requires an explanation of the mechanisms that allow a fibrous media to capture particles that are orders of magnitude smaller than the pores or openings between fibers. For a greater understanding of these mechanisms, papers on fibrous filters[1] and filter media modeling[2] and Dr. Davies' book on filtration theory[3] are all suggested. The major forces include interception, inertia, diffusion, and brownian motion.

In very general terms, interception simply means that when particles moving in a laminar flow pattern through the filter come in direct contact with a filter fiber, van der Waals forces and electric charge keep the two bodies in contact, effecting capture and filtration. Air flowing through fibrous media must follow a tortuous path due to the random pattern of fibers in the media. The multiple directional changes allow inertia to aid the interception process by overcoming the forces of the laminar flow pattern which would carry the particle past a fiber. This permits contact and capture primarily of larger particles moving at high velocities through open-structure medias.

Diffusion is the phenomenon where particles in an airstream will move in such a manner as to create a uniform distribution throughout, moving from areas of high concentration to areas of low concentration. As particles contact the fibers in the filter media and become attached, the concentration of particles in the immediate vicinity of the fiber becomes less than that in the airstream in general. The forces of diffusion then attempt to reestablish the uniformity of concentration, bringing other particles into closer proximity to the fiber. Brownian movement, random motion of particles suspended in a fluid, is the primary force that produces diffusion (see Fig. 9.1).

As particles continue to collect on a media fiber, the particles also tend to build on themselves, generating matrices or chainlike growth. The photograph shown in Fig. 9.2, taken with the help of a scanning electron beam microscope, graphically shows the results of particu-

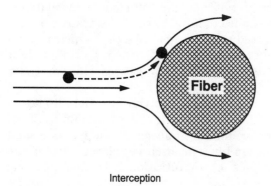

Interception

Figure 9.1 Diagrams depicting the particle collection mechanics of interception, inertia, and diffusion. (*Courtesy of Torit Products, Donaldson Company, Inc.*)

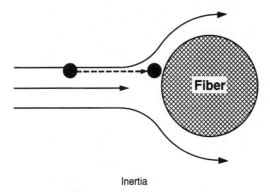

Inertia

Figure 9.1   (*Continued*)

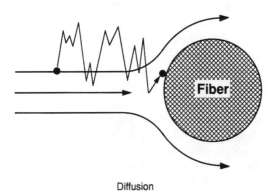

Diffusion

Figure 9.1   (*Continued*)

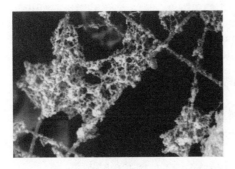

Figure 9.2   Scanning electron beam generated microphotograph, taken at a magnification ratio of 10,000:1. The microphotograph shows the agglomeration of very small (submicrometer) particles and a branching treelike growth pattern, as particles collect on filtration fibers. (*Courtesy of Torit Products, Donaldson Company, Inc.*)

late agglomeration. In this case, the particulate developed a dendritic or branching treelike formation on the submicrometer-diameter filter fibers. As this process continues, the agglomerated particles form a layer covering the media (the dust cake mentioned earlier). Once established, this dust cake does most of the filtration.

The agglomeration of particles provides the basis for cleanable, reusable filter materials. Individual particles have too little mass and/or cross section to allow normal cleaning mechanisms to readily remove them from filter fibers. Agglomerations of particles can be removed by vibration, reverse airflow, pulsing reverse airflow, washing, and other cleaning mechanisms, since agglomerates will act much the same as a large-sized single particle.

On a practical level, scientists can manipulate many variables to provide effective filtration, including the size of the fiber, the shape and texture of the fiber, the density of the fibers in the filter matrix, the choice of binders, and the surface finish. In general, the finer the fiber in the filter, the more efficient the filter will be. This can be easily visualized if you think about a much larger scale. Consider the density and filtration ability of a filter made of wood pencils thrown into a box at random, compared to the density and filtration characteristics if the box contained wood pulp fibers.

The shape and texture of the fibers contribute desired characteristics to the filter media. If the fiber is humanmade, it usually has a smooth cylindrical shape, allowing easy release of sticky particulate. In contrast, natural cellulosic fibers typically have a very irregular shape and rough texture which may help build a stable dust cake but might create problems with particulate that does not release easily. On a microscopic scale, this is similar to the difference between a smooth, hard surface floor and a carpet. The hard, smooth surface allows easy movement of particulate deposited on it; the carpet has an irregular surface that tends to hold the deposited material until it is forcefully removed. Depending on the application, these characteristics can be used to advantage. The natural fiber may increase the efficiency of a filter on a particulate that is difficult to collect due to spherical shape, small size, or uniformity of size. The synthetic fiber may give the ability to collect and release sticky particulate that might plug a natural fiber filter media.

If the airstream or the particulate is chemically active, or the temperature extreme, the choice of fiber and other components in the filter media may need to be based on chemical resistance to degradation, and/or the ability to survive the temperatures involved. The density and thickness of the filter media and the diameter of the fibers used in the media affect the size of the openings or pores, the amount of fiber surface area, and the length of the tortuous path, consequently affecting most of the primary filter mechanisms.

The performance of a filter used in an industrial application will be determined by several factors. The filter media chosen for a dust collecting system will affect the efficiency, the pressure loss (the energy used to force the airflow through the media), and the useful life of the filter media. Much of the effort of research and development in this field is to balance these three factors in order to develop the best possible combination of performance, initial cost, and total cost.

To further complicate the quest for optimum performance, the operating conditions from application to application will vary greatly in terms of the particulate loading; particle size, shape, density, and chemical characteristics; airstream velocity through the media; temperature; moisture content; and gaseous contamination that may affect the media. Any of these factors can seriously detract from the desired blend of efficiency, pressure loss, and useful life.

9.5   General Categories of Media

The majority of industrial filtration medias generally fit into three product groups: needle-punched felts, woven fabrics, and filter paper. Other types that are used less often will also be covered.

9.5.1   Needle-punched felt

Most needle-punched felts have a woven support or scrim with a nonwoven, loose fiber layer driven back and forth through the scrim by barbed needles until it forms a dense mat on both sides of the scrim. There are *scrimless* felts for some specialty applications. Felt medias most often contain synthetic or humanmade fibers, with wool about the only natural fiber commonly available.

9.5.2   Woven media

Woven media come in all the synthetic fibers, plus cotton. The natural fibers provide the best efficiency, while synthetics provide an advantage in release characteristics and chemical resistance, but at the cost of decreased efficiency. The woven medias also employ several weave styles, produced by the same weaving technique and equipment as other textile fabrics. The fibers that run the length of the fabric are called *warp*. The fibers that run from side to side at right angles to the warp are called *fill*. By varying the relationship of the two fibers, the mill can generate the different weaves.

In plain weaves, the fill fiber alternately goes over one and under one warp fiber. This makes for many voids between fibers, large pore sizes, and lower filtration efficiency than the other choices.

Sateen weaves have a characteristic smooth face which is created by having the fill fibers go under one warp fiber and then over four or more warp fibers, or the warp fiber will pass under one fill fiber and over four or more fill fibers. This weave has fewer large pores created when adjacent fibers pass to opposite sides of the intersecting fibers. This weave gives the best filtration efficiency, but the same weave characteristics that give sateen its efficiency lessen its strength and durability.

Twill weaves have a very obvious diagonal pattern. The angle of the pattern will vary depending on the over or under pattern utilized. Typical weaves include the fill fiber passing over one and under two, over one under three, over two under two, and over two under three. Twills have greater durability than sateen and better efficiency than plain weaves.

9.5.3 Filter paper

Filter paper is a nonwoven media made from natural fibers, synthetic fibers, or blends of both types, held in place by a cured resin. It usually resembles a heavyweight, roughly textured paper, is pleated, and is installed in a supporting cage or cartridge. This filter paper is used in what is commonly referred to as a cartridge collector.

9.5.4 Plastic media

One product that does not fit into the three categories above has a filter fabricated of a sintered polyethylene core made from 50- to 100-μm-diameter granules, thermally fused to make a porous substrate. This substrate supports a surface layer of polytetra fluoroethylene (PTFE). The coating increases the ease of release of collected materials and increases the filtration efficiency. The substrate has a convoluted cross section to create more surface area in the same volume.

9.5.5 Ceramic filter candles

Ceramic filter medias have the ability to withstand high temperatures (1400°F [760°C]) that would destroy most filter media, as well as being almost inert to chemical attack except for alkali attack at high temperatures. Ceramic filter media has been tested in fly ash collection at temperatures up to 900°F [500°C].

9.5.6 Sintered metal

Metal-fiber medias are manufactured by sinter-bonding either powdered metal or very fine metal fibers into a thin mat. Metal-fiber filters operate in a temperature range similar to ceramics. Stainless-steel construction permits temperatures up to 800°F [425°C], and INCONEL (a registered trademark of International Nickel Company) provides satisfactory performance up to 1200°F [650°C]. Sintered metal filters are highly abrasion resistant and are largely unaffected by solvents, acids, salts, and high humidity.

9.6   Surface Treatments

To increase the odds of satisfactory performance, many different surface treatments have become available to modify industrial filter medias. The surface treatments include mechanical modification of the fiber, sprayed-on finishes, and laminations of different products to enhance the filtration process.

9.6.1   Singeing

Singeing consists of passing the media under an open flame to melt back fibers, making the surface a little smoother and less prone to accumulate particulate that does not clean off readily. It is most commonly used on synthetic fiber felts.

9.6.2   Glazing

Glazing uses a combination of heat and physical contact to melt and smear the surface to increase surface loading and improve dust cake release by closing a portion of the pore space at the filter media surface. It is most commonly used on synthetic fiber felts.

9.6.3   Sprayed-on coatings

These coatings are a wide array of compounds designed to improve dust cake release, lessen the effects of abrasion, improve resistance to acid and chemical attack, reduce blinding or media plugging, improve water repellency, reduce hydrolysis, and add oleophobic (oil repelling) characteristics and flame retardants.

9.6.4   Membranes

A microscopic layer of porous material such as PTFE can be laminated onto a substrate media to improve surface loading characteristics, efficiency, and dust cake release.

9.6.5   Multilayer construction

A base or substrate filter media can provide the physical strength and support, while a layer of submicrometer-sized fibers bonded to the dirty side of the media improves the surface loading, efficiency, lower pressure drop, and dust cake release.

9.6.6   Abraded surface

Abrading the surface of the fibers in the media provides a nappy finish, intended to improve the efficiency and loading capability of the media. Primarily this is a treatment for woven medias.

9.7   Filters and Selection Criteria

There are two families of media filters: static filters not designed for repeated loading cycles and dynamic or cleanable filters that are designed to be loaded and cleaned for hundreds or thousands of cycles.

9.7.1   Static filters

The term *static filter* implies a filter that is designed to be used one time and then thrown away. The filter will have a pressure drop that rises from the clean value to a point that prevents adequate airflow. At that time it must be replaced. In a typical system without automated flow control, airflow changes continuously, always decreasing, as the pressure drop rises. This filter system design works best with applications that are relatively insensitive to flow, that handle very low concentrations of solids, or where there are contaminants that cannot be cleaned from a filter in a self-cleaning system. Examples would be a vent filter on an extruder day bin handling plastic pellets which generate very small amounts of dust; flux and fume from soldering that are so corrosive, sticky, and agglomerative that they cannot usually be removed during the cleaning process; and safety or HEPA filters downstream of a primary filter system, rarely cleaned because of the possibility of degrading the filter efficiency.

The media used in the static filter family is designed to collect particulate, developing an ever thicker and denser dust cake, until the pressure drop becomes so high that the system blower can no longer draw an adequate volume of air through the filter. The filter is then removed from the system and thrown away. The best examples of the static filters include roll and panel filters used in HVAC systems and HEPA and ULPA filters.

9.7.1.1   Static filter efficiency measurements.   ASHRAE Standard 52.1-1992[4] and the earlier Standard 52-76 provide a widely accepted means to compare the expected performance of air filters, proscribing the test procedures, test equipment, and standardized reporting. The tests interrelate attributes important to filter performance: restriction to airflow, efficiency, arrestance, and dust holding capacity. In greatly simplified form, the test is described below.

Since airflow volumes and the static pressure necessary to generate that airflow are critical to measurement of system performance, test reports include a static pressure versus flow curve at varying flow rates. The tests report both initial or clean filter performance as well as the final or dirty filter results.

The dust-spot efficiency test determines the ability of the subject filter to remove particles by measuring the staining or darkening of a separate test filter by atmospheric dust. The test involves drawing samples of air upstream and downstream of the subject filter at generally equal flow rates but not equal total volumes. The sample flows pass through very fine fiber filters, and, at the conclusion of the test period, the test filters undergo optical comparison of light transmission. Efficiency of the subject filter is determined by comparing the ratio of air volumes necessary to generate equal staining. Staining is a measurement of the combined cross-sectional area of the particles collected on the test filters.

The dust arrestance test utilizes a standardized test dust consisting of mineral dust, cotton lint, and carbon. Arrestance measures the percentage of solids retained by the subject filter, compared to the solids that penetrated the subject filter and collected on an absolute filter installed downstream of the subject filter.

Dust-holding capacity measures the amount of dust that a filter can hold by comparing the weight gain of the filter to the rise in pressure drop during the test period. The test dust is fed until the filter reaches the maximum rated pressure drop or until on two successive runs the arrestance value measures less than 85 percent of the maximum measured efficiency.

ASHRAE-rated filters are available in many physical forms, typically rated by efficiency with a range such as 30 to 35 percent though 90 to 95 percent. The efficiency rating provides only part of the information needed to make a competitive decision. Knowing the remaining information helps predict the filter life and the economics of your choice.

Thermal dioctyl phthalate (DOP) tests are used to challenge HEPA filters. The test is based on Military Standard 282, Method 102.9.1 and uses a monosized 0.3-μm aerosol generated by vaporization and condensation of dioctyl phthalate as the test dust. A photometer that senses light scatter measures particle penetration of the filter. ULPA filters have a minimum efficiency of 99.999 percent on particles of 0.12 μm and larger. This test utilizes a particle counter based on dual lasers obtaining readings upstream and downstream of the filter. An atomizer injects a solution of dioctyl phthalate and alcohol, mineral oil in hexane, or other solution that has a particle size range with a particle mean size distribution in the 0.1- to 0.2-μm range. The laser-based particle counter stores data on many size ranges, but the efficiency data is based on evaluation of particle counts in the appropriate size range.

9.7.2   Dynamic or cleanable filters

The media filter is the basis of many types of cleanable or self-cleaning filter systems in today's industry. Most applications require the filter systems to be cleanable or self-cleaning in order to control the labor and material costs associated with frequent changes of filters. The use of self-cleaning systems also minimizes the airflow fluctuations caused by the loading cycle of static filters.

A self-cleaning filter must not only collect the particulate and release the material during the cleaning sequence but must also prevent the fine particulate from bleeding through the media. All media filters are not created equal, but the best will operate at a stable pressure drop with a narrow pressure drop hysteresis band, using one of several cleaning mechanisms to maintain the pressure drop. Under these dynamic conditions they will test at 99.9 to 99.999 percent based on particle counts upstream and downstream of the filters.

Self-cleaning filters are highly efficient, but their efficiency is rated slightly differently than a static HEPA or ULPA filter. Rather than a test at clean conditions using monosized 0.3- or 0.12-μm particles, the dynamic filter test more typically will be based on an element that has reached a stable operating pressure drop, using a standardized test dust with a wide range of particle sizes.

If media intended for a dynamic system were to be tested by the method used in evaluation of an HEPA filter, the filter would exhibit an initial clean condition efficiency of near 0 to the 30 percent range depending on the type of media and manufacturer. Conversely, an HEPA-style media used in a dynamic cleaning system would likely fail to maintain its rated efficiency, since cleaning the media would cause leakage due to the mechanical stresses imposed on the media. Obviously, the filters are both very efficient, but they do not perform the same, must be rated by appropriate tests, and typically are not interchanged in their respective functions.

Self-cleaning filter systems typically maintain a relatively constant, stable pressure drop once they have been in service a few hours or days. This stable condition will be maintained for a period of months or years. Accurate indication of the long-term performance of a self-cleaning filter requires testing at equilibrium conditions through many loading and cleaning cycles.

The most accurate indication of performance also requires identification of the particle size distribution of the penetrating dust. Modern test equipment, including optical detectors and computers, counts particles in more than 1000 size ranges. This allows extremely accurate indication of a filter's ability to remove particulate across the entire size distribution range or in any segment of the range you wish to examine. This is termed *fractional efficiency*.

9.7.3 Air-to-media ratio

The air-to-media ratio is a measurement of the velocity of the air passing through the filter media. The ratio definition is the volume of air expressed in cubic feet (per minute) or cubic meters (per hour) divided by the filter media area expressed in square feet or square meters.

$$\frac{20{,}000 \text{ ft}^3 \text{ of air (per minute)}}{2000 \text{ ft}^2 \text{ of media}} = 10{:}1 \text{ air-to-media ratio}$$

$$\left[\frac{20{,}000 \text{ m}^3 \text{ of air (per hour)}}{2000 \text{ m}^2 \text{ of media}} = 10{:}1 \text{ air-to-media ratio} \right]$$

Table 9.1 contains representative air-to-media ratios. The values in this table provide general information only. The actual ratio used in any project must reflect many values that are specific to the individual application. The values you use must reflect the moisture content of the airstream, the solids loading expected, the particle sizes and shapes generated, the brand-specific characteristics of the equipment purchased, and many other factors. The values used to determine the ratios in Table 9.1 reflect nuisance dust concentrations, not process or pneumatic conveying concentration levels.

9.8   Cleaning Mechanisms for Filter Media

9.8.1   Manual cleaning

The least expensive systems rely on manual cleaning of the media. In the case of media filters used in dust collectors, envelope filters might be brushed down by hand or filter tubes with one end closed with a zipper can be opened and emptied. Washing the filters at intervals may extend the useful life. The most common examples of manual cleaning include secondary filters for cyclones and the small unit filters used in woodworking shops.

9.8.2   Mechanical shaker

The next step up in complexity and versatility is the mechanical shaker style of collectors. To clean the filter elements or bags, the collector must be shut down since the shaker style does not have sufficient cleaning energy to allow reliable on-line cleaning. The cleaning cycle typically consists of multiple shake cycles, allowing time for solids to drop away between cycles. The shaking flexes the fabric, breaking up the dust cake and allowing the agglomerated particulate to fall into the hopper or solids collection container.

TABLE 9.1 Air-to-Media Filtration Ratios

Description— application or dust	Cartridge filter	Envelope and baghouse filter, shaker	Baghouse low pressure, high volume	Baghouse high pressure, low volume
		Collector type		
Abrasive blasting	1.6	3	8	6
Activated carbon	2	2	6	5
Alumina	2	3	9	8
Ambient air ≤ 0.01 gr/ft³ [23 mg/m³]	2	4	15	10
Asbestos	—	2	10	10
Bauxite	1.5	2.5	6	5
Bentonite	1.2	2	4–6	3–5
Beryllium	1.5	—	8	6
Bin venting	Size at ½ of the dust's listed air-to-media ratio			
Buffing and polishing	—	2.5	8	7–10
Calcium carbonate	1.2	2	4–6	5
Carbon black				
Fused (heated and compressed)	1			
Sintered (powdered)	1.3			
Cement	1.5	2	6–8	6
Ceramic	1.6	2.5	7	6
Clay, brick, and marble	1.6	2.5	7	6
Coal	1.5	2	7	6
Cocoa	1.5	2.5	6	5
Coffee (not roasting)	1.5	2.5	9	7
Coke	1.5	2.5	9	7
Composites	—	2	7	6
Cutting				
Laser (metals only)	8–1.1			
Oxyacetylene	1.4–1.7			
Plasma	1.0			
Diatomaceous earth	1.6	2.5	7	6
Dyes	1.0	—	5	3–5
Fertilizer	1.5	2	6	5
Fiberglass	—	2	8	6
Flour	1.5	2	10	6
Fly ash (not flue gas)	1.5	3	6	5
Frit	1.6	2	8	6
Fumes (metallurgical)	1.2	—	—	2.5
Graphite	1.6	1.5	7	5
Grinding				
Aluminum	1.5	2	8	6
Brake shoe	—	2.5	7	4–5
Cast iron	1.5	2.5	8	5
Composites	—	2	7	4–5
Rubber (precleaner recommended)	—	2.5	10	6
Steel	1.6	2	8	6–8
Titanium	0.5			
Gypsum	1.8	2	8	6
Iron oxide (rust)	1.2	2	5–8	4–5

TABLE 9.1 Air-to-Media Filtration Ratios (*Continued*)

Description— application or dust	Cartridge filter	Envelope and baghouse filter, shaker	Baghouse low pressure, high volume	Baghouse high pressure, low volume
Kaolin	1.2	2	4–6	5
Lead powder	1.2			
Leather	—	3	10	8
Lime	1.8	2.5	7	6
Lime, hydrated	1.2	2	6	5
Limestone	1.8	3	6–8	6
Lignite	1.5	2	7	6
Metal, powdered	1.8	2	8	6
Metallizing				
Electric arc spray	0.5			
Plasma arc spray	1.0			
Flame spray, powder, and wire	1.0			
Milk solids (powdered)	—	2	8	6
Paint pigments	0.9	2	7	6
Pharmaceuticals				
Dry powder	1.2	2	7	6
Mixing and coating	0.5–0.8	2	5	3
Plaster	1.8	2	7	5
Powder coating				
Black	0.8			
White and colors	1.7			
Teflon	1.3			
Rock, mineral, ore, and stone	2.0	2.5	10	8
Rubber	—	2	10	6–8
Salicylic acid	1.3	2	6	5
Salt (mineral)	—	2	8	6
Sand (nonfoundry)	1.8	3	9	8
Sand (foundry)	1.4	2	8	6
Silica	1.8	2	9	8
Silica, fumed	0.5			
Silicates	1.7	2	8	7
Soda ash	1.7	2	7	5
Starch	1.7	2	10	6
Surgical starch	0.5			
Sugar	1.5	2	8	6
Talc	1.5	2	8	6
Talcum powder	1.5	2.5	8	6
Titanium dioxide	1.7	—	5	4
Tobacco	—	3	10	8
Toner	1.0			
Weld fume, source capture				
Laser and plasma	1.5			
All others	2.0			
Weld fume, ambient				
Laser and plasma	1.7			
All others	2.2			

TABLE 9.1   Air-to-Media Filtration Ratios (*Continued*)

Description— application or dust	Collector type			
	Cartridge filter	Envelope and baghouse filter, shaker	Baghouse low pressure, high volume	Baghouse high pressure, low volume
Woodworking				
Sanding	—	2	8	6
High-speed cutting	—	3	10	8
Low-speed cutting and planing	—	3	12	12

SOURCE: Courtesy of Torit Products Div., Donaldson Company, Inc.

The mechanical shaker cleaning mechanism for envelope filters consists of a shaker bar that strikes the metal clips on the bottom edge of the filter bags, raising the bags, deflecting the media, providing manual cleaning on demand, or automatically going through a cleaning cycle every time the collector shuts down (Fig. 9.3). Mechanical shaker baghouses typically use filter tubes with the lower end fixed and the top end attached to a movable support (Fig. 9.4). Movement of the upper support causes bag flexing and dust cake discharge. The cartridge filter can be cleaned by mechanical shaking. This is a small part of the market, typically utilized in unit collectors with flow rates below 3000 ft³/min [5100 m³/h].

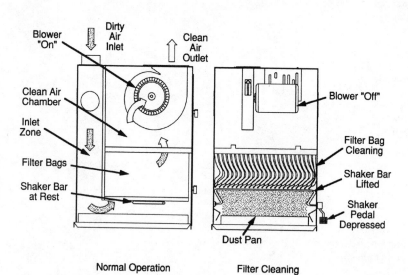

Figure 9.3 Cabinet filter cleaning mechanism. The Shaker bar, powered by hand, foot, or electric motor, lifts and deflects envelope filter to break up the dust cake. (*Courtesy of Torit Products, Donaldson Company, Inc.*)

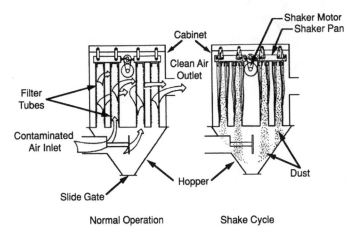

Figure 9.4 Mechanical shaker style of baghouse; shows both filtering and bag cleaning modes. The dust cake breaks up and falls away as the bags flex due to movement of the upper bag support. (*Courtesy of Torit Products, Donaldson Company, Inc.*)

9.8.3 Pulse cleaning

Pulse cleaning of the fabric filter uses a combination of two mechanisms: reversing the airflow direction through the filter media and physical movement of the media to break up the dust cake, allowing the agglomerated dust to drop into the hopper or dust storage area below. There are two distinct groups in this cleaning style: low pressure, high volume and high pressure, low volume.

9.8.4 Low-pressure, high-volume pulse cleaning

This style of cleaning system applies only to baghouse filters (Fig. 9.5) and is most frequently housed in a cylindrical body using a rotating manifold to distribute the cleaning air. (The low pressure and high volume require large tubing to deliver the air for cleaning the filter bags. Multiple fixed-position cleaning-air delivery tubes would be very difficult to fit into a clean air plenum.) The cleaning system utilizes a positive displacement blower, a reservoir, and a solenoid-actuated diaphragm valve to deliver short bursts of air to the filter bags. The positive displacement blower compresses air into the reservoir until it reaches a pressure of 6 to 8 lb/in^2 [0.42 to 0.56 kg/cm^2]. At that point a timer assembly energizes a solenoid valve, which in turn allows a large diaphragm valve to open between the reservoir and the cleaning manifold. This timing function can be random, with no geometric relationship to bag placement, firing at an interval timed to allow the pressure to build to the chosen set point. Another design uses a trigger

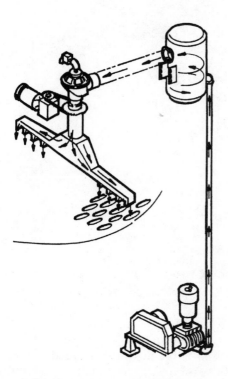

Figure 9.5 Cleaning mechanism for a low-pressure, high-volume pulse-cleaning system. Includes positive displacement cleaning-air pump, rotating manifold, diaphragm, and solenoid valve. (*Courtesy of Torit Products, Donaldson Company, Inc.*)

mechanism to open the diaphragm valve at preset points in the rotation. Both systems deliver pulse cleaning to the bags with a frequency that provides a stable pressure drop across the filter bags.

The diaphragm valve is a fast-acting valve with the ability to allow a large volume of air to pass through the valve in a few milliseconds. This pulse of pressurized air enters the throat of the filter bag, momentarily stopping the clean air flow since it has greater pressure than the system pressure forcing the air through the bag during the filtering cycle. The pulse of pressurized air reverses the airflow direction and momentarily inflates the bag, fracturing the dust cake and allowing it to fall free.

9.8.5 High-pressure, low-volume pulse cleaning

High-pressure pulse cleaning is among the most common cleaning mechanisms for cartridge filters, panel filters, sintered filters, and baghouses and is also used for some envelope filter systems. This cleaning cycle works in much the same manner for all pulse-cleaned collectors. The typical system utilizes a solid-state timer with relays that sequentially energize a series of solenoid valves. The solenoid valves, in turn, actuate diaphragm valves which release pulses of compressed air into an element, into a pair of elements end to end, or into a blowpipe which distributes the pulse into several elements, a row of filter bags, or envelope filters.

The high-pressure, pulse-jet cleaning style (Fig. 9.6) requires a source of clean, dry compressed air at a pressure of 80 to 120 lb/in^2 [5.6 to 8.4 kg/cm^2], depending on the manufacturer's parameters. The cleaning action is similar in nature to the low-pressure, high-volume system described in the preceding section.

The delivery system consists of a compressed air reservoir and multiple solenoid valves, diaphragm valves, and frequently blowpipes, one pipe for each row or column of bags, cartridge filters, or envelope filters in the collector. An electronic timer actuates the solenoid valves in sequence, and the solenoid valve allows the diaphragm valve to open, pressurizing a blowpipe, when used, or discharging directly into a filter.

The blowpipe is simply a pipe with orifices located such that they direct a jet of high-pressure air down the middle of each filter in the row. Variations of the blowpipe and the orifice design include drilled openings, pierced openings, drilled holes fitted with tubular nozzles, and converging or diverging nozzles designed to give greater cleaning performance. Many systems also use a venturi at each filter tube sheet opening to improve the cleaning characteristics.

The high-velocity air jet induces additional air with it, helping overcome and stop the normal flow of air through the filter. The compressed air expands as it decreases in pressure, inflating the filter, reversing the airflow direction through the media, breaking up the dust cake, and allowing the dust to drop into the hopper. Some of the

Figure 9.6 Cleaning mechanism for a high-pressure, low-volume pulse-cleaning system. Depicts blowpipe for compressed air delivery, venturi, and bag. (*Courtesy of Torit Products, Donaldson Company, Inc.*)

dust will reentrain and end up back on the same filter or nearby filters; however, most of the dust removed from the surface of the filter will be highly agglomerated and will end up in the hopper.

The cleaning of pleated filter media uses the same mechanisms as a baghouse, but there are a few differences in the actual application. The solids must be removed from between the pleats and out of the cartridge frame, rather than dropping free from a cylindrical surface like a filter bag. The amount of cleaning air in relationship to the square footage of filter area is also significantly different. A pulse of air that would clean up to 60 or 80 ft² [5.6 to 7.4 m²] of bag area in a baghouse may clean 150 to 350 ft² [13.9 to 32.5 m²] of media in a cartridge filter.

9.8.6   Reverse air cleaning

As in the case of low-pressure, high-volume cleaning, the reverse-air-cleaning mechanism generally applies only to baghouses. As its name implies, this version of the baghouse uses reversal of the airflow through the filter media during the cleaning cycle to effect cleaning of the bags. The energy to provide the reversal of flow and the delivery methods may take many forms. Collectors with multiple compartments piped in parallel allow isolation of a single compartment by closing the clean air outlet and opening a vent to atmosphere. If the dirty air duct is operating under negative pressure, clean air from the atmosphere will be drawn through the filter bags from the clean air side to the dirty air side, reversing the direction of flow. This version requires stoppage of the normal flow before the cleaning cycle begins. A filter cleaning system that requires shutting down the airflow means interruption of the process or redundant capacity allowing isolation of a portion of the system to facilitate cleaning.

An early attempt to provide on-line reverse air cleaning used traveling rings that encircled the bags, traveled up and down the length of the bags, and blew air through the bag fabric from the clean air side to the dirty side. Although this mechanism worked, its popularity faded due to the frequent maintenance of the many moving parts.

The next generation of on-line cleaning baghouses had a cylindrical housing with the bags installed in concentric rings. A rotating manifold in the clean air plenum had multiple compartments each with a valve operated by cams located on the tube sheet (the plate that separates the clean air plenum from the dirty air plenum) (Fig. 9.7). The cams were located such that they opened the cleaning valves only when the manifold compartment was directly over a filter bag. The manifold also had a trailing plate that effectively blocked the openings to and the flow through the preceding row of bags, so that the material displaced during cleaning would not be drawn back to the

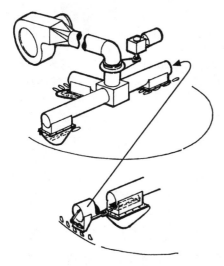

Figure 9.7 Cleaning mechanism for a low-pressure, high-volume non-pulse-cleaning system. Depicts pressure blower for cleaning air, rotating manifold with butterfly valve, cams to open valves, and trailing plate. (*Courtesy of Torit Products, Donaldson Company, Inc.*)

previously cleaned bag. The cleaning energy came from a pressure blower mounted on the collector. This style provided very good filtration performance but also suffered from the complexity of the mechanical components needed for the cleaning and the amount of service required to keep them fully operational.

The most current models of reverse air, low-pressure, high-volume cleaning systems use a delivery system essentially the same as the one shown in Fig. 9.5 with the exception that they do not utilize the diaphragm valve—solenoid valve assembly to create a cleaning pulse. This style is mechanically simple, with few moving parts; however, the tradeoff is they need greater horsepower to generate sufficient cleaning energy.

9.9 Housing Styles

The typical filter system consists of an enclosure or housing, inlet area and baffle, solids collections area (flat dustpan or hopper), the filter section, cleaning mechanism, clean air plenum, and an internal or external primary blower. The function of the housing is self-explanatory. The three general styles are an open bottom or bin vent unit to vent an enclosure or vessel, a flat-bottom style containing a removable dustpan, or a leg-mounted style that has a hopper. Depending on hopper style, an external dust container or a conveying device may be used for final collection, removal, and disposal of the collected solids.

The filtration system enclosure shape, round or square, reflects the intended use and the pressure rating of the collector. Square collectors are typically a little less expensive to build and lend themselves to compact packaging and the most economical layout of multiple rows of filter bags (Figs. 9.6 and 9.12). Square collectors can be reinforced for moderate operating pressures; however, about 2.9 lb/in^2 [0.20 kg/cm^2] seems to be a practical limit. Above that point, the reinforcing becomes quite expensive. Square construction lends itself to the manufacturing capabilities of many fabricators and tends to generate a smaller percentage of scrap than round construction.

Round collectors have some real advantages as well. The natural hoop strength allows higher pressure ratings with little reinforcement. This inherent strength has value even on systems with low operating static pressures when explosion vents are required.

Cylindrical collectors also lend themselves to having precleaning capability built in by use of tangential or volute inlets (Fig. 9.8) above the hopper. This has been very popular with the woodworking, metal machining, and other industries that produce very heavy solids loading and large chip sizes.

Round, conical hoppers also present less opportunity for hopper buildup of materials that have a tendency to bridge and not flow readily. The valley angles in rectangular and square hoppers are more apt to collect material than the smooth contours of a cone.

Job site modification of round collectors tends to be more difficult when inlet or outlet openings need to be cut and adapted to ducting. Layout, accurate cutting, and fit-up for welding can create a challenge once the collector is on the job site.

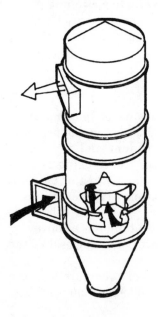

Figure 9.8 Volute inlet design for cylindrical baghouse. Brings airstream into the baghouse on a tangent to the body. (*Courtesy of Torit Products, Donaldson Company, Inc.*)

The type of cleaning system utilized can dictate the body shape. A high-volume, low-pressure system (Fig. 9.5) requires a cleaning flow delivery system that uses a large manifold, in motion, to deliver cleaning to many filter bags from one source. Rotational motion is a natural choice for that system and consequently a round body will be used. A high-pressure, low-volume cleaning system (Fig. 9.6) uses small-diameter piping which does not appreciably interfere with the primary airflow, and the piping is in fixed relationship to the bag locations. Piping is much easier to arrange in straight rows than in circular patterns or radial lines so rectangular or square collectors are the obvious first choice for this cleaning system.

9.10   Inlet Style and Location

9.10.1   Hopper inlets

The most common inlet area for the vertically oriented collectors is the hopper (Figs. 9.4, 9.8, and 9.12). If the material being collected contains large quantities of very large particulate, gravity and the directional change required for the air to reach the filter area provide some inertial separation. The inlet design must include a means of diffusion to prevent solids from ricocheting off the sloped wall of the opposite side of the hopper and eroding the ends of the filters.

Round collectors lend themselves well to tangential or volute inlets (Fig. 9.8). This design allows the air to enter on a tangent to the hopper. Then the air follows the curve of the body and hopper, centrifugal force moves the heavier solids to the wall of the collector, and gravity carries the material down into the hopper discharge. The semiclean air then moves on to the filter section for final cleaning. A flow straightener section between the hopper and filter section limits bag sway in baghouse collectors by converting rotational air motion to vertical. This also improves bag wear life when handling abrasive products or heavy loading.

9.10.2   Side inlets

One of the major factors in the lack of success or outright failure of collectors in applications with fine or light, fluffy particulate is the upward velocity of the air. This is referred to as updraft, axial velocity, or *can* velocity. During the on-line cleaning of a filter, the particulate displaced by the cleaning cycle comes off the filter in agglomerates rather than individual particles. This material must fall to the hopper or collection area below the filters. If the updraft of incoming air overcomes the gravitational settling forces (Fig. 9.9b), the material stays in the filter area, reentrains into the incoming airstream, and

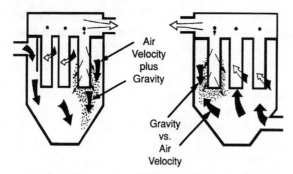

Figure 9.9 Comparison of the hopper entry, up flow pattern versus top or high entry, horizontal or down flow designs, and the effect on light and fluffy material discharged from filters during the cleaning cycle. (*Courtesy of Torit Products, Donaldson Company, Inc.*)

is then pulled back on the filters. If the collected solids settling rate does not equal the rate of additional particulate coming into the filter bag area, the concentration of particulate in the airstream will continually rise as will the pressure drop across the filters. The cleaning system must be operated with reduced or no primary airflow to recover from this condition.

In many applications, the use of a high inlet, eliminating much of the updraft velocity, allows the material pulsed off the filters to settle into the hopper (Fig. 9.9a). Good practice usually requires a baffle positioned between the high inlet and the bags or filters. This minimizes the abrasion and erosion of filters due to the typically high transport velocities present at the collector inlet.

If any combination of particulate shape, small size, and low density suggests a problem with updraft, consider a collector with the inlet at or near the top of the filter area. The airflow pattern of a top or high inlet design allows for the airflow direction and inertial forces to reinforce the gravitational pull on the particulate rather than setting the forces to work against each other, greatly reducing the effects of updraft or can velocity.

9.10.3 Top inlet

Body designs that do not have a clean air plenum on the top of the collector allow use of a top inlet. When the clean air plenum is not above the filters and dirty air plenum, the dirty air can enter from the top and flow down. Top inlet designs aid in the movement of the material pulsed off the filters into the hopper for discharge to an even greater extent than the high side inlet. This style also significantly reduces reentrainment of solids, lowers particulate concentration in the dirty air plenum, and consequently helps maintain a lower stable pressure drop across the filters.

9.11   Filter Replacement Service

9.11.1   Service from the dirty air plenum

For the majority of the vertical format collectors, including baghouses and cartridge filters, it is necessary to enter the dirty air plenum of the collector to service the filters. For small collectors, it may only be necessary to reach in from an access door, but the majority are too large to completely service without physically entering the enclosure. If the dust being collected is inert or only a nuisance, the only drawback may be the discomfort of the maintenance person. If the dust is hazardous, the process may require anyone entering the collector to wear a respirator or even a full isolation suit.

A work platform, either permanently installed below the filters or put in place during service, is usually required for worker safety. If the collector is mounted on a storage vessel, safety rules require a permanent safety grid be added to prevent personnel from falling into the storage space. OSHA's Confined Space Standards, 1910.146, appear to define both the dirty air and clean air sides of dust collector housing as a "permit-required confined space."

One of the advantages of this style of collector is that additional height is not needed to service the bags in a baghouse. This may dictate servicing from the dirty side when a baghouse is installed inside a building with limited headroom.

9.11.2   Service from the clean air plenum

Servicing the filters from the clean air side is a great improvement in terms of personnel comfort, exposure, and safety. This design is not available in envelope filters or in the mechanical shaker filters. Unfortunately, during a bag or filter change, both a walk-in–style, and an access-door–style clean air section would still likely be considered a permit-required confined space by OSHA due to the exposure to the dirty air side caused by the filter removal. To allow clean air side service in pulse-cleaned baghouses, the bag cage design changes to one of two common styles that allow insertion of the bag and cage from the clean side of the tube sheet.

One style of bag and cage uses an O-ring sewn into the cuff at the top of the bag. The bag drops into the hole in the tube sheet, and a wire cage with a formed collar at the top slips into the bag. The formed collar uses retention bolts to compress the O-ring and filter fabric between the collar and the tube sheet. This style provides very positive grounding of the frame to the tube sheet, essentially eliminates misalignment of the assembly, and minimizes bag sway. Misalignment can allow bags to touch each other or the collector body, providing a place for product buildup and/or filter bag damage from impact or friction. Bag sway can cause similar damage due to impact.

Another style of bag and cage uses a snap band in the cuff to position and secure the bag in the hole through the tube sheet. The bag drops into the hole, and the band is deformed into the center of its area until it can be inserted into the hole. The snap band cross section straddles the tube sheet providing retention. Once the bag is in place, the cage drops inside and, when seated properly, further compresses the fabric against the edge of the opening in the tube sheet. The advantage of this style of assembly is the speed of filter bag installation. The disadvantages include the need for a separate grounding method for the bag and cage, less positive alignment, and a greater chance of bag sway in turbulent airflow conditions. In either case, proper service technique is to remove the bag cage, leaving the bag itself in the dirty air plenum. The dirty bag can then be pushed through the tube sheet and allowed to drop into the hopper for later removal. This technique keeps the contamination of the clean air side of the collector to a minimum.

Cartridge collector designs exist that allow service from the clean air plenum. The most common design has a clean air plenum chamber that allows individuals to remain standing while inside the chamber and protection from the weather. Cartridge filters typically cannot be removed without bringing them into the clean air plenum. This means a thorough clean up of the clean air plenum of the collector after the filter replacement.

9.11.3 Outside access

Horizontal formal collectors, both baghouses and cartridge collectors as well as most envelope filters, allow access and service from outside of the collector. The round cartridge filters in a horizontal or inclined format allow access through a series of round porthole covers (Fig. 9.10). This arrangement allows service with very limited exposure to the inside of the collector. The envelope filters, V-pack filters, and sintered element filters also allow service from the outside, through access doors that typically open the major part of one or more sides of the collector.

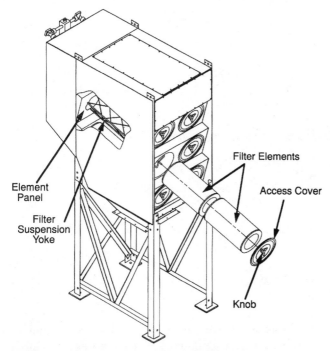

Filter Elements

Access Cover

Element
Panel

Filter
Suspension
Yoke

Knob

Figure 9.10 The external service capability of the horizontal- or inclined-axis cartridge dust collector. Maintenance personnel can service elements without entering the body of the collector. (*Courtesy of Torit Products, Donaldson Company, Inc.*)

9.12   Media Filter Styles

9.12.1   Cabinet envelope filter

Cabinet filters use a flat envelope of woven filter media (see Fig. 9.11). Woven media is typically used in systems that use mechanical shaking for cleaning, while systems that clean with compressed air can utilize felt fabrics. The envelope consists of two pieces of fabric sewn together around three sides, with the fourth side open to allow passage of the filtered air to the clean air plenum. Open cell foam, wire mesh, or a similar material placed between the two pieces of fabric spaces them apart to allow a channel for airflow. In the case of mechanical shaker systems, the bottom edge has a metal clip that runs the full length of the bag, stiffening the bag edge. This allows a bar to strike the bag edge, lifting and dropping the bag and dislodging collected dust. The top edge of each bag, or set of bags, has a frame to support the media and seal the dirty air side from the clean air side once installed.

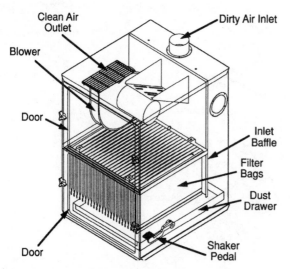

Figure 9.11 Phantom view of a typical cabinet-style envelope filter, and dust collector. (*Courtesy of Torit Products, Donaldson Company, Inc.*)

Because of the low cleaning energy, a mechanical shaker collector must be shut down to be cleaned. It cannot be cleaned satisfactorily with the blower operating. There will be very little reduction in the pressure drop, and the filter will experience loss of efficiency and bleed through. However, pulse-cleaned envelope filters using felt fabrics can be cleaned with the collector in use and the blower operating.

Mechanical shaker unit collector models typically have an airflow range of 100 to 8000 ft³/min [170 to 13,500 m³/h] and commonly have the blower mounted internally or directly to the collector. The pulse-cleaned envelope collectors can be designed in a modular fashion to provide systems of almost any flow, using internal, direct-mounted, or remote blowers. Many accessories add to the versatility of the collector: optional filter medias, special materials of construction, electrical control for the motor(s), pressure gauges, drum lids, explosion vents coupled with sparkproof construction, silencers, casters, safety filters, and weatherization kits.

The mechanical shaker unit collector applications include grinding, polishing, bag dump stations, and many other applications that do not require continuous operation, allowing the opportunity to periodically shut down for cleaning. Good applications have such characteristics as low to moderate airflows that are not extremely sensitive to variation and low to moderate solids loads that are not wet or oily. Weld fume and other very fine particulate applications are not recommended.

The pulse-cleaned version of the envelope filter with felt media allows a wide variety of applications. As with any media filter, wet or oily airstreams and product may cause problems. Weld fume and other similar applications are not recommended for this style of collector. Since pulse cleaning and felt fabrics allow continuous operation, cleaning during the normal loading cycle allows higher dust loading without undue service requirements.

9.12.2 Sintered filters

The sintered element is typically installed in a housing similar to that used for cabinet envelope filter systems. Because of the rigid nature of the filter construction, the filter does not require a frame or cage for support.

The cleaning mechanism is a pulse-jet compressed air system essentially the same as that used for cartridge filter and baghouse filter systems. The system of diaphragm and solenoid valves, blowpipes, and solid-state control systems provides periodic pulses of high-pressure compressed air to break up the dust cake and discharge the collected material into the solids collection container.

The sintered filter has a greater capacity to handle heavy solids loading, a low risk of filter rupture or collapse, high efficiency, and easy product release. Because of the nature of the media, the pressure drop across the filter is significantly higher than that of other media choices, requiring higher-horsepower blowers to overcome the added restriction.

9.13 Baghouses

The term *baghouse* encompasses an entire family of collectors with several types of filter bag shapes, cleaning mechanisms, and body styles (see Fig. 9.12). The bag filter would be more properly called a tube. The filter bag or tube is available in a variety of shapes, sizes, and materials of construction. Shapes include round and oval cross sections. Sizes range from diameters of about 4 to 12 in [100 to 300 mm] and lengths of 3 to 20 ft [1 to 6 m] long. The most common lengths for industrial use are 8, 10, and 12 ft [2.4, 3, and 3.7 m]. Bag styles include one end cuffed, designed to clamp to a collar, with the opposite end sewn shut; both ends open with clamping cuffs; cuffs with O-rings designed to seal by compression; and cuffs with spring steel snap bands designed to force the bag tightly into the openings in the center plate (tube sheet) separating the clean and dirty plenums.

The bags may be fabricated from either woven or felt fabrics. Usually the felt fabrics must be cleaned with some type of air-pulse

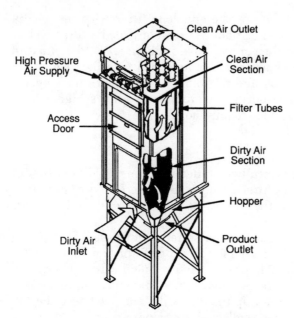

Figure 9.12 Cross section of a baghouse-style dust collector. (*Courtesy of Torit Products, Donaldson Company, Inc.*)

system. Specify woven filter tubes or lightweight shaker felts if the system is cleaned by shaking. As always, there are exceptions. The choice of filter material will depend on the operating conditions: temperature, humidity, material compatibility with both the solids collected and the airstream, physical characteristics of the solids, and the efficiency desired. Probably the most common felt fabric is polyester, a relatively inexpensive product with good resistance to acids, fair resistance to alkaline products, excellent tensile strength, and resistance to abrasion. Polyester felt can be used on applications up to temperatures of 275°F [135°C] with moderate moisture levels.

9.13.1 Manually cleaned systems

The least expensive style of baghouse is the manually cleaned filter section. This type is most frequently used as an after-filter for a cyclone or other inertial separator. There are several variations of housing styles including a dirty air plenum without any other enclosure, a dirty air plenum with side wall enclosure and an open bottom, or a complete enclosure with a hopper or other solids collection arrangement. Cleaning is done either strictly by hand or by manual shake, with periodic replacement and/or laundering. Cleaning is always at no-flow conditions.

9.13.2   Mechanical shaker collectors

Mechanical shaker baghouses usually use a hopper inlet. The dirty air enters the hopper, losing some of the larger particulate due to deceleration and directional change. The semiclean air flows into the open end of a filter bag with a flow pattern from inside to outside. Solids build up on the inside of the bag until the restriction to flow forces a cleaning cycle.

Mechanical shaker baghouses typically require use of woven bags rather than felt bags. The stiffness of felts designed for pulse cleaning prevents the proper flexing of the fabric to break up the dust cake. This limits the versatility of this style, since the woven fabrics lack the performance characteristics needed for applications with very fine particulate. There are shaker felts specifically designed for mechanical shaker collectors, providing a compromise with greater efficiency than woven medias and greater flexibility than felts intended for pulse cleaning.

Mechanical shakers are a good choice for applications that use the filter tube as a substrate to support a filter aide or precoat that actually does the filtration. In that application, the airflow begins without contaminant and the precoat filter aide is added to the airstream until the bag surface is completely coated. The contaminant is then introduced into the airflow and operates until the pressure drop exceeds the acceptable level. At this point, the filter must be shut down and a cleaning cycle initiated, displacing both the collected contaminant and the filter precoat material. The filter is then recoated prior to being put back on-line. This precoat and filter cycle works especially well on applications with sticky product and with applications that ingest sparks since the filter bag has a protective layer of noncombustible material.

The mechanical shaker style is a good option for installations lacking compressed air at the site. The only utility required is electric power for the blower motor, the shaker motor, and for the timer controlling the shake cycle.

The mechanical mechanism located inside the collector housing means the collector cannot be made explosionproof at a reasonable cost. The mechanism is on the clean air side of the filter element, and, except for mechanical failure, the probability of reaching the lower explosive limit of suspended solids is remote, but the risk is certainly higher than with other styles of collectors available. Typical accessories include a selection of fabrics to match the characteristics of the airstream and particulate, blowers, sound attenuation, flow control devices, ladders and platforms, sprinkler systems, explosion vents, and devices for handling the solids collected such as slide gates, rotary air locks, screw conveyors, and pneumatic conveying systems.

Good applications for shaker-type baghouses include grinding, woodworking, plastics, resins, gypsum, sandblast, mixer venting, lime and limestone, and other applications with moderate loading and the ability to interrupt the airflow to periodically clean the filter tubes. The shaker-style baghouses can handle heavier loading and larger chips than an envelope-style collector. Applications involving very fine particulate such as weld fume are poor prospects for this style. For units that have shaker drive components mounted inside, air temperature limitations are dictated by the capability of the motor, gearbox, and other components to operate at elevated temperatures.

9.13.3 Pulse-cleaned baghouses

In most cases, both the high-pressure cleaning and low-pressure cleaning versions of this style of collector use a wire cage to support the filter bag. The pulsing action lifts the bag away from the frame, and then, as the pressure diminishes, the bag snaps back against the support wires due to the pressure differential that exists between the clean and dirty air plenums. This action also helps break up the dust cake. Both round and oval cross-section bag designs are used, with an advantage claimed for the oval cross-section bag due to greater utilization of the snap action during the cleaning cycle.

The high-pressure, low-volume baghouse provides the cleaning energy needed for difficult applications with sticky, agglomerative, and hygroscopic product such as milk products, fertilizers, clay, spices, grain meal, applications that have very heavy loading, and other difficult conditions. Less attractive applications are those that primarily collect very fine particulate and have light solids loads such as weld fume, metallizing, powder paint, and ambient air filtration.

The low-pressure, high-volume style of collector uses air supplied by a positive displacement pump rather than plant-compressed air for cleaning. This eliminates the problems associated with the condensation that is so commonly found in compressed air systems. This is especially useful in systems installed outdoors in areas of the country that routinely must face temperatures far below zero. It is also desirable for handling hygroscopic or soluble products such as portland cement, fertilizer, salt, and sugar.

Other very common application areas include grain, flour, starch, paper, wood products, coal handling, and fly ash. Accessories for both groups of collectors are similar to those listed for the mechanically cleaned baghouses. A wide range of fabrics, blowers, sound attenuation, flow control for both the clean and dirty side ducting, ladders and platforms, sprinkler and explosion vent systems, and solid collection devices such as rotary air locks and feeder valves, screw conveyers, and pneumatic conveying systems are available for this style of collector.

9.14 Cartridge Filters

Cartridge filters, with a few exceptions, use nonwoven filter paper rather than the woven media or felt medias common to the envelope and baghouse style collectors. This nonwoven media has a rigidity that allows it to take and permanently hold a pleat. The pleated media usually is packaged in flat panels, V-shaped packs constructed of two pleated flat panels, or cylindrical filter cartridges. One of the significant features of cartridge filter systems is their compact size. Comparing the airflow against the housing volume shows that a baghouse requires four times as large a dirty air plenum for the same airflow as a cartridge filter.

Cartridge filters have an important efficiency advantage over baghouse filters for particulate in the respirable size range. The increase in efficiency is due to the density of the filter media and the size of the fibers in the media. Penetration, the inverse of efficiency, may be many times greater in a baghouse or envelope filter than in the top-of-the-line cartridge filters.[5] The exception to this rule is filter bags that have PTFE membrane surface treatment.

An important variation on the design of the cartridge collector places the elements mounted on either a horizontal or inclined axis, rather than the typical vertical arrangement (Fig. 9.13). The key to the success of this style is the increased placement flexibility of the filter components. Since the clean air plenum is at the back of the col-

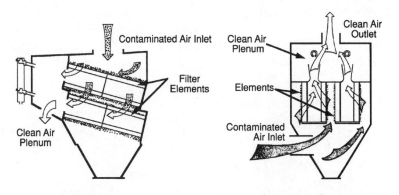

Horizontal or Inclined Arrangement   Vertical Element Arrangement

Figure 9.13 Cross sections of a vertical-style cartridge filter compared to an inclined- or horizontal-style cartridge filter. (*Courtesy of Torit Products, Donaldson Company, Inc.*)

lector rather than on top, the dirty air inlet can enter from the hopper, either side, front, or top. If the application requires an explosion vent, it can be mounted on the front, either side, or on the top. The top installation of the explosion vent adds versatility by directing the exhaust blast upward and away from people. It eliminates overturning loads presented by an explosion vent that discharges horizontally and allows ducting the vent through the roof without any turns in the ducting.

Another significant feature of this style is the ability to remove and replace each filter through an access in the wall of the housing. The ability to change a set of filters without entering the filter housing eliminates the problems inherent in meeting the requirements of the OSHA Confined Work Space standard. With the addition of a collar around the element service access openings, the dirty filters can be put directly into a plastic bag, minimizing contamination of the service area and people. This is very important when dealing with hazardous materials and severe nuisance dusts such as dyes and pigments. With special bags, the filters can be changed without intermingling air from inside and outside the collector. This requires use of a modified version of the *bag-in, bag-out* procedures established for changing HEPA filters in nuclear applications.[6]

9.15   V-Pack Filters

V-pack filters consist of two flat panel filters installed in a common frame that holds them at a shallow angle to each other (Fig. 9.14). Multiple V-packs typically are installed in a collector in a manner quite similar to horizontal cylindrical filters. This filter style offers the same flexibility in choosing the inlet and explosion vent locations. The filter media installed in a V-pack would be similar to that used in a round cartridge, a pleated nonwoven media that allows large areas of filter media to be put into a single frame.

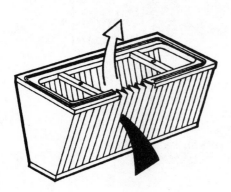

Figure 9.14 The construction of a V-pack style of pleated media filter. (*Courtesy of Torit Products, Donaldson Company, Inc.*)

As in all pulse-cleaned self-cleaning systems, the airflow must be from the outside to the inside. This allows the solids discharged during the cleaning cycle to fall freely to the hopper rather than being trapped inside the V-pack.

The V-packs are typically arranged in vertical columns served by blowpipe arrangements similar to those used in baghouses. The blowpipes have diaphragm and solenoid valves as in all other pulse jets and frequently use venturi systems to aid the cleaning process.

9.16   Basic Selection Criteria

There are several factors to consider when choosing a filtration system. Among the most important are airflow, several characteristics of the solids in the airstream, temperature, efficiency requirements, physical space constraints, duty cycle, explosion or fire risk, system static pressure requirements, and the availability of a compatible media.

9.16.1   Airflow

Before beginning the process of dust collector selection, you must have an idea of the airflow requirements. Only general information is needed. The different types of media filtration systems have overlapping airflow ranges, so this criteria only serves to steer the search when the flows are either very high or very low. Very high flows eliminate most of the envelope filters with mechanical shake cleaning, due to the complexity of providing the mechanical cleaning hardware. Almost all other types are available, with differing purchase and operating costs. The other end of the spectrum, very low flow, eliminates system types only by economic consideration, since even baghouses are available in small sizes with short bag lengths and small numbers of bags.

9.16.2   Particulate characteristics

The size range, particle density, shape, loading, chemical, and other physical characteristics of the particulate will greatly affect the equipment choice. The particulate size range can vary from nearly 100 percent submicrometer fumes from a metallizing job to large chips and pieces in a woodworking application. The equipment choice has to reflect the particle size range expected. Particle size rarely affects static filters if the appropriate efficiency range is chosen; however, when choosing a self-cleaning system, be aware of the efficiency requirements needed for the fine particulate but also consider the system's ability to separate and discharge the material at the other size extreme. Examine the design for impediments to the free flow of product into the solids collection container or removal system and satisfy yourself that the system will handle the expected loading.

The bulk density or apparent density can be important in systems that require that the collected solids fall through a rising airstream to be discharged from the collector. Low-density material may simply not be able to reach the hopper or other collection area. Light, fluffy materials such as leather dust and textile lint have a 90° angle of repose, making it very difficult to get the collected material to flow out of a hopper. Particle shape can cause a problem at times. For example, some starches have particles that are nearly spherical in shape and fall into a very narrow size range. This type of dust will not build a stable dust cake, resulting in bleed-through and leakage problems. Material such as steel grinding fines generated while producing flat surfaces may resemble miniature lathe turnings. This particulate can agglomerate into a steel wool mat which clings tenaciously to many filter fabrics. Choose a felt fabric with a surface finish designed for easy release in this case. Cartridge filters with very tight pleat spacing will not clean well when handling fibrous or linty solids. Open-pleat construction designed for this type of application will solve the problem.

Continuously high loading or a solids loading cycle that includes extremely high spikes may suggest the need for a high-pressure, low-volume or a low-pressure, high-volume pulse-cleaned baghouse. Envelope filters and cartridge filters are a little less forgiving of the extremes of loading. Mechanical shakers and other intermittent duty collectors do not work well in heavy loading situations because of the frequent shutdowns needed for filter cleaning. Chemical and physical characteristics of the particulate might require making allowances for abrasion and choosing materials of construction and filter materials that are compatible with chemically active solids.

9.16.3 Temperature

Temperature normally is not the deciding issue in equipment selection. At temperatures exceeding 350°F (175°C), there are limitations in the choice of media in cartridge filters, although current development continues to raise the upper limits. Baghouses have a wide variety of medias readily available that perform well even at very high temperatures. Mechanical shaker and other collectors that have drive components in the airstream limit the high temperature extremes by the inability of the drive components to survive and function at elevated temperatures. At low temperatures, such as might be expected in outdoor installations in intemperate climates, collectors that do not use compressed air for cleaning may offer an advantage. They avoid the problems possible with solenoid and diaphragm valve freeze-up and with condensation that occurs during the depressurization and expansion of the compressed air downstream of the diaphragm valve.

9.16.4   Efficiency requirements

The efficiency requirements of a dust collection system frequently form the basis of the selection process or have a major impact on the choice. The efficiency requirements may be related to the desire to recycle the air back into the building, the particulate emission may be regulated due to health concerns, or the particulate may be inert but very fine. The most critical situations involve particulate that must be controlled due to toxicity.

Fabric filters that utilize woven fabric are the least efficient, although the weave, fiber choice, and coatings can provide differing levels of efficiency. Shaker felts are more efficient than woven fabrics, and the felt fabrics intended for use with pulse cleaning are better yet. Within the family of felts, there are many choices of fiber, denier of the fiber, the weight of fabric, and special coatings and finishes that can affect the efficiency. PTFE-membrane-surfaced felts and the PTFE-coated sintered plastic media filters share the honors of highest efficiency with the multilayer and PTFE-membrane-laminated paper medias used in cartridge filters. Cartridge filters maintain an edge due to their ability to squeeze large areas of media into compact packages, allowing low air-to-media ratios. This further improves the performance of the paper-media-based systems.

9.16.5   Physical space constraints

With any project, finding the space to fit the equipment into the plant can be a real problem. The wide variety of equipment available makes the job a little easier. If the dust collector can go outside beside the building, there is usually little concern with physical size. If the equipment will be mounted on the roof, the overall height may cause concern due to wind load factors or zoning requirements. Inside installation may limit the overall height or require use of a minimum of floor space. Prior to deciding on a style of equipment, determine if it will physically fit in the space available.

9.16.6   Duty cycle

Continuous duty versus noncontinuous or intermittent duty forms one of the major divisions of media filter types. This means that if the application permits regular opportunities to shut the equipment down to clean the filters and service the dust collection containers, intermittent duty collectors will work satisfactorily. On the other hand, if the process requires a nonstop flow of air or does not tolerate wide variations in airflow to operate properly, the choices include a continuous duty collector that can clean itself at full flow or a compartmentalized intermittent collector with provisions for isolating each compartment for off-line cleaning. Intermittent collectors, generally speaking, are the mechanical shaker or reverse-air-cleaned filters. Continuous duty collectors include some pressurized reverse air filters, and pulse-cleaned filters.

9.16.7   Fire and explosion risk

On applications that have a high risk of fire and explosion due to the flammable nature of the dust or the airstream, equipment choices need to reflect that risk. Solid materials that normally pose little fire risk may be very flammable or capable of generating a deflagration when finely divided, as is the norm with materials collected in dust collectors. Aluminum, titanium, and magnesium dusts are very explosive, and even steel wool will burn! Collectors should not have moving parts in the airstream that could cause metal-to-metal contact, creating a spark hazard. Round collector bodies with their inherent pressure containing strength means you need smaller or fewer explosion vents. Collectors with a horizontal format might allow the use of explosion vents that discharge through the roof. Plan ahead for the location and be sure there is direct, close access to an outside wall or roof if you must install a collector with explosion vents indoors since the vents must be ducted to a safe outside location.

9.16.8   Static pressure

Rectangle or square construction may limit the ability to withstand high static pressures needed for some systems. Round collectors have built-in hoop strength that permits greater pressures, both positive and negative, with a minimum of reinforcement.

9.17   Installation

9.17.1   Typical components

A typical media filtration system consists of a hood to funnel the airborne solids into the system at the collection point; ducting to provide containment and transportation of the solids to the dust collector; the dust collector to separate the solids from the airstream; the blower to provide the energy to move the air and solids inside the system; the dust take away system to remove the collected solids, isolate them from the airstream, and provide storage; and the utility connections. For a successful system, all the components must complement each other.

9.17.2   Choosing the location

The choice of installation location is rarely easy. The first hurdle will be locating a space large enough to fit a dust collector into the existing layout. The scope of the search can include inside floor space, ceiling, outside ground level, and even the building roof.

Inside or outside, both have advantages and disadvantages. Many dust collectors end up outside since they are frequently large and do not need the constant attention of other equipment. This section will touch on some logical reasons to pick or reject a location. As an example, in cold winter weather, pumping warm moist air into a cold collector mounted outside the building may cause condensation. In the summer, the sun might make the collector a significant heat load if you wish to recyle air-conditioned air back into the building.

The first consideration should be access for the ductwork, both for ducting the dirty air to the dust collector and returning the air to the workplace and, when required, for ducting between any explosion vents and the outside of the building. While electric wiring and other utilities can make sharp and multiple turns, the ducting must have the absolute minimum number of turns and be as short as practical. Every elbow in the dirty airstream ducting system causes loss of energy and, consequently, increases the horsepower requirements and energy cost. Also, the physical size of ductwork makes it more difficult to route than other utilities. If the system requires explosion vents, the recommended practice is a straight run, no abrupt turns, and not over 3 m long. Just remember, explosions do not turn corners readily.

If the dust collector requires explosion vents and cannot be located outdoors, the products of combustion must be vented out of the building and in a safe direction. If the combustion gases are not directed outside, the shock wave that results from the explosion frequently dislodges the dust that accumulates on ledges, light fixtures, and equipment. The fireball that leaves the explosion vent ignites the dust cloud, and the secondary explosion can be much more violent than the original, at times leveling entire buildings.

With this in mind, the ideal choice is to locate any collector with explosion vents outdoors, either at grade level or on the building roof. Even when located outdoors, care should be taken to aim the vents in a safe direction and avoid discharging toward walkways, widows, walls, ducting, or flammable materials. Maintain enough distance between the vent and any obstructions so that the pressure wave has an opportunity to dissipate. An obstruction in close proximity to the vent may provide enough restriction to flow that the collector may see a dangerous and unplanned overpressure condition.

When the explosion-vent–equipped dust collector must be located indoors, locate the dust collector against an outside wall and provide a duct from the vent through the wall. When located in a single-story building, a second possibility is to choose a dust collector designed to allow the explosion vents to be mounted on top of the collector and ducted through the roof. Both installations require a means of preventing the entry of birds and small animals. A roof penetration requires a rain cover designed to prevent liquid from entering the ducting. In areas of the country subject to snow, pitch the rain cap sufficiently to prevent snow buildup which could cause dangerous overpressure in the ducting and the dust collector.

No matter what location you choose, or how good the quality of the system you install, service will be necessary. Nonroutine service of moving or wear parts and electrical components, periodic replacement of filters, and routine servicing of the dust collection containers will be required. If drums or tote bags are employed, the system must have easy access for a forklift or pallet mover to carry full containers away.

Noise can be a significant factor in locating filtration systems. Large-horsepower blowers without proper silencing or attenuation equipment can be very loud, seemingly amplified by the reflection of noise off building walls, ceilings, and other equipment. Pulse-cleaned systems have periodic noise that can startle or irritate those in the mediate area. When located outdoors, nighttime operation may disturb nearby residents. Corrective actions such as sound barriers, walls to direct the noise up rather than horizontally, and pointing blower discharges away from residential areas all may have to be considered if there are nearby neighbors.

Building codes or local building restrictions may cause problems when locating equipment outdoors. The overall height of baghouses may cause problems if mounted on the roof. Screening requirements, clearance to lot lines, and other zoning requirements must be explored.

9.17.3 Mounting pads

Designing a concrete pad to mount a dust collector outdoors must take into consideration the normal loads such as the dead load of the equipment; the expected wind load based on location, height above grade, and other buildings in the area; and seismic loading when appropriate. If the dust collector has a sprinkler system, the weight of the collector should be based on the assumption that the collector could partially fill with water in the event the sprinkler system discharges. It should also be considered that the collector would partially plug with particulate. If the collector housing has explosion vents, do not ignore the overturning loads imposed on the structure and mounting pad.

9.17.4 Utilities

The utilities required for a dust collector will obviously depend on the style of collector chosen. Literally all systems will require electric power to operate the blower motor, plus the timer, solenoid valves, and other devices. Complex systems may require an electrical system capable of controlling hopper heaters, rotary air lock valves, screw conveyors, and other components.

Compressed air for pulse-cleaned systems must be both clean and dry to ensure a long-term troublefree life of the cleaning mechanism. Winter operation is very difficult without a quality dryer for the compressed air line. For systems located inside a temperature-controlled building, a refrigerant dryer system that can achieve a dew point of about 32°F (0°C) should be sufficient. If you are located where temperatures routinely fall below freezing and the collector is located outdoors, use a dryer system designed to provide compressed air with a dew point about −40°F (−40°C). A separate dryer dedicated to the collector, located indoors but as close as possible to the collector, increases the chance of troublefree operation. The air used to pulse clean the filters should also be oilfree. If not, the oil accumulates on the inside of the filter tubes or cartridges, causing premature failure. Filling an air line oiler with antifreeze may keep the valves from freezing but also will saturate the filters, leading to very short filter life. A solids filter at the discharge of the dryer traps the dirt, scale, desiccant, and other matter that can lodge in solenoid or diaphragm valves.

9.17.5 Sprinkler systems

Dust collectors almost never start fires; however, they collect the finely divided flammable material that becomes the fuel in a fire, and they also collect any sparks or burning materials that enter the ducting system. The finely divided material, airflow, and turbulence typical of a dust collector lead to serious fires, once started. If the solid material in the dust collector is flammable, seriously consider the services of a trained professional to design and install a sprinkler or other fire and explosion suppression system.

9.17.6   Ducting installation

Ducting can range from one hood connected to a unit collector to complex systems that draw from many hoods, using a network of ducting to connect them all to a single collector. In all systems, the shorter and straighter the better. Friction between the air and the ducting wall requires energy, and every change of direction increases the energy consumption.

Clean air discharged back into the building can save on heating and air-conditioning costs. As long as the air quality exceeds regulatory requirements, it is a sound economic decision to recycle. This may not be true if the airstream contains nonsolid contaminants such as solvent fumes, strong odors, and high levels of water vapor or if the temperature is high. It is common practice to design the return ducting so the air can be returned to the building when it is advantageous and diverted outside at other times. An example would be to route warm air back into the building during the heating season and discharge it outside in the summer when the building benefits from the ventilation. If the system collects flammable materials, provide fast-acting isolation valves to prevent propagating fires back into the building through the ductwork.

9.17.7   Collector installation

Physically installing the dust collector usually involves unloading from a truck and either erecting a completely welded unit or attaching legs, hopper, ladder and platform, railings, and other external parts of the system. Carefully read and follow the manufacturer's instructions for rigging and slinging the collector during unloading and erection. The center of gravity may not be symmetrical. When installing, do not forget to plan for access for service of the filters and the solids collection container. Follow all building code requirements that apply. Be sure to anchor the collector solidly to prevent accidental overturning. Face blower discharges away from occupied areas. Simply aiming the discharge ducting up or away from people will lower the apparent sound level.

9.17.8   Dust take-away system

The choice of a method of collecting and disposing of the solids collected by the dust collector depends to a great extent on the volume involved. Small dust collectors on nuisance dust applications or even large systems on fume collection may accumulate solids slowly. This system will have little need for extensive equipment to deal with the solids collected. Unit collectors may come equipped with an internal dust drawer or with an external container ranging from a small pail to a drum.

Larger collectors with hoppers may use any of a number of methods of disposal. In all cases, the key to success is allowing the solids to leave the collector without allowing appreciable amounts of air to enter the hopper discharge. Large amounts of air entering the collector through the hopper discharge opening reentrain the dust and put it right back on the filters.

Large collectors frequently use one or more drums connected to the hopper(s) with a flexible hose and a drum lid adapter with gasket designed to provide an airtight seal at the drum rim. If the drums need to be emptied or replaced while the collector is running, a slide gate installed between the hopper and the hose and drum lid assembly allows service without interruption of airflow.

A slight variation to the drum arrangement substitutes the large sacks commonly used for storing and transporting chemicals and granular products. Caution must be used to prevent the soft bags from being drawn up into the dust collector when operating under negative pressure. A support frame, designed to work with tie-down loops on the bag, will keep the bag in place until enough solids collect to hold it in place.

Rotary air locks operate much like a revolving door. A rotor with multiple blades rotates inside a tight-fitting housing. As each blade passes the open end facing the hopper opening, dust can drop into the wedge-shaped compartment between that blade and the following blade. As each of these compartments rotates around the shaft, the compartments will empty their load of dust though the opening facing the solids collection container below the valve. The empty compartment then continues back to the dust collector side allowing only a small volume of air to enter the hopper. This type of valve is covered in more detail in another portion of this book (see Chap. 4).

Another version of discharge valve is called a double dump or trickle valve. This operates by opening a trapdoor above an airtight chamber, allowing any collected solids to enter that chamber. The trapdoor then closes, and a similar trapdoor at the bottom of the chamber then opens, discharging the solids to the container below. A single trapdoor version of this assembly is also referred to as a *trickle* valve. The single-valve version allows more airflow to circulate when the system has a pressure differential across the valve.

A third method of discharging solids utilizes a soft flexible version of the trickle valve. This valve has essentially a flattened elastomer tube held shut by the atmospheric pressure on the outside. As solids collect in that tube, the weight overcomes the air pressure holding the tubular section flat, allowing the solids to flow out the bottom. Once the solids leave the tube section, air pressure again flattens the tube section, closing the path between the outside and the inside of the collector.

Any of these valves allows for the use of your choice of closed or open containers set below the discharge device. A screw conveyor, located such that it accepts the solids from one or more hopper discharges, can transport solids to a point away from the dust collector. At the end of the screw conveyor, a single discharge valve isolates the system from outside air. A pneumatic conveying system can provide transportation when large amounts of solids need to be moved away from the dust collector. They work well over long distances and allow multiple directional changes. A typical use of a pneumatic conveying system would be to deliver the material to the top of a storage silo or into a closed truck, or other collection container.

When the dust collector installation handles toxic dusts, be sure that the dust removal system protects the people that will service the collection containers. It makes little sense to carefully collect the dust and then allow it to reenter the airstream at a different point in your facility or to protect one group of workers, only to contaminate others.

At times, chemicals collected in the systems may react in storage containers. Some buffing by-products can cause fires due to spontaneous combustion. Some chemicals can also ignite on contact with oxygen or produce exothermic reactions when exposed to moisture either from going through the dew point or from rain leakage into the system.

Make the choice of a dust removal system based on volume and convenience in your circumstances. Avoid collection containers that require too frequent service. If access is limited, do not use containers that, when loaded, become too heavy to move easily. Anticipate the problems; loaded drums are often too heavy to move without mechanical help.

9.18   Troubleshooting Media Filters

Although there are any number of possible problems associated with installation and operation of a filtration system, the four most common areas of concern are high pressure drop, low efficiency, premature filter failure, and insufficient airflow. Since there are a variety of filter media choices, cleaning mechanisms, and housing types, not all items below will apply to any one system.

9.18.1   High pressure drop

Possible cause	Checks and remedies
Inappropriate air-to media ratio	Check table of representative air-to-media values in this chapter and for the recommendations of the collector or filter manufacturer
	Close some branches of the ducting system, remove part of the attached load

	Improve hood designs to allow lowering flow requirements Adjust blower damper or slide gates
Plugged filter media	*End of filter life.* Replace with new filters *Excess loading.* Add precleaner, operate filter cleaning system with reduced or no airflow (downtime cleaning) *Wet, agglomerative, sticky, hygroscopic dusts.* Change filter to appropriate design and material *Oily airstream.* Difficult to remedy; precoating with absorbent material such as mason's lime may extend life *Wet or oil compressed air.* See below
Incorrect choice of media or surface treatment	Work with collector or filter supplier to determine best choice of filter media to meet system needs
Wet filter media	Eliminate source of water, check for collector body rain leaks, check spark extinguishing system, check for water in compressed air Eliminate dew point excursions, and use start-up and shutdown procedures to dry media
Cleaning mechanism not working	Check electrical, mechanical, and/or pneumatic systems for correct operation and replace or repair as needed
Loss of compressed air and/or low pressure	Monitor compressed air and ensure adequate volume and pressure for all shifts the collector operates. Increase compressed air line size or install dedicated compressor. Malfunctioning solenoid or diaphragm valve may be allowing continuous air loss
Wet or oily compressed air	Add filter or dryer to system, or service existing system Reduce dew point to less than lowest expected operating temperature
Frozen pulse valves	Add enclosure heater to solenoid enclosure Add filter or dryer to compressed air line, or service existing system
Pulse pipes missing, installed upside down, or not properly aligned with filters	Check mechanical system for proper installation and alignment
Dust storage full or plugged	Service dust storage or take-away system on regular schedule. Do not use collector hopper for dust storage
Instrument error	Check for dust-filled, plugged, kinked, cracked, crushed, or missing pneumatic lines Check all fittings for leaks Check gauge for evidence of dust contamination Check for mechanical or electrical instrument failure Check gauge calibration

9.18.2 Low efficiency

Possible cause	Checks and remedies
Air-to-media ratio too high	Check table of representative air-to-media values in this chapter, and for the recommendations of the collector manufacturer Close some branches of the ducting system

	Improve hood designs to allow lowering flow requirements Adjust blower damper or slide gates
Damaged filter media	Check for mechanical damage of filters, tears, split seams, damage due to bag sway impact with other bags or collector body, installation damage, bent or dented end caps or liners Test with fluorescent powder and black light Replace any damaged filters
Improper filter installation	Check installation information. Look for airflow directional arrows on filters. Look for gasketed filters installed backwards
Loose filter retention	Check band clamps, cranks, hand knobs, bolt downs, snap bands, or other filter retention mechanisms for proper tightness. Tighten as directed by installation instructions Test with fluorescent powder and black light
Damaged or missing filter gaskets	Check for gaskets that are torn or detached. On cartridge filters with two open end caps, check to be sure gasket faces the correct way
Tube sheet leaks	Check for cracks, broken welds, or missing fasteners Test with fluorescent powder and black light
Warped tube sheet	Test flatness of tube sheet; look for leaks at filter gasket areas. Common problem after fires Test with fluorescent powder and black light
Missing filters	Check all filters. Install any that are missing
Abrasion of filter media	High velocity inside collector, unequal air distribution, duct directional changes close to collector inlet, high solids loading, solids collection container full or plugged Add precleaner or abrasion reduction inlet box. Redesign inlets to lower velocities. Remove elbows too close to collector inlet. Service solids collection containers on regular schedule
Fire-damaged filters	Look for filters with burn holes or for missing media. Test with fluorescent powder and black light Replace any damaged filters. Eliminate all sources of ignition or burning embers. Add spark and fire suppression system
Excessive cleaning frequency	Overcleaning filters reduces efficiency; look for very low pressure drop across filters Install pressure-drop-based cleaning control. Reduce frequency of cleaning
Excessive cleaning energy	Check operating instructions, regulate cleaning air pressure at collector manufacturer's recommended value
Chemical attack	Check the chemical compatibility of the solids and entrained vapors with the filter media. May be affected by concentration and temperature. Choose media suited to application
Temperature	Measure airstream temperature and compare to media rating. Install filters with proper temperature rating

9.18.3 Premature filter failure

Possible cause	Checks and remedies
Air-to-media ratio too high	Check table of representative air-to-media values in this chapter and for the recommendations of the collector manufacturer Close some branches of the ducting system Improve hood designs to allow lowering flow requirements Adjust blower damper or slide gates
Abrasion of filter media	High velocity inside collector, unequal air distribution, duct directional changes close to collector inlet, high solids loading, solids collection container full or plugged Add precleaner or abrasion reduction inlet box. Redesign inlets to lower velocities. Remove elbows too close to collector inlet. Service solids collection containers on regular schedule
Chemical attack	Check the chemical compatibility of the solids and entrained vapors with the filter media. May be affected by concentration and temperature. Choose media suited to application
Temperature	Measure airstream temperature and compare to media rating. Install filters with proper temperature rating
Incorrect choice of media or surface treatment	Work with collector or filter supplier to determine best choice of filter media to meet system needs
Wet or oily compressed air	Add filter or dryer to system, or service existing system. Reduce dew point to less than lowest expected operating temperature
Dust storage full or plugged	Service dust storage or take-away system on regular schedule. Do not use collector hopper for dust storage
Oily airstream	Difficult to remedy; precoating with absorbent material such as mason's lime may extend life

9.18.4 Insufficient airflow

Possible cause	Checks and remedies
Blower rotating backward	Check rotation arrows on blower housing. Change rotation if necessary
Blower malfunction	Eroded fan wheel, misadjusted-adjusted inlet cone, slipping drive belt Repair or replace as required
Collector or ducting openings not closed	Look for hopper discharges, inspection doors, access doors, unused duct branches left open. Close off any unnecessary openings on both clean and dirty side of collector and in ductwork
Ducting collapsed or plugged	Check all affected portions of system for plugged ducting
Improper duct design	Have knowledgeable person check duct design for proper static and flow balance and duct design

Excess flexible ducting

Flex duct requires more static pressure due to irregular interior surface and direction changes. Calculate losses, increase duct diameter, increase static available, or reduce length of run

Improper transitions or elbows

Check for transitions facing into flow rather than with flow. Use minimum angle between trunk line and branch inlet. Use long-radius elbows

Blast-gate settings

Check all blast gates for proper settings. Readjust any set out of tolerance

Blower system effects

Elbows close to blower inlet, improper inlet box without turning vanes, improperly designed or no-discharge duct, directional change close to blower discharge, improper rain hood. Consult trained duct designer or blower supplier, correct ducting errors. Speed up blower

Blower inlet or exhaust dampers

Check settings of all flow dampers in system

Blower motor speed

Check with blower supplier to ensure blower and motor speeds are compatible

High pressure drop

Any fault that causes high pressure drop can also cause low airflow. See Sec. 9.18.1

References

1. B. Y. H. Liu and K. L. Rubow, "Air Filtration by Fibrous Media," in R. R. Raber (ed.), *Fluid Filtration: Gas, vol. I, ASTM STP 975,* American Society for Testing and Materials, Philadelphia, 1986, pp. 1–12.
2. D. R. Monson, "Key Parameters Used in Modeling Pressure Loss of Fibrous Filter Media," in R. R. Raber (ed.), *Fluid Filtration: Gas, vol. I. ASTM STP 975,* American Society for Testing and Materials, Philadelphia, 1986, pp. 27–45.
3. C. N. Davies, *Air Filtration,* Academic Press, Inc., London, 1973.
4. ASHRAE Standard 52.1-1992, "Gravimetric and Dust-Spot Procedures for Testing Air-Cleaning Devices Used in General Ventilation for Removing Particulate Matter," American Society of Heating, Refrigerating, and Air-Conditioning Engineers, Inc., Atlanta, 1992.
5. M. H. Bergin, D. Y. H. Piu, T. H. Kuehn, and W. T. Fay, "Laboratory and Field Measurements of Fractional Efficiency of Industrial Dust Collectors," *ASHRAE Transactions, vol. 95, part 2,* American Society of Heating, Refrigeration and Air-Conditioning Engineers, Atlanta, 1989, pp. 102–112.
6. C. A. Burchsted, J. E. Kahn, and A. B. Fuller, *Nuclear Air Cleaning Handbook, ERDA 76-21,* Energy Research and Development Administration, 1976, Oak Ridge National Laboratory, Oak Ridge, TN.

Chapter 4

Particle Scrubbing

William L. Heumann
and
Venkatesh Subramania

Fisher-Klosterman, Inc.
Louisville, KY

10.1 Introduction

Particulate scrubbing using water is possibly the oldest method of air pollution control. A part of the earth's natural pollution control system is the cleaning of the lower atmosphere by rain. This is most evident by the freshness of the air following a rainstorm. Because of the simplicity of scrubbing in its most basic form (i.e., spraying water into a gas stream) and the relatively high particulate collection efficiencies achievable, scrubbers have been used within industry extensively since the early 1900s.

Most pollution control problems are solved by the selection of equipment based upon two simple questions. These are (1) Will the equipment meet the pollution control requirements? and (2) Which selection will cost the least? With this premise for selection, there are a number of distinct advantages and disadvantages associated with particulate collection by liquid scrubbing as compared to dry methods. Advantages of wet collection are

1. High-temperature gases may be economically quenched with water and subsequently scrubbed.

2. Scrubbers as basic pieces of equipment are generally less expensive in initial capital cost than comparable baghouses or electrostatic precipitators.

3. Gaseous and particulate contaminants may be collected in the same device and odors may be reduced.

4. Many combustible and explosive materials are safely collected and neutralized.

5. Many sticky and/or hygroscopic materials may be handled.

6. Systems are usually more compact in size compared to dry systems.

Disadvantages of wet collection are

1. The cost of treatment of liquid and solids prior to discharge may be high.

2. Energy consumption may be higher than with other types of equipment.

3. Accurate and detailed design data (i.e., aerodynamic particle size distribution) is required for proper design.

4. Corrosion problems not experienced on dry systems may occur.

5. Systems to prevent liquids from freezing in cold climates may be required.

6. Exhaust gas will be saturated with water, often resulting in a steam plume and/or condensation in the stack.

For any given industrial application, the first economic consideration is the reliability of the technology. In many instances scrubbers are favored in this consideration because they generally are of simple design and provide reliable operation for particulates that may be extremely troublesome in other devices. Since a pollution control system is typically a regulated addition to a production facility and adds little or no value back to the process, the worst possible outcome is significant downtime and loss of production caused by the pollution control system. For this reason the engineer should carefully analyze the expected downtime and maintenance for any equipment selected.

10.2 How Scrubbers Work

The basic principle of all particulate scrubbers is to generate easily collected large particles by combining liquid droplets with relatively small dust particles. In this process the dust particles are grown into larger particles by several methods. These methods are combining them with relatively large droplets, absorption of water by the dust particles and subsequent increases in mass, or formation and growth of condensable particles by relatively cold temperatures within the scrubber.

Of the particle growth methods listed above, the first is by far the most significant collection method used within scrubbers in most applications. The methods of contacting a dust particle with a water droplet and the subsequent wetting of the dust particle are explained in the following text.

10.2.1   Inertial impaction

If small particles are dispersed throughout a flowing gas in which there is an obstruction (see Fig. 9.1), inertia will cause the particles to break through the flow stream that travels around the obstacle. Some of the particles will strike the obstruction. The likelihood of this occurrence is dependent on a number of variables, most significantly the amount of inertia placed upon the particle and the size of the obstruction. In scrubbers, inertial impaction occurs between small dust particles and relatively large droplets. The most common mechanical design used to achieve inertial impaction is shown in Fig. 10.1. Here dust particles and water droplets are present in a moving gas stream. The mixture enters a *contraction* section where the cross-sectional area is reduced, thereby increasing the gas velocity. The relatively large water droplets require some time to accelerate, while the small particles do not (due to the relative inertial properties of each mass). In this section inertial impaction tends to make the dust particles run into the slow-moving water droplets. The mixture then travels through a throat section and into a diverging section where the process is reversed. As the cross-sectional area increases, the gas slows as do the small particles. The liquid droplets with their greater mass and inertia tend to remain at higher velocities and overtake dust particles. This contraction throat and divergence design is commonly called the venturi or contactor section of the scrubber.

Although venturis are the most common method utilized for inertial impaction scrubbing, there are other methods available. One of these methods is the use of spray towers of various design (cocurrent, current flow, cross flow, etc.). These scrubbers are effectively used in applications in which the desired collection efficiency may be achieved at low energy consumption due to the relatively large particle sizes or low efficiency requirements.[1]

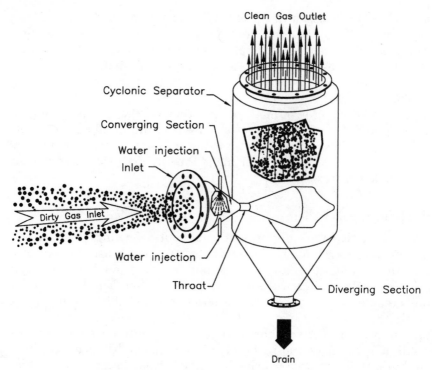

Figure 10.1   Conventional hourglass venturi. (*Illustration by Alex Donnenberg.*)

10.2.2 Interception

If a small particle is moving around an obstruction in the flow stream, it may come in contact with that object due to the particle's physical size (see Fig. 9.1). An easy way to visualize this effect is to imagine yourself as the obstacle when you are driving an automobile. As you drive, objects enter the flow stream over the hood. If the object is in a flow stream that passes within one-half of its diameter of the hood of the car, it is likely to strike.

10.2.3 Diffusion

Small particles with aerodynamic particle diameters of less than 0.3 μm (at specific gravity = 1) are primarily collected by diffusion as they have little inertial impaction due to their small mass and interception is limited by their reduced physical size. Small particles move from areas of high concentration to areas of low concentration in a process called diffusion (see Fig. 9.1). Diffusion is largely an effect of brownian motion, which is the erratic zigzag movement of small particles as they are hit by surrounding gas molecules and other small particles. As these particles are collected into a liquid droplet, the concentration of particles in the immediate vicinity is reduced and once again small particles move from a higher concentration toward the lower concentration in the vicinity of the collecting body.

10.2.4 Condensation

If the thermodynamic characteristics of the flowing gas stream are controlled in a manner in which condensation occurs within the gas stream, small particles will serve as the seed of growth for the condensate. The then liquid-coated particles are more easily collected by the main collection mechanisms listed above.[2,3] Common methods of achieving condensation are by contraction of the vapor and gas from a low pressure to a higher pressure, introduction of steam into the saturated flow stream, and/or direct cooling of the flow stream.

10.2.5 Electrostatic charging

When a different electrostatic charge between particles and droplets is created, an increased efficiency in contacting dust particles to the droplets is achieved. This principle is described in greater detail in Chap. 11 concerning electrostatic precipitators.

10.2.6 Other methods

There are numerous mechanisms of relatively small magnitude which may add to the ability to collect small particles in certain scrubber applications. Since the magnitude of these effects is relatively small and their effect is difficult to verify quantitatively, we will not mention them in depth here. Some of these mechanisms are particle agglomeration, thermophoresis, and diffusiophoresis.[2]

10.2.7 Mist eliminator

Once a particulate has been contacted and absorbed into a liquid droplet, all that remains is to separate the droplet from the gas stream. Generally speaking, droplets generated by mechanical processes (nozzles, air atomization, etc.) are 50 μm in diameter and larger. Typical scrubber droplet size distributions are shown in Figs. 10.2 and 10.3. This enables the relatively easy removal of droplets from the gas stream by a number of proven methods. The methods most commonly used are

1. Cyclonic separators
2. Chevron mist eliminators

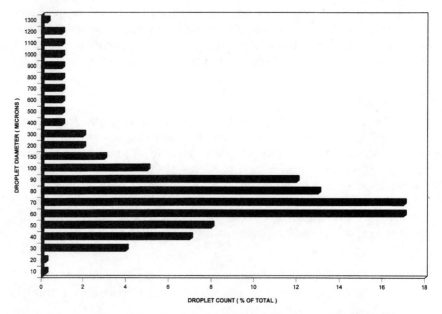

Figure 10.2 Typical scrubber droplet distribution.[4] (*Courtesy of Fisher-Klosterman, Inc.*)

3. Mist filters

One interesting aspect of scrubbing installations is that the capital outlay required in the relatively simple task of separation of liquid and gas often exceeds that associated with the actual scrubbing process. The same may be said for the space required for installation. Many problems associated with improper scrubber installation will become evident as failure of the separation of liquid and gas, resulting in the emission of dirty liquid droplets through the scrubber exhaust. For these reasons, equal, if not greater, care should be given to the method of liquid separation as to that of particle and liquid contact.

10.3   Types of Scrubbers

Basic scrubber designs consist of three variable components. Each of these design components can be uniquely arranged to provide a wide array of equipment available in the marketplace today. Component arrangement combined with individual manufacturers' design modifications provide the engineer with an almost infinite array of equipment configurations to choose from.

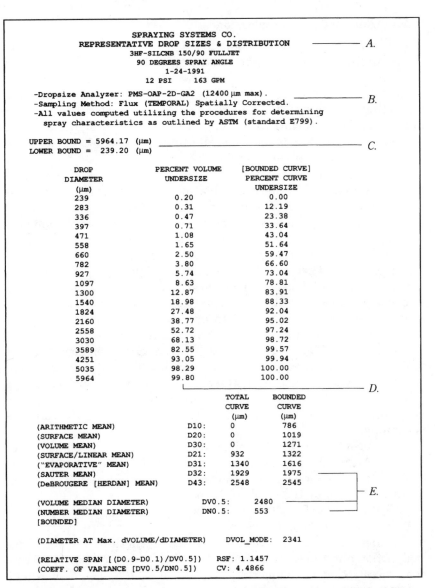

```
                    SPRAYING SYSTEMS CO.
         REPRESENTATIVE DROP SIZES & DISTRIBUTION         ————— A.
                  3HF-SILCNB 150/90 FULLJET
                   90 DEGREES SPRAY ANGLE
                        1-24-1991
                  12 PSI     163 GPM

    -Dropsize Analyzer: PMS-OAP-2D-GA2 (12400 µm max).
    -Sampling Method: Flux (TEMPORAL) Spatially Corrected.   ————— B.
    -All values computed utilizing the procedures for determining
     spray characteristics as outlined by ASTM (standard E799).


    UPPER BOUND = 5964.17 (µm)   ———————————————————————————————
    LOWER BOUND =  239.20 (µm)                                    C.


            DROP            PERCENT VOLUME      [BOUNDED CURVE]
          DIAMETER          UNDERSIZE          PERCENT CURVE
           (µm)                                 UNDERSIZE
            239                0.20                0.00
            283                0.31               12.19
            336                0.47               23.38
            397                0.71               33.64
            471                1.08               43.04
            558                1.65               51.64
            660                2.50               59.47
            782                3.80               66.60
            927                5.74               73.04
           1097                8.63               78.81
           1300               12.87               83.91
           1540               18.98               88.33
           1824               27.48               92.04
           2160               38.77               95.02
           2558               52.72               97.24
           3030               68.13               98.72
           3589               82.55               99.57
           4251               93.05               99.94
           5035               98.29              100.00
           5964               99.80              100.00
                            └————————————————————————————————————— D.
                                    TOTAL       BOUNDED
                                    CURVE       CURVE
                                    (µm)        (µm)
    (ARITHMETIC MEAN)         D10:    0          786
    (SURFACE MEAN)            D20:    0         1019
    (VOLUME MEAN)            D30:    0         1271
    (SURFACE/LINEAR MEAN)     D21:   932        1322
    ("EVAPORATIVE" MEAN)     D31:  1340        1616
    (SAUTER MEAN)            D32:  1929        1975   ———┐
    (DeBROUGERE [HERDAN] MEAN) D43: 2548        2545   ———┤
                                                            E.
    (VOLUME MEDIAN DIAMETER)      DV0.5:    2480    ———┤
    (NUMBER MEDIAN DIAMETER)      DN0.5:     553    ———┘
    [BOUNDED]

    (DIAMETER AT Max. dVOLUME/dDIAMETER)   DVOL_MODE:  2341

    (RELATIVE SPAN [(D0.9-D0.1)/DV0.5])    RSF: 1.1457
    (COEFF. OF VARIANCE [DV0.5/DN0.5])     CV: 4.4866
```

A. Nozzle designation, nominal spray angle, discharge pressure and flow rate.

B. Sampling method and statement of conformance to ASTM-E799 (Revision 1987).

C. Minimum and maximum drop diameters of accumulated (test) volume.

D. Per ASTM-E799-87; section 4.5, less than 1.0% of the cumulative volume is contained in the largest drop measured.

E. Three principal means of reporting drop size are commonly requested. All other values shown (arithmetic mean, surface mean, etc.) are provided for conformance to the ASTM-E799-87 standard.

*PMS® is a trademark of Particle Measuring Systems, Inc.

Figure 10.3 Lower-pressure full cone spray nozzle distribution. (*Courtesy of Spraying Systems Co.*)

10.3.1 Liquid and particle contactors

The physical process of scrubbing may be considered to take place in the part of the scrubber where contacting of dust particles and liquid droplets occurs. In most scrubber systems this is where the majority of the contacting power is consumed. There are numerous designs for contacting particles with liquid droplets. It is important to note that despite the many variations in design the basic performance in terms of particle collection is similar for all contactor designs at equal inlet conditions and energy consumption.[4]

10.3.1.1 Venturis. The most common devices for liquid and particle contactors are venturis. Within the general category of venturis there is a wide variety of designs to choose from. The most simple and popular form of venturi is the hourglass venturi (Fig. 10.1). This venturi derives its name from its hourglass shape. It may have a circular or rectangular cross section. The advantages of this type of design are simple construction and relatively good accessibility for maintenance. The disadvantages are that this design proves relatively cumbersome if an adjustable venturi is desired and throat parts are particularly susceptible to erosion if an abrasive particulate is present.

Venturi rods (Fig. 10.4) offer a more economical construction that still provides a method of venturi adjustment. The rods may be readily replaced when worn. Adjustment of scrubber pressure drop is a simple matter of the addition or subtraction of rods while the device is off-line. This device is not commonly adjustable while the scrubber is running.

Slot venturis (Fig. 10.5) combine the benefits inherent in the prior designs. Although they may be more expensive than a venturi rod design, they provide an economical method of adjusting scrubber pressure drop while the scrubber is operating.

Orifice venturis (Fig. 10.6) are simply made using perforated plates or grids which are readily available in numerous construction materials. Orifice venturis offer the shortest axial length (along the direction of flow) and are economical when the process allows for orifice construction that is commercially available. These designs will plug up easily with certain particulates, are not easily adjustable, and may be expensive for special materials of construction.

10.3.1.2 Nonventuri contact methods. There is an assortment of additional methods available for particle and liquid contact that are less common. One of these types is the simple spray tower (Fig. 10.7). These devices are most suitable for applications with low energy requirements (relatively light loading and/or large particulate).

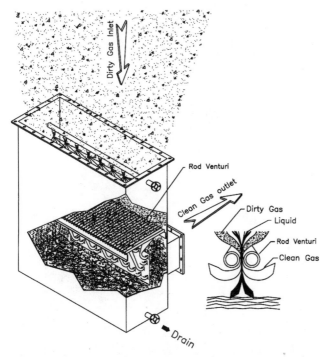

Figure 10.4 Venturi rod scrubber. (*Illustration by Alex Donnenberg.*)

Wet wall cyclones or straight-flow cyclones (Fig. 10.8) use the inherent centrifugal force of a cyclonic device with a wetted wall to ensure particles that reach the walls of the collector are not reentrained. This is essentially a cyclonic separator with liquid injection (Fig. 10.9).

Bed scrubbers include gravel beds and packed towers (see Chap. 12). These devices utilize large contact areas, long residence or contact times, and tortuous paths for gas flow to cause particles to contact a wetted surface.

Wetted media scrubbers (Fig. 10.10) use a wetted collection media (usually fabric) to contact particles to liquid. As the gas passes through the wetted media, collection occurs by the same mechanism as for bed scrubbers (above) and the small gaps between fibers of the media serve as small venturis.

10.3.2 Liquid distributors

The method of liquid distribution is more important to the real operating efficiency of a scrubber than it is to the calculated collection effi-

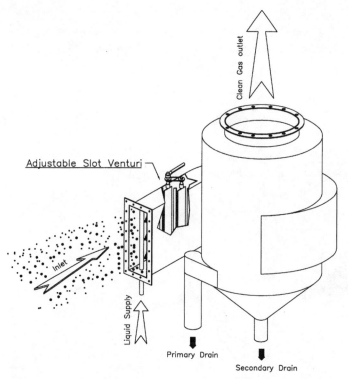

Figure 10.5 Slot venturi scrubber. (*Illustration by Alex Donnenberg. Courtesy of Fisher-Klosterman, Inc.*)

ciency of the device. The real effect of the liquid distribution method is evident when it fails, causing a loss of collection efficiency and/or excessive maintenance. In most scrubbers the amount of contacting power devoted to liquid injection is relatively low and there are numerous methods of trading liquid injection for contacting power and vice versa. Therefore, from a practical operating perspective, the liquid injection method is vital if one is to realize the rated performance of the scrubber.

The main function of any liquid distributor is to provide an adequate liquid supply to all required parts of the venturi or other contacting device. If part of the contacting device is left dry, then collection efficiency will deteriorate. Although the required form of the liquid (liquid film, droplets, etc.) varies from device to device depending on the contactor design, all devices require the liquid to be fully distributed over the contactor section.

Spray nozzles (Figs. 10.11 to 10.13) provide a simple method of widely dispersing the liquid over the required area and atomizing the liquid into suitably sized droplets. There are a wide variety of nozzles

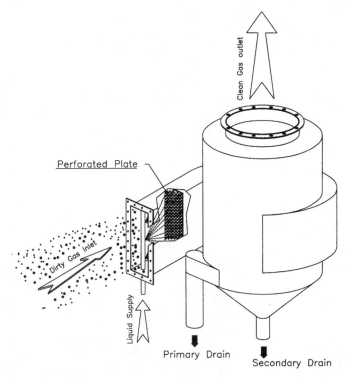

Clean Gas outlet

Perforated Plate

Dirty Gas Inlet

Liquid Supply

Primary Drain

Secondary Drain

Figure 10.6   Orifice venturi scrubber. (*Illustration by Alex Donnenberg. Courtesy of Fisher-Klosterman, Inc.*)

available for use within scrubbers, and with proper selection they can provide for reliable operation and performance of the scrubber. Spray nozzles utilize energy, usually in the form of water pressure, to atomize and disperse the water into droplets. In most cases this results in a reduced cross section at some point in the nozzle when compared to the feed pipe.

The reduced cross section or orifice diameter will be the point of nozzle plugging under most circumstances. With recycled liquid streams it is advisable to use nozzles with the largest orifice diameter and free passage size possible and to keep the suspended solids in the recycle liquid to a minimum.

Injection tubes (Fig. 10.14) inject the liquid at relatively low pressure (compared to spray nozzles) as a liquid stream. The water is then atomized by the venturi, and the energy consumed in this process is largely measured as scrubber gas pressure drop. Injection tubes are relatively free from being plugged up and can be designed to be easily serviceable. Designs utilizing injection tubes require

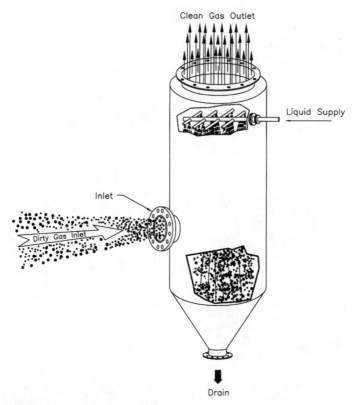

Figure 10.7 Spray tower scrubber. (*Illustration by Alex Donnenberg.* *Courtesy of Fisher-Klosterman, Inc.*)

greater design expertise to ensure the contactor section is fully wetted (Fig. 10.15).

Liquid pools (Fig. 10.16) rely on a pool of liquid through which the gas passes generating its own atomized droplets at the contactor section. These designs provide reliable wetting of the full contactor section and are relatively maintenance-free since the liquid supply system is greatly simplified. Often these designs utilize the liquid sump as a settling tank. Most liquid pool systems are utilized in relatively low energy (low efficiency) scrubbers due to problems that are encountered with using this mechanism at higher pressure drops. They are also subject to losses in efficiency if the liquid level drops or solids build up to a high level in the pool. By nature of the liquid pool design, scrubber pressure drop is not readily adjusted aside from the narrow range achievable by slight variations in liquid level.

Sieve trays (Fig. 10.17) utilize perforated troughs and/or overflow weirs to spread the liquid over the contactor. The liquid may be injected at low pressure, and generally these designs are reasonably free

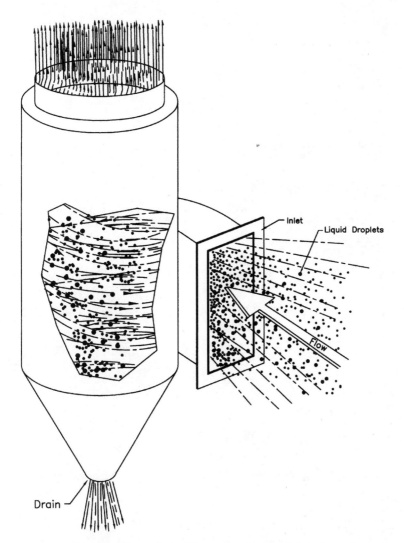

Figure 10.8 Straight-flow cyclone. (*Illustration by Alex Donnenberg. Courtesy of Fisher-Klosterman, Inc.*)

from being plugged up. They are most often used as liquid distributors within bed scrubbers and are generally not used in high-energy venturi scrubbers.

10.3.3 Mist elimination

The final stage of most scrubbers usually comprises a device to remove the dirty liquid droplets from the gas stream. Although the actual scrubbing process is largely complete prior to this section, the

Figure 10.9 FKI WL low-energy wet wall Cyclone scrubber at asphalt plant. (*Courtesy of Fisher-Klosterman, Inc.*)

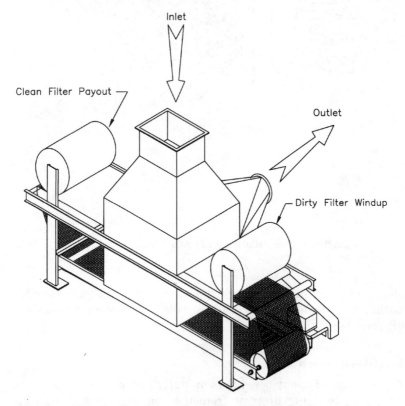

Figure 10.10 Wetted roll media scrubber. (*Illustration by Alex Donnenberg.*)

Figure 10.11 Spiral nozzles offer a relatively plug-free nozzle for recycled liquids. Spraying Systems Co. offers a broad line of special jet nozzles in a wide variety of materials. (*Courtesy of Spraying Systems Co.*)

Spray Patterns

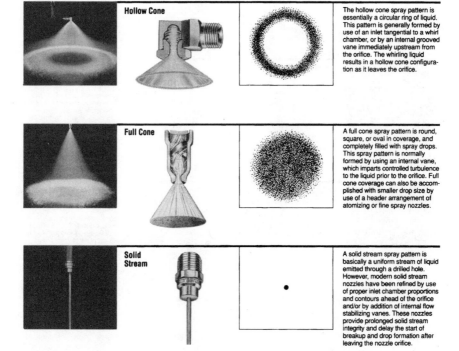

Hollow Cone

The hollow cone spray pattern is essentially a circular ring of liquid. This pattern is generally formed by use of an inlet tangential to a whirl chamber, or by an internal grooved vane immediately upstream from the orifice. The whirling liquid results in a hollow cone configuration as it leaves the orifice.

Full Cone

A full cone spray pattern is round, square, or oval in coverage, and completely filled with spray drops. This spray pattern is normally formed by using an internal vane, which imparts controlled turbulence to the liquid prior to the orifice. Full cone coverage can also be accomplished with smaller drop size by use of a header arrangement of atomizing or fine spray nozzles.

Solid Stream

A solid stream spray pattern is basically a uniform stream of liquid emitted through a drilled hole. However, modern solid stream nozzles have been refined by use of proper inlet chamber proportions and contours ahead of the orifice and/or by addition of internal flow stabilizing vanes. These nozzles provide prolonged solid stream integrity and delay the start of breakup and drop formation after leaving the nozzle orifice.

Figure 10.12 Spray nozzles. (*Courtesy of Spraying Systems Co.*)

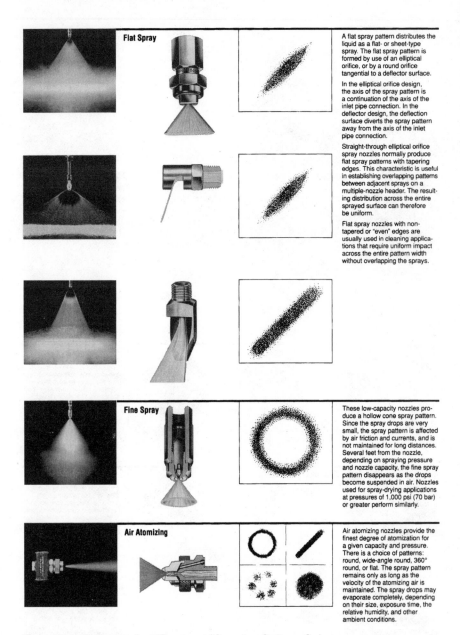

Flat Spray

A flat spray pattern distributes the liquid as a flat- or sheet-type spray. The flat spray pattern is formed by use of an elliptical orifice, or by a round orifice tangential to a deflector surface.

In the elliptical orifice design, the axis of the spray pattern is a continuation of the axis of the inlet pipe connection. In the deflector design, the deflection surface diverts the spray pattern away from the axis of the inlet pipe connection.

Straight-through elliptical orifice spray nozzles normally produce flat spray patterns with tapering edges. This characteristic is useful in establishing overlapping patterns between adjacent sprays on a multiple-nozzle header. The resulting distribution across the entire sprayed surface can therefore be uniform.

Flat spray nozzles with non-tapered or "even" edges are usually used in cleaning applications that require uniform impact across the entire pattern width without overlapping the sprays.

Fine Spray

These low-capacity nozzles produce a hollow cone spray pattern. Since the spray drops are very small, the spray pattern is affected by air friction and currents, and is not maintained for long distances. Several feet from the nozzle, depending on spraying pressure and nozzle capacity, the fine spray pattern disappears as the drops become suspended in air. Nozzles used for spray-drying applications at pressures of 1,000 psi (70 bar) or greater perform similarly.

Air Atomizing

Air atomizing nozzles provide the finest degree of atomization for a given capacity and pressure. There is a choice of patterns: round, wide-angle round, 360° round, or flat. The spray pattern remains only as long as the velocity of the atomizing air is maintained. The spray drops may evaporate completely, depending on their size, exposure time, the relative humidity, and other ambient conditions.

Figure 10.13   Spray nozzles. (*Courtesy of Spraying Systems Co.*)

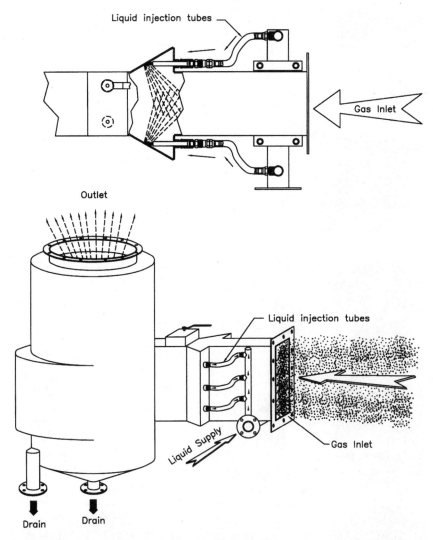

Figure 10.14 Liquid injection tubes. (*Illustration by Alex Donnenberg. Courtesy of Fisher-Klosterman, Inc.*)

actual performance, operation, and capital costs of most scrubbers are often centered in this portion of the device. Any failure or lack of performance in mist elimination results in particulate emissions as the liquid carryover contains collected particulate. This will be measured in an emission test as particulate as surely as the airborne particles.

It is important that the mist eliminator function properly for operational reasons also. A small percentage of liquid carryover emitted from a stack may cause extremely severe safety, operational, and

Figure 10.15 High-energy adjustable venturi with liquid injection via tubes. Injection tubes have reaming device to prevent pluggage. (*Courtesy of Fisher-Klosterman, Inc.*)

maintenance problems in the vicinity of the stack. The increased mass flow caused by water carryover through a blower can cause the fan motor to overload and/or cause severe balance and vibration problems. Mist eliminators should remove at least 99.0 percent (by weight) of the free water generated in the scrubber, and frequently separation efficiencies of more than 99.9 percent are achieved using any of the technologies discussed herein. Applications requiring very high levels of particulate collection and/or low emission levels must allow for the emissions that occur in liquid carryover from the mist eliminator.

Operators and engineers utilizing scrubbers should be careful to not confuse liquid droplets formed by condensation of the gas in the scrubber exhaust with mist carryover. Scrubbers are frequently used as the control technology on "hot systems" resulting in a saturated gas stream at elevated temperatures exiting the scrubber. This steam plume is visible on the exhausts of many scrubbers. In many systems, this saturated gas cools after exiting the scrubber resulting in noticeable amounts of condensation. Often these droplets will contain visible particles since the uncollected particles will serve as condensation nuclei (remember that no scrubber is 100 percent efficient).

In practice, selection of the mist elimination method is based upon an evaluation of cost, physical size restrictions, and operability.

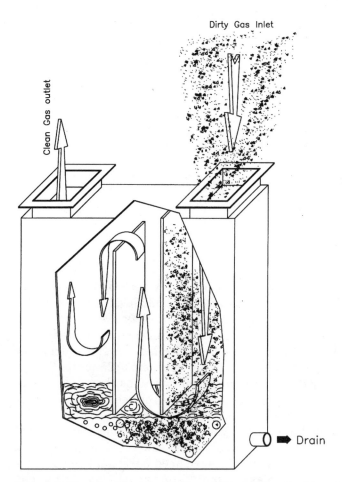

Figure 10.16 Liquid pool scrubber. (*Illustration by Alex Donnenberg. Courtesy of Fisher-Klosterman, Inc.*)

Usually, any of the devices discussed will provide an adequate level of performance for an application if operating properly.

Cyclonic separators are straight-flow cyclones with tangential inlets (see Chap. 8). The typical configuration for this device is shown in Fig. 10.8. Between manufacturers, numerous variations in design and sizing are evident. Some designs require the separator to operate at average axial velocities below 600 ft/min (3 m/s) with a separator height of three to four diameters, while other effective designs allow average axial velocities of more than 1200 ft/min (6.1 m/s) and are less than two diameters tall. Within the limitations of the particular cyclonic separator, these devices effectively remove most liquid droplets generated mechanically during the scrubbing process.

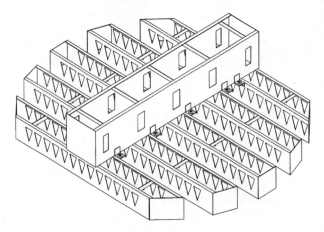

Figure 10.17 Sieve tray. (*Courtesy of ACS Industries, Inc.*)

Droplets formed by nonmechanical processes such as condensation may be too small to be collected at an acceptable level with a cyclonic separator alone. Mechanically generated droplets are generally 50 μm in diameter and larger, while droplets formed by condensation may be less than 10 μm in diameter.[3] Typical pressure drops in cyclonic separators for this service range between 1 (25 mmH$_2$O) and 6 in w.c. (152 mmH$_2$O). In most basic configurations there are no significant internal parts, so these devices are the most maintenance-free of mist elimination devices. A typical efficiency curve for droplet removal is shown in Fig. 10.18. Typical overall droplet removal efficiencies are + 99.5 percent for mechanically generated droplets.

Chevron mist eliminators are zigzag-shaped parts that remove liquid droplets from a gas stream by inertia and impaction. As the gas flows between the chevrons, it must follow the zigzag course set by the chevrons. As the gas changes directions, the water droplets impact on the chevron surface due to their inertia. On the surface of the blade they then coalesce with other collected droplets and run down. These devices may be fabricated from a variety of materials and may be installed into a compact housing. Variations in designs allow for operation with vertical or horizontal installations and axial velocities between 300 and 1200 ft/min. Typical pressure drops for these devices range from 1 (25 mmH$_2$O) to 3 in w.c. (76 mmH$_2$O). A typical fractional efficiency curve is shown in Fig. 10.19. The resulting overall collection efficiencies on mechanically generated droplets will exceed 99.5 percent by weight.[5] Chevron-type mist eliminators are more subject to problems associated with solids buildup than cyclonic separators. Often, wash-down nozzles are installed to remove any buildup that may occur on the chevron surfaces.

GAS FLOW RATE:	10000	ACFM
GAS DENSITY:	0.075	LB/FT^3
DROPLET DENSITY:	62.4	LB/FT^3
PRESSURE DROP:	2.6	IN. W.C.

DROPLET SIZE (MICRONS)	% COLLECTION (BY WEIGHT)
0.8	5.89
1.0	11.35
1.2	17.99
1.4	25.36
1.6	33.02
1.8	40.66
2.1	46.15
2.3	50.75
2.5	55.06
2.7	59.07
3.0	64.56
3.3	69.41
3.6	73.68
4.0	78.55
4.4	82.58
4.9	86.63
5.6	90.83
6.7	94.99
9.6	99.02
10.8	99.51

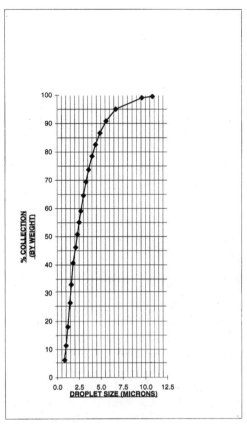

Figure 10.18 Cyclonic separator performance. (*Courtesy of Fisher-Klosterman, Inc.*)

Mesh pads consist of a bed of fibrous material that the gas flow must pass through before exiting to atmosphere. The liquid droplets impinge on the fibers where they coalesce with other droplets forming a mass that is then large enough to drain down.

There are a wide variety of materials in use as mesh mist elimina-tors ranging from fiberglass and polypropylene to high alloy steel fibers. Efficiencies in mesh pads are a function of the fiber diameters and spacing of fibers as well as the depth of the mesh pad. Typical efficiencies for 4-in-thick mesh pads exceed 99.5 percent with pres-sure drops of less than 2 in w.c. (Figs. 10.20 and 10.21).

Mesh pads are the most susceptible of all the methods of mist elimi-nation discussed to the effects of solids buildup and plugging. For this

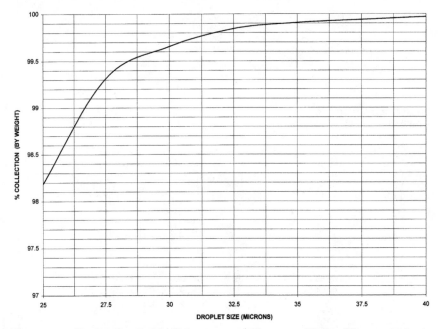

Figure 10.19 Chevron fractional efficiency curve.[4] (*Courtesy of Fisher-Klosterman, Inc.*)

reason wash-down sprays are often utilized with mesh pads to prevent the buildup of solids and subsequent plugging. The advantage of mesh pads is that they may provide very high levels of mist elimination even on droplets less than 20 μm in diameter while using relatively little space and pressure drop. High-efficiency mesh pads can provide up to 99.9 percent removal on 1-μm-diameter liquid droplets.[6] It is not uncommon to utilize multiple layers of mesh pads if very high levels of liquid separation are required.

10.3.4 Scrubber selection

Depending on the application, one can choose a variety of scrubbers ranging from venturi scrubbers, packed tower scrubbers, tray towers, and mist scrubbers or combinations of different scrubbers. Typically, when particulate scrubbing has to be accomplished along with gas absorption, a combination of venturi and packed tower scrubbers is used. The smaller the diameter of the particulate (below 3 μm), the more likely that a venturi scrubber would be appropriate. If the particulate diameter is larger than 3 μm and the process has to be accompanied with gas absorption, one might select a tray tower. Tables 10.1 and 10.2 indicate the type of scrubbers used in relation to particulate size and application.

PRESSURE DROP

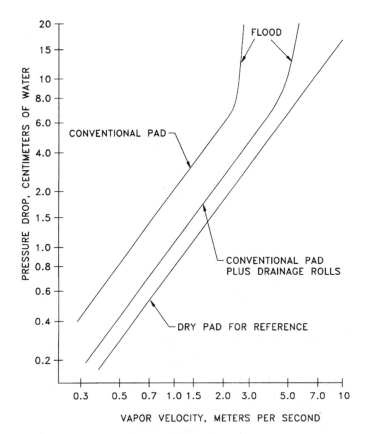

Figure 10.20 Mesh pad pressure drop. (*Courtesy of ACS Industries, Inc.*)

10.4 Scrubber Efficiency

Scrubbers, as well as any other inertial device, collect aerodynamically larger particles more readily than small ones. The continuous curve that shows the removal efficiency of a device versus particle size is the fractional efficiency curve. If the particulate entering a scrubber is large, we would expect very high collection efficiencies, while if the particulate is predominately submicrometer size, we would expect a much lower rate of collection because the particles we are trying to collect are of such obvious difference in size. This property is true of all inertial separators and is the reason that their ability to collect particulate is expressed as size efficiency or fractional efficiency. This denotes that the ability of an inertial device to collect par-

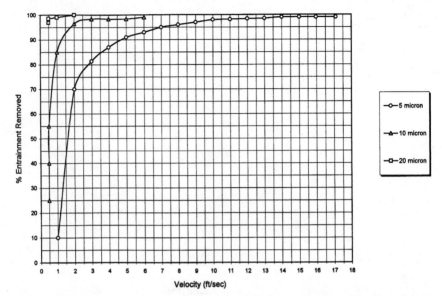

Figure 10.21 Separation efficiency of water droplets in air. (Passing through 48 layers of 0.01-in-diameter wire. Droplet sizes of 5, 10, and 20 microns.) (*Courtesy of Fisher-Klosterman, Inc.*)

TABLE 10.1   Type of Scrubber Used

Particulate diameter, μm	Particulate loading	Type of scrubber
0.3–1000	High	Centrifugal separator
0.01–100	2–3 gr/ft^3	Venturi scrubber
0.01–1000	<1 gr/ft^3	Packed tower scrubber
0.01–1000	1–2 gr/ft^3	Tray tower scrubber
0.1–100	2–3 gr/ft^3	Venturi rod scrubber

ticulate is dependent on the aerodynamic particle size of the particles entering it. Typical fractional efficiency curves for scrubbers at several operating conditions are shown in Figs. 10.22 to 10.26.

Fractional efficiency curves will generally appear as S-shaped curves and, when plotted on normal scales, will become asymptotic toward 0 and 100 percent efficiencies, never actually reaching either value. It is proper to represent fractional efficiency curves in either a tabular or graphical form as shown.

In a normal industrial application the user has no real concern for the fractional efficiency of the scrubber but rather a total collection efficiency for that particular application. If we use a scrubber with the fractional efficiency curve as expressed above for a given application, it will have a total collection efficiency that is dependent on the

TABLE 10.2 Type of Scrubber Used Based on Application

Application	Source of gas	Particulate size, μm	Type of scrubber
Particulate removal	Iron, coke, and silica dust	0.1–10	Venturi scrubber
Acid fume	Sulfuric acid mist	—	Venturi with cyclonic spray
Acid and fume	Titanium dioxide, hydrogen chloride fumes	0.5–1	Venturi scrubber
Cement particulate	Cement dust	0.5–55	Venturi scrubber
Incinerator exhaust	Fly ash, hydrochloric acid, sulfur dioxide	0.05–1	Venturi-packed tower
Ammonia removal	Ammonia, ammonium chloride	—	Packed tower
Paper industry	Hydrogen sulfide, particulate, sodium carbonate, carbon dioxide	0.5–50	Venturi with packed tower and tray tower
Chemical industry	Acetic acid fume, particulate	0.5–10	Tray tower

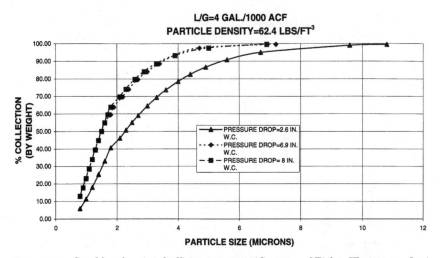

Figure 10.22 Scrubber fractional efficiency curves. (*Courtesy of Fisher-Klosterman, Inc.*)

particle size distribution of the particulate entering. For this reason it is vitally important prior to selecting a scrubber for a given application that extreme care is taken in accurately measuring the characteristics of the particulate that is to be collected, especially the size distribution.

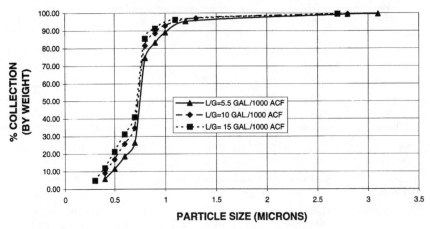

Figure 10.23   Scrubber fractional efficiency curves. (*Courtesy of Fisher-Klosterman, Inc.*)

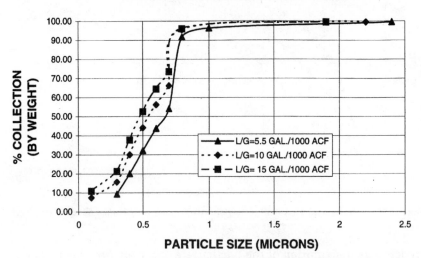

Figure 10.24   Scrubber fractional efficiency curves. (*Courtesy of Fisher-Klosterman, Inc.*)

PRESSURE DROP=40 IN. W.C.
PARTICLE DENSITY=62.4 LBS/FT³

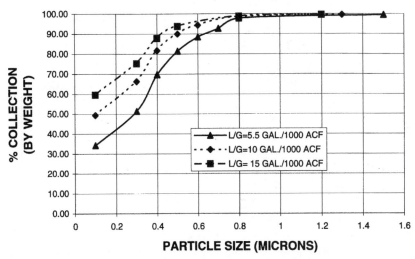

Figure 10.25 Scrubber fractional efficiency curves. (*Courtesy of Fisher-Klosterman, Inc.*)

The aerodynamic particle size distribution should be determined by a method which is consistent with that used to develop the fractional efficiency curve. The engineer may then calculate a total collection efficiency E_T by the equation

$$E_T = \int_{d=0}^{d=\infty} \frac{\partial g(d)}{\partial d} \cdot f(d) \partial d \cdot 100 \qquad (10.1)$$

where $g(d)$ is the function which describes the particle size distribution and $f(d)$ is the function which describes the fractional efficiency curve.

In most practical applications the calculation of total collection efficiency is performed utilizing a computer to perform the series equivalent of this calculation instead of using integral calculations directly. In practice it is common to perform the above operation as a series calculation, since the integral is not apparent or readily known in many cases for the product $f(d) \cdot \partial g(d)/\partial y$. In this case the generalized form of the calculation of total collection efficiency is

$$E_T = \sum_{d=0}^{d=\infty} \left\{ [g(d)_{N+1} - g(d)_N] \cdot f\left[\frac{(d)_n + (d)_{N+1}}{2} \right] \right\} \cdot 100 \qquad (10.2)$$

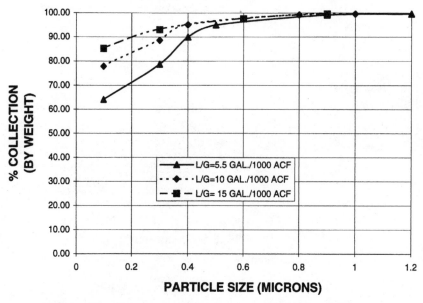

Figure 10.26 Scrubber fractional efficiency curves. (*Courtesy of Fisher-Klosterman, Inc.*)

Example Based upon the design data shown in Figs. 10.27 and 10.28, calculate the total collection efficiency E_T and emissions E_F.

solution The tabular method of utilizing Eq. (10.2) is shown in Fig. 10.29. This calculation is identical for that shown in Chap. 8, and a step-by-step description of the solution is discussed there.

The most important aspect in understanding scrubber operations is the realization that scrubber collection efficiency is most significantly dependent on energy consumption for a given inlet condition. This is called the contacting-power theory and has been well-proven by testing and operation.[4] In other words, for a given application with proper and complete description of the inlet conditions, all venturi scrubbers should have a similar energy consumption to achieve a given emission standard. It is important to keep in mind that energy consumption in a scrubber system may take numerous forms and there may be real advantages to one form of energy consumption over another in any given application based upon the cost of various forms of energy. Some of the forms of energy consumption that should be realized in system analysis are given in the following table.

Item	Common form of energy consumption
Scrubber pressure drop	Fan horsepower
Liquid supply rate	Pump horsepower
Liquid supply pressure	Pump horsepower
Cooling or condensation	Pump horsepower and energy used to cool the gas. May include thermal losses to atmosphere if evaporative cooling is used to cool water which in turn is used to cool the gas
Condensation by steam injection	Steam heat content

PARTICLE SIZE MICRONS	DISTRIBUTION % LESS THAN BY WEIGHT	COLLECTION EFFICIENCY % LESS THAN BY WEIGHT
0.05	0.00590	8.14000
0.15	0.01705	12.35000
0.25	0.03171	18.88000
0.35	0.04924	29.17000
0.45	0.06928	45.86000
0.55	0.09157	58.87000
0.65	0.11592	69.11240
0.75	0.14218	89.43920
0.85	0.17024	97.85940
0.95	0.20000	98.11920
1.05	0.23137	98.30728
1.15	0.26428	98.47655
1.25	0.29867	98.62890
1.35	0.33448	98.76601
1.45	0.37166	98.88941
1.55	0.41016	99.00047
1.65	0.44994	99.10042
1.75	0.49097	99.19038
1.85	0.53321	99.27134
1.95	0.57663	99.34421
2.05	0.62120	99.40979
2.15	0.66688	99.46881
2.25	0.71366	99.52193
2.35	0.76150	99.56973
2.45	0.81039	99.61276
2.55	0.86031	99.65148
2.65	0.91123	99.68634
2.75	0.96313	99.71770
2.85	1.01600	99.74593
22.05	20.42752	100.00000
22.15	20.55348	100.00000
22.25	20.67954	100.00000
22.35	20.80570	100.00000
22.45	20.93196	100.00000
22.55	21.05831	100.00000
22.65	21.18476	100.00000
22.75	21.31130	100.00000

Figure 10.27 Example design data.

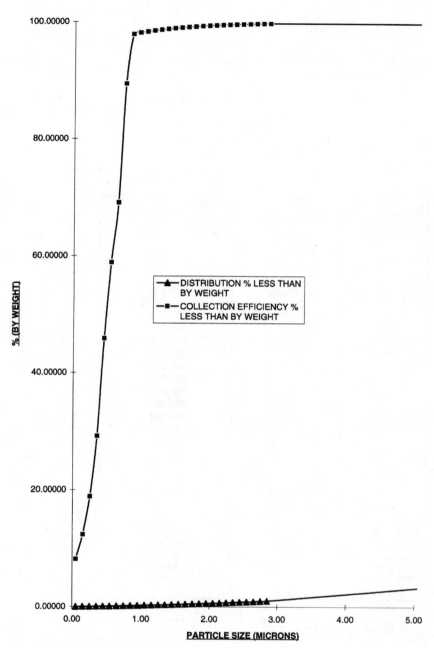

Figure 10.28   Design example data graph.

1	2	3	4	5	6	7
PARTICLE SIZE(MICRONS) RANGE			MASS			
LOW	HIGH	MID	FRACTION	$E_{MID}/100$	CATCH	EFFLUENT
0.00	0.10	0.05	0.00005903	0.08140000	0.00000481	0.00005423
0.10	0.20	0.15	0.00011148	0.12350000	0.00001377	0.00009771
0.20	0.30	0.25	0.00014659	0.18880000	0.00002768	0.00011892
0.30	0.40	0.35	0.00017535	0.29170000	0.00005115	0.00012420
0.40	0.50	0.45	0.00020038	0.45860000	0.00009189	0.00010849
0.50	0.60	0.55	0.00022287	0.58870000	0.00013121	0.00009167
0.60	0.70	0.65	0.00024349	0.69112400	0.00016828	0.00007521
0.70	0.80	0.75	0.00026264	0.89439200	0.00023490	0.00002774
0.80	0.90	0.85	0.00028060	0.97859400	0.00027459	0.00000601
0.90	1.00	0.95	0.00029757	0.98119200	0.00029198	0.00000560
1.00	1.10	1.05	0.00031371	0.98307280	0.00030840	0.00000531
1.10	1.20	1.15	0.00032911	0.98476552	0.00032410	0.00000501
1.20	1.30	1.25	0.00034388	0.98628897	0.00033917	0.00000471
1.30	1.40	1.35	0.00035809	0.98766007	0.00035367	0.00000442
1.40	1.50	1.45	0.00037178	0.98889406	0.00036766	0.00000413
1.50	1.60	1.55	0.00038503	0.99000466	0.00038118	0.00000385
1.60	1.70	1.65	0.00039785	0.99100419	0.00039427	0.00000358
1.70	1.80	1.75	0.00041030	0.99190377	0.00040698	0.00000332
1.80	1.90	1.85	0.00042240	0.99271340	0.00041932	0.00000308
1.90	2.00	1.95	0.00043417	0.99344206	0.00043132	0.00000285
2.00	2.10	2.05	0.00044565	0.99409785	0.00044302	0.00000263
2.10	2.20	2.15	0.00045684	0.99468807	0.00045441	0.00000243
2.20	2.30	2.25	0.00046777	0.99521926	0.00046554	0.00000224
2.30	2.40	2.35	0.00047846	0.99569733	0.00047640	0.00000206
2.40	2.50	2.45	0.00048891	0.99612760	0.00048702	0.00000189
2.50	2.60	2.55	0.00049915	0.99651484	0.00049741	0.00000174
2.60	2.70	2.65	0.00050918	0.99686336	0.00050758	0.00000160
2.70	2.80	2.75	0.00051901	0.99717702	0.00051755	0.00000147
2.80	2.90	2.85	0.00052866	0.99745932	0.00052732	0.00000134
22.00	22.10	22.05	0.00125858	1.00000000	0.00125858	0.00000000
22.10	22.20	22.15	0.00125961	1.00000000	0.00125961	0.00000000
22.20	22.30	22.25	0.00126062	1.00000000	0.00126062	0.00000000
22.30	22.40	22.35	0.00126161	1.00000000	0.00126161	0.00000000
22.40	22.50	22.45	0.00126259	1.00000000	0.00126259	0.00000000
22.50	22.60	22.55	0.00126355	1.00000000	0.00126355	0.00000000
22.60	22.70	22.65	0.00126449	1.00000000	0.00126449	0.00000000
22.70	22.80	22.75	0.00126542	1.00000000	0.00126542	0.00000000
+	22.80	24.00	0.78688696	1.00000000	0.78688696	0.00000000
			1.00000000		0.99921861	0.00078139

$E_T =$ **99.92%**
$E_E =$ **0.08%**

Figure 10.29 Total collection efficiency calculation.

It is certainly true that other characteristics of scrubber design and/or the thermodynamics of the system may affect scrubber operation and performance, but the previously listed effects are the most significant. In some designs, techniques commonly used for gaseous scrubbing are used in the collection of particulate. These designs which utilize increased surface area for the contact between the particulate and scrubbing liquor may operate at lower energy levels to achieve the same level of collection as a conventional venturi. The penalty, of course, is the cost of construction and the increased potential for plugging or fouling which accompanies these designs. Designs which utilize increased contact surfaces are generally most suitable for large particulate (+0.7 μm) that is readily soluble in the scrubbing liquor.

10.4.1 Pressure drop relation

The scrubber pressure drop depends on particle diameter, density of the particle and scrubbing liquid, viscosity of the scrubbing liquid, liquid and gas rate, and velocity across the throat. This can be expressed in the following functional form:

$$E_F = f(D_L, D_p, \eta_L, U_G, \rho_p, \frac{Q_L}{Q_G}, \Delta P, Y) \qquad (10.3)$$

where
E_F = fractional efficiency
D_L, D_p = diameters of liquid and particulate, respectively
η_L = liquid viscosity
U_G = gas velocity
ρ_p = particulate density
Q_L/Q_G = liquid-to-gas ratio
ΔP = pressure drop across venturi
Y = surface tension

Scrubber efficiency increases with

1. Increasing pressure drop
2. Increasing liquid-to-gas rate
3. Increasing particle and liquid droplet diameter
4. Increasing pressure drop across the throat
5. Increasing particle density
6. Increasing number of droplets in contactor section

As seen in Eq. (10.3), the variables are numerous and complex. For this reason, it is extremely difficult to obtain reliable and accurate design data. Through a number of field and/or laboratory tests it is possible to compile a good set of data to predict the scrubber efficiency. Various research work has been done to adequately predict the scrubber efficiency as a function of pressure drop for a given particle size and properties and liquid-to-gas rates.[7-12] Assuming that all the energy is used to accelerate the liquid droplets to the throat velocity of the gas, Calvert derived the following equation relating the pressure drop to liquid-to-gas rate and throat velocity:[8]

$$\Delta P = 1.65 \times 10^{-3} \cdot U_{\text{TR}} \cdot \frac{Q_L}{Q_G} = \left[1.03 \times 10^{-3} \cdot U_{\text{TR}} \cdot \frac{Q_L}{Q_G} \right]$$

where U_{TR} = throat velocity, ft/s [cm/s]

Q_L/Q_G = liquid-to-gas ratio, gal/1000 ft^3 [L/m^3]

ΔP = pressure drop, in w.c. [cmH$_2$O]

Example   Assume U_{TR} = 200 ft/s = [6096 cm/s] and $\frac{Q_L}{Q_G}$ = 10 gal/1000 ft^3 = [1.3365 L/m^3]. Then the pressure drop is

$$\Delta P = 1.65 \times 10^{-3} \cdot 200 \cdot 10 = 3.3 \text{ in w.c.}$$

$$= [1.03 \times 10^{-3} \cdot 6096 \cdot 1.3365 = 8.39 \text{ cmH}_2\text{O}]$$

There are two general approaches for the development of a predicted fractional efficiency curve for a new application. These are (1) to use previously measured fractional efficiency curves for a scrubber operating at a similar pressure drop and liquid rate and (2) to use a generalized method for calculation. Of the two methods listed above, utilizing empirical data is more simple and reliable if the design conditions are similar to those shown. We have included fractional efficiency curves for scrubbers operating over a wide range of pressure drops and liquid rates (Figs. 10.22 to 10.26). These curves may be used effectively as long as the gas density, gas viscosity, and particle density pressure and liquid rate (Q_L/Q_G) are similar to that specified for the given curve.

Calvert also presents a collection efficiency equation related to collection mechanism (impaction, diffusion, etc.), slip, drag, size of atomized droplets, and uniformity of the atomized droplets in terms of particle penetration $P(D)$.[8] The particle penetration is defined as $1-E_{dF}$ and is related to the inertial impaction by the following relationship[12]:

$$P(D) = 1 - exp^{\left(-K \cdot \frac{Q_L}{Q_G} \cdot \psi_L \right)} \tag{10.5}$$

where Q_L/Q_G = liquid-to-gas rate in gal/1000 acf

K = system parameter (or inertial impaction correlation factor, approximately 0.12 to 0.2)

ψ_L = impaction parameter [see Eq. (10.8)]

For example, if Q_L/Q_G = 10 gal/acf, ψ_L = 1, and K = 0.12, then penetration $P(D)$ is

$$P(D) = 1 - e^{(0.12 \cdot 10 \cdot 1)} = 0.6988$$

Based on Eqs. (10.4) and (10.5), Hesketh studied the variation of venturi pressure drop for various particle diameters and liquid rates and formed a generalized correlation that related the venturi pressure drop to the throat velocity and liquid-to-gas rate for particles with diameters less than 5 μm.[7] This correlation is

$$P(D) = 3.47\Delta P^{-1.43} = [13.16\Delta P^{-1.43}] \qquad (10.6)$$

where ΔP is the pressure drop, in w.c. [cmH$_2$O], and is related to throat velocity and liquid-to-gas ratio by Eq. (10.4). For example, if the pressure drop is 5 in w.c. [12.7 cm w.c.], then penetration $P(D)$ calculated per Eq. (10.6) is

$$P(D) = 3.47(5)^{-1.43} = 3.47(0.1) = 0.347$$

$$= [13.16(12.7)^{-1.43} = 0.347]$$

It is apparent from Eqs. (10.4) to (10.6) that in order to collect a particle of know diameter we have to produce the liquid droplet of a certain diameter that is necessary for impaction and collection of these particles. The diameter of the liquid droplet is a function of the pressure drop, liquid-to-gas ratio, and properties of the gas and liquid. In other words, to produce a liquid droplet of a known diameter, we have to provide enough pressure drop through the venturi. This pressure drop can be provided by increasing the liquid-to-gas rate or by increasing the throat velocity (or decreasing the throat area). Higher scrubbing efficiencies of smaller particles are achieved by large pressure drops across the venturi. In other scrubber systems (nonventuri type), fine liquid droplets may be produced using a high-pressure nozzle (dual phase nozzles). Therefore, when choosing the mechanism of scrubbing by impaction, clearly it is necessary to determine the size of the liquid droplet that is required to collect these particles, the properties of the gas and liquid, particle diameter, liquid-to-gas rates, and throat velocities. The experimental and theoretical study of droplet sizes is shown in Refs. 2 and 9. The term *Sauter mean diameter* is predicted for liquid droplets in venturi scrubbers and is determined by the Nukiyama-Tanasawa equation given below.[10,11] (Sauter mean diameter is defined as the average ratio of the volume to the surface [i.e., the average surface diamter] of the liquid droplet in pneumatic sprays.)

$$D_L = \frac{585}{U_G} \cdot \left(\frac{\sigma}{P_L}\right)^{0.5} + 597 \cdot \left[\frac{\mu_L}{(\sigma \cdot P_L)^{0.5}}\right]^{0.45} \cdot \left(1000 \cdot \frac{Q_L}{Q_G}\right)^{1.5} \qquad (10.7)$$

where D_L = liquid droplet diameter, μm
σ = liquid surface tension, dyn/cm
μ_L = liquid viscosity, P

$$P_L = \text{liquid density, g/cm}^3$$
$$Q_L/Q_G = \text{liquid/gas ratio, L/m}^3$$
$$U_G = \text{relative velocity between gas and liquid, cm/s}$$

Example If the relative velocity between the gas and liquid is 6096 cm/s, P_L is 1 g/cm³, μ_L is 0.001 P, σ is 70 dyn/cm, Q_L is 1000 cm³/s, and Q_G is 10^6 cm³/s, what is $D_{L?}$

solution

$$D_L = \frac{585}{U_G} \cdot \left(\frac{\sigma}{P_L}\right)^{0.5} + 597 \cdot \left[\frac{\mu_L}{(\sigma \cdot P_L)^{0.5}}\right]^{0.45} \cdot \left(1000 \cdot \frac{Q_L}{Q_G}\right)^{1.5}$$

$$= \frac{585}{6096} \cdot \left(\frac{70}{1}\right)^{0.5} + 597 \cdot \left[\frac{0.001}{(70 \cdot 1)^{0.5}}\right]^{0.45} \cdot (1000 \cdot 0.001)^{1.5}$$

$$= 0.803 + 597 \cdot 0.017 \cdot 1$$

$$= 10.95 \ \mu\text{m}$$

The impaction parameter ψ_L is then related to the particle diameter by the relation

$$\psi_{L1} = C(\rho_p - \rho_G) \frac{D_P^2 \, u_p}{18 \, \mu_L \, D_L} \tag{10.8}$$

where u_p = relative velocity between particle and droplet, cm/s
ρ_p = particle density, g/cm³
ρ_G = gas density, g/cm³
D_P = particle diameter, cm
μ_L = liquid viscosity, P
D_L = droplet diameter, cm [see Eq. (10.7)]
C = Cunningham slip correction factor

The Cunningham slip correction factor in the viscous flow region is a function of the particle diameter and the mean free path of gas molecules. It is given by the relation[2]

$$C = 1 + \frac{2\lambda}{D_P} \left(1.23 + 0.41 \, E^{(-0.44D_p/\lambda)}\right) \tag{10.9}$$

where D_P is the particle diameter and λ is the mean free length of the gas molecule in micrometers.

The mean free length can be calculated for a given gas using the relation

$$\lambda = \frac{\mu_G}{0.499\, \rho_G u} \tag{10.10}$$

where μ_G = viscosity of gas, P
ρ_G = gas density, g/cm^3
u = average velocity of gas molecules, cm/s [expressed by Eq. (10.11) and based on kinetic theory of gases].

$$u = \sqrt{\frac{8\,RT}{\pi M}} \cdot 100 \tag{10.11}$$

where R = universal gas constant (8.314 J/mol · K)
T = temperature, K
M = molecular weight of gas (29 for air)
π = constant (3.14159)

Example If air at 20°C [293 K] with a viscosity μ_G of 1.81×10^{-4} P and density ρ_G of 0.001 293 g/cm^3 contains 0.2-μm-diameter particles, what are u, λ, the Cunningham slip correction factor C, and the inertial impaction factor ψ_L?

solution u is calculated per Eq. (10.11) to be

$$\sqrt{\frac{(8)(8.314)(293)}{(3.14159)(29)}} \cdot 100 = 1462.5 \text{ cm/s (or 47.96 ft/s)}$$

The mean free length λ is calculated by Eq. (10.10) to be

$$\lambda = \frac{(1.81 \times 10^{-4})}{(0.499)(0.001293)(1462.5)} = 1.918 \times 10^{-4} \text{ cm} = 1.918\ \mu\text{m}$$

Therefore, the Cunningham slip correction factor C is calculated using Eq. (10.9) to be

$$C = 1 + \left(2 \cdot \frac{1.918}{0.2}\right) \cdot \left[1.23 + 1.41 \cdot \exp\left(-0.44 \cdot \frac{0.2}{1.918}\right)\right] = 32.1$$

The slip correction factor is less than 1 percent for 20-μm-diameter particles in ambient air, about 5 percent for 5-μm-diameter particles, and 16.7 percent for 1-μm-diameter particles. Outside the viscous flow region the experimental values overestimate the slip correction estimates by at least 0.2 percent. Hence, the inertial impaction parameter is calculated using Eq. (10.8) to be

$$\psi_L = \frac{(32.1)(1-0.001293)(0.2 \times 10^{-4})^2(6096)}{(18)(0.001)(10.95 \times 10^{-4})} = 3.996$$

Therefore, using Eqs. (10.5) and (10.7) to (10.11), one can compute the actual pressure drop required to collect a particle of known diameter. It is also very clear from Eq. (10.8) that scrubbing action is extremely dependent on the properties of the liquid and particle and that pressure drop is needed to produce droplets based on known particle diameters. The above equation also provides insight to the fact that the addition of detergent provides improved particle adherence due to reduced surface tension. Based on Eqs. (10.4) to (10.8) and knowing the fact that the sum of E_F and $P(D)$ must equal 1 at any particle size, then

$$E_F = 1 - P(D)$$

One can then develop fractional efficiency curves for a given liquid-to-gas rate, particle and liquid properties, and pressure drop. Figure 10.30 shows a typical correlation between experimental and theoretical data (Calvert's model) from a venturi scrubber used for removal of coal particulate (fly ash). Figure 10.31 predicts the penetration of submicrometer particulate using Calvert's model.[12]

10.5 Applications

Scrubbers are commercially suited for a variety of industries and applications like the paper and pulp industry, boiler industry, chemical plants, flue gas desulfurization, and explosive manufacturing. In this chapter two cases have been given for illustration:

1. Coal fired utility boiler application
2. Lime and limestone flue gas desulfurization

It is advised to the reader that choosing a method of scrubbing depends on the process and a full understanding of the process. The two examples were chosen to demonstrate the wide application range of scrubbers and also to emphasize that a combination of scrubbing systems can be chosen to provide the most suitable system for any process.

10.5.1 Coal fired utility boilers

A variety of scrubbing systems and other particulate control devices can be used for this application, some of which are highlighted in Ref. 13. For this process, venturi scrubbers are mostly used because of the nature of the process and particulate distribution. Typical particle size distribution data based on the different types of boilers is shown in Table 10.3.

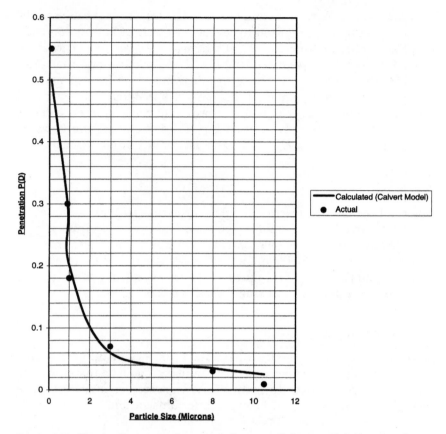

Figure 10.30 Penetration comparison—actual versus Calvert model. Pressure drop = 27 in w.c.; particulate is fly ash.

The typical pressure drop used in this type of system is on the order of 20 in w.c. or more. The liquid-to-gas ratios are between 10 to 15 gal/1000 acf. Most of the particulate is 0.1 to 10 μm. The specific gravity of the fly ash (typical in a coal fired boiler) is 1. Therefore, we can calculate (as described in Sec. 10.4.1) the fractional efficiency curve for the scrubber (Table 10.4). The measured test data for the process is as follows:

Liquid-to-gas rate = 9 gal/1000 acf = [1.45 L/m^3]

Pressure drop = 28 in w.c. = [71 cmH$_2$O]

Throat velocity = 217 ft/s = [66.1 m/s]

Sauter mean diameter (using 10.7 for air and water system) = 111.35 μm

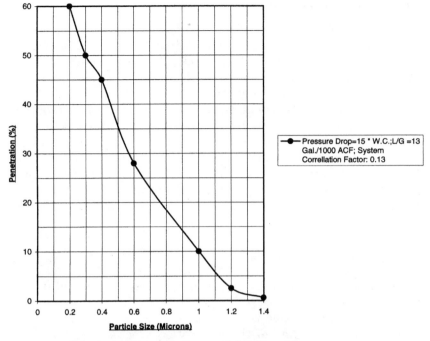

Figure 10.31 Submicrometer penetration per Calvert's model.

TABLE 10.3

Boiler	Loading	Particle size	Combustible content in fly ash, %
Stoker	Low to medium, depending on coal ash	Coarse, 20%<10 μm	40–60
Cyclonic	Low	Fine, 80%<10 μm	10–30
Pulverized coal	High to medium, depending on coal ash and type bottom	Medium, 50%<10 μm	<5

TABLE 10.4

Particle size, μm	Measured penetration (1−removal efficiency), %	Inertial impaction [using Eq. (10.8)]	Calculated penetration [using Eq. (10.5)], %
0.2	60.2	0.73	45
0.6	28.6	0.33	24
0.9	16.4	1.49	20
1.5	5.37	3.1	4
3	0.33	4.96	1
7	0	225,466	0

10.5.2 Limestone scrubbing

Limestone scrubbing is used in applications where sulfur dioxide and particulate must be scrubbed. A typical application would be flue gas desulfurization for a utility power boiler. The typical scrubber that is used for such a process is a venturi with packed tower system. Figure 10.32 shows a typical flow diagram for such a process.[14] The venturi is used for scrubbing particulate and to start the initial absorption process. The absorption process is for removing sulfur dioxide, and the scrubbing liquid is lime (calcium hydroxide). For lime and limestone scrubbing, the gas-phase absorption or chemical reaction rate, or both, may be rate-controlling resistances. Hence, it is necessary to maximize the surface area of the calcium-contributing reactant and to minimize the resistance of the gas-to-liquid interface at minimum cost. Slurry recirculation on the order of 20 to 80 gal/1000 ft^3 would be required depending on the scrubber used. A solids content of 5 to 15 percent by weight is typically used in the slurry. Here venturis with a pressure drop of 10 to 15 in w.c. at a liquid rate of 10 to 15 gal/1000 acf are used. The venturis used in this system have an adjustable throat that allows the user to change the pressure drop. The removal efficiency of sulfur dioxide is on the order of 80 to 90 percent. For a complete understanding of the absorption process we recommend reading Chap. 12 of this book. Care must be taken to account for the heat of reaction between the lime and sulfur dioxide. Owing to the corrosive nature of the gas, typical materials of construction for the venturi are alloys like Inconel and Hastelloy. In addition, 316L stainless steel is sometimes used for the process. Secondly, owing to the

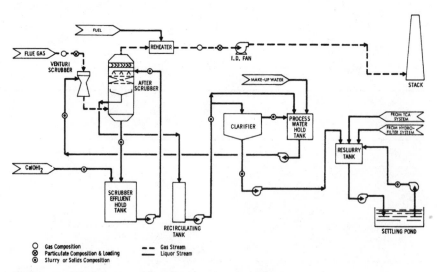

Figure 10.32 Typical process flow diagram for venturi system.

abrasive nature of the particulate, abrasion-resistant materials must be used. Also, the material should withstand the highest temperatures encountered during normal operation and upset conditions.

10.6   Installation

10.6.1   Site selection

Scrubber installation will vary depending on the type of scrubber and the location of the desired installation. If the scrubber installation is outdoors, the site must be inspected to determine that the soil at the site will be sufficient for the bearing load and that site drainage is adequate should a system upset occur. Depending on local weather conditions, freeze protection may be required for pipes, tanks, and pumps. Scrubber and support design calculations must not only include the appropriate equipment design loads and earthquake loads but must also include wind loads (Fig. 10.33). Indoor installations are usually free from concerns of freezing and wind loads, but fitting many scrubber systems into existing plant space may be difficult.

Adequate access to all mechanical components and tanks is important in site selection. Over the life of a scrubber system it should be assumed that

1. Each electric component (motors, instruments, etc.) will require replacement

2. Pump impellers will require replacement

3. Spray nozzles or injection tubes will require frequent cleaning or replacement

4. Any recycle or settling tanks handling suspended solids will need to be fully drained and manually cleaned frequently

5. Scrubber internals will need to be accessed, cleaned, and/or repaired on a periodic basis

Figure 10.33 Installation of four FKI WL low-energy scrubbers at fiberglass manufacturing plant. (*Courtesy of Fisher-Klosterman, Inc.*)

The necessity of providing good access to all parts of a scrubber system that *may* require service cannot be overemphasized. Failure to do so and the resulting lack of required service on the system will eventually result in losses in efficiency, system downtime, and possibly a loss of plant operation for operational or regulatory reasons.

10.6.2 Support

As with any mechanical equipment, scrubber support should take into account all appropriate loads with adequate safety factors. Static loads should be considered with the scrubber flooded with water to as high an elevation as is physically possible. It is common for scrubber drain lines to become plugged.

Piping, ductwork, blowers, and pumps should not be supported from the scrubber unless specifically included in the vendor's design. All ductwork and connections to the scrubber should have vibration isolation sleeves if they are connected to equipment with moving parts. In hot gas systems, adequate allowances must be made for thermal expansion to prevent the transmission of excess loads to the scrubber.

10.6.3 Inlet and outlet ducting

The ducting considerations mentioned in Chap. 8 on cyclones are similar to those required for scrubber installation. In general, however, scrubber performance is somewhat less sensitive to inlet duct design than is cyclone performance. The basic rules to follow are

1. Duct velocities should be sufficiently high to keep dust particles airborne.

2. Horizontal elbows in the inlet ductwork should not be close-coupled to the scrubber but should be at least two duct diameters away from the scrubber inlet if possible. If a horizontal elbow leads up to the scrubber inlet, it should be in the same rotation as the scrubber (if appropriate).

3. Vertical upward elbows into a scrubber inlet should be at least two duct diameters away from the scrubber inlet and of such a design to ensure that eddy currents do not allow a solids buildup in the scrubber inlet.

4. Inlet transitions should be designed so that solids do not build up on the scrubber inlet. Included angles should not exceed 30°.

5. Outlet ducting should be designed to provide minimum system effects if the fan is downstream.

6. Outlet ducting should have provisions for access and maintenance as there is still remaining particulate in the gas stream and mist

surfaces downstream of the scrubber will be moist and accumulate a buildup of solids over time.

10.7   Operation

The basic requirements for successful operation of a properly designed scrubber are to maintain the proper airflow and liquid flow to the scrubber. Most commonly, a failure of a scrubber to meet its expected efficiency is the result of

1. Improper design and/or incorrect design data
2. A failure in operation as a result of incorrect gas flow rate, incorrect liquid supply, or inadequate mist elimination

Often the original design data used to select a scrubber is erroneous due to the difficulty involved with accurate data acquisition and/or the methods selected for obtaining the data. It is vital for proper scrubber design that not only is the gas flow measured accurately and its components identified properly but that the particulate loading and aerodynamic size distribution be ascertained. Chapter 3 discusses this issue more fully. An operational failure often occurs in conjunction with more than one of the above causes as they are integrated in most scrubber designs.

10.7.1   Gas flow rate

The gas flow to the scrubber is largely controlled by factors aside from the scrubber (ventilation requirements, fan capacity, etc.). Ventilation and fan system design are discussed in detail in Chaps. 4 and 5, respectively.

Many scrubbers are *fixed* devices, meaning their pressure drop (energy consumption) is dependent on the gas flow rate. Should the gas flow rate be less than the design, the pressure drop will be lower and so will the scrubber collection efficiency. Flows that are higher than the original design will result in higher scrubber pressure drop and subsequent efficiencies but may exceed the static pressure allowed for that device and result in less than desired flow in the rest of the system. Gas density and gas viscosity also affect scrubber performance but are largely outside of the designer's control and must be dealt with as they occur within the system.

If flow rates are uncertain or variable, it is often wise to use an adjustable venturi scrubber. This allows the operator to maintain the design pressure drop over a wide range of flow rates. Often the venturi is automatically controlled to maintain a given pressure drop and collection efficiency.

Scrubbers are frequently the technology of choice on the exhaust of hot gas systems. It is important to cool the gas prior to entering the scrubber if the gas temperature will result in a significant evaporation of water. Most scrubber manufacturers require quenching if the inlet gas temperature exceeds 300 to 500°F. Quenching is a simple and economical method of cooling the gas in a scrubber system. If hot gases are not quenched, scrubber performance will be compromised as liquid is evaporated in the scrubber contactor section.

System designers should be aware of the changes that will occur in the gas characteristics as the gas passes through the quench and/or scrubber. The gas will change in volume, density, and mass as follows:

1. The system pressure decreases causing an increase in volume

2. The mass and volume increase due to the evaporation of water

3. The volume decreases due to a reduction in temperature

For virtually all systems where the fan follows the scrubber, the scrubber inlet conditions are significantly different than those for the fan.

10.7.2 Liquid supply

Maintaining the design liquid flow rate to the scrubber is an absolute requirement in maintaining performance and may be vital in reducing maintenance and/or damage to parts of the scrubber and downstream equipment. The full aspects of proper design of liquid systems are beyond the scope of this work, and problems in some part of this system are most frequently the cause of scrubber failure. The designer of the liquid supply system should not only be familiar with the characteristics of the specific process under consideration but should also be familiar with proper piping and tank design practices, pumps, instrumentation, and techniques to separate liquids and solids. If this expertise is not readily available, the scrubber should be purchased as a complete system from a reliable vendor and expertise must be developed internally or by the use of outside design consultants.

In particulate scrubbers there are three general liquid supply arrangements. These are

1. Once-through systems

2. Recycle systems

3. Closed systems

Once-through systems are extremely rare in industrial applications as the cost and difficulty associated with using large quantities of fresh water and disposing of large quantities of dirty water are often prohibitive. Once-through systems utilize fresh water from some source as the liquid supply. All the liquid discharge is sent to the wastewater system in the sewer. Some industrial operations have their own large liquid reservoirs that may be used for liquid supply and disposal. For practical consideration, these applications resemble once-through systems due to the relatively vast size of the reservoirs compared to the scrubber flow. Once-through liquid supply systems are the most maintenance-free as it is a relatively simple matter to pump fresh water without creating line or nozzle pluggage, impeller erosion, or instrument fouling, and scrubber internals will better resist fouling. This is also the simplest liquid supply system, so maintenance is reduced by having fewer components that require upkeep or that can fail.

Recycle systems are the most common of liquid supply systems. Recycle systems reduce the amount of liquid that must be supplied to and disposed from the scrubber by using the same water many times. In a recycle system most of the dirty water exiting the scrubber is used as the liquid supply to the scrubber. A small amount of liquid is removed continuously from the system so that the pollutant concentration in the liquid remains at an acceptable level. Fresh water is added to the system to provide for evaporation losses and the water that is bled off from the system. The main disadvantages of recycle systems are

1. The complexity of the system makes operation and maintenance more extensive and difficult.
2. The solids and pH level of the liquid may result in shorter equipment life and increased maintenance.

In most industrial applications, the disadvantages listed above are far outweighed by the advantages and some form of a recycle system is used.

Closed systems (Fig. 10.34) are generally nothing more (or less) than a recycle system with periodic bleed-off. Many small and some large industrial applications will recycle the liquid in a closed loop (except for the makeup of evaporation losses) for a given operating cycle. At the end of that cycle, collected solids, in the form of sludge, are removed from the system. This is a reasonable approach to liquid supply design if

1. The liquid supply reservoir is large enough to operate through a full cycle without the solids concentration and/or pH exceeding allowable levels.
2. The operating cycle period is set short enough that maintenance and operational problems do not occur.

Figure 10.34 FKI PS Scrubber system with skid-mounted recycle tank, pump, and instrumentation. (*Courtesy of Fisher-Klosterman, Inc.*)

3. Adequate provisions are made for removal and disposal of the relatively large mass and volume of sludge at one time.

10.7.3 Mist elimination

The basic mechanisms for failure of any of the mist eliminator designs discussed herein are (Fig. 10.35)

1. Too high a velocity through the mist eliminator which may be caused by
a. Too high a gas flow rate
b. Buildup of solids or other material on the surfaces of the eliminator resulting in a reduced cross-sectional area and increased velocity through the section
2. Buildup of solids or other material over a functional part of the mist eliminator such as a liquid trap or drain
3. Loss of a functional part of the mist eliminator such as a liquid trap, mesh pad, or drain by corrosion, erosion, or some other mechanism
4. Excess liquid entrainment beyond the capacity of the mist eliminator, usually a result of a failure in the design or operation of the drain piping

Figure 10.35 Liquid removal—looking down into a FKI MS cyclonic separator. (*Courtesy of Fisher-Klosterman, Inc.*)

The exact arrangement, design, and specific components of a scrubber system vary from application to application as the style of scrubber and plant requirements change. Although it is not practical within the scope of this work to provide a full presentation of the technologies involved in these systems, we offer the following descriptions and usages of the most common components.

10.7.4 Supply piping

All piping runs should be sized adequately to convey the specified liquid and suspended solids without excessive pressure drop or solids settling. Good plumbing practices should apply to the design, construction, and installation of all piping. Clean-out access ports, plugs, or unions should be provided at locations likely to require maintenance.

10.7.5 Pumps

All pumps should be sized to supply the required liquid rate at a sufficient pressure (or head) to overcome all nozzle, pipe, and lift losses. The pump must be designed so that liquid is free-flowing to the inlet at the specified flow rate or such that the pump net positive suction head (NPSH) is adequate to overcome the appropriate inlet losses.

10.7.6 Fans and blowers

Fans and blowers are covered in detail in Chap. 4. Some special considerations should be given to the selection and construction of air moving equipment associated with scrubber systems. One of the primary decisions that must be made is whether to locate the fan upstream or downstream of the scrubber. Upstream fans will often operate at lower gas flow volumes if the gas is reasonably cold prior to reaching the scrubber due to the pressure-volume relationship (Boyle's law) of the gas and may have fewer problems from corrosion and solids accumulation within the fan. Corrosion problems may be fewer in an upstream fan since usually there is no free water in this area to form acid. Care must be taken in evaluating the likelihood of corrosion, though in many cases the ions that form acids when water is present are corrosive even in the absence of water. Chloride ions, which form hydrochloric acid (HCL) when combined with water, are a prime example.

Downstream fans have the advantage of operating on a relatively cool gas, free from high particulate loadings. In systems where the particulate is abrasive, this becomes a significant consideration. Operating on a gas that has been cooled to saturation as it passes through the scrubber may allow for the use of fiberglass reinforced plastic (FRP) or coated carbon steel construction in some applications where corrosion is a concern, but high temperatures exist upstream of the scrubber. The gas volume may be reduced after the scrubber due to the temperature-volume relationship (Charles' law) and the cooling of the gas as it passes through the scrubber.

One misconception that is often incorporated into the decision to locate the fan downstream of the scrubber is that the gas is free of particulate. Although the loading and average particle size are greatly reduced after the scrubber, there are still some particulate emissions. Consider a scrubber operating at 20,000 ft^3/min with an emission level of 0.01 gr/ft^3 which is far below visible levels and most emission standards. This outlet loading equates to 1.7 lb/h or 288 lb/week of particulate emissions. Ductwork, fans, and stacks downstream of the scrubber are frequently damp from condensation, and some of the particulate emissions will accumulate on these damp surfaces. In the case of fans, this may eventually result in balance problems.

If the fan is installed downstream of the scrubber, it should include a drain, access door, and other provisions that will allow for easy cleaning. Stacks should not be arranged directly over the fan outlet so water condensing in the stack does not drain back into the fan. A recommended fan and stack arrangement is shown in Fig. 10.36.

10.7.7 Instrumentation

There is a wide variety of instrumentation that is commonly used on scrubber systems. Instruction concerning the proper selection and

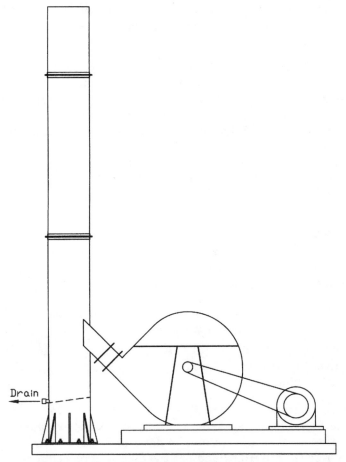

Drain

Figure 10.36 Recommended fan and stack arrangement. (*Illustration by Alex Donenberg. Courtesy of Fisher-Klosterman, Inc.*)

usage of this instrumentation is beyond the scope of this work, but we can provide some basic guidelines. The two guidelines for selection of where and what to measure are

1. Measure the operating parameters that are critical to scrubber performance, usually scrubber pressure drop and liquid flow rate. On many scrubber installations this level of instrumentation is a requirement of the appropriate environmental regulatory agency since if either of these conditions is not maintained at design levels, performance will suffer.

2. Measure those operating parameters for the scrubber and related equipment that will ensure operability at the conditions required for scrubber operation and obtain adequate troubleshooting informa-

tion. This category may also include gas flow rate, scrubber inlet and outlet temperatures, fan and pump motor amperage, pH, and water temperature.

In general, the greater the level and sophistication of the instrumentation incorporated into a scrubber system, the more simple it is for the operators to keep the system running properly and troubleshoot system problems when they occur. It should be well understood, however, that the instrumentation itself adds numerous additional components to the system that may fail and/or require maintenance. The cost of the instrument package must also be considered in the selection process. It is common for the cost of instruments and controls to exceed the cost of the basic scrubber on small systems.

10.7.8 Drain piping

Care must be exercised in the selection, location, and installation of drain piping. The following precautions are recommended when designing a proper drain system for a scrubber:

1. The scrubber must be elevated above the highest expected liquid level of your recirculation tank or waste liquid system.

2. There must be a gas-tight seal at the bottom of the drain pipe extension, usually created by submerging the pipe beneath the lowest expected liquid level.

Elevating the scrubber sufficiently above the high liquid level will ensure that the liquid is not sucked back into the scrubber, causing carryover into the exhaust gas stream. To determine what is a sufficient elevation, refer to Fig. 10.37 and use the following procedures.

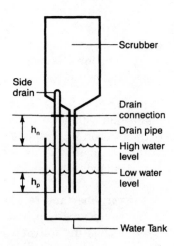

Figure 10.37 Scrubber drain design. (*Courtesy of Fisher-Klosterman, Inc.*)

The minimum elevation with negative pressure (when the fan is located downstream of the scrubber) is determined by

$$h_n = \frac{P_2}{0.83} + 12 \quad \text{and} \quad h_p = 12$$

$$= \left[\frac{P_2}{0.83} + 31 \quad \text{and} \quad h_p = 31 \right] \quad (10.12)$$

where P_2 = scrubber outlet static pressure, in w.c. [cmH$_2$O]
$\quad h_n$ = scrubber minimum elevation above maximum water level, in [cm]
$\quad h_p$ = drain pipe minimum submergence below minimum water level, in [cm]

Example Assume a scrubber outlet pressure of -42 in w.c. [107 cmH$_2$O]. What is the minimum scrubber elevation above the tank water level h_n and the minimum submergence h_p?

solution Using Eq. (10.12),

$$h_n = \frac{42}{0.83} + 12 = 62.6 \text{ in} = \left[\frac{107}{0.83} + 31 = 129 \text{ cm} \right]$$

$$h_p = 12 \text{ in} = [31 \text{ cm}]$$

The minimum elevation with positive pressure (when the fan is located upstream of the scrubber) is determined by

$$h_p = \frac{P_1}{0.83} + 12 \quad \text{and} \quad h_n = 0$$

$$= \left[\frac{P_1}{0.83} + 31 \quad \text{and} \quad h_n = 0 \right] \quad (10.13)$$

where P_1 = scrubber inlet static pressure, in w.c. [cmH$_2$O]
$\quad h_p$ = drain pipe minimum submergence, in [cm]
$\quad h_n$ = scrubber minimum elevation above water level, in [cm]

Example Assume a scrubber inlet static pressure of $+ 32$ in w.c. [81 cmH$_2$O]. What is the minimum submergence h_p and the minimum scrubber elevation above the tank water level h_n?

solution Using Eq. (10.13),

$$h_p = \frac{32}{0.83} + 12 = 50 \text{ in} = \left[\frac{81}{0.83} + 31 = 129 \text{ cm} \right]$$

$$h_n = 0$$

10.8   Maintenance

Scrubbers and their related equipment and systems often receive less than adequate maintenance since they are not commonly value-added production equipment. This attitude, although short-sighted, is all too prevalent within many facilities. The eventual result of this approach is a loss in production due to a complete failure of the scrubber system and subsequent shutdown and/or excessive emissions while operating, which may result in fines from regulatory agencies, loss of public and employee goodwill, and real damage to the surrounding environment. Scrubber systems generally provide the most economical service for the owner when they are maintained in a regular manner that ensures their continued reliable operation.

Periodic inspection of major system components can help to prolong the life of the system and avoid unexpected failures. A regularly scheduled inspection and checklist are recommended. External system components (pumps, fans, instruments, etc.) will each have a recommended inspection and maintenance procedure that should be followed closely. Internal scrubber components (nozzles, trays, packing, etc.) should be inspected visually on a regular basis that will vary depending on the application. This inspection should occur at least on an annual basis, and for many systems should occur monthly. This inspection must be tailored specifically to the scrubber selected, but some of the key points follow.

10.8.1   Spray nozzle maintenance

Scrubber performance can be compromised and rendered ineffective by eroded, damaged, or plugged nozzles. It is important to perform regular inspections of the spray nozzle manifold to maintain scrubber system efficiency. All nozzles are selected to deliver a liquid volume at a particular design pressure. If this flow is not being delivered, the nozzles themselves are a logical source of the problem. Nozzle performance degradation can be caused by several things; the most common are discussed below.

10.8.2   Erosion, wear, or corrosion

Gradual enlargement and distortion of the nozzle orifice usually causes an increase in scrubbing liquid flow. Pressure may decrease, the spray pattern may become irregular, and spray droplets may become larger. These conditions cause a very definite drop in collection efficiency. A breakdown of the nozzle material by chemical action from the sprayed liquid will cause performance problems similar to those caused by erosion. Chemical action can also cause severe damage to the exterior surfaces of the nozzle.

10.8.3 Clogging or caking

Recirculating solid particles may be of such size that they block the orifice (clear passage), causing a restricted or blocked spray pattern that hampers uniformity of the scrubbing action.

10.8.4 Temperature effect

Prolonged high temperatures may cause material breakdown or deformity, which could tend to alter spray patterns. If your system must operate with a hot process flow, careful monitoring of the nozzle's condition is essential.

10.8.5 Improper reassembly after cleaning

Misaligned or damaged gaskets, O-rings, or internal vanes may cause an improperly overhauled nozzle to leak or produce a faulty spray pattern. Take special care when reassembling; do not overtighten nozzle caps onto bodies.

10.8.6 Spray manifold alignment

Frequently, many nozzles are mounted into a common header or manifold. Usually nozzle bases are designed to fit into pipe fittings welded directly to the manifold and no adjustment is necessary. If, after maintenance, the manifold is reinstalled backward (facing the wrong direction), scrubber efficiency will be hampered. The scrubber design should prevent any misalignment. Visually check the directional alignment at installation and after nozzle cleaning or repair. If the alignment is not correct, remove and restore to the proper location.

10.8.7 Adjustable venturi throat

The internal parts of an adjustable venturi throat should be checked periodically for wear and corrosion. These parts, the shaft and damper, may be directly in the gas stream and can experience heavy wear due to process materials that are abrasive or corrosive.

If the venturi is adjustable, the external operator mechanism (Fig. 10.38) should be cleaned and relubricated periodically. Check and replace shaft packing if needed. Use a high-quality general service compression packing material suitable for service conditions.

10.8.8 Interior surfaces

Periodic checks of the scrubber interior should be made to determine if there is solids buildup on the inside surfaces. Inspection will determine if an abrasive and/or corrosive condition exists. This could

Figure 10.38 Manually adjustable venturi with turnbuckle adjustment. (*Courtesy of Fisher-Klosterman, Inc.*)

result in degradation of the scrubber to a point where it is not structurally or functionally sound. Determine that leaks located at gasketed joints are not due to wear or deterioration of internal surfaces. Clean away any excess of solids buildup on the scrubber inside surfaces (thicker than 1/8 in). This should be done with regard to materials of construction, exercising care to prevent damage. High-pressure water sprays or industrial cleaners are preferred. Chisels may be used, but avoid the use of hammers which may cause dents in the walls of the scrubber.

It may be possible to patch weakened areas with field-welded steel patches if action is taken soon after severe wear or corrosion is detected. If immediate repair measures are not taken, there may not be sufficient solid metal to support patching at a later date. Complete replacement may become necessary to maintain particulate collection control.

10.8.9 Abrasion and corrosion

The presence of any severe abrasion or corrosion is an indication of an ongoing process that may result in serious deterioration. Close monitoring is recommended. The best solution to any abrasion or corrosion problem is in the initial selection of (*a*) the proper scrubber design and (*b*) material of construction. Material selection should always be based upon the owner's experience in handling process gases and solids.

10.8.10 Troubleshooting

Table 10.5 is a typical checklist that is to be used while troubleshooting a scrubber system.

TABLE 10.5   Troubleshooting guide

Item number	Description	Possible cause
1	Too-high pressure drop	*a.* High gas rate as compared to design *b.* High liquid-to-gas rate *c.* Small venturi throat (high throat velocity). If damper is used, damper could be in too fully closed position *d.* If packed tower is used, check liquid rates for tower flooding
2	Solids buildup	*a.* Not enough blowdown (and therefore fresh makeup) *b.* High inlet loading *c.* Not enough scrubbing liquid *d.* If packed tower, more solids are formed in reaction than estimated. Check process flow
3	Corrosion of material	*a.* High chlorides content *b.* Wrong material of construction
4	Abrasion	Needs abrasion-resistant liner in the inside
5	High outlet temperature	*a.* Scrubber is not saturating the gas. Review process saturation calculation *b.* Check if system is designed to have the right amount of liquid required for saturation
6	Water mist out of the scrubber stack (or carryover)	*a.* Verify stack velocity (stack velocity should be less than 3000 ft/min with a recommended range of 2000 to 2500 ft/min) *b.* Mist eliminator is inefficient (or clogged if carryover has salts buildup) *c.* Too high a liquid rate in packed tower scrubbers
7	Liquid overflows or pump cavitates	Check integral sump design. A minimum of 3-min retention time is recommended
8	Instrumentation defects	Check individual instrumentation and controls based on process and instrumentation design
9	Fan not running properly	*a.* Check fan design characteristics *b.* Check fan balancing *c.* Check if the fan is running at its operating curve *d.* Check if fan is mounted properly and bearings are greased properly *e.* Check fan alignment *f.* Check the drive guards
10	Recycle pump failure	*a.* Check pump characteristics including pump curve *b.* Check pump alignment

TABLE 10.5   Troubleshooting guide (*Continued*)

Item number	Description	Possible cause
10	Recycle pump failure	*c.* Check seals and see if they need service
		d. Check seal water flow and pressure characteristics (certain manufacturers require seal water)
		e. Check if nozzles are plugged up resulting in higher static pressure than required

References

1. W. Licht, *Air Pollution Control Engineering,* Marcel Dekker, New York, NY, 1988.
2. W. Strauss, *Industrial Gas Cleaning,* Pergamon Press, Oxford, 1975.
3. Gad Hetstroni (ed.), *Handbook of Multiphase Systems,* Hemisphere Publishing Corp., Washington, D.C., 1982.
4. Manfred Wicke, "Collection Efficiency and Operation Behavior of Wet Scrubber," *Proceedings of the Second International Clean Air Congress of the International Union of Air Pollution Prevention Association,* December 6–11, 1970.
5. C. C. Manners, "Removal of Entrained Moisture from Exhaust/Scrubber System," *Proceedings of the Second World Filtration Congress,* 1979.
6. Steve Franke and Kanti Patel, "Improving Gas Dehydrator Efficiency," *Pipeline and Gas Journal,* July 1989.
7. Howard E. Hesketh, "Fine Particle Collection Efficiency Related to Pressure Drop, Scrubbant and Particle Properties, and Contact Mechanism," *Journal of Air Pollution Control Association,* vol. 24, no. 10, 1974.
8. S. Calvert, J. Goldsmith, G. Leith, and D. Mehta, "Scrubber Handbook," Natl. Tech. Info Service, Springfield, VA, 1972.
9. D. S. F. Atkinson and W. Strauss, "Droplet Size and Surface Tension in Venturi Scrubbers," *Journal of Air Pollution Control Association,* vol. 28, no. 11, 1978.
10. S. Nukiyama and Y. Tanasawa, "An Experiment on Atomization of Liquid by Means of an Air Stream," *Trans. Society of Mechanical Engineers (Japan),* vol. 4, Summary Section, 14, 1938.
11. S. Nukiyama and Y. Tanasawa, "An Experiment on Atomization of Liquid," *Trans. Society of Mechanical Engineers (Japan),* vol. 5, Summary Section, 14, 1939.
12. S. Calvert, "How to Choose a Particulate Scrubber," *Chemical Engineering,* August 29, 1977.
13. Michael F. Szabo and Richard W. Gerstle, "Operation and Maintenance of Particulate Control Devises on Coal-Fired Utility Boilers," EPA Technology Series, EPA-600/2-77-129, July 1977.
14. I. A. Raben, "Status of Technology of Commercially Offered Lime and Limestone Flue Gas Desulfurization Systems," presented at the EPA Flue Gas Desulfurization Symposium, New Orleans, Louisiana, May 14–17, 1973.

Electrostatic Precipitators

Samuel G. Dunkle

Fisher-Klosterman, Inc.

11.1 Introduction

Electrostatic precipitation has been utilized to collect solid and/or liquid particulate matter in dirty gas and/or air streams since the early 1900s. The electrostatic precipitator is a simple and effective device for controlling particulate emissions from coal fired power stations, cement kilns, pulp and paper mills, metallurgical process plants, steel mills, incinerators, and other process plants. The electrostatic precipitator has been successfully applied with overall collection efficiencies exceeding 99.90 percent on particles with diameters larger than 10 μm and has been proven to be a very effective device for removing particulate matter less than 10 μm in diameter.

The two major categories into which precipitators can be divided are dry and wet. The dry electrostatic precipitator is the most common. It collects and removes the particulate matter in a dry state, such as cement and fly ash. The wet electrostatic precipitator collects the particulate matter in a wet or moist state and removes the particulate matter by water-washing the inside of the precipitator.

The basic electrostatic precipitator comprises a casing with inlet transition, gas distribution devices, outlet transition, dust collection hoppers, and support steel. The casing encloses and supports the internal equipment such as collection surfaces, discharge electrodes, and rapping devices. A typical dry electrostatic precipitator is shown in Figs. 11.1 and 11.2. A typical wet electrostatic precipitator is shown in Fig. 11.3. The most common material of construction for the casing and internal equipment is mild steel. Stainless steel, fiberglass-reinforced plastic (FRP), and/or polyvinyl chloride (PVC) materials are also used in special applications.

This chapter will focus on the single-stage high-voltage dry electrostatic precipitator most commonly used in industrial processes today. The principles discussed are also applicable to wet and multistage electrostatic precipitators.

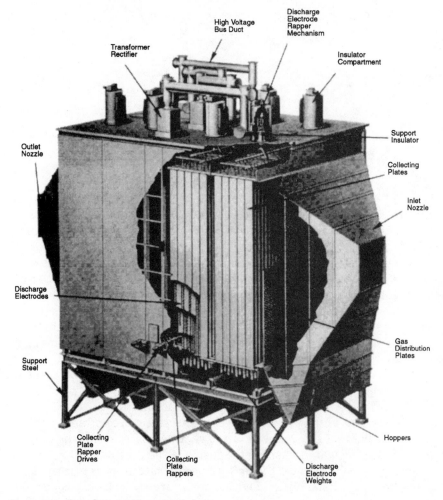

Figure 11.1 Dry electrostatic precipitator.

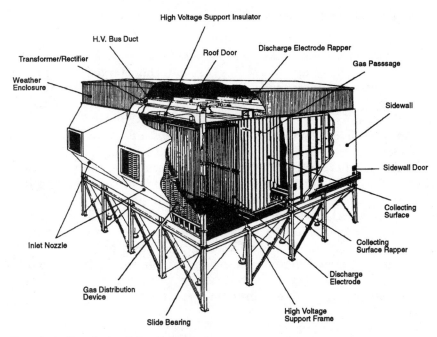

Figure 11.2   Dry electrostatic precipitator.

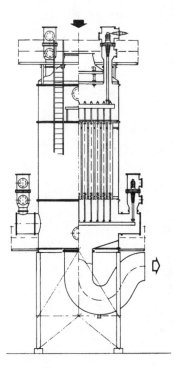

Figure 11.3   Wet electrostatic precipitator.

11.2   Fundamental Principles

The fundamental principle of collection is defined as using electric forces to impart a charge which is usually negative to the particulate matter in the dirty air and/or gas stream. The charged particulate matter migrates and becomes attached to a collecting surface of opposite polarity. Cleaning the particulate from the collecting surface is the final step.

The process of electrostatic precipitation, although still under investigation, is well-documented in publications, such as Dr. H. J. White's book *Industrial Electrostatic Precipitation.*[1] The purpose of this section is to develop a general knowledge of the precipitation process as it relates to the common industrial electrostatic precipitator, which operates with negatively charged discharge electrodes and positively charged collecting surfaces. The actual electrostatic precipitation process occurs in the space between the discharge electrode and collecting surface. The operating voltage can vary between 15,000 and 100,000 V or higher depending on the configuration of the precipitator. Refer to Fig. 11.4 for a graphic presentation of the precipitation process.

The first step in particulate collection is to impart a charge to the particulate matter. This is accomplished by applying a high-voltage, pulsating direct current to the discharge electrode system. Actual field measurements have shown that the best precipitator performance is obtained using pulsating direct current, because higher voltages and currents can be achieved when intermittent voltages are used under the conditions normally found inside the precipitator.

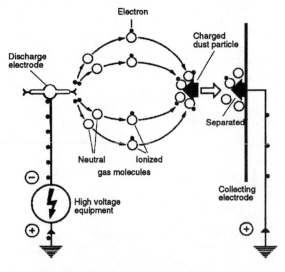

Figure 11.4   Electrostatic precipitation process.

There is a direct relationship between the precipitator operating voltage and the precipitator performance. Increasing the operating voltage will increase the collection efficiency of a precipitator provided that a high-resistivity condition does not exist. Particle charging occurs when the particles pass through an intense electric field. This field is developed when the voltage is applied at a value which causes a molecular breakdown of the gas at the discharge electrode and is referred to as *corona discharge* or *gas ionization.* Corona discharge or gas ionization occurs when the electric field potential across the space between the discharge electrode and collecting surface reaches a value where conductivity or the flow of free electrons (current flow) has been achieved. The current flow will continue to increase until there is a catastrophic electric discharge consisting of sparks or arcs. The corona discharge is the primary effect which imparts the driving force or electric charge into the particulate matter. Maximum precipitator performance is achieved when the highest effective charging current and voltage are applied to the precipitator. Field charging and diffusion charging are the two main mechanisms which affect the charging of the particulate inside the precipitator. Field charging, also known as impact charging, will dominate particles larger than 1.0 μm by impaction of free negative ions when they bombard and collide with the suspended particulate. Diffusion or ion diffusion charging will dominate particles below 0.3 μm. Ion diffusion depends on the thermal energy or velocity of the ions, not the electric field. The negative ions will collide with the suspended particulate due to the thermal motion of the ion. Collision of the particle and the ion will impart the charge on the particle. Both field and diffusion charging affect the particles between 0.3 and 1.0 μm.

The particulate matter, after being charged, migrates and attaches itself to the oppositely charged collecting surface and subsequently loses its charge. The collected particulate is then removed from the collecting surface by shaking, vibrating, or mechanical rapping and is deposited into a collection hopper. The most common removal method is mechanical rapping whereby the collecting surface is suddenly struck by a hammer which in turn dislodges the particulate. Particles tend to agglomerate as they attach themselves to the collecting surface. When rapped, the particulate layer is removed as a sheet of agglomerated particulate. In the case of a wet electrostatic precipitator, the collected particulate is usually removed by water-washing.

11.3 Precipitator Arrangement

The single-stage electrostatic precipitator is made up of a gas-tight casing which houses longitudinally mounted parallel rows of collecting surfaces which form gas passages. The discharge electrodes are suspended in the center of the gas passages. Externally mounted rapping systems are provided for the collecting surfaces and discharge electrodes. Inlet and outlet transitions are used to connect the precipitator casing to the industrial process. The transitions with the appropriate gas distribution devices are designed to provide uniform gas and dust distribution through the electrostatic precipitator. Collection hoppers located below the casing are used to collect the particulate matter prior to evacuation to a central storage location. Special *antisneakage* baffles are located throughout the precipitator to prevent untreated gas from bypassing the active treatment zone.

The maximum efficiency and highest reliability can be achieved if each individual discharge electrode has an independent power supply. This is not often a practical arrangement because a small industrial precipitator can have more than 500 discharge electrodes, and the cost of installing a power supply for each discharge electrode would be prohibitive. The most cost-effective precipitator would have one power supply for the entire precipitator. However, using one power supply is not practical because reliability and performance are sacrificed. The solution is to provide a sufficient number of power supplies to maximize the performance of the electrostatic precipitator yet maintain reasonable reliability and cost-effectiveness. As a result, suppliers of electrostatic precipitators divide the electrostatic precipitator into many mechanical and electrical sections. The mechanical sectionalization is based on the number of longitudinal chambers and the configuration of the collecting surface rapping systems called fields. The electrical sections are divided into independent areas called bus sections. The bus section is the smallest area of the precipitator which can be independently deenergized by subdivision of the high-voltage system and arrangement of support insulators (refer to Fig. 11.5). The power supplies utilized for an electrostatic precipitator must be designed with special consideration being given to the electrical transients (extreme variations in the voltage and current) and stresses which are created by the electrical disturbances inside the precipitator.

The high-voltage, pulsating direct current is applied to the discharge electrodes by transformer-rectifiers and associated control panels. The typical power supply will consist of a single-phase high-voltage transformer-rectifier set, a transformer-rectifier control panel, and an automatic voltage control (AVC) (Fig. 11.6). The control panel regulates the voltage and current on the primary side of the high-voltage transformer-rectifier, which is designed to meet the normal service voltage characteristics, i.e., 480 volts alternating current (V ac), 1 phot (ph). The transformer steps up the service voltage (400 to 575 V ac) to the high voltages necessary in electrostatic precipitation (15 to 100 kV). The rectifier converts the alternating current to direct current.

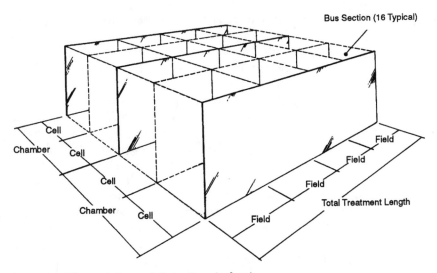

Figure 11.5 Electrostatic precipitator (terminology).

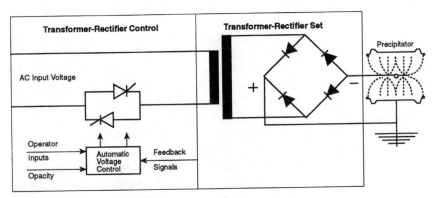

Figure 11.6 Typical power supply.

The quantity of transformer-rectifiers and controls depends on the size of the precipitator and number of bus sections. The transformer-rectifiers are usually mounted on the roof of the precipitator but can be located in any remote area near the precipitator. The control panel provides regulation and indication of the electric power going into the precipitator. Automatic voltage controls perform the function of automatically optimizing the power going into the precipitator under various operating conditions.

A dedicated AVC is provided for each transformer-rectifier. The AVC is usually located in the same panel as the power regulation equipment consisting of silicon controlled rectifiers (SCRs), firing circuit, control transformers, contactors, circuit breaker, and miscellaneous electrical equipment. The primary purpose of the AVC is to monitor the operation of the specific bus section and automatically adjust the operating parameters to maximize the precipitator performance.

The AVC monitors and utilizes feedback signals from the primary current and voltage as well as the secondary current and voltage. The current technology uses microprocessor-based AVCs to evaluate and analyze the feedback signal and respond by automatically adjusting the voltage and current levels to preset parameters. A well-designed AVC will incorporate a sophisticated spark-sensing circuit to determine when sparks or arcs occur inside the precipitator. Controls are also provided for the collecting surface and discharge electrode rapping systems to optimize the performance of the electrostatic precipitator by allowing the adjustment of intensity and/or frequency of rapping.

Electrostatic precipitator designs are normally classified by the type of discharge electrode support and the type of rapping system employed to clean the collecting surfaces. Types of discharge electrodes include the rigid frame, rigid discharge electrode, and the wire/weight design.

The rigid frame design, sometimes referred to as the European design, was introduced into the United States in the late 1960s. The rigid frame design (Fig. 11.7) consists of a rigid support frame made of structural tubing or pipe with the discharge electrodes located between the support frame. The length of the discharge electrode varies between 6 and 9 ft with the entire frame exceeding 50 ft. Typical electrodes (Fig. 11.8) consist of small-diameter wires (<0.125 in), coiled wires, flat rectangular shapes, flat rectangular shapes with spikes, barbed wires, square twisted rods, star wires, and rigid tubes with spikes. A variety of discharge electrode shapes are available depending on the supplier. The rigid frame is suspended from the roof of the electrostatic precipitator by four support insulators. The major advantage of the rigid frame design is the increased reliability and reduced maintenance when compared to the wire/weight design.

The rigid discharge electrode (RDE) design was introduced into the United States in the 1960s from Europe and became standard equipment for most American manufacturers in the mid- to late 1970s. The rigid discharge electrodes are individually suspended by a rugged structure from roof-mounted support insulators (Fig. 11.9). The quantity of support insulators varies between two and four. One unique characteristic of the RDE is the evenly spaced emitter sources over the entire length of the electrode. The length of an RDE can exceed 50 ft; however, depending on the design, a center support may be required to stabilize and prevent swinging. The major advantage of

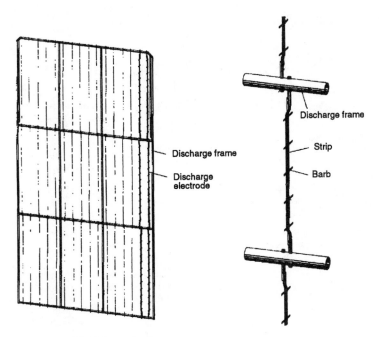

Figure 11.7 Rigid frame design.

the RDE is the increased reliability and reduced maintenance when compared to the weighted-wire design and reduced erection time when compared to the rigid frame design.

The weighted-wire design, sometimes referred to as the American design, has been used in the United States since the 1920s. This design consists of a small-diameter (approximately 0.125 in) wire individually suspended by a structure from roof-mounted support insulators (Fig. 11.10). The wire is stabilized by a weight at the bottom usually made of cast iron within a lower stabilizing frame. The maximum length of the weighted-wire discharge electrode is approximately 36 ft. The major advantage of this design is its low cost.

The type of rapping systems normally employed to clean the collecting surfaces consist of electric or pneumatic vibrators, impulse rappers, drop rods, and electromechanical rotating drop hammers. The electric and pneumatic vibrators, impulse rappers, and drop rods are normally used to clean dust deposits by imparting a rapping force at the top of the collecting surface (Fig. 11.11). The electromechanical rotating hammers usually impart a rapping force at the bottom of the collecting surface (Fig. 11.12).

The rapping control systems and rappers are designed to allow adjustment of intensity and frequency to permit optimization of pre-

SPIKE TYPES

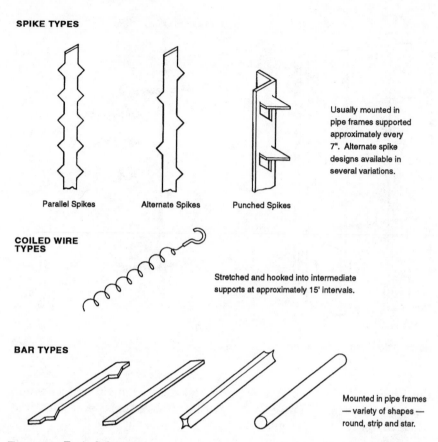

Usually mounted in pipe frames supported approximately every 7". Alternate spike designs available in several variations.

Parallel Spikes   Alternate Spikes   Punched Spikes

COILED WIRE TYPES

Stretched and hooked into intermediate supports at approximately 15' intervals.

BAR TYPES

Mounted in pipe frames — variety of shapes — round, strip and star.

Figure 11.8 Typical electrodes.

cipitator performance. One very important and often overlooked detail of the rapping system is the arrangement and support of the collecting surfaces. The top-rapped collecting surfaces typically consist of a series of roll-formed sections which are welded, interlocked, or crimped to make one continuous collecting surface (12 ft long by 50 ft high) firmly supported at the top, making a single rigid collecting surface (Fig. 11.11). The bottom-rapped collecting surfaces typically consist of a series of independently supported roll-formed (noninterlock) plates approximately 18 in wide which are mechanically assembled to create one collecting surface assembly (Fig. 11.12). The entire collecting surface assembly consisting of 5 to 12 panels (7.5 to 18 ft long by 50 ft high) is firmly supported at the top. Utilizing bottom-rapped collecting surfaces could be an advantage when the particulate being collected has characteristics which make it difficult to remove from the collecting surface, e.g., sticky dust.

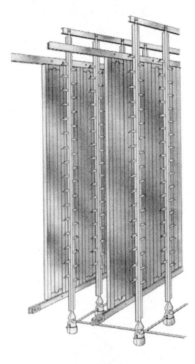

Figure 11.9   Rigid discharge electrode (RDE) design.

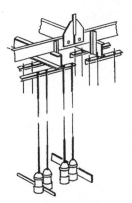

Figure 11.10   Weighted-wire (W/W) design.

11.4   Design Parameters

The electrostatic precipitator must be a reliable and dependable part of the process plant to meet the current emission standards and owner expectations. A properly designed electrostatic precipitator will be able to operate reliably with routine maintenance and continuously meet the required emission standards.

The primary measure of performance for an electrostatic precipitator is the overall collection efficiency. The total collection efficiency

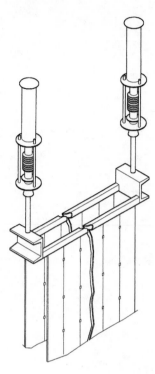

Figure 11.11 Top rapping with interlocked panels.

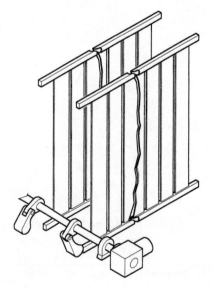

Figure 11.12 Bottom rapping with independent panels.

can be calculated when the inlet and outlet dust quantities are known by the equation

$$E_T = \frac{W_1 - W_2}{W_1} \cdot 100 \tag{11.1}$$

where E_T = total collection efficiency, %
W_1 = inlet particulate loading, gr/acf [g/m³]
W_2 = outlet particulate loading, gr/acf [g/m³]

Example 11.1 Determine the total collection efficiency when inlet and outlet particulate loadings are known.

solution

W_1 = inlet particulate loading = 5.0 gr/acf = [11.465 g/m³]

W_2 = outlet particulate loading = 0.01 gr/acf = [0.0229 g/m³]

$$E_T = \frac{W_1 - W_2}{W_1} \cdot 100$$

$$= \frac{5.0 - 0.01}{5.0} \cdot 100 = \left[\frac{11.465 - 0.0229}{11.465} \cdot 100 \right]$$

$$= 99.80\%$$

The collection efficiency of a specific precipitator is related to the effective collecting surface area, gas flow through the precipitator, and the particle migration velocity. The simplest way to understand the relationship between collection area, gas flow, and migration velocity is to review the fundamental electrostatic precipitator sizing equation known as the Deutsch-Anderson equation. This equation or derivatives thereof are used extensively throughout the precipitator industry.[2]

$$E_F = 1 - e^{(A/Q) \cdot \omega} \tag{11.2}$$

where E_F = collection efficiency (as a decimal)
A = effective collecting surface area, ft² [m²]
Q = gas volume or flow (actual conditions) through precipitator, ft³/s [m³/s]
e = base of natural logarithm = 2.718
ω = particle migration velocity at specific plate spacing, ft/s [m/s]

Example 11.2   Determine the migration velocity when the precipitator arrangement, gas volume, and overall collection efficiency are known. The precipitator arrangement is

30 gas passages spaced 12 in apart
36-ft effective collecting surface height
4 fields in direction of gas flow
10-ft effective field length

$$\text{Gas volume } Q = 3500 \text{ ft}^3/\text{s} = [99.09 \text{ m}^3/\text{s}]$$

$$\text{Collection efficiency } E_F = 0.9980 \text{ or } 99.8\%$$

solution

The collecting surface area is

$$A = 30 \times 2 \times 36 \times 4 \times 10 = 86{,}400 \text{ ft}^2$$

The number of gas passages is multiplied by 2 since each gas passage has two sides.

Solving Eq. (11.2) for migration velocity

$$\omega = \frac{-\ln(1 - E_F)}{A/Q} \tag{11.3}$$

Then

$$\omega = \frac{-\ln(1 - 0.9980)}{86{,}400/3500} = \frac{6.2146}{24.6857} = 0.2518 \text{ ft/s}$$

In the example above, the gas volume was measured and/or calculated. The precipitator collection surface area was calculated based on the physical dimensions of the equipment. The efficiency was calculated based on the measured inlet and desired outlet loading. The migration velocity was calculated using the volume, area, and efficiency and can be used to predict performance or to estimate the size of a precipitator installed on a similar application with the same efficiency requirement. However, it must be recognized that the Deutsch-Anderson equation is scientific in nature and the results are theoretical. The equation makes several assumptions which must be evaluated in the final conclusions. They are: uniform gas flow exists for all particles in the gas stream, no rapping reentrainment occurs, and particle size of the particulate in the gas stream is consistent.

The particle migration velocity ω is sometimes referred to as the drift velocity or the precipitation rate. The migration velocity represents the parameter at which a group of dust particles in a specific process can be collected in a precipitator and is usually based on empirical data. The particle migration velocity can also be calculated for any given application as shown above. Therefore, the precipitator sizing specialist will most often use the migration velocity based on the actual experience in a full-scale electrostatic precipitator to determine the collection area required for a given efficiency and gas volume.

The above example identifies the methods used to establish the effective collecting surface area for a specific precipitator. The actual or physical precipitator arrangement or configuration (i.e., the number of gas passages, the plate height, and the number and length of fields) can be significantly affected by the gas velocity, gas and dust distribution, treatment time, aspect ratio, and the geometry of discharge electrodes and collecting surfaces. These design parameters and their relationship to the precipitator size are discussed below.

The gas velocity is determined by dividing the total gas flow rate by the effective cross-sectional area of the precipitator. The effective cross-sectional area is calculated by multiplying the effective height of collecting surfaces times the number of gas passages times the width of the gas passage. When calculating the gas velocity, it is assumed that uniform gas distribution is present across the face of the precipitator, although this is frequently not the situation.

To meet current efficiency and emission requirements, the design gas velocities will be in the range of 3 to 5 ft/s [0.9 to 1.7 m/s]. Utilizing a velocity within this range will reduce the effects of reentrainment. Reentrainment occurs when collected particulate is reintroduced into the gas stream, requiring it to be collected again and thus reducing the overall performance of the precipitator. A high gas velocity will cause sweeping (reentrainment) of the dust from the collecting surfaces, while too low a gas velocity will result in poor gas and dust distribution.

Uniform gas and dust distribution within the precipitator treatment zone is a very important parameter to ensure a high removal efficiency of the precipitator. The most widely accepted standard for the uniform gas distribution is the Institute of Clean Air Companies, Inc., (ICAC) criteria. The ICAC defines the uniform gas distribution within the treatment zone near the inlet and outlet faces of the precipitator collection chamber. An acceptable velocity pattern will have a minimum of 85 percent of the velocities not more than 1.15 times the average velocity and 99 percent of the velocities not more than 1.40 times the average velocity. The treatment zone at the inlet and outlet is divided into equal areas, and the actual velocity is measured at the center of the selected area. The selected area is based on the number of gas passages and the collecting surface height. Usually velocity measurements are taken in every other gas passage and at 2-ft intervals of the plate height.

Uniform gas distribution is achieved by using transitions (inlet-outlet nozzles) that create a slow expansion and contraction of the gas as opposed to sudden changes. Also, perforated distribution plates with special deflectors are often applied to improve the gas distribution.

Another important design parameter which is common sense but is often overlooked is the importance of ensuring that all the dirty gases pass through the treatment zone inside the precipitator. This is achieved by installing antisneakage baffles in the top, bottom, hoppers, and sides of the precipitator casing. These baffles prevent dirty gases from bypassing the treatment zones.

Treatment time or residence time denotes the theoretical time that the dirty gases are exposed to the treatment zone within the precipitator. Treatment time is usually expressed in seconds and is calculated by dividing the effective precipitator length (ft [m]) by the gas velocity (ft/s [m/s]). This parameter is very useful when comparing different electrostatic designs for a similar application. The treatment time will vary depending on the required efficiency.

The aspect ratio is determined by the ratio of the effective length to the effective height of the precipitator. To achieve efficiencies greater than 99.0 percent, it has been generally recognized that an aspect ratio greater than 1.0 is desirable. Efficiencies greater than 99.99 percent have been achieved with an aspect ratio less than 1.0, but special consideration must be given to the gas velocity, gas and dust distribution, and treatment time.

The geometry of the collecting surfaces and discharge electrodes can have an effect on the electric field inside the precipitator. This will affect the arrangement of the equipment in the precipitator and subsequently the overall performance. Theoretically, the geometry of the collecting surfaces and discharge electrodes should vary from application to application and from the first field to the last field.

Many different types of discharge electrodes and collecting surfaces have been experimented with and used in practice. The RDE and panelized collecting plate (interlocked or noninterlocked) have received the widest acceptance in the United States. The shape, size, number, and location of the corona points on the RDE are selected based on the corona current characteristics and the mechanical parameters. In an application where high dust loadings are expected, the electric field will be suppressed. To increase the current flow characteristics, a special RDE design will be used in the first fields which incorporates many corona generating points. The arrangement and quantity of rappers is often adjusted based on how easily a specific dust can be removed (rapped) from the collection surfaces and/or discharge electrodes. A difficult-to-remove dust will result in the use of additional rappers installed on the precipitator.

The plate spacing (the distance between collecting surfaces) can have a significant impact on the cost and performance of the electrostatic precipitator. These spaces typically range from 6 to 18 in [150 to 500 mm]. Twelve-inch [300-mm] spacing has become the most common spacing. However, 16-in [400-mm] spacing has been utilized industrially with good results (provided the precipitator was designed for the higher voltages) and is rapidly becoming the norm.

It is sometimes difficult for those unfamiliar with electrostatic precipitators to understand the relationship between efficiency and plate spacing. The following example will help in understanding this relationship.

Example 11.3 Precipitator A has 9-in [225-mm] plate spacing with 260,000 ft^2 [24,164 m^2] of collecting surface area, and precipitator B has 16-in [400-mm] plate spacing with 146,000 ft^2 [13,569 m^2] of collecting area. Which precipitator, A or B, will result in a higher efficiency?

solution Reviewing the Deutsch-Anderson equation, it would initially seem as if precipitator A will have a higher efficiency than precipitator B because it has more collecting area. However, in practice, migration velocity is also affected by plate spacing. The general rule of thumb is that the migration velocity increases or decreases proportionally with the plate spacing provided the transformer-rectifiers are correctly sized. The reason for this is that the charging voltage may be increased for the wider plate spacing thus increasing the migration velocity. Migration velocity is directly proportional to the charging voltage. The result is that both precipitators A and B will achieve the same collection efficiency.

There is no one parameter that can be used to predict the best arrangement and highest precipitator performance. All the parameters must be considered based on their importance and the effect they may have on the electrostatic precipitator performance. The most significant process parameters which can affect the size or quantity of collecting surface area of the electrostatic precipitator are

- Particulate loading, gr/acf [g/m^3]
- Gas temperature, °F [°C]
- Electrical resistivity of dust, ohm · cm (Ω · cm)
- Particle size, μm
- Surface of the particles (spheres or leaves)
- Adhesion ability of particulate
- Cohesion ability of particulate
- Intensity of electrical field (current density)
- Gas composition

- Dew point of gases (H_2O and SO_3 dew point)
- Degree of cleanliness of discharge electrode and collecting electrode

The quantity of particulate entering the electrostatic precipitator affects the efficiency of the precipitator. The higher the quantity of particulate, the greater the collecting area required for a specific application and emission level. Inlet loading will affect the type of inlet transition and gas distribution devices required to ensure uniform gas and dust distribution. When inlet loading exceeds 25 to 50 gr/acf [50 to 100 g/m^3], special consideration must be allowed for the dust fallout in the inlet transition of the precipitator. Gas flow disturbances or drift caused by the action of dust fallout caused by gravity should not be overlooked when sizing the electrostatic precipitator. To compensate for this phenomenon, special inlet baffles, inlet transitions, deflector plates, and distribution devices may be used. Special materials of construction must be used for abrasive dust to reduce the effects of erosion.

The effect of gas temperature on the performance of the precipitator brings to light many areas of interest. The most important are the effect on the electrical characteristics and the effect of particle adhesion and cohesion. The actual effects due to gas temperature are related to the moisture content and chemical composition of the particulate. In general, precipitator performance will improve as the operating temperature approaches the moisture dew point temperature. Table 11.1 shows the dew point temperature as it relates to the moisture content (percentage by volume). Further data on the saturation temperature is covered in App. 1B and App. 1C. Operating at or close to the dew point (acid or moisture) of the gas will improve the precipitator performance. However, severe corrosion of the mild steel parts may occur and result in adverse effects on the life of the precipitator.

The electrical resistivity of the particulate is a very important parameter. However, its effect on precipitator sizing must be evaluated with all the other parameters. The resistivity is defined as the inverse of conductivity and is a measurement of the ability, or more properly the inability, of the particulate matter to pass an electric current. The unit of measurement is ohm-centimeters and is determined by measuring the electric resistance of a dust sample 1 cm^2 in cross-sectional area by 1 cm thick. Typical resistivity curves for cement dust at various moisture contents are shown in Fig. 11.13.

Particulates with low resistivities (10^4 to 10^8 $\Omega \cdot cm$) will be difficult to collect because they take on and lose electric charge easily, sometimes giving up the charge before they are attached to the collecting surface. This process of charging, discharging, and recharging occurs very quickly in the treatment zone with the result being significant

TABLE 11.1

H$_2$O, % by volume	Approximate dew point	
	°C	°F
1	7	45
2	17.5	64
3	24	75
4	29	84
5	33	91
6	36	97
7	39	102
8	42	108
9	44	111
10	46	115
12	50	122
14	53	127
16	55.5	132
18	58	136
20	60.5	141
25	65	149
30	69.4	157
35	73	163
40	76	169
45	79	174
50	81.5	179
55	84	183
60	86	187
65	88	190

reentrainment of particulate into the gas stream and reduced precipitator performance.

Particulate matter having normal resistivities (10^8 to 10^{11} $\Omega \cdot$ cm) will be easy to collect. The particulate will take on an electric charge, and upon attachment to the collecting surface a slow consistent discharge will occur allowing a layer of particulate to be built up on the collecting surface. The particulate layer is dislodged by the rapping system and collected in a hopper or similar collection device.

High-resistivity particulate (10^{11} $\Omega \cdot$ cm and above) is difficult to collect for two reasons; first it is difficult to impart a charge, and, second, once the particulate has taken on a charge, it is very slow to discharge. The time required to impart a charge on high-resistivity particulate is much longer as compared to normal- or low-resistivity particulate. Thus, the precipitator must be increased in size. After the particulate has been charged and is collected, the slow consistent discharge that is typical with a normal-resistivity dust does not occur, and, as a result, a layer of particulate remains on the collecting surface which has a high field potential similar to the discharge electrode. This phenomenon is referred to as *back corona*. The occurrence

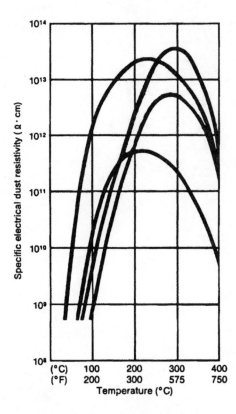

Figure 11.13 Resistivity curves.

of back corona results in a significant reduction in the collection efficiency of the precipitator.

The size of the particulate will affect the migration velocity and the collection efficiency of the precipitator. Measuring the effect which the particle size has on the collection efficiency is very complex and is made more complicated by the various shapes which can occur in a single dust sample. Shapes can include large or small spheres (hollow or solid), jagged or star shapes, irregular shapes like leaves, and pins or needles. A general rule in the process industries is that the smaller the particles, the larger the electrostatic precipitator will have to be to achieve a specific emission.

11.5 Applications

Electrostatic precipitators have been utilized to reduce emissions, regenerate materials, collect dust, etc., on a wide variety of industrial processes. Some examples include coal-, wood-, and oil-fired boilers; cement kilns (wet and dry); lime rotary kilns; coal dryers; coal mills; coke ovens; clinker coolers; magnesite kilns; municipal solid-waste incinerators; sludge incinerators; chemical plants (catalytic cracking); alumina calcining kilns; blast furnaces; bauxite dryers; black liquor boilers (sulfate and/or sulfide); smelting and roasting furnaces; scarfing machines; sintering plants; glass furnaces; electric aluminum furnaces; and so on.

The table below outlines many electrostatic precipitator applications and the dust materials which are collected.

Industry	Materials collected
Electric power generation	Fly ash—coal-fired boilers
	Oil ash—oil-fired boilers
	Fly ash—large incinerator boilers
Pulp and paper	Salt cake—black liquor recovery boilers
	Fly ash—wood and bark boilers
	Lime kiln
Chemical	
Sulfuric acid	Mist
Phosphorus	Rock dust and acid mist
Plastics	H_2SO_4 mist from acid sludge combustion
Carbon black	Precipitator as agglomerator followed by mechanical or bag collector
Rock products	
Portland cement	Clinker cooler dust
	Cement dusts, alkalis
	Kiln dust—wet and dry
	Mill dust—raw material
Gypsum	Dry gypsum particles
Trona	Dry trona (sodium carbonate)
Iron and steel	
Coke oven gas	Tar
Sintering machines	Iron oxide and fluxing dusts
Blast furnace gas	Dusts—secondary collector following wet washers
Open hearth furnace	Fumes and dusts
Basic oxygen furnace	Fumes and dusts
Scarfing machines	Scale, fumes and dusts
Electric arc furnace	Fumes and dusts
Zinc roasters	Fumes and dusts
Municipal incinerators	Fly ash
Petroleum	
Acetylene gas manufacture	Tars, carbon, oil mist
Oil shale	Tars, oil mist
Catalytic crackers	Catalyst dust, carbon
Coal-oil conversion	Tars, carbon
Nonferrous metals	
Aluminum	Alumina dust pot line fumes, carbon
Copper, lead, zinc, tin refining, and ore roasting	Oxide fumes, dusts
Precious metals refining	Mists and fumes
Other	
Glass manufacturing	Fine dusts and fiber material

Typical inlet conditions which might exist in various applications are shown in Table 11.2.

11.6 Troubleshooting

The troubleshooting guidelines listed below are general in nature to assist the precipitator specialist to find the source of trouble and develop a plan to quickly resolve the situation and restore the precipitator to maximum operating performance. Specific troubleshooting techniques will have to be developed for each plant based on the type of precipitator used and the plant maintenance and safety procedures. The troubleshooting suggestions are based on the assumption that an adequate power supply or source is available and all primary and secondary meters are in proper working order.

1. *Symptoms:* No secondary voltage, low primary voltage, no sparking, control trips on undervoltage.

   Possible causes: Short or ground inside precipitator, broken wire or discharge electrode, dust bridging inside collecting surfaces, hopper full of dust, broken transformer-rectifier ground switch, safety ground jumper in place, foreign material between discharge electrode and collecting surface, transformer secondary shorted.

   Remedies: Inspect components in high-voltage circuit, i.e., inside precipitator, and correct situation; remove broken wire; remove dust buildup; repair ground switch; remove ground jumper; remove foreign material; replace transformer.

TABLE 11.2

Application or process	Temperature of flue gas, °F	Inlet dust concentration, gr/acf	Resistivity, $\Omega \cdot cm$
Coal-fired boiler			
High sulfur, >4%	350–400	2–5	10^7–10^9
Medium sulfur, > 1% and <4%	325–375	3–7	10^8–10^{10}
Low sulfur, <1%	300–350	3–9	10^{10}–10^{13}
Lignite-fired boiler	300–400	5–15	10^8–10^{12}
Incinerator	450–550	2–8	10^8–10^{11}
Black liquor boiler	260–375	5–14	10^8–10^9
Clinker cooler	300–650	2–6	10^9–10^{12}
Lime kiln	450–650	8–14	10^8–10^{10}
Preheater cement kiln	300–360	5–20	10^9–10^{10}
Cement mill (raw material)	150–200	50–300	10^9–10^{10}
Catalytic cracking unit	450–750	1–3	10^9–10^{11}
Copper smelter	300–650	1–10	10^9–10^{10}

2. *Symptoms:* Power levels increase normally with sudden loss of secondary voltage and current approaches limit, control trips on undervoltage, excessive sparking may also occur.

   Possible causes: Broken support insulator or bus duct insulator.

   Remedies: Replace broken insulator.

3. *Symptoms:* Contactor energized, vent fan on, no primary voltage or current, no secondary voltage or current.

   Possible causes: AVC malfunction, soft start circuit not working; SCR (primary) fuses blown; SCRs opened.

   Remedies: Replace AVC, correct soft start circuit, replace fuses and/or SCRs.

4. *Symptoms:* Primary voltage at 50 percent, current level at 50 percent, full conduction of SCRs, noisy current limiting reactor.

   Possible causes: Open SCR (one), firing circuit malfunction, high-voltage rectifier failure.

   Remedies: Replace bad SCR, replace firing circuit and/or AVC, replace rectifier(s).

5. *Symptoms:* High sparking, trips on overcurrent.

   Possible causes: Overload circuit or relay malfunction, AVC malfunction.

   Remedies: Reset overload circuit, replace overload circuit or relay, replace AVC.

6. *Symptoms:* High primary current, no precipitator current, no primary or secondary voltage.

   Possible cause: Short circuit in primary.

   Remedies: Inspect and megger primary power wiring; megger current limiting reactor and transformer-rectifier primary, replace as necessary.

7. *Symptoms:* High spark rate meter indication, low primary voltage and current, no sparking observed by meter movement.

   Possible causes: Spark-sensing circuit malfunction, line noise in spark-sensing circuit, failure of spark-sensing unit or circuit.

   Remedies: Replace spark-sensing unit, ground control circuits.

8. *Symptoms:* Unstable voltage and current, no control setback apparent, meters indicate sparking but no control response.

   Possible causes: Spark-sensing circuit malfunction or disconnected.

   Remedy: Connect and/or replace spark-sensing circuit.

9. *Symptoms:* Higher than normal voltage with lower than normal current with little or no sparking, high corona onset voltage, current suppression.

Possible causes: Excessive dust buildup on discharge electrodes and/or collecting surfaces, rappers not working, excessive moisture in flue gas.

Remedies: Increase rapping intensity and/or frequency, inspect rappers and repair as necessary, reduce moisture in flue gas, power-off rapping may assist in cleaning.

10. *Symptoms:* Control operates under normal condition for short period of time (1 to 2 h), control trips on undervoltage.

    Possible cause: Dust buildup inside precipitator.

    Remedies: Empty hoppers, increase rapping intensity and/or frequency, verify rapper operation, verify purge air system operation.

11. *Symptom:* Intermittent (sudden burst) sparking.

    Possible causes: Swinging discharge electrode, loose high-voltage connection.

    Remedies: Remove electrode, tighten high-voltage connection.

11.7   Improving the Performance of an Electrostatic Precipitator

Because of increasingly more stringent particulate emission regulations, plant maintenance personnel have often been given the responsibility to maintain a continuously clear stack by achieving the maximum performance with an electrostatic precipitator. Reduced maintenance budgets and corporate downsizing may make achieving this objective difficult. Industries such as chemical, pulp and paper, utility, cement, and metallurgical have been affected. The purpose of this outline is to present a practical approach to the precipitator maintenance specialist for improving and maintaining the overall performance of an electrostatic precipitator. Maximum performance is defined as minimizing emissions, increasing reliability, and reducing maintenance cost. The practical approach to improve the precipitator performance begins with determining what were the original design conditions of the precipitator and what is the current performance level. The precipitator maintenance specialist must assemble for review and evaluation, if available, the most recent operation and maintenance logs, emission test reports, gas distribution tests, dust analysis reports, resistivity curves, inspection reports, the original design specification, maintenance manual, drawings, and any other documents that relate to the operation and/or performance of the precipitator. This information will be used as reference source or database to assist the precipitator maintenance specialist during the performance evaluation.

The performance improvement program is divided into four major activities or steps:

Step 1. *Mechanical* inspection and repair

Step 2. *Electrical* inspection and repair

Step 3. *Process* review and adjustment

Step 4. *Optimization* of precipitator performance

The program described below is to be used as a guideline and should be adjusted to suit specific plant operating and maintenance procedures. To achieve the best results, it is recommended that the program sequence be followed as described. It is very important for the precipitator maintenance specialist to document the date and time when any adjustment and/or changes are made to the precipitator. This will permit the maintenance specialist to monitor the cause and effect the adjustments and/or changes have had on the precipitator performance.

11.7.1   Step 1. Mechanical inspection and repair

The precipitator must be taken out of service. An internal inspection must be performed to ascertain the mechanical condition of the precipitator. The mechanical condition and location of discrepancies must be documented in an inspection logbook or similar file, such as a precipitator general arrangement drawing. Discrepancies between the actual mechanical condition and the manufacturer's recommended tolerances must be corrected. Areas to be inspected include

- Collecting surfaces and discharge electrode alignment
- Collecting surfaces rapping system
- Discharge electrode rapping system
- High-voltage support insulator system
- Collecting surfaces for bows and bends
- Discharge electrodes for bows, missing electrodes
- Corrosion of collecting surfaces
- Corrosion of discharge electrodes
- Gas distribution screens (plugged holes, missing screens)
- Gas distribution of casing, hoppers, inlet and outlet transitions
- Dust removal system

- Locations at which air in-leakage can occur
- Corrosion of casing

Discrepancies such as broken discharge electrodes, bowed plates, and excessive corrosion are usually a result of other problems within the process operation and/or the precipitator. Potential problems such as operating the precipitator below the acid dew point temperature and/or mechanical fatigue can be identified at this time and corrective action taken prior to a catastrophic failure. Recommended repairs and corrective action should be supervised by a qualified electrostatic precipitator technician to ensure that the mechanical condition of the electrostatic precipitator is in accordance with the approved standards.

11.7.2   Step 2. Electrical inspection and repair

The mechanical condition and the electrical operating properties of the transformer-rectifier, transformer-rectifier controls, insulator heaters, rapper (coils or motors), rapping controls, etc., must be inspected to ensure that the operation is in accordance with the specifications. Documenting the mechanical condition and the electrical readings in an inspection logbook is very helpful in identifying future problems. The electrical equipment should be operated at design parameters, electrical data recorded, and replacement and repairs made where discrepancies to the specifications are noted. Equipment to be inspected and operated includes

- Transformer-rectifiers and controls
- Rapper controls
- Rapper motors, coils, vibrators
- Key interlock system
- Hopper heaters
- Insulator heaters
- Insulator compartment and penthouse ventilation system
- Data management systems

A very important tool to the precipitator maintenance specialist is the development of current and voltage curves for each electoral bus section under air load (static) and gas load conditions. These curves can be used for future reference and evaluation to identify grounds, dust buildups, full hoppers, and high resistivity and to assist in troubleshooting or diagnosing operating problems.

The purpose of documenting the findings is to establish a baseline for precipitator operating parameters, mechanical condition, and electrical condition. This baseline data can be compared to information which is obtained in the future to continue to evaluate precipitator performance.

11.7.3   Step 3. Process review and adjustment

The operating conditions of the process must be evaluated, documented, and compared to the original precipitator design conditions. A simple spreadsheet is recommended so comparisons can be made readily and accurately. Also, an evaluation of the actual performance versus the expected performance of the precipitator must be completed. Data on gas volume, gas temperature, dew point, inlet dust loading, resistivity, particle size, and the gas and dust flow distribution must be collected, recorded, and evaluated. Historical information and performance data, such as emission test reports, gas distribution tests, dust particle size, resistivity curves, inspection reports, and process changes which were assembled for review, can be utilized at this time to evaluate the changes in precipitator performance.

The specific application or process conditions can have a significant impact on the performance of the precipitator. The process variables must be reviewed and evaluated to determine their effect on the precipitator performance. The objective of the process review is to find areas in which the process variables can be optimized to reduce the gas volume into the precipitator, increase the moisture dew point of the flue gas, adjust the temperature to the optimum resistivity, and reduce the dust loading into the precipitator. Optimizing the process will not only stabilize the electrostatic precipitator performance but can reduce production and operating cost. This is also a good time to interview the plant operators to discuss what changes in the plant operation have affected the precipitator performance.

When the precipitator operates under a high negative pressure (more than −15 in w.g.), air in-leakage can cause significant operating problems such as corrosion, dust buildup, and reduced production. Unnecessary air in-leakage must be prevented in areas such as access doors, expansion joints, screw conveyors, and penetrations in the casing. Reducing or eliminating air in-leakage will reduce gas volume in the precipitator, reduce emissions, increase temperature, reduce corrosion, reduce power consumption, and increase process production rates. The final result is that the process runs more efficiently and the precipitator works better.

11.7.4   Step 4. Optimization of precipitator performance

Precipitator performance optimization can take place only after the mechanical and electrical equipment has been inspected, maintained, and reconditioned in accordance with specifications and after the process review has been completed and the process optimized. The objective of step 4 is to compare the baseline data to the future operating data to determine what effect (positive or negative) adjustments to the system will have on the overall performance of the precipitator. After determining what changes affect the precipitator performance, positive steps can be taken to improve performance and reduce emissions. Precipitator performance can be improved by

- Changing product and/or process variables (quantity and characteristics)
- Improving gas and dust distribution
- Retrofitting with state-of-the-art AVCs (microcomputer controls sensing secondary voltage spark detection, semipulse energization, multiple mode operation)
- Optimizing the rapper timing intensity and sequence
- Substituting power-off rapping or reduced power rapping with standard rapping to decrease the effects of high-resistivity fly ash
- Installing a precipitator management control system (supervisory control)
- Additional field sectionalization (additional bus sections)
- Installation of performance enhancement equipment, e.g., SO_3 injection, evaporation coolers, NH_3 injection, H_2O injection

11.7.5   Summary

In summary, to achieve the best performance from the electrostatic precipitator requires maintaining the mechanical and electrical components, tuning the process, and observing what changes had the most significant effect on the precipitator performance. Using this knowledge and taking positive proactive steps will result in a well-maintained precipitator that has reliable and consistent performance. In addition, the results of an optimization program can be used to develop future plans for maintenance and upgrade alternatives.

11.8   Inspection Checklist

The precipitator performance can be easily compromised by overlooking what is considered to be a seemingly small detail. As an example, maintaining proper electrical clearance throughout the internals of the precipitator is of prime importance. The performance of any precipitator can be significantly degraded if a single electrical clearance between the collecting surface and a discharge electrode falls outside of the specifications. A close clearance can create a situation where premature sparking occurs and reduces the power into the precipitator, resulting in an increase in emissions. One method used by many precipitator maintenance specialists is to document the mechanical and electrical condition of the precipitator by using a *checklist*. A checklist is another tool to assist the precipitator maintenance specialist to improve reliability, reduce maintenance time and cost, and reduce long-term operating expenses.

Remember, prior to entering the confined space of the precipitator all safety procedures must be followed to ensure that it is safe to do so. The general checklist shown below should be used for daily, weekly, quarterly, and annual inspections. The items on the list can be deleted and/or new items added.

Daily inspections _____ Initial _____ Date

_____ Ventilating fans on

_____ Rappers on

_____ Vibrators on

_____ Dust removal system on

_____ Hopper heaters on

_____ Insulator heaters on

_____ Transformer-rectifier control power level documented

AC amps _____ A ac

AC volts _____ V ac

DC amps _____ mA dc

DC volts _____ kV dc

Spark rate _____ s/min

_____ Opacity _____ %

_____ Load data _____

Weekly inspections _____ Initial _____ Date

_____ Hopper level detector operation

_____ Transformer-rectifier oil level

_____ Transformer-rectifier oil temperature

_____ °F   _____ °C

_____ Key interlocks covered and weather-protected

_____ Verify no air leakage around door gaskets and casing penetrations

Quarterly inspections   _____ Initial   _____ Date

_____ Complete current and voltage curves

_____ Clean control cabinets and rapper control panel

_____ Verify control set points

Undervoltage _____

Overcurrent _____

_____ Document rapper wear and alignment

_____ Replace filters in ventilation system

Annual inspections   _____ Initial   _____ Date

The annual inspection should be done when the process is shut down or is off-line and the precipitator can be entered safely.

Top of precipitator

_____ Oil sample from transformer-rectifier for dielectric testing

_____ Verify key interlock operation

_____ Alignment of rappers (tightness, wear, etc.) and overall conditions

_____ Verify operation of vent fans, doors, louvers, etc.

_____ Verify ground switch operation

_____ Remove bus duct inspection covers, inspect and clean insulators

_____ Verify ground straps are in place and utilized

_____ Inspect and clean support insulators inside and outside

_____ Clean ventilation ports

_____ Tighten loose rapper assemblies

_____ Clean penthouse and insulator compartment floor, if more than $\frac{1}{8}$ in of accumulated dust

_____ Verify operation of insulator or penthouse heaters

_____ Verify tightness of high-voltage connection

Inside precipitator

_____ Check collecting surface and discharge electrode alignment of each bus section: top, bottom, and midpoint

_____ Remove all dust buildups

_____ Inspect condition of upper and lower high-voltage support elements

_____ Inspect condition of collecting surface: top, bottom, and midpoint support elements

_____ Verify placement of rappers

_____ Inspect all collecting surfaces for bows, bends, and distortions

_____ Check corrosion reviews and documentation, i.e., for casing, collecting surfaces, discharge electrodes

_____ Verify discharge electrodes are secure and tight

_____ Check perforated plates and turning vanes for dust buildup and binding

_____ Verify antisneakage baffles are in place

_____ Verify condition of antisway insulators

_____ Verify placement of internal doors on casing, hopper baffles, perforated plates, etc.

Air load precipitator

_____ Verify control operations: A ac, V ac, mA dc, kV dc

_____ Verify rapper operation: intensity, lift, and frequency

_____ Insulator heaters operate at _____ A _____ V

_____ Hopper heater operates at _____ A _____ V

_____ Hopper level detectors operate at _____ A _____ V

_____ Current and voltage curve documentation

_____ Dust removal system operation

11.9   Advantages and Disadvantages

The question is asked frequently, What are the advantages and disadvantages of using an electrostatic precipitator compared to a fabric filter? The answer to the question is outlined below. Each application must be evaluated based on the customer's priorities and plant conditions and the overall economics associated with either selection. The following list should help in this evaluation.

11.9.1   Increase in inlet grain loading with no increase in air volume

Fabric filter.   Would necessitate an increase in cleaning cycles to maintain a constant pressure drop across the system. The increase in cleaning cycles would reduce bag life.

Precipitator.   Overall precipitator efficiency should not change, but outlet dust emissions may increase if the precipitator were sized for lower inlet grain loadings.

11.9.2   Process variable—moisture content

Fabric filter.   Depending on other constituents and temperature, increased moisture will affect the dew point and could promote acid corrosion of bags and collection internals. Increased moisture could also increase filter drag (pressure drop).

Precipitator.   Increases in moisture could improve precipitator efficiencies but also increase the need for effective rapping systems due to the increased stickiness of the dust. It can result in hopper bridging in either device and in the need for hopper heating and/or vibrating. In addition, increased moisture could promote acid corrosion of collector internals.

11.9.3   Process variable—fuel and chemical composition of particulate

Fabric filter.   There will be no major effect on collector performance, but increases in sulfur or fluorine contents with no associated increase in temperature can lead to acid corrosion that can be harmful to fabric. (This is especially true when operating temperatures are in the range of 255 to 300°F [107 to 149°C].)

Precipitator.   The precipitator design (size) is based on specific fuel specifications. Changes in fuel may affect performance. Acid corrosion caused by low temperatures can be harmful to the steel parts. Low-sulfur coal will require a larger electrostatic precipitator size than that for a high-sulfur coal for a specific application. Increases in sulfur content will affect the life of the fabric collector or precipitator platework if the systems operate below the flue gas dew point.

11.9.4 Process variable—other ash constituents: Calcium, silica, sodium, magnesium, SO₃

Fabric filter. These elements could affect the cleaning ability of a fabric collector since the dusts can become sticky or charged to a point where they will not be easily released from the bag. This could lead to excessively high pressure drop or excessive cleaning of the bags which could adversely affect bag life.

Precipitator. Different combinations of these elements can have a definite effect on the precipitator efficiency. These elements not only affect the resistivity of the dust entering the precipitator but have a dramatic influence on the ability to clean the precipitator electrodes and collecting plates since certain combinations can make dusts very difficult to remove.

11.9.5 Erosion possibilities

Fabric filter. If the collector is *not* properly designed, it can have an adverse effect on bag life due to the abrasion occurring at the tubesheet connection and the lower one-third of the bag or due to uneven particulate distribution between hoppers.

Precipitator. There will be very little, if any, effect on the precipitator because of inherently low velocities. Inlet ductwork and gas distribution devices must be properly designed and fitted with abrasion-resistant steel as necessary for applications where erosion can occur for both devices.

11.9.6 Gas temperature

Fabric filter. Low temperatures can affect the flue gas dew point which can cause bag blinding and attack on the bag finish. Also, low temperatures could affect dust stickiness. High temperatures will adversely affect the life of the bag.

Precipitator. Temperature can affect dust resistivity and overall efficiency. It can also affect the stickiness of the dust. In these instances, the need for a very positive, proven rapping system is essential. High temperatures will not affect the precipitator structure unless design temperatures are exceeded. In either device, low gas temperatures reaching the dew point can cause dust bridging in the hoppers.

11.9.7   Volume and flexibility of operation beyond design flow conditions

Fabric filter.   Air-to-cloth ratios would increase causing increased pressure drop and the need to increase the cleaning cycle. The effect on efficiency would be minimal. There would be a slight increase in emissions due to increased cleaning, but this effect is usually negligible.

Precipitator.   Specific collecting plate area is decreased with addition to the gas flow, while precipitator efficiency would decrease; thus, emissions would probably increase.

11.9.8   Flue gas distribution

Fabric filter.   Gas distribution devices are suggested for inlets and under tubesheets. Unequal gas distribution would probably lead to a high pressure drop across certain compartments, bag pluggage, and/or excessive bag wear in certain areas of the compartment.

Precipitator.   The flue gas distribution is very critical especially in high-efficiency units. Low velocities through the electrostatic precipitator cannot compensate for higher velocities in other areas. Therefore, efficiency of the unit falls off. Inlet and outlet gas distribution devices will increase the effect on overall pressure drop.

11.9.9   Changing load capabilities

Fabric filter.   This will not be a major problem, except the temperature must be maintained above the dew point to prevent acid attack of the bag coating and possible bag blinding. Maximum design temperature of bag material must not be exceeded or bag life will be reduced.

Precipitator.   The need for a reliable rapping system that produces significant accelerating energy for each collecting plate and electrode becomes more important as gas temperature drops below the dew point. The system can be sensitive to upset conditions that cause high resistivity and lower efficiency. As the gas temperature goes below the dew point, corrosion of shell, ductwork, etc., of both systems does occur.

11.9.10 Power requirements

Fabric filter. The fabric collector will typically have a pressure drop of between 4 and 6 in w.c. [102 to 152 mmH$_2$O] (flange to flange); therefore, more fan horsepower will be required than for a precipitator. A reverse air fan is required for a reverse air collector. A compressor for pneumatic operation or electric power will be required for damper operation.

Precipitator. Pressure drop through the precipitator can vary from 0.5 to 1.5 in w.c. [13 to 38 mmH$_2$O)] depending on the necessary gas distribution devices. The power needed to operate the transformer rectifiers should be added to the power requirements. Hopper heating, if used, should be very similar for both devices. The precipitator will consume less power than a fabric collector for a given removal efficiency.

11.9.11 Control and control systems

Fabric filter. A properly designed collector system will operate in an automatic mode.

Precipitator. The precipitator AVC operates in an automatic mode. Troubleshooting of controls will require a qualified technician. Controls can be integrated into a plant management control system.

11.9.12 Start-up and shutdown

Fabric filter. Proper start-up and shutdown of the collector are extremely critical with regard to bag life. Bypass operation may be required (if allowed) if upset conditions occur during start-up and shutdown.

Precipitator. The precipitator can essentially be energized any time after the flue gas temperature is above the moisture dew point. The precipitator can be energized at temperatures below the dew point, but good rapping systems become critical to maintain clean plates and electrodes.

11.9.13 System overall collection efficiency

Fabric filter. The fabric filter has been proven to be very reliable. An increase in volume or dust loading usually will increase the pressure drop across the collector and can normally be handled by increasing the cleaning cycle. Emissions will remain relatively constant. Opacities well below 10 percent should be normal, assuming tight damper shut-offs and no broken bags. Submicrometer particulate can be collected by the collector, but the dust cake on the bag must be sufficient.

Precipitator.   Properly sized and powered, the precipitator is very reliable and will have a constant collection efficiency. Reentrainment of dust particles during the rapping cycle can be a problem when very low outlet particulate loadings are required. Opacities below 10 percent can be easily obtained. Submicrometer particulate can be collected by the precipitator.

11.9.14   Maintenance requirements

Fabric filter.   The collector bags are the heart of the fabric collector system. The frequency of complete bag replacements will vary depending on various factors; 3 to 5 years of operation with glass bags can be expected. Detection of broken bags can be difficult, but a properly designed collector will permit a broken bag to be replaced while the collector is in full operation. Inlet and outlet module dampers do require preventive maintenance, but properly designed dampers are very reliable and parts can be replaced while the system is in operation.

Precipitator.   With rigid electrodes and hammer rapping, increased power sectionalization, more reliable controls, and better sizing criteria, the overall maintenance requirements have been substantially reduced. These improvements have significantly increased the precipitator's reliability. Since internal mechanical items cannot be repaired during operation unless redundant fields are installed, a mechanical failure may lead to an unscheduled shutdown. Normally, major maintenance repairs will coincide with outage schedules especially when rigid electrode precipitators have been installed.

11.9.15   Space

Based on current-day emission levels (99.99 percent efficiency), the space needed for the precipitator is usually somewhat greater than the baghouse.

11.9.16   Erection

Erection can be performed by the existing labor force. However, qualified erection supervisors must be present for both devices during installation of critical components including bags and discharge electrodes.

11.9.17   Experience

Fabric collectors and precipitators have a proven track record in most applications.

11.9.18   Evaluation

Capital cost, operating cost (power consumption), and maintenance cost (bag replacement) must be evaluated on a case-by-case basis. The evaluation must take into account long-term operating expenses as well as customer experience with maintenance and operating personnel.

11.10   Terminology for Electrostatic Precipitators

The terminology for electrostatic precipitators has been standardized by the ICAC. The terms most used are defined below and are shown on Figs. 11.1, 11.2, and 11.11.

antisneakage baffles   Internal baffle elements within the precipitator used to prevent the gas from bypassing the active field or causing hopper reentrainment.

arc   A discharge of substantial magnitude of the high-voltage system to the grounded system, of relatively long duration, and not tending to be immediately self-extinguishing.

aspect ratio   The ratio obtained by dividing the effective length of the precipitator by the effective height.

automatic power supply   The automatic regulation of high-voltage power for changes in precipitator operating conditions utilizing feedback signals. It is sometimes referred to as an automatic voltage control (AVC).

auxiliary control equipment   Electrical components required to protect, monitor, and control the operation of precipitator rappers, heaters, and other associated equipment.

bolted plate   A cover provided with sufficient bolts to ensure tight closure where occasional accessibility is required.

bus section   The smallest portion of the precipitator which can be independently deenergized (by subdivision of the high-voltage system and arrangement of support insulators).

cable   An oil-filled cable or dry cable for transmitting high voltages.

casing   An enveloping structure to enclose the internal components of the precipitator. A rectangular or cylindrical configuration is used. The casing includes the gas-tight roof, side walls, and end walls and hoppers and/or bottoms. A gas-tight dividing wall is used to separate chambers. A non-gas-tight load-bearing wall is used to separate bus sections. The casing is sometimes called the housing or shell.

cell (in width)   An arrangement of bus sections parallel to gas flow. *Note:* The number of cells wide times the number of fields deep equals the total number of bus sections.

chamber   A gas-tight longitudinal subdivision of a precipitator. A precipitator with a single gas-tight dividing wall is referred to as a two-chamber precipitator. *Note:* Very wide precipitator chambers are frequently equipped with non-gas-tight load-bearing walls for structural considerations. These precipitators by definition are single-chamber precipitators.

collecting surface area   The total flat projected area of collecting surfaces exposed to the active electric field (effective length×effective height×2×number of gas passages).

collecting surface rapper   A device for imparting vibration or shock to the collecting surface to dislodge the deposited particles or dust.

collecting surfaces   The individual elements which make up the collecting system and which collectively provide the total surface area of the precipitator for the deposition of dust particles.

collecting system   The grounded portion of the precipitator to which the charged dust particles are driven and to which they adhere.

collection efficiency   The weight of dust collected per unit time divided by the weight of dust entering the precipitator during the same unit time expressed as a percentage.

control damper   A device installed in a duct to regulate the gas flow by degree of closure, for example, butterfly or multilouver.

corona power (kW)   The product of secondary current and secondary voltage. Power density is generally expressed in terms of (1) watts per square foot of collecting surface or (2) watts per 1000 acfm of gas flow.

current density   The amount of secondary current per unit of precipitator collecting surface. Common units are milliamperes per square foot and nanoamperes per square centimeter.

discharge electrode   The part which is installed in the high-voltage system to perform the function of ionizing the gas and creating the electric field. Typical configurations are rigid frame (RF), weighted-wire (W/W), and rigid discharge electrode (RDE).

discharge electrode rapper   A device for imparting vibration or shock to the discharge electrodes in order to dislodge dust accumulation.

doors   A hinged or detached cover provided with a hand-operated fastening device where accessibility is required.

dust or mist concentration   The weight of dust or mist contained in a unit of gas, e.g., pounds per thousand pounds of gas or grains per standard dry cubic foot (the temperature and pressure of the gas must be specified if given as volume).

effective cross-sectional area   The effective width times the effective height.

effective height   The total height of the collecting surface measured from top to bottom.

effective length   The total length of the collecting surface measured in the direction of gas flow. The length between fields is to be excluded.

effective width   The total number of gas passages multiplied by the spacing dimension of the collecting surfaces.

field (in depth)   A field is an arrangement of bus sections perpendicular to gas flow that is energized by one or more high-voltage power supplies.

gas distribution devices   Internal elements in the transition or ductwork used to produce the desired velocity contour at the inlet and outlet of the precipitator, for example, turning vanes or perforated plates.

gas distribution plate rapper   A device used to prevent dust buildup on perforated plates.

gas passage   A passage formed by two adjacent rows of collecting surfaces, measured from collecting surface centerline to collecting surface centerline.

high-voltage bus   A conductor enclosed within a grounded duct.

high-voltage conductor   A conductor used to transmit the high voltage from the transformer-rectifier to the precipitator high-voltage system.

high-voltage power supply   The supply unit to produce the high voltage required for precipitation, consisting of a transformer-rectifier combination and associated controls. Numerous bus sections can be independently energized by one power supply.

high-voltage power supply control equipment   Electrical components required to protect, monitor, and regulate the power supplied to the precipitator high-voltage system. Regulation of the primary voltage of the high-voltage transformer-rectifier is accomplished by one of the following devices: (1) saturable core reactor (a variable impedance device), (2) variable auto-transformer control, and (3) silicon-controlled rectifier (SCR) (electronic switch for voltage regulation).

high-voltage structure   The structural elements necessary to support the discharge electrodes in their relation to the collecting surface by means of high-voltage insulators.

high-voltage system   All parts of the precipitator which are maintained at a high electric potential.

high-voltage system support insulator   A device to physically support and electrically isolate the high-voltage system from ground.

hopper capacity   The total volumetric capacity of hoppers measured from a plane 10 in below the high-voltage system or plates, whichever is lower.

impedance devices   (1) linear inductor or current limiting reactor required to work with SCR-type controllers, (2) a transformer with a specially designed high-impedance core and coils, (3) saturable core reactor, and (4) resistors.

insulator compartment   Enclosure for the insulators supporting the high-voltage system (may contain one or more insulators, but not enclosing the roof as a whole).

isolation damper   A device installed in a duct to isolate a precipitator chamber from process gas.

lower weather enclosure   A non-gas-tight enclosure at the base of the precipitator to protect hoppers from wind and/or detrimental weather conditions.

manual power supply   Manual regulation of high-voltage power based on precipitator operating conditions as observed by plant operators.

migration velocity   A parameter in the Deutsch-Anderson equation used to determine the required size of an electrostatic precipitator to meet specified design conditions. Other terms used are *ω-value* and *precipitation rate*. Values are generally stated in terms of feet per minute or centimeters per second.

penthouse   A weatherproof, gas-tight enclosure over the precipitator to contain the high-voltage insulators.

precipitator current   The rectified or unidirectional average current to the precipitator measured by a multimeter in the ground leg of the rectifier.

precipitator gas velocity   A figure obtained by dividing the volume rate of gas flow through the precipitator by the effective cross-sectional area of the precipitator. Gas velocity is generally expressed in terms of feet per second.

precipitator voltage   The average dc voltage between the high-voltage system and the grounded side of the precipitator.

primary current   The current in the transformer primary as measured by an ac ammeter.

primary voltage   The voltage as indicated by an ac voltmeter across the primary windings of the transformer.

rapper insulator   A device to electrically isolate discharge electrode rappers yet mechanically transmit forces necessary to create vibration or shock in the high-voltage system.

rapping intensity   The gravity force measured at various points on collecting or discharge electrodes. Measured forces should be specified as longitudinal or transverse.

safety grounding device   A device for physically grounding the high-voltage system prior to personnel entering the precipitator. (The most common type consists of a conductor, where one end is grounded to the casing and the other end is attached to the high-voltage system using an insulated operating lever.)

silicon rectifier   A rectifier consisting of silicon diodes immersed in mineral oil or silicone oil.

single precipitator   An arrangement of collecting surfaces and discharge electrodes contained within one independent casing.

spark   A discharge from the high-voltage system to the grounded system which is self-extinguishing and of short duration.

specific collecting area (SCA)   A figure obtained by dividing the total effective collecting surface area of the precipitator by the gas volume, expressed in thousands of actual cubic feet per minute.

transformer-rectifier   A unit comprising a transformer for stepping up normal service voltages to voltages in the kilovolt range and a rectifier operating at high voltage to convert alternating current to unidirectional current.

transition   An aerodynamically designed inlet or outlet duct connection to the precipitator. Transitions are normally included as part of the precipitator and are sometimes referred to as inlet-outlet nozzles.

treatment time   A figure, in seconds, obtained by dividing the effective length, in feet, of a precipitator by the precipitator gas velocity figure.

turning vanes   Vanes in ductwork or transitions used to guide the gas and dust flow through the ductwork in order to minimize pressure drop and to control the velocity and dust concentration contours.

upper weather enclosure   A non-gas-tight enclosure on the roof of the precipitator used to shelter equipment (transformer-rectifier sets, rappers, purge air fans, etc.) and maintenance personnel.

References

1. H. J. White, *Industrial Electrostatic Precipitation*, Addison Wesley Publishing, 1963.
2. Sabert Oglesby, Jr. and Grady Nichols, *A Manual of Electrostatic Precipitator Technology*, Southern Research Institute, 1970.
3. ICAC—Institute of Clean Air Companies, Inc.

Control Technologies: Gaseous

Absorption
Separators

Jerry A. Maudlin

Fisher-Klosterman, Inc.

William L. Heumann

Fisher-Klosterman, Inc.

Venkatesh Subramania

Fisher-Klosterman, Inc.

12.1 Introduction

Many pollutants, natural and synthetic, are gaseous in nature and subsequently cannot be removed from a flow stream by the technologies used for particulate separation. Gaseous industrial pollutants include acid gases (HCl, H_2SO_4, HBr, HCN, H_2S, HF, etc.), other inorganic vapors (SO_x, NO_x, NH_3, Cl_2, etc.), and organic vapors (formaldehyde, ethylene, benzene, etc.). Absorption is one of the main mechanisms utilized within the industrial air pollution control industry to remove gaseous pollutants.

It is the intent of this chapter to describe in simple and easily understood terms a very complex process and yet provide enough detail for those who desire insight into some of the mechanisms involved. Absorption separators include a wide array of devices which utilize a common process, absorption, to remove a gaseous pollutant from a gaseous flow stream. Absorption separators utilize a solvent to absorb the pollutant. This solvent may be either a liquid or a dry bulk solid.

The term *absorption* literally means "a taking in" of something, and everything that will be described later will deal with this basic concept albeit in more complex terms. The reverse process, desorption, means literally "a giving up" of something. The material that absorbs is called the *solvent*, and the gas that is to be absorbed is called the *solute*. Absorption and desorption describe the processes whereby something is going from a gas to a solvent and from a solvent to a gas, respectively.

12.2   How Absorption Separators Work

In order for either of the above processes to occur, there has to be some driving force present. Also, there must be sufficient contacting area between the gas and solvent as well as some length of time to ensure that the gas and solvent have had an opportunity to reach the chemical and physical equilibrium of the process.

It is interesting to note that the most common device used for absorption, the packed tower, is also commonly used to cool gases. There are two basic methods of absorption: chemical and physical. Within any given process, it is important to understand that either or both of these methods may be playing a significant role in absorption. Physical absorption describes the process by which a gas will reach equilibrium with a solvent without changing its chemical properties. A common example of physical absorption is the collection of gaseous HCl into water. The HCl gas is very soluble in water and is readily absorbed. Within the water, liquid HCl is formed. The main factors that will drive the rate of physical absorption are

1. The difference between the solvent's equilibrium molar concentration and the molar concentration of the gas prior to absorption into the solvent. The greater the difference between the actual and equilibrium concentrations, the greater the rate of absorption.

2. The contact area between the solute and solvent. The greater the contact area, the higher the rate of absorption.

In chemical absorption, the solute is chemically reacted with solvent to form a new compound. Although chemical absorption does not alter the mechanisms of physical absorption (mass transfer), it does affect the equilibrium curve. As previously described, the rate of absorption is directly related to the difference between the concentration of the solute within the bulk gas and the concentration the solvent will contain at equilibrium. By chemical reaction, we are able to generate a new equilibrium curve. This means that the solvent can absorb greater quantities of gas, thus increasing the difference in concentrations, which increases the rate of absorption. The chemical reactions used in absorption are relatively rapid when compared to the time required for mass transfer and, therefore, can usually be assumed to be instantaneous.

An important prerequisite to understanding the process of absorption is a basic understanding of equilibrium. A simple equilibrium curve (Fig. 12.1) plots the steady-state concentration of a given compound in a bulk gas versus the steady-state concentration of that compound in a bulk gas where there is perfect mixing of the two and an infinite amount of time. In practice, equilibrium can be achieved with less than perfect mixing and relatively short lengths of time. For any given temperature and pressure, an equilibrium curve like the one shown in Fig. 12.1 could be drawn showing a plot of the empirically determined values of equilibrium at various gas concentrations.

Gas absorption can also be described as a process in which a gas mixture is contacted with a liquid for the purpose of preferentially dissolving one or more of the gas components into the liquid. The transfer of material from the solute to the solvent is mass transfer. Mass transfer is also the movement of molecules induced by some form of potential energy or driving force. A common example is HCl vapor in a flue gas dissolving into water. In some cases, it is desirable to add chemicals to speed up the process, to end up with a desired product of reaction, and/or to make the process feasible.

The most widely used theory for designing mass transfer devices is the two-film theory. The physical laws describing mass transfer, using the two-film theory, have the same form as those for heat transfer or the flow of electricity. The rate that the solute gases are taken into

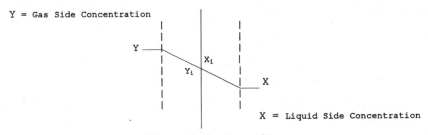

Y = Gas Side Concentration

Y

X_i

Y_i

X

X = Liquid Side Concentration

Boundary of the Two Films

Figure 12.1 Boundary of the two films. (*Courtesy of Fisher-Klosterman, Inc.*)

the solvent is a function of the potential energy divided by the resistance. In the two-film theory, the potential energy is caused by the difference in partial pressure resulting from the difference in concentration of a given solute between the two phases (gas and solvent). The resistance that is offered by the two films is due to the rates of diffusion and the solubility of the solute into the solvent. The rates of diffusion and the solubility of a solute into a solvent are dependent on the solute concentration in the surface layer of the solvent.

12.3   Principles of Mass Transfer

In equation form, the law of mass transfer is

$$\text{Flow} = \frac{\text{potential energy}}{\text{resistance}} \qquad (12.1)$$

where   flow = rate of mass transfer
potential energy = difference in gradients of partial pressure of solute and solvent films
resistance = combination of diffusion and solubility and varies with the type of liquid and gases under consideration

Many models have been presented to describe the phenomenon of mass transfer along the boundary of the process gas and solvent. The models serve to reduce experimental data into useful forms and to predict the performance of absorption systems. The simplest and best known are

1. The stagnant film model
2. The penetration model
3. The surface renewal model

Although theoretical and beyond the scope of this book, introduction to these models is helpful to the understanding of mass transfer. In addition, the engineer must be careful to ensure that the model used is valid for the case under consideration.

12.3.1   The stagnant film model[1]

The stagnant film model is depicted by the diagram shown in Fig. 12.1. The model was first proposed by Nernst and later was used by Whitman to develop the two-film theory. It is the most common model used for modern absorption equipment design. The theory states that the mass transfer process is represented by a concentration gradient on the gas side Y and a concentration gradient on the solvent side X which occur across the gas and solvent films, respectively. The gradients represent the change from the bulk concentrations (of the component undergoing mass transfer) in the gas and solvent. Variables Y_i and X_i represent the equilibrium saturation values corresponding to the concentration of the solute at the interface.

A major weakness of the stagnant film model is that it is not usually possible to measure the concentrations at the interface of the two films. To overcome this difficulty, it is more practical to use overall coefficients based upon the overall driving force between Y and X, i.e., the concentration in the bulk of the gas and the bulk of the liquid, respectively. These coefficients are

$$K_{OG} = \text{coefficient of overall gas}$$

$$K_{OL} = \text{coefficient of overall solvent}$$

$$K_G = \text{gas film coefficient}$$

$$K_L = \text{liquid film coefficient}$$

The value of the stagnant film model is much greater if the slope of the equilibrium line is constant or nearly constant. However, many applications have a nonlinear slope, and this severely limits the use of this model. (High solute concentrations, for example, result in nonlinear equilibrium lines because the gas being removed is a significant portion of the entering gas volume.)

Situations in which the slope of the equilibrium line is constant show that the overall mass transfer coefficients are a function of the slope of the equilibrium line. They also describe a particularly important case involving the situation when there is a chemical reaction between the solute and a solvent. When a chemical reaction is also taking place, no partial pressure exists in the solvent film at the interface. The absence of this partial pressure reduces the solvent resistance to zero, which means that the slope is zero and the overall mass transfer coefficient is then equal to the gas film mass transfer coefficient. In the instance where the gas is infinitely (or nearly so) soluble in the solvent, the slope would also be zero (e.g., NH_3 in water). Therefore,

$$\frac{1}{K_{OG}} = \frac{1}{K_G} \quad \text{or} \quad K_{OG} = K_G \tag{12.2}$$

Another special case that presents itself is the situation where gases have very low solubility, and the slope of the equilibrium line then approaches infinity. In this case, the solvent film mass transfer coefficient will describe the process. For this special case,

$$\frac{1}{K_{OL}} = \frac{1}{K_L} \quad \text{or} \quad K_{OL} = K_L \tag{12.3}$$

Although the stagnant film model has limitations, it demonstrates the essential features of real systems. The concept of the slope of the equilibrium line controlling the rate of mass transfer demonstrates the real problem of resistances that can dramatically affect the design of a mass transfer device. A solute must get into the bulk of the solvent by molecular diffusion and dissolution before it can be transferred away from the film area by convection. The results of the use of this model are in many cases identical, or nearly identical, to the predictions by more sophisticated models. Its value is in its simplicity, and for this reason it is the most widely accepted model for the development of design correlations from experimental data.

12.3.2 Penetration model[2]

The basic concept of this model is that the physical elements of the solvent at the gas-solvent interface are replaced at intervals of time by other elements from the bulk of the solvent, which have the mean bulk composition. The concept further assumes that while an element is exposed to the gas at the interface, it absorbs the solute as though the element was in a quiescent zone and was infinitely deep.

The rate of mass transfer is then described to be a function of the time of exposure for both the gas and solvent films. The time of exposure is very seldom known. Consequently, the main value of this model is in predicting the change in the rate of mass transfer when some conditions are changed in the absorption equipment while others are held constant.

It is generally accepted that the penetration theory provides a more accurate description of the absorption process when data is available to allow for its use.[4] The penetration model and hybrids of the stagnant film and penetration models most closely model gas absorption in nonisothermal cases, that is, where there is a temperature change as a result of a chemical reaction between the solute and solvent.

12.3.3 Surface renewal theory[4]

This is an extension of the penetration theory. Higbie's penetration theory assumed that the exposure time was the same for all the exposed elements. The surface renewal theory assumes a wide spectrum of exposure times and then averages the varying degrees of penetration. The practical application is limited because the fraction of the interface area of the gas-liquid surface that is replaced in an interval of time is usually not known. The useful applications again are those where a process is changed and the goal is to predict the effect of the change.

12.4   Design of Absorption Equipment

12.4.1   Phase equilibria

Gas-solvent equilibria play a major role in the design of absorption equipment. If chemical enhancement is not possible or desirable, the role is of even higher importance. The equations which are listed below are useful in defining the equilibrium conditions.

12.4.1.1   Ideal solutions.   Raoult's law applies for the liquid side of mass transfer between a solute in gas and a liquid solvent. Raoult's law says that at equilibrium the partial pressure of a gas containing the solute over a liquid solvent is equal to the vapor pressure of the pure solute at the given temperature times the mole fraction of the liquid component in the mixture. A limitation to the usage of Raoult's law is that the mixture must be an ideal solution. Ideal solutions are defined as solutions where all liquids in a mixture are independent of each other.

For ideal solutions, the partial pressure is the controlling factor of the mass transfer that is possible. Raoult's law may be expressed as

$$p = P_S \cdot X \tag{12.4}$$

where p = partial pressure of solute in gas phase, psia [mmHg]
P_S = vapor pressure of pure solvent component at operating temperature and pressure, psia [mmHg]
X = mole fraction of component in liquid, mol solute/mol solvent

Example   Given methanol (CH_3OH) at 0.36% B.U. in air with a vapor pressure P_S of 23.6 psia [1221.7 mmHg] at 77°F [25°C], what is the equilibrium concentration in water at atmospheric pressure?

solution

$$P = P_S \cdot X$$

Solving for X,

$$X = \frac{P}{P_S}$$

Per Dalton's law,

$$P = \frac{0.36}{100} \cdot 14.7 = 0.053 \text{ psia} = \left[\frac{0.36}{100} \cdot 760 = 2.7 \text{ mmHg} \right]$$

$$P_S = 23.6 \text{ psia} = [1221.7 \text{ mmHg}]$$

$$X = \frac{0.053}{23.6} = \left[\frac{2.7}{1221.7} \right]$$

$$= 0.0022 \text{ mol methanol/mol water}$$

If P_S is not known, it may be calculated using Antoine's equation:

$$\ln P_S = A - \frac{B}{T + C} \tag{12.5}$$

where P_S = saturation vapor pressure, bar
    A, B, C = constants from Table 12.1
        T = gas temperature, K

Example   Given bromine (Br_2) at 300 K, what is P_S?

solution

$$\ln P_S = A - \frac{B}{T + C}$$

Solving for P_S

$$P_S = e^{[A - B/(T + C)]}$$

From Table 12.1,

$$A = 9.2239 \qquad B = 2582.32 \qquad C = -51.56$$

$$P_S = \exp\left[9.2339 - \frac{2582.32}{300 + (-51.56)} \right]$$

$$= 0.3103 \text{ bar} = 4.5 \text{ lb/in}^2 \text{ [23 mmHg]}$$

TABLE 12.1   Constants[6] for Calculating P_S

Type of gas	Temperature range, K	A	B	C
Biomine	259–354	9.2239	2582.32	−51.56
Silicon tetrachloride	238–364	9.1817	2634.16	−43.15
Carbon tetrafluoride	217	9.4341	1244.55	−13.06
Formic acid	271–409	10.3680	3599.54	−26.09
Ethylene glycol	364–494	13.6299	6022.18	−28.25
Isobutanol	293–388	10.251	2874.73	−100.3
Acrylonitrile	7300	9.3051	2782.21	−51.15

The above concept becomes very important whenever there is a recirculation of liquid to a packed tower. If the partial pressure of the solute in the gas phase on the liquid side exceeds the partial pressure of the solute in the gas phase on the solvent side, the packed tower becomes a stripper; i.e., the solvent gas actually removes the solute from the liquid.

12.4.1.2 Moderately soluble gas. Henry's law applies in quantifying how much of a gas will dissolve in a liquid solution. A set of gas constants has been developed by several researchers, referred to as Henry's constants after the initial researcher. Henry's law states that the mole fraction in the gas phase of a solute in a solvent is a function of the solute's molar concentration in the solvent and the total pressure of the system. This is expressed as

$$p = H \cdot X \qquad (12.6)$$

Therefore,

$$Y = \frac{p}{P} = \frac{H \cdot X}{P} \qquad (12.7)$$

where Y = mole fraction of solute in gas phase
P = total pressure, lb/in^2 [mmHg]
H = Henry's law constant, atm/mol fraction
p = partial pressure of solute in gas phase, lb/in^2 [mmHg]
X = mole fraction of solute in solvent

Example How much hydrogen can be dissolved in 1 mol of water at a pressure of 1 atm when the temperature is 68°F [20°C] and the partial pressure of hydrogen in the gas phase is 3.868 psia [200 mmHg]?

solution Rearranging the equation:

$$X = \frac{p}{H}$$

where H is 6.83×10^4 atm/mol fraction of solution in liquid (see Table 12.2) and P is 14.7 lb/in^2 [760 mmHg].

$$p = 3.868 \text{ psia} = \frac{3.868}{14.7} \text{ atm} = 0.263 \text{ atm}$$

$$X = \frac{0.263}{6.83 \times 10^4} = 0.000\ 003\ 85 \text{ mol } H_2/\text{mol } H_2O$$

For many applications, Henry's gas constants are a function of temperature but are relatively independent if the partial pressure in the

TABLE 12.2   Henry's Constants[7]

Solute	Henry's law constants between solute and water, bar · m³/mol
Formaldehyde	3×10^{-7}
Methyl bromide	6.3×10^{-3}
Ethylene oxide	1.2×10^{-4}
Acrylic acid	3.2×10^{-7}
Methyl chloride	2.4×10^{-2}
Ethyl benzene	8.5×10^{-3}

gas phase is less than 1 atm. In ionic solutions of inorganic salts, Henry's law constant is also a function of ionic strength.

12.4.1.3   Organic compounds in water.   The equilibrium condition for the situation where the temperature is a constant and the organic compound concentration is low (less than 2% by volume) offers a unique solution to the problem of defining the slope of the equilibrium curve. If γ is equal to the activity coefficient for the solute (see Table 12.3), it can be used to determine other constants by using equations for the excess Gibbs energy.

The equation for estimating the activity coefficient for dilute systems is

$$\log \gamma = \alpha + \varepsilon N_1 + \frac{\xi}{N_1} + \frac{\sigma}{N_2}$$

where N is the number of carbon atoms in molecules 1 and 2.

Example   Given ethanol in water at 140°F [60°C] where the saturation pressure P_S is 4.18 lb/in² [216 mmHg]. Ethanol is a primary alcohol and $N_1 = 2$. What is the activity coefficient for ethanol?

TABLE 12.3   Correlating Constants for Activity Coefficients at Infinite Dilution

Solute	Solvent	Temperature, °C	α	ε	ξ	σ
n-Acids	Water	25	−1.0	0.622	0.490	0
		50	−0.80	0.590	0.290	0
		100	−0.620	0.517	0.140	0
n-Primary alcohols	Water	25	−0.995	0.622	0.558	0
		60	−0.755	0.583	0.460	0
		100	−0.420	0.517	0.230	0
m-Ketones	Water	25	−1.475	0.622	0.500	0
		60	−1.040	0.583	0.330	0
		100	−0.621	0.517	0.200	0

solution From Table 12.3,

$$\alpha = -0.755 \qquad \varepsilon = 0.583 \qquad \xi = 0.460 \qquad \sigma = 0$$

$$\log \gamma = -0.755 + 0.583.2 + \frac{0.460}{2} + 0 = 0.641$$

Therefore, $\gamma = 4.38$.

In equation form, the equilibrium curve can be stated as

$$y = m \cdot X \tag{12.8}$$

For dilute concentrations

$$m = \frac{\gamma P_S}{P} \tag{12.9}$$

Therefore,

$$y = \frac{\gamma P_S}{P} \cdot X \tag{12.10}$$

where Y = mole fraction of solute in gas phase
$\quad\quad\;\; X$ = mole fraction of solute in liquid phase
$\quad\quad\;\; P_S$ = saturation pressure of pure solute at given temperature, lb/in² [mmHg]
$\quad\quad\;\; P$ = absolute pressure, lb/in² [mmHg]

Then,

$$P_S = 4.18 \text{ lb/in}^2 \text{ [216 mmHg]}$$
$$P = 14.7 \text{ lb/in}^2 \text{ [760 mmHg]}$$
$$m = \frac{4.38 \times 4.18}{14.7} = \left[\frac{4.38 \times 216}{760}\right]$$

and

$$Y_i = 1.25 \, X_i$$

12.4.2   Absorption equipment design

All absorption equipment design involves the following basic steps:

1. The gas and solvent must be brought together in some sort of contacting device.

2. The gas and solvent phase must reach equilibrium if the results are to be predictable.

3. Following absorption contact, the cleaned gas and solvent must be separated.

Whatever absorption device is chosen, the rate at which the solute is transferred from the gas to the solvent will depend on the contaminant concentration in each phase, the mass transfer coefficients in each phase, the solute's solubility in the solvent, whether or not there is enhancement due to a chemical reaction between the solvent and solute, and the solvent-solute interfacial area made available by the absorption device.

There are two basic types of absorption devices:

1. Stagewise as in tray scrubbers

2. Continuous differential as in packed bed scrubbers and fluid bed reactors

12.4.2.1 Stagewise absorption. This type of absorption uses devices where the gas and liquid are allowed to reach equilibrium in separate stages, or steps, until the desired total absorption and mass transfer have been attained. Devices that are commonly used to achieve stagewise absorption include bubble cap trays, sieve trays, and ballast trays. The design of these devices is based on the efficiency of each tray. The diagram shown in Fig. 12.2 depicts the absorption for a single tray or stage:

Point a is the point before the contacting stage or tray.

Point b is the point after the contacting stage or tray.

Theoretically point b should be equal to point c. However, the diagram is presented in this way to illustrate that experimental data is required to correctly design *stages* for absorption. In practice, ab/ac is less than 1, or, in other words, the equilibrium line is not reached due to the physical variations in the device (such as uneven gas velocity, uneven temperature, and uneven liquid distribution). Furthermore, the unique physical characteristics of various compounds can lead to less than ideal results.

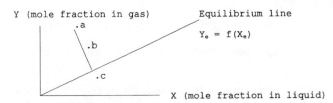

Figure 12.2 Gas-liquid equilibrium line. (*Courtesy of Fisher-Klosterman, Inc.*)

12.4.2.2 Continuous differential contact devices. A packed tower or column is an example of a continuous differential sketch contact device. The difference between a stagewise and a continuous differential contactor is the method for providing mass transfer. The continuous differential contactor uses a subdivided solid, fluidized solid or packing to provide the interfacial surface. In a packed tower, liquid flows over and through the packing, while the gas simultaneously passes through the solid-liquid matrix.

The design of the packed tower, which is the most widely used continuous differential contact device for gas scrubbing, is based on one of four methods which can be stated briefly as follows:

1. *Single-phase coefficient,* K_G *or* K_L. Should be used if the concentrations and the partial pressure at the interface between the two films are known (or can be readily determined) and for predicting changes to some variables while holding others constant. The integral equation for the packing height follows later.

2. *The height of a single-phase transfer unit* H_G *or* H_L. In many cases, gas phase and/or liquid phase absorption data is reported in terms of H_G or H_L. This method is the most convenient if the data is available. Again, the main value is in predicting results for changes. The design equations will not be dealt with in this book.

3. *The overall coefficient* K_{OG} *or* K_{OL}. If the two-film interface concentrations and partial pressure are not known, the overall coefficients can be used. As described in Sec. 12.3.1, the overall coefficients are based on the component concentrations in the bulk of the gas and liquid, respectively, and thus may be available.

4. *The height of overall mass transfer unit* H_{OG} *or* H_{OL}. This method is used where overall coefficients are more convenient than single-phase coefficients or where it is impractical to determine interfacial values. It is the most common method used. The design equations will be examined (and to some extent method 3 above also), and a numerical example will be presented as well.

Single-phase coefficient design. In the earlier discussion of the stagnant film model, the rate of mass transfer was stated in words. The equation for single-phase coefficients where P is a function of pressure and D is a function of the diffusivity is

$$N_A = K_G P(Y - Y_i) = K_L D(X_i - X)$$

where N_A = mass transfer rate, lb $\cdot$ mol/(h $\cdot$ ft^2) [g $\cdot$ mol/(s $\cdot$ cm^2)]
$\quad P$ = system pressure, atm
$\quad K_G$ = gas film mass transfer coefficient
$\quad Y$ = concentration of solute in bulk gas (solvent)

Y_i = concentration of solute at interface

K_L = liquid film mass transfer coefficient, lb · mol/(h · ft²) [g · mol/(s · cm²)]

D = liquid molar density, lb · mol/ft³

X = concentration of solute in bulk liquid

X_i = concentration of solute at interface

However, in continuous differential devices such as packed towers, the mass transfer is not linear. In this situation, a volumetric mass transfer expression is often used.

$$N_A a_V = K_G a_V P(Y - Y_1) = K_L a_V (X_1 - X)D \qquad (12.11)$$

where N_A = mass transfer rate, lb · mol/(h · ft²) [g · mol/(s · cm²)]

a_V = interface area, ft² per unit volume

K_G = gas film mass transfer coefficient, lb · mol/(h · ft²) [g · mol/(s · cm²)]

Y = concentration of solute in bulk gas, mol/mol

Y_1 = concentration of solute in gas at point 1

X_1 = concentration of solute in liquid at point 1

D = liquid molar density, mol/ft³

K_L = liquid film mass transfer coefficient, lb · mol/(h · ft²) [g · mol/(s · cm²)]

P = system pressure, atm

Once the rate of absorption is known, the height of the packing Z is determined using the following equation:

$$Z = \int_{Y_2}^{Y_1} \frac{G_M \, dy}{K_G a_V \rho_G (1 - y)(y - y_1)} \qquad (12.12)$$

where G_M = molar gas velocity, lb · mol/(h · ft²) [g · mol/(s · cm²)]

K_G = single-phase gas film transfer coefficient, mol/(h · ft² · atm) [g · mol/(s · cm²)]

a_V = packing interfacial area, ft²/ft³ [m²/m³]

y = mole fraction of gas

ρ_G = gas density, lb/ft³ [g/mL]

Design based on overall coefficients K_{OG} and K_{OL}, or overall heights of transfer units H_{OG} or H_{OL}. As previously stated, the use of single-phase coefficients is sometimes inconvenient or impractical. The use of overall coefficients that are based on hypothetical interfacial compositions (gas-liquid films) is a method to overcome the difficulties of single-phase coefficients. Overall coefficients for gases and liquids, respectively, are based on the driving forces $Y - Y^*$ for a gas, and $X^* - X$ for a liquid.

The mass transfer rates are
For a gas

$$N_A = K_{OG}(Y - Y^*) \tag{12.13}$$

For a liquid

$$N_A = K_{OL}(X^* - X) \tag{12.14}$$

The following equations for dilute gas systems (less than 2% by volume) are the important points of our consideration:

$$H_{OG} = H_G + \frac{m \cdot G_m}{L_m} \cdot H_L \tag{12.15}$$

On the liquid side of the mass transfer consideration,

$$H_{OL} = H_L + \frac{L_m}{m_1 \cdot G_m} \tag{12.16}$$

The decision as to whether to base the design on the liquid side [Eq. (12.16)] or the gas side [Eq. (12.15)] depends on the value of mG_m/L_m. If mG_m/L_m is less than 1, the design should be based on the gas side [Eq. (12.15)], and if it is greater than 1, the most economical design is the one based on the liquid side [Eq. (12.16)].

12.4.3 Design procedure for a dilute gas concentration with a straight equilibrium line

In some cases, it is reasonable to assume that the equilibrium curve and the operating line are linear over the range in which they are used for the design of a packed bed. In these cases, the logarithmic mean of the terminal potentials is theoretically correct. If Y^* tends to go to 0 due to chemical reaction or dissociation of the gas (as occurs in dilute solutions of HCl or H_2SO_4), then

$$N_{OG} = \ln \frac{Y_1}{Y_2} \tag{12.17}$$

and the efficiency equals

$$E_T = (1 - e^{-N_{OG}}) \cdot 100 \tag{12.18}$$

The treatment of N_{OG} evaluations for highly concentrated gases (with straight or curved equilibrium lines) is not covered in this chapter. Standard reference textbooks should be consulted for this situation.

Example Consider a soluble acid gas (HCl) that will react instantly with a reactant in the liquid phase (NaOH). The HCl concentration is less than 2% by volume, so $m = 0$. The particulars are

Saturated gas volume = 1414 acfm [0.26833 actual m^3/s]
Saturated temperature = 150°F [65.56°C]
Gas density = 0.053 lb/ft^3 [0.84874 kg/m^3]
Inlet gas pressure = 14.7 psia [760 mmHg]
HCl = 138.9 lb/h [.0175 Kg/s]

solution Choose a tower diameter such that the gas velocity is between 450 and 550 ft/min [137 and 168 m/min) (see Table 12.4). The final selection normally is predicated on using a standard-size pipe or mandrel in order to incur minimal fabrication costs. In this case, a 24-in-diameter packed tower gives a velocity of 450 ft/min. (Keep in mind that the less the solubility of the gases, the lower the velocity that should be chosen.)

Next, choose a liquid recirculation rate. For highly soluble gases, a liquid ratio (L/G) of 6 gal/1000 ft^3 gas flow [.8 L/m^3] is normally adequate. The less soluble the gas, the higher the L/G that will be necessary. Once an L/G is chosen, check to ensure that there is at least 2.5 GPM/ft^2 [102 LPM/M^2] to assure complete wetting of the packing. Lower values are permissible in some cases that are beyond the scope of this discussion. In our example, an L/G of 6 gives 1.414×6 which is approximately 8.5 GPM [32.2 L/M]. Since the area of the bed is 3.14 ft^2 [.29 M^2], our average coverage is 8.5/3.14 [32.2/.29] or 2.71 GPM/ft^2 [111.0 LPM/M^2], which is greater than 2.5 [102]; it is therefore adequate.

The choice of the packing to be used needs to be considered at this point. In general, packing manufacturers recommend smaller size packing for small-diameter packed towers and larger size packing for larger-diameter towers. The packing manufacturer's recommendations should be followed. For this example, we will choose 2-in Jaeger Tripaks.

Next, calculate G (in English units of pounds per hour per square foot of area for the gas) and L (pounds per hour per square foot of area for the recycle liquid). Do not neglect to check the packing for flooding. The equations for the flooding conditions are supplied by the packing vendors and vary with the packing used. For the packing selected above, $G_F = (1.24 \times 10^{-6}) L^{-0.617}$ in English units.

TABLE 12.4

Description	Typical velocities recommended, ft/min
Low solubility gas	300–400
High solubility gas	450–550
High reacting gas	450–550
Very low solubility gas (recycle loop)	200–350

$$G = \frac{60Q_S\rho_S}{A} \ \text{lb/(h} \cdot \text{ft}^2)$$

where Q_S = saturated volume, acfm [m³/s]
ρ_S = saturated density, lb/ft³ [kg/m³]
A = tower cross-sectional area, ft² [m²]

In our example, the numerical result is

$$G = 1431.3 \ \text{lb/(h} \cdot \text{ft}^2) = [1.9409 \ \text{kg/(m}^2 \cdot \text{s})]$$

Next, we solve for L:

$$L = \frac{500 \ Q_L}{A} = 1352.8 \ \text{lb/(h} \cdot \text{ft}^2) = \left[\frac{3.2996 \ Q_L}{A} = 8.92 \ \text{kg/(m}^2 \cdot \text{s}) \right]$$

where Q_L is the liquid volume (gal/min [kg/s]).

In our example, $H_{OG} = H_G$ and the derivation of Eq. (12.15) for the Tripak results in the following form:

$$H_{OG} = (2.4717 \times 10^8) \ \frac{G^{0.33}}{L^{0.48} \times D_V^{0.5} \times T_S^{3.0575}}$$

$$= \left[(7.5337 \times 10^7) \ \frac{G^{3.615}}{L^{1.5815} \times D_V^{0.5} \times T_S^{3.3662167}} \right]$$

The diffusivity of HCl in air D_V is 0.166 cm²/s. *Note:* The units conversion to English and SI equations is accounted for in the equation constant and is used because the diffusivity values are readily available in SI units. The saturation temperature $T_S = 460 + 150 = 610°R$ [338.756 K]. Therefore,

$$H_{OG} = .639 \ \text{ft} \ [.195 \ \text{m}]$$

This represents the physical height of one transfer unit in this case.

Now suppose that we require 99 percent removal of HCl.

$$N_{OG} = \ln \frac{1}{1-\text{eff}} \qquad \text{where the efficiency is in decimal form}$$

$$= 4.605$$

which is the number of transfer units required. The required packing height is Z. For our example,

$$Z = H_{OG} \times N_{OG} = 0.639 \times 4.60517 = 2.943 \ \text{ft} \ [.8974 \ \text{m}]$$

In practice, the actual bed height is normally increased by 20 percent to allow for gas and liquid distribution irregularities, so the bed height Z' to be used is

$$Z' = \frac{2.943}{0.8} = 3.679 \ \text{ft} \ [1.121 \ \text{m}]$$

and we would use 4.5 ft (1.372 m). Unless the 10 percent is accepted in all cases, a better description of settling bleed rates is required here. The remainder of the design process includes material selections compatible with the chemistry involved and mechanical strength analysis to determine wall thickness, etc.

12.5   Absorption Equipment

In the broadest sense, any device that is used to bring gas containing a solute and solvent (into which the solute will be absorbed) into intimate contact is an absorber. A discussion of when to use various types of equipment will be made later in Sec. 12.6. A brief description of the most common types of equipment will be considered in this section. There are two broad categories of absorption equipment. These are wet and dry. Wet absorption is by far the most common within industry although the usage of dry absorption systems has been increasing significantly.

12.5.1   Wet absorption systems

12.5.1.1   Stagewise equipment.   A group of horizontal metal plates, or trays, arranged in a vertical series, usually in a cylindrical housing, comprises stagewise absorbers. Each horizontal plate is one stage. The plates can be sieves, bubble type, or ballasts. The gas flow is countercurrent to the liquid flow in all cases.

1. *Bubble type.* (See Fig. 12.3.) The gas flows through chimneys set in holes under each bubble cap and out through the serrated rims of the caps. This action produces massive quantities of bubbles and foam in the liquid in the tray. The bubbles provide the interfacial area for mass transfer. Bubbles in foam scrubbers are suitable for contacting large amounts of process gas containing dilute concentrations of the solute with small quantities of liquid solvent.[4]
2. *Sieve tray type.* Similar to the bubble type, except the bubble cap is omitted and there are many smaller-diameter holes in the plate which disperse the gas and create bubbles in the liquid above the plate. Again, the bubbles formed provide the interfacial area.
3. *Ballast tray type.* (See Fig. 12.4.) This is similar to the sieve tray type, except that the ballast tray contains floating plugs which are lifted up by the air bubbling through the holes. In the case of a shutdown, the plugs drop down and plug the holes.

In all the types listed above, liquid is maintained on the tray surface by a dam at an entrace to a downcomer or sealed conduit that allows overflow liquid to pass to the tray below.

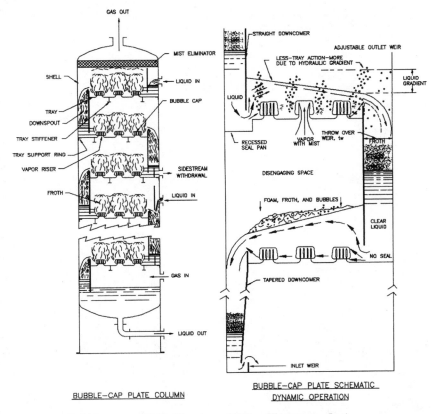

Figure 12.3 Bubble cap scrubber. (*Courtesy of Fisher-Klosterman, Inc.*)

Stagewise absorbers have positive mixing between the gas and liquid streams. One main disadvantage, when compared to packed columns, is that the trays create a much higher pressure drop. It should also be noted that the tray design readily lends itself to cooling by coils which can be placed in the trays. This is of major importance in the production of concentrated nitric, hydrochloric, and sulfuric acids where exothermic heats of formation are occurring.

12.5.1.2 Packed columns or towers. (See Fig. 12.5.) Cylindrical, vertical packed towers are the most common, although horizontal rectangular towers are also being employed. Horizontal packed bed scrubbers (see Fig. 12.6) are very valuable whenever headroom is important, and they are the most economical for a wide range of applications. Limitations in liquid rates and some structural problems are potential negatives. Also, when chemical reaction enhancement is involved, the quantity of reagent per unit volume of the liquid

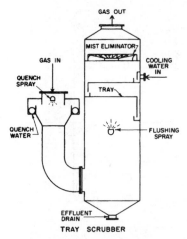

OPERATING CHARACTERISTICS

- Low to moderate pressure drop (approx. 4-1/2" wc)
- Effective over wide range of gas throughputs
- 1-10 gal of scrubbing liquor per 1000 cf of gas
- Low-pressure liquor feed

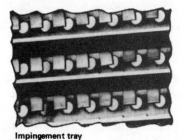

Impingement tray

Figure 12.4 Ballast tray scrubber. (*Courtesy of Fisher-Klosterman, Inc.*)

varies continuously from the top to bottom of the packing. (Liquid is introduced at the top and flows crosscurrent to the horizontal gas flow.) Vertical columns generally have the freshest absorbent at the top where the gas exits and the liquid is introduced. The liquid flows countercurrent to the gas by gravity in the most common designs.

The different shapes of packings available are too numerous to name. They can be generalized into two groups, structured and random packings. Structured packing has more interfacial surfaces per unit volume and higher mass transfer, but it induces a larger pressure drop and is more susceptible to plugging than random packing.

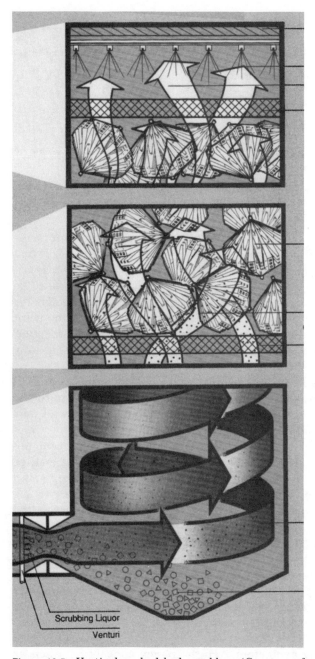

Figure 12.5 Vertical packed bed scrubber. (*Courtesy of Fisher-Klosterman, Inc.*)

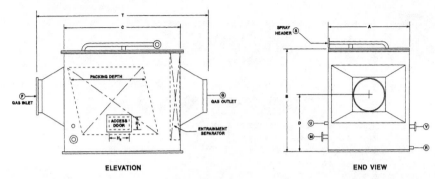

ELEVATION   **END VIEW**

Figure 12.6 Horizontal packed bed scrubber. (*Courtesy of Celicote Air Pollution Control.*)

A recent series of new packed bed scrubbers has come into some use. They employ a fluidized bed of plastic spheres (see Fig. 12.7) constrained between two coarse horizontal screens several feet apart. This style allows higher gas flow rates without flooding, but the allowable range of gas flow is limited to that which keeps the spheres from resting on the lower screen but does not carry them to the point that they are bunched up against the upper screen. A nearly constant gas flow is required for these to be practical. Secondly, the packing tends to break often during the process of fluidization.

The packing size for a given application for any type of packed tower is very important. If the ratio of the tower diameter to the packing size is less than 8:10, the flow of the gas along the wall may be excessive and the efficiency will not be maximized.

Two characteristics of packing that are extremely important are *flooding velocity* and *holdup*. The subject of flooding velocity will be covered in Sec. 12.7. Liquid holdup has two elements, dynamic and static. Dynamic holdup is the liquid that drains out of the packing after gas flow is stopped. Static holdup is the liquid that stays in the packing when flow is stopped.

The better packings provide large surface areas per unit volume, have high flooding velocities, and have a low operating pressure drop. The tradeoff between random and structural packing is in general that the structured packing requires less height but induces a higher pressure drop. Also, as previously noted, structured packing is much more susceptible to pluggage by particulate.

12.5.1.3 High-energy contactors. High-energy contactors, such as venturi scrubbers, have some value as absorbers. The use of the penetration model to treat the data is required to determine the performance of this type of contactor. The theory will not be presented here,

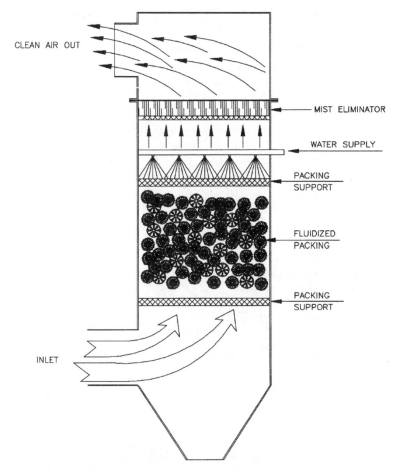

Figure 12.7 Fluidized packed bed scrubber. (*Courtesy of Fisher-Klosterman, Inc.*)

but the mass transfer concepts are very similar to packed beds. The mass transfer surface area is provided by the droplets generated by the pressure drop across the venturi. The size of the droplets becomes smaller as the pressure drop increases. Therefore, the total surface area of the droplets for a given quantity of water increases with pressure and the mass transfer also increases.

12.5.2 Dry absorption systems

By nature, dry absorption systems are similar to chemical reactors that utilize dry bulk chemicals. There are three common types of dry absorption systems.

12.5.2.1   Dry scrubbers.   (See Fig. 12.8.) These absorption systems usually involve the injection of a dry solvent directly into a process gas stream. Frequently a mixing chamber or a long run of ductwork will be placed after the solvent injection to allow adequate contact time for the absorption to occur. This will be followed by an adequate collection device for the removal of the particulate from the gas stream. The selection of the particulate collection device will depend on the characteristics of the particulate and the required emission levels. Baghouses, cyclones, and electrostatic precipitators are most commonly used in industry for this purpose, e.g., SO_2 injection absorption using lime.

12.5.2.2   Spray dryers.   (See Fig. 12.9.) These utilize injection of a wet solvent into a hot gas stream. The liquid is evaporated leaving a dry solvent in contact with the process gas. After the absorption has taken place, the particulate may be filtered from the gas stream by the most appropriate technology. Baghouses and cyclones are most commonly used for this purpose.

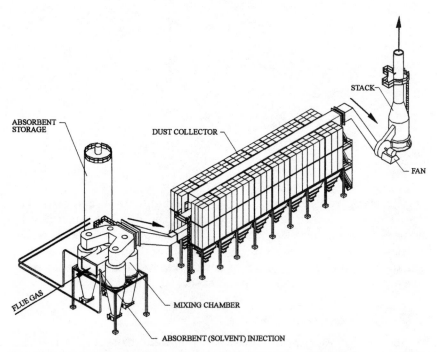

Figure 12.8 Dry scrubber absorption system. (*Illustration by Ronda White. Courtesy of Fisher-Klosterman, Inc.*)

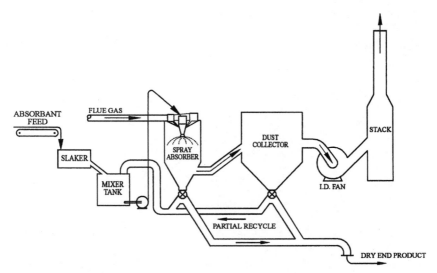

Figure 12.9 Spray dryer absorption process. (*Illustration by Ronda White. Courtesy of Fisher-Klosterman, Inc.*)

12.5.2.3 Fluid bed reactors. (See Fig. 12.10.) These utilize a bed of granulated solvent that is fluidized within a vessel. The process gas within the solvent flows through the fluidized bed coming in contact with the solvent. Most of the solvent is retained in the bed by the design characteristics of the fluidized bed and/or return of eletruiated solvent by cyclones to the bed.

12.6 Basic Selection Criteria

The selection of which type of contactor (absorber) to be used depends on a very large number of factors including space, the process or application on which the absorber is to be used, and economics. The following is a list of considerations to help determine which device is best-suited to accomplish the desired goals.

The factors that will determine which type of absorption equipment is most appropriate for any given application are primarily economical considerations. In the broadest sense, the consideration of whether absorption equipment itself is the best choice for an application is one of economics. Once it has been determined that absorption provides the best methods of removal of the solute, the engineer must choose between wet and dry methods.

In cases where the solute does not create a usable product when absorbed with the solvent, dry systems may provide significantly lower disposal costs resulting in lower overall operating costs. Wet

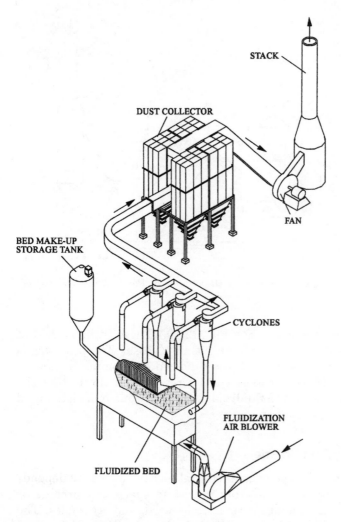

Figure 12.10 Fluidized bed absorption system. (*Illustration by Ronda White. Courtesy of Fisher-Klosterman, Inc.*)

systems typically provide significantly higher collection efficiencies over a broader range of solute than dry systems. Table 12.5 shows some typical removal efficiencies that may be achieved utilizing different absorption equipment in various solutes. Tables 12.6 and 12.7 list some of the pros and cons for various absorption equipment.

In many cases, release of a gaseous pollutant into the atmosphere is controlled because it is malodorous. Odor control is most frequently accomplished by absorption or adsorption. In applications where the odorous gas is reasonably soluble into a solvent, absorption will be a likely method of control.

TABLE 12.5   Typical Absorption Efficiencies, %

Device	Solute			
	HCl	SO$_2$	NO$_x$	Soluble VOCs
High-energy contactor	<90	<90	50	<70
Stagewise absorber	Up to 99.9	Up to 99.9	NA*	Up to 99
Packed tower .	Up to 99.9	Up to 99.9	60–80	Up to 99.9
Spray tower	70–85	60–75	55	5–45
Dry scrubber	75–85	80–95	NA	70–80
Spray dryer	80–90	85–98	NA	75–85
Fluid bed reactor	85–95	90–99	NA	80–90

NA = not applicable.

TABLE 12.6   Wet Absorbers

Equipment type	Pros	Cons
Stagewise absorbers	1. Can handle high particulate load 2. Can provide high levels of absorption 3. Can operate with low liquid rates	1. Expensive 2. Moderately high pressure drops
High-energy contactor	1. Can handle high loadings 2. Moderately high level of absorption on highly reactive gases 3. Low cost	1. High energy costs 2. High water usage
Packed tower	1. Very high levels of absorption 2. Low energy consumption 3. Moderate cost	1. High water consumption 2. Cannot handle high particulate loading
Spray tower	1. Low pressure drop 2. Can handle high particulate loading	1. High water usage 2. Low rates of absorption

In applications involving odor control, the first step is to determine the chemical composition of the compound causing the odor. Since the detection of a malodorous substance is a subjective matter, definition of the expectations of the pollution control system can be very difficult. Most odors are designated by the concentration of the compound at which 50 percent of people on a test board can detect that odor. The 50 percent threshold for odor detection is quite variable among difficult compounds (see Table 12.8). Careful analysis of the process is requested for proper design since the olfactory detection of compounds is so variable. For instance, the 50 percent threshold for H$_2$S (hydrogen sulfide) is at a level that is over 2000 times lower than that of SO$_2$ (sulfur dioxide).

TABLE 12.7   Dry Absorption Equipment

Equipment type	Pros	Cons
Dry scrubber	1. Can utilize hot flue gas 2. Can handle high particulate loadings 3. Very simple equipment and system 4. Reliable operation	1. Moderately high pressure drop 2. Low levels of absorption
Spray drier	1. Can utilize hot flue gas 2. Can handle high particulate loading	1. Moderately high cost 2. Maintenance problems associated with atomization nozzles
Fluid bed reactors	1. Thermal stability 2. High absorption rates 3. Reliable operation	1. Cost 2. High pressure drop

TABLE 12.8   Odor Threshold of Common Materials

Chemical compound	50% Threshold, ppb by volume*
Acetaldehyde	210
Acetic acid	210
Acrolein	100
Aniline	1000
Benzyl chloride	10
Benzyl sulfide	10
Butyric acid	0.5
Carbon disulfide	100
Chloral	47
p-Cresol	0.5
Dimethylamine	21
Dimethyl sulfide	1
Diphenyl sulfide	2.1
Ethyl acrylate	0.1
Ethyl mercaptan	0.5
Formaldehyde	1000
Hydrogen sulfide	0.2
Methyl mercaptan	1
Methyl methacrylate	210
Monochlorobenzene	210
Nitrobenzene	4.7
Phenol	21
Phosgene	470
Pyridine	10
Styrene	47
Sulfur dioxide	470
Toluene diisocyanate	210
Trimethylamine	0.2

*ppb = parts per billion.

TABLE 12.9 Typical Odor Control Applications[5]

Industry	Containment	Absorbant	Oxidant
Brass foundry	Phenol formaldehyde	Na_2CO_3 solution	$KMnO_4$
Varnish plant	Alkyd resin	NaOH solution	$KMnO_4$
Iron foundry	Amines	$NaHSO_4$	$KMnO_4$
Silver plating	Cyanides	NaOH solution	NaOCl
Food processing	Vegetable oil	H_2SO_4	$KMnO_4$
Pulp mill	Organic sulfides	H_2SO_4	ClO_2
Hardboard	Linseed oil	NaOH solution	$KMnO_4$
Rendering	Ham fat	NaOH solution	NaOCl

Care must also be taken in the design of adsorption systems to ensure that the solute that is stripped from the solvent does not exceed the detection threshold as it leaves the device. Since these values are so low, usually a chemical additive is used that *fixes* the solute (keeps it from being stripped) and/or the solvent is used only one time. In the case of odor-causing compounds that are inorganic (e.g., sulfur dioxide, hydrogen chloride, nitric oxide, hydrogen sulfide, hydrogen cyanide, ozone, chlorine, and ammonia), the absorption process may be relatively straightforward utilizing a packed tower with a caustic solution.

In general, the removal of organic odors is more difficult than that of inorganic odors. For most successful applications utilizing absorption of organic odors, a two-step process is used. In this process, the odor-causing compound is first absorbed into the solvent, and then it is oxidized into a less malodorous substance. Careful control of pH and oxidant concentrations is necessary for successful operation. Table 12.9 shows some successful applications utilizing absorption on organic compounds.

12.7 Installation

The following few key points should be considered when installing absorption systems.

1. *Saturation of gas.* For wet absorption, the gas to be treated should always be saturated if maximum efficiency is to be realized. If the gas entering the device is not saturated, the efficiency is less because the system has not reached an equilibrium condition. The loss in efficiency is directly related to the temperature and to the degree of saturation if the gases are not saturated.

2. *Gas and solvent distribution.* Uniform distribution of both the gas and solvent liquid is absolutely essential if predictable results are to be achieved. The gas entry should be designed so that a full, uniform distribution has been achieved prior to entry into the absorbing

devices. The solvent distribution is no less important. In the case of random packed towers, for example, packing suppliers recommend the collection and redistribution of the liquid if the packing height is too great. The recommended heights between redistribution points vary, and the packing supplier's recommendations should be followed.

3. *Gas and liquid flow rates.* Random and structured packing have operational limits where flooding will occur. This condition is the result of reaching the point where the gas velocity will hold up the liquid to the point where the liquid no longer will flow through the packing. The flooding rates of flow vary depending on the packing used, and packing suppliers will provide data that can be used to avoid this condition.

4. *Material selection.* The absorbing device and all the internals must be constructed of materials that will withstand the chemical and temperature conditions of the process.

12.8   Troubleshooting Absorption Towers

There are many possible causes for improper operation of absorption systems. The following is a listing of the most common problems encountered for packed towers.

12.8.1   Gas volume and gas distribution

The presence of a high molar gas velocity expressed in terms of volume is the most common problem. If the velocity exceeds about 600 to 650 ft/min [3 to 3.3 m/s], the residence time within the tower can be too short for the mass transfer to take place. Also, the gas velocity and the liquid rate must be considered simultaneously in order to assure that there is adequate gas-to-liquid contact time.

In addition to the axial gas velocity, the distribution of the gas must be considered. For example, if the entering gas velocity is too high, the distance from the gas entry point to the packing must be increased in order to achieve a uniform gas flow across the bed. In general, the maximum axial gas velocity variation should not exceed 5 percent. Uniform gas distribution can be achieved by distance or by the introduction of some device that causes a pressure drop to occur. Such a device, while effective in producing uniformity in the gas flow, will also add to the operating cost of the tower by using more fan horsepower. A convenient method to use is to have the packing support serve as a gas flow distributor also. Simple grating works fine if the gas stream velocity at the entry point is less than 2500 ft/min [12.7 m/s] and the tower cross section is such that the average gas velocity is 450 to 650 ft/min [2.3 to 3.3 m/s]. The grating opening size should be based on a maximum of about 0.1 in w.c. loss across the grating and be small enough to prevent the packing from dropping through.

12.8.2   Liquid volume and distribution

The next most common problem is the liquid flow rate which can be either too low or too high. If the liquid rate is too high, a flooding condition can occur whereby it is impossible for the quantity of liquid to flow through the packing. The packing vendor gives the flooding condition parameters, and good design is to have the liquid rate be about 80 percent of the flooding condition. Normally the liquid rate results in an L/G of 6 to 20 gal/1000 acfm saturated [801 to 2670 L/actual m³]. The higher the liquid rate, the higher the pressure drop induced per foot of packing depth. In addition to the above, it is possible to have too little liquid flow. For example, if a chemical reaction is being counted on to keep the slope of the equilibrium curve equal to zero (i.e., no resistance to mass transfer), then sufficient solute must be present at each point of the packing to react with the gas mass transfer rate. A computerized program allows the designer-troubleshooter to look at many points of a packed bed and to analyze the solute concentration at each point.

If there is not any chemical reaction and the mass transfer is simply from the gas phase to a dissolved state in the liquid, the partial pressure of the liquid state must also be compared to the partial pressure in the gaseous state. Please be aware that *if the liquid is recirculated,* the bleed rate of the liquid must be sufficient so that the gas partial pressure of the contaminant (*at the design efficiency*) at the top of the packing is not exceeded by the contaminant partial pressure in the liquid at the same point in the tower. *If this is not considered, the packed bed will become a liquid stripper instead of a gas absorber.*

Liquid distribution can also become a problem if the packing height is too great or if the liquid distributor used allows too much of the liquid to flow down the tower walls instead of through the packing. In the latter case, if a nozzle-type distributor is used and the bed is too far below the nozzles, a large portion of the spray will go to the walls and will not contribute to the mass transfer. If the packing height is too great, a liquid redistributor is required. The packing will gradually cause more and more liquid to migrate to the vessel walls as it flows down through the packing. The packing height where liquid distribution is required varies from packing to packing, so it is recommended that this point be discussed with the packing vendors.

12.8.3   Collapsed packing

Packing (particularly random style) over time will collapse due to the weight of the packing above it with the liquid. Obviously, operating temperature upsets can produce this instantaneously. Periodic checks of the packing condition should be made. A continuous monitoring of the pressure drop across a packed bed will highlight any impending problem of packing collapse or packing plugging due to some upset condition resulting from particulate formation from chemical reaction or foreign matter being introduced to the system. Replacement of the packing with one that is made of more temperature-resistant materials will many times dramatically increase packing life before collapse. Chemical compatibility with the packing material must also be considered if the reagent or the gas being absorbed has changed.

12.8.4 Operating condition changes

The initial system design criteria should be carefully checked against the current conditions. Factors such as volume, temperature, and incomplete gas saturation, as well as the concentration of the contaminant gas, are the most common items to be checked.

12.8.5 Plugged packing

Generally the pressure drop across the bed will indicate when this has occurred. Foreign particulate or reaction products can be the source of the plugging. In some cases, hard water can be the problem. Calcium will react with acid gases and produce a solid even when the reagent used produces a soluble salt. An analysis of the makeup water should always be made when solids plugging of the packing is occurring and there is not any particulate in the gas stream.

References

1. Nernst, Walther Hermann, *The New Heat Theorem, Its Foundations in Theory and Experiment.* Translated from the second German edition by Guy Barr, New York, Dover Publications, 1969.
2. Higbie, R., *Trans. AICHE,* vol. 31, p. 365, 1935.
3. Danckwerts, P. K., *Ind. Eng. Chem.,* vol. 43, p. 1460, 1951.
4. Siddharth G. Chatlerjee and Elmar R. Altwicker, "Film and Penetration Theories for a First-Order Reaction in Exothermic Gas Absorption," *Canadian Journal of Chemical Engineering,* vol. 65, June 1987.
5. Biswas, Asolekar, and Keimer, "Effect of Surface Resistance Arising Due to Surfactant on Gas Absorption Accompanied by Chemical Reaction in a Foam-Bed-Reactor," *Canadian Journal of Chemical Engineering,* vol. 65, June 1987.
6. Ralf F. Strigle, Jr., *Random Packings and Packed Towers (Designs and Applications),* Gulf Publishing Co., Houston, TX, 1987, pp. 50–51.
7. Robert C. Reid, John M. Prausnitz, and Bruce E. Pounh, *The Property of Gases and Liquids,* 4th ed., McGraw-Hill, Inc., 1987, pp. 656–732.
8. David R. Lide, ed., *Handbook of Chemistry and Physics,* 72d ed., CRC Press, Inc., 1992, pp. 16–22 and 16–24.

Adsorption

Jerry A. Maudlin

Fisher-Klosterman, Inc.

William L. Heumann

Fisher-Klosterman, Inc.

Venkatesh Subramania

Fisher-Klosterman, Inc.

13.1 Introduction

Adsorption, like absorption, is a mass transfer operation. The process can be described as one in which a porous solid is brought into contact with either a liquid or gaseous fluid stream to selectively remove unwanted contaminants by depositing them on the solid. The removal of moisture from an airstream utilizing a silica gel is an example of one type of adsorption. Adsorption takes place because molecular forces at the surface of the gas or liquid are in a state of imbalance with the solid material. Within this book, the contaminant will be referred to as the *adsorbate* and the collecting media will be called the *adsorbent*.

Adsorption devices are used commonly within industry for the removal of certain organic and inorganic compounds and elements.

Since adsorption captures instead of chemically altering or destroying the adsorbate, it is often used in applications where recovery of the adsorbate is desired. The adsorbent may capture the adsorbate by both physical and chemical means.[1] In physical adsorption, the adsorbate molecules adhere to the adsorbent material. There is a physical bonding force (primarily van der Waals forces). When a chemical bond is formed between the adsorbate and adsorbent, the process is referred to as chemisorption. Chemisorption usually occurs at elevated temperatures where there is adequate available energy to make or break chemical bonds. Within the industrial air pollution control field, adsorption is used for

1. Purifying intake, circulation, or exhaust gases from toxic gases, odors, and other noxious gases.

2. Solvent (VOCs) recovery from air leaving an evaporation chamber or process. Examples are spray painting and dry cleaning.

3. Fractionization of gases.[2]

In practical usage, adsorption normally takes place within a fixed bed of adsorbent through which the process gas passes. The adsorbent bed then has a limited life before it must be replaced and/or regenerated. To regenerate an adsorbent, the process of adsorption is reversed and is called desorption. In desorption, the adsorbate is stripped away from the adsorbent usually by a reduction in pressure and/or an increase in temperature.

Successful adsorption of a contaminant gas requires certain physical and, in some cases, chemical properties of the adsorbent. By far, the most common adsorbent used industrially is activated carbon. Other common adsorbents include activated alumina, silica gel, zeolite, and clays.

Chemisorption, or activated absorption, is essentially the same process as described for catalytic oxidation (see Chap. 14) within industrial air pollution control. For that reason, within this chapter, we will deal with physical adsorption only.

13.2   How Adsorption Works

In physical adsorption, particular gas molecules are attracted and attached to the absorbent due to molecular forces.[3] The primary molecular force is van der Waals force, and physical adsorption is sometimes referred to as van der Waals adsorption.[3] The adsorbate will condense on the surface of the adsorbent releasing heat. The amount of heat released is somewhat greater than the latent heat of vaporization for a given adsorbate and will more closely match the heat of sublimation. If the adsorbent is highly porous, there is a greater surface area for the adsorbate to attach to. Furthermore, the vapor pressure of a small-radius concave liquid surface is lower than that over a flat surface. This results in an increase in adsorption. In all cases of adsorption at equilibrium, the partial vapor pressure of the adsorbed gas equals the vapor pressure of the contacting gas phase, and by lowering the partial vapor pressure of the gas phase or by increasing the temperature of the adsorbate, the process can be reversed (or desorption takes place). In industrial applications the dependence of reversibility is an important factor as gases adsorbed are subsequently desorbed and recovered or reused. It is important to note that the process of desorption is not confined to gases but is applicable to liquids as well.

As previously stated, at ordinary temperatures adsorption is based on intermolecular attractive forces. The molecular forces are imbalanced at the interface of the gas and solid and up to a depth equal to a molecular thickness at the boundary of the gas and solid. Because of this fact, a large boundary surface area is necessary for significant adsorption to occur.

The discussion of how adsorption devices work requires consideration of the following:

1. Physical criteria such as size, shape, and the method of packing the solid adsorbent (or nature of adsorbent)

2. Equilibrium

3. Stoichiometric capacity

4. Rates controlling the separation

13.2.1 Nature of adsorbents

Virtually all solid materials have some adsorptive capacity. However, there are very few solids that have adequate adsorptive properties to make them useful for industrial gas cleaning. One common characteristic of all industrial adsorbents is a large surface area per unit volume. This is evidenced on a microscopic level by irregular surface shapes and pores. Some of the common physical characteristics of adsorbents are shown in Table 13.1 and in the following sections. Most adsorbents are in granular form with individual particle diameters ranging from 50 μm to ½ in.[3] The solids must be chosen such that they have a high rate of adsorption (depending on the gases adsorbed), offer low pressure drop, and are able to sustain their individual particle shape (or surface area) when subjected to their own weight in bulk. There are a number of adsorbents that are commonly used in industry that satisfy the above criteria. They are listed below.

TABLE 13.1 Physical Properties of Adsorbents[2]

Material and uses	Shapes of particles	Internal porosity, %	Typical bulk dry density, kg/L	Surface area, km²/kg	Sorptive capacity, kg/kg (dry)
Aluminas					
Low-porosity (fluoride sorbent)	Granules, spheres	40	0.70	0.32	0.20
High-porosity (drying, separations)	Granules	57	0.85	0.25–0.36	0.25–0.33
Desiccant, $CaCl_2$ coated	Granules	30	0.91	0.2	0.22
Activated bauxite	Granules	35	0.85		0.1–0.2
Chromatographic alumina	Granules, powder, spheres	30	0.93		~0.14
Silicates and aluminosilicates (molecular sieves)	Spheres, cylindrical pellets, powder				
Type 3A (dehydration)		~30	0.62–0.68	~0.7	0.21–0.23
Type 4A (dehydration)		~32	0.61–0.67	~0.7	0.22–0.26
Type 5A (separations)		~34	0.60–0.66	~0.7	0.23–0.28
Type 13X (purification)		~38	0.58–0.64	~0.6	0.25–0.36
Mordenite (acid drying)			0.88		0.12
Chabazite (acid drying)			0.72		0.20
Silica gel (drying, separations)	Granules, powder	38–48	0.70–0.82	0.6–0.8	0.35–0.50
Magnesium silicate (decolorizing)	Granules, powder	~33	~0.50	0.18–0.30	
Calcium silicate (fatty-acid removal)	Powder	75–80			
Clay, acid-treated (refining of petroleum, food products)					
Fuller's earth (same)			0.80	~0.002	
Diatomaceous earth	Granules, powder		0.44–0.50		
Carbons					
Shell-based	Granules	60	0.45–0.55	0.8–1.6	0.40
Wood-based	Granules	~80	0.25–0.30	0.8–1.8	~70

Petroleum-based	Granules, powder	~80	0.45–0.55	0.9–1.3	0.3–0.4
Peat-based	Granules, powder, spheres	~55	0.30–0.50	0.8–1.6	0.5
Lignite-based	Granules, powder	70–85	0.40–0.70	0.4–0.7	0.3
Bituminous-coal-based	Granules, powder	60–80	0.40–0.60	0.9–1.2	0.4
Organic polymers					
Polystyrene (removal of organics, e.g., phenol; antibiotics recovery)	Spheres	40–50	0.64	0.3–0.7	
Polyacrylic ester (purification of pulping wastewaters; antibiotics recovery)	Granular, spheres	50–55	0.65–0.70	0.15–0.4	
Phenolic (also phenolic amine) resin (decolorizing and deodorizing of solutions)	Granular	45	26.2	0.08–0.12	0.45–0.55

1. *Activated carbon.* As previously stated, activated carbon and related forms of carbon such as charcoal are the widely used industrial adsorbents. Activated carbon is produced by the carbonization of wood, coal, nutshells, coconut shells, peat, fruit pits, etc. Carbonization involves exposing the base material to a high temperature in the absence of oxygen. The carbon products, if produced from coal, may then be further activated by steam treatment at a high temperature. Activated carbon is available in granulated and pellet form and may be used for the recovery of solvent vapors, odor control, hydrocarbon control, and the removal of certain inorganic vapors (e.g., mercury). Activated carbon is easily regenerated or reactivated by steam or other techniques which evaporate the collected adsorbate from the carbon. Activated carbon is regenerated at temperatures of 212 to 284°F [100 to 140°C] and may be operated at temperatures up to 302°F [150°C].

2. *Activated alumina.* Activated alumina is hydrated alumina oxide which is activated by high temperatures which evaporate the free moisture. Activated alumina is available in pellet, granule, and powder forms. The primary industrial usage of activated alumina is as a desiccant to dry process gases.

3. *Silica gel.* Silica gel is a neutralized form of sodium silicate. The term *gel* arises from the soft consistency of the material during one of the stages of its production. As an industrial adsorbent, silica gel is available in granular and pellet forms. It may be used at temperatures up to 752°F [400°C] and is regenerated at temperatures of 248 to 482°F [120 to 250°C]. Silica gel is used principally for the dehydration of air and other gases, fractionalization of hydrocarbons, and flue gas desulfurization.

4. *Fuller's earth.* Fuller's earth is natural clays comprised of magnesium aluminum silicates. It is available in powder and granular forms. The main industrial usages of Fuller's earth are in the purification and decoloring of petroleum and animal and vegetable oils. The material may be regenerated by washing and/or burning off the collected organic matter.

5. *Molecular sieves.* The term *molecular sieve* refers to adsorbents of a number of different chemical compositions, primarily compounds of alumina silicates. The adsorbent itself is crystalline in nature, and the diameters of the passageways may selectively adsorb molecules by physical size. The pore and/or passage diameter can filter or sieve out molecules that are too large, preventing their adsorption (Fig. 13.1). Molecular sieves may be used for temperatures up to 1025°F [552°C] and are regenerated at temperatures of 250 to 500°F [120 to 260°C].

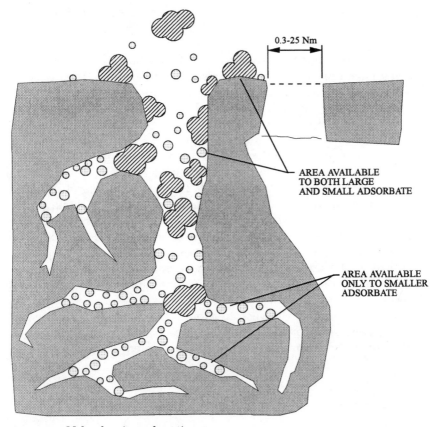

0.3-25 Nm

AREA AVAILABLE
TO BOTH LARGE
AND SMALL ADSORBATE

AREA AVAILABLE
ONLY TO SMALLER
ADSORBATE

Figure 13.1   Molecular sieve adsorption.

13.2.2   Theory of adsorption process

The discussion that follows will be limited to processes that are isothermic, meaning that the heat generated by the exothermic adsorption process is small and the simplification involved in assuming the heat generated is negligible does not introduce a large error into the analysis. There are three basic mathematical expressions that describe vapor adsorption equilibria. These equations are referred to as the

Langmuir equations

Benauer-Emmett-Teller (BET) equations

Freundlich isotherms

None are universally applicable, and, unless prior research has been done, it cannot be predicted as to which will apply to a specific case.

13.2.2.1 Vapor adsorption equilibria.

The equilibrium relationship is called an *isotherm* which means constant temperature. The definition of an isotherm is the amount of adsorbate adsorbed as a function of the concentration (partial pressure) of the adsorbate in the gas phase at a constant temperature. Different gases are adsorbed to different extents under similar conditions. Figures 13.2 to 13.4 provide the vapor equilibrium curves for various types of organic vapors for the activated carbon, molecular sieve (artificial zeolites), and silica gel adsorbents. As can be seen in Figs. 13.2 to 13.4, the adsorption capacity decreases with an increase in temperature for the same partial pressure. This is due to the desorption taking place where the equilibrium partial pressure of the gas phase is lowered (see the beginning of Sec. 13.2). It can also be clearly seen that increasing the partial pressure of the adsorbate causes more gases to be adsorbed and, vice versa, decreasing the partial pressure causes more gases to be desorbed.

There are a number of ways to express the equilibrium concentration of an adsorbate on the adsorbent in relation to the gas phase concentration of the adsorbate. The general statement is

$$q^* = f(p) \tag{13.1}$$

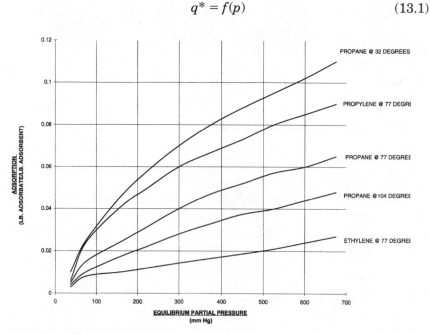

Figure 13.2 Silica gel adsorption.[4]

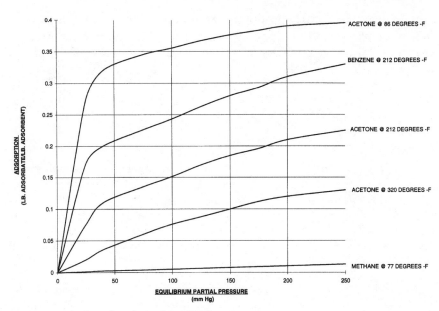

Figure 13.3   Activated carbon adsorption.[3]

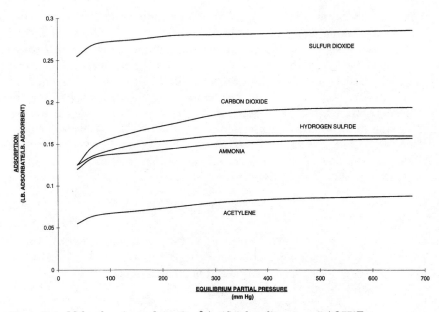

Figure 13.4   Molecular sieve adsorption.[3] Artificial zeolites, type 5.A@77°F.

where q^* is the equilibrium concentration in moles per unit weight of adsorbent and p is the phase concentration of the adsorbent in moles per unit volume of gas. One equation for the expression of isotherms is the Langmuir equation:

$$\frac{q^*}{q_m} = \frac{K_L p}{1 + K_L p} \tag{13.2}$$

where q^* = equilibrium concentration, mol/g
$\quad q_m$ = amount of adsorbate needed to cover a single molecular layer per gram of adsorbent, g/g
$\quad K_L$ = Langmuir's constant, mL/mmol
$\quad p$ = pressure, mmHg

Example If K_L is 375 mL/mmol for anthrocene onto alumina and p is 5 mmHg,

solution

$$\frac{q^*}{q_m} = \frac{(0.375)(5)}{1 + (0.375)(5)} = .6522 \text{ g/g}$$

13.2.2.2 Adsorption capacity. The adsorption capacity (also called stoichiometric capacity) of a column describes the volume of a fluid (or gas) that can be treated before breakthrough occurs. *Breakthrough* is defined as the point at which the removal rate of the adsorbate begins to decrease below the designed efficiency value. This capacity is a function of the nature of the compound being adsorbed (or the adsorbate) and the type of adsorbent. Reference 5 provides a mathematical expression and correlation constants for the adsorption capacity of activated carbon for different organic vapors. Some of these are highlighted below. The correlation equation for the adsorption capacity of activated carbon is given as

$$\log_{10} Q = A + B \log_{10} y + C(\log_{10} y)^2 \tag{13.3}$$

where $\quad y$ = concentration of adsorbate, in parts per million by volume (ppmv) at 77°F [25°C] and 14.7 lb/in^2 [1 atm]
$\quad Q$ = grams of compound per 100 g of carbon
$\quad A, B, C$ = constants (see Table 13.2)

13.2.2.3 Determining rates of adsorption. Adsorption is a molecular scale process. There are two basic mechanisms that require definition and control the rate of adsorption.

TABLE 13.2   Adsorption Correlation Constants

Name	A	B	C	Min, ppm	Max, ppm	Q@10 ppmv	Q@100 ppmv	Q@1000 ppmv
Styrene	1.35701	0.13495	−0.01451	10	8,044	30.02	37.06	42.78
p-Xylene	1.31115	0.14069	−0.01458	10	10,000	27.37	34.22	40.00
Ethyl benzene	1.30444	0.14449	−0.01493	10	10,000	27.16	34.17	40.14

1. *Fluid phase external diffusion process.*   This mechanism deals with the rate of material transfer to the particle, i.e., the adsorbate rate of transfer from the fluid into the adsorbent, or solid. The process can be expressed as: The rate of accumulation of the adsorbate on the solid per unit time is equal to the fluid phase mass transfer coefficient times the surface area of the solid per the unit of volume times the ratio of voids between the solid particles per unit density times the driving force. The driving force in this case is simply the difference between the adsorbate concentration in the incoming fluid and the equilibrium adsorbate concentration.

2. *The solid phase internal diffusion process.*   The adsorbate must be transferred from the outer surface of the particles (adsorbent) at the same rate that they are deposited externally to the particle surface for equilibrium to exist. An example which illustrates the process is the adsorption of water vapor from a gas stream using activated carbon or a silica gel. A sample calculation for a silica gel column will be given later in this section.

In a fixed bed, initially, the exterior of the adsorbent particle and up to one molecule deep absorbs the adsorbate where it condenses into liquid form. This first step would be the end of the process if it was not for the capillary action, whereby the condensed adsorbate is transferred to molecules deeper and deeper inside the adsorbent particle. As the initial adsorbate travels deeper into the particle, additional adsorbate condenses on the surface. Over time, the process will continue until the particle is saturated. All the adsorbent particles in a layer of the bed are being acted upon at the same time and will become saturated at roughly the same time. Subsequently, another layer is similarly affected and so forth until the whole usable portion of the bed has become saturated. The migration from one layer to another is referred to as a *wave*. In the case of a moving bed, the medium is continuously replenished so the process is continuous rather than being a patch process.

The adsorption process can be a single stage or multistage. The following discussion will be limited to a single stage which can be either a continuous or a batch type of process. It is similar to a single-stage absorption process where the solid adsorbent is analogous to the solvent and the adsorbate is analogous to the solute.

For purposes of our discussion, think of a gas stream with a mass of contaminant to be absorbed in a single-stage tower. The gas stream is the solvent and the contaminant is the solute. The concentration of the solute in the solvent is Y lb solute/lb solvent. The solid adsorbing medium is the absorbent, and the amount of solute absorbed is X lb solute/lb solvent. As the contaminated gas flows through the adsorbent, X and Y can be plotted as an equilibrium curve.

Further, if G is the mass of solvent flowing through a mass of adsorbent designated as S, the solute removed from the solvent is equal to the solute picked up by the adsorbent so that

$$G \cdot (Y_0 - Y_1) = S_S \cdot (X_1 - X_0) \tag{13.4}$$

where G = solvent mass flow, lb/(h · ft²) [kg/(h · m²]
$\quad\quad X_0$ = initial solvent concentration, lb adsorbate/lb adsorbent
$\quad\quad X_1$ = final solvent concentration, lb adsorbate/lb adsorbent
$\quad\quad S_S$ = adsorbate-free adsorbent amount, lb/(h · ft²) [kg/(h · m²]

The value of Eqs. (13.4) to (13.10) will be seen in the example that follows and the equilibrium curve that demonstrates the concepts introduced.

If Eq. (13.4) is plotted (see Fig. 13.5), it is a straight line with a slope of $-S_S/G$ and is referred to as the operating line. Expressions can also be developed for multistages, but these are beyond the scope of this study.

The *Freundlich equation* (Eq. 13.5) can be applied for many situations that we are considering here.[3] If we let $m = S/G$, then we can state

$$Y^* = m \cdot X^m$$

and at equilibrium conditions,

$$X_1 = \left(\frac{Y_1}{m} \right)^{1/n} \tag{13.5}$$

Further, if we assume that the adsorbent does not contain any adsorbate initially, at the entry point, then $X_0 = 0$ and

$$\frac{S_S}{G_S} = \frac{Y_0 - Y_1}{(Y_1/m)^{1/n}} \tag{13.6}$$

For a *one-component solute*, this computational process can be clearly explained by considering it as analogous to gas absorption where a

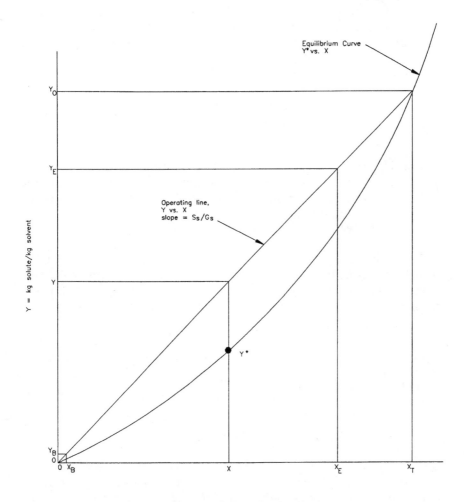

X = kg adsorbate/kg absorbed

Figure 13.5 Adsorption equilibrium curve.

solid absorbent replaces a liquid solvent. Using this concept, Eqs. (13.7) to (13.9) are derived to compute the height of the carbon bed.

If the overall gas mass transfer coefficient can be described in terms of the outside surface of the particles making up the adsorbent, the rate of mass transfer over a segment of the tower with a total height Z is

$$S_S \, dx = G_S \, dy = K_Y a_p (Y - Y^*) \, dZ \qquad (13.7)$$

where a_p is the outside surface of the particles and Y^* is the equilibrium gas composition at point X with a given adsorbate composition.

The driving force for the adsorption process is $Y - Y^*$ which is the vertical distance between the straight line (operating line) and the equilibrium curve as shown in Fig. 13.4.

The equation above, after integration, yields

$$N_{OG} = \int_{Y_2}^{Y_1} \frac{dy}{Y - Y^*} \times = \frac{K_Y a_p}{G_S} \int_0^Z dZ = \frac{Z}{H_{OG}} \qquad (13.8)$$

where $H_{OG} = G_S/K_Y a_p$ = height of a transfer unit, ft [m], given here
N_{OG} = ordinarily integrated graphically and is equal to area under equilibrium curve shown in Fig. 13.1 and is the number of transfer units required
$Z = H_{OG} \times N_{OG}$ = overall bed height, ft [m]

It must be understood that the use of an overall coefficient of mass transfer requires that the mass transfer within the discrete solid particles can be described by a mass transfer coefficient k_S for each particle of surface area a_p. In equation form, this is stated as

$$\frac{G_S}{K_y a_p} = \frac{G_S}{k_y a_p} + \frac{mG_S}{S_S}\left(\frac{S_S}{k_S a_p}\right) \qquad (13.9)$$

so that

$$H_{OG} = H_G + \frac{mG_S}{S_S} H_S \qquad (13.10)$$

The variable k_S in Eq. (13.9) assumes a linear driving force based on the difference in concentration. The linear driving force assumption is true if the mass transfer coefficient of the adsorbate is controlling the process. Cases where the resistance to diffusion within the discrete particles of the adsorbent controls the process are not covered here, but there are many instances where this occurs.

Example Assume a fixed bed of silica gel removing a *small* quantity of water from a stream of air such that it is reasonable to assume an isothermal process. The following conditions and assumptions are given.

1. The inlet temperature of the gas stream T = 80°F [26.7°C].
2. The humidity of the gas stream = 0.00267 lb water/lb air [0.00267 kg/kg].
3. The bed depth = 2 ft [0.61 m].
4. The superficial mass velocity of the air = 95.536 lb air/(h · ft²) [0.1295 kg/(m² · s)].
5. Assume the breakthrough point Y_B = 0.0001 lb water/lb dry air [0.0001 kg/kg]; the inlet humidity Y_0 = 0.00267 lb water/lb dry air [kg/kg], and the

bed exhaustion point $Y_E = 0.0024$ lb water/lb dry air [kg/kg]. The mass transfer coefficients determined by Eagleton and Bliss in English units are $k_y a_p = 31.6G^{0.941}$ lb water vapor/(ft$^3 \cdot$ h $\cdot \Delta Y$) [$31.6G^{0.55}$ kg/(m$^3 \cdot$ s $\cdot \Delta y$)] and $k_S a_p = 216.94$ lb water vapor/(ft$^3 \cdot$ h $\cdot \Delta X$) [0.965 kg/(m$^3 \cdot$ s $\cdot \Delta x$)]. *Note:* The work by Eagleton and Bliss was done on a silica gel with an apparent bed density of 41.91 lb/ft^3 [671.2 kg/m^3], an average particle diameter of 0.068 in [1.727 mm], and a particle external surface area of 0.067 ft^2/lb [2.167 m^2/kg].

6. $X_T = Y_0 G_S/S_S = 0.0858$ lb water vapor/lb gel [0.0858 kg/kg] and $m = \Delta Y/\Delta X$
   $= 0.0185$.

Estimate the time required for the bed to reach the breakthrough point.

solution   Calculate the mass transfer coefficients.

$k_y a_p = 31.6G^{0.941} = 31.6 \cdot (95.536)^{0.941} = 2308$ lb water vapor/(ft$^3 \cdot$ h $\cdot \Delta Y$)
$= [10.27$ kg/(m$^3 \cdot$ s $\cdot \Delta y$)]

$k_S a_p = 216.94$ lb water vapor/(ft$^3 \cdot$ h $\cdot \Delta X$) $= [0.965$ kg/(m$^3 \cdot$ s $\cdot \Delta x$)]

Calculate the adsorbate mass S_S.

$$S_S = \frac{Y_0 G_S}{X_T}$$

where Y_0 = inlet humidity
  G_S = superficial air mass velocity
  X_T = lb water vapor/lb gel at point Y_0 on the equilibrium line.

Adding in the values,

$$S_S = \frac{0.00267 \cdot 95.536}{0.0858} = 2.973 \text{ lb gel/(h} \cdot \text{ft}^2) = [3.92 \times 10^{-3} \text{ kg/(s} \cdot \text{m}^2)]$$

Calculate mG_S/S_S. Note: m is determined using graphical methods; plotting Y versus X, $m = 0.0185$.

$$\frac{mG_s}{S_S} = \frac{0.0185 \times 95.536}{2.973} = 0.5945 \text{ ft} = \left[\frac{0.0185 \times 0.1295}{3.92 \times 10^{-3}} = 0.61 \text{ m} \right]$$

Calculate the height of a gas transfer unit, H_g.

$$H_g = \frac{G_S}{k_y a_p} = \frac{95.536}{2308} = 0.0414 \text{ ft}$$

$$= \left[\frac{0.1295}{10.27} = 0.0126 \text{ m} \right]$$

Calculate the height of a solid (gel) transfer unit H_S.

$$H_S = \frac{S_S}{k_S a_p} = \frac{2.973}{216.94} = 0.0137 \text{ ft}$$

$$= \left[\frac{3.92 \times 10^{-3}}{0.965} = 4.062 \times 10^{-3} \text{ m} \right]$$

Calculate the overall mass transfer unit H_{OG}.

$$H_{OG} = H_G + \frac{mG_S}{S_S} \cdot H_s = 0.0414 + 0.5945 \cdot 0.0137 = 0.0495 \text{ ft}$$

$$= [0.0126 + 0.61 \cdot (4.062 \times 10^{-3}) = 0.0151 \text{ m}]$$

Determined by graphical methods, N_{OG} is equal to 9.304 (the area under the equilibrium curve). Calculate the required bed height Z_a.

$$Z_a = H_{OG} \cdot N_{OG} = 9.304 \cdot 0.0495 = 0.46 \text{ ft} = [9.30 \cdot 0.0151 = 0.14 \text{ m}]$$

Calculate the degree of saturation. At the breakthrough point,

$$\frac{Z - f \cdot Z_a}{Z}$$

where $f = 0.53$. The degree of saturation at the breakthrough point is the physical property of the gel to absorb solute.

$$\frac{Z - f \cdot Z_a}{Z} = \frac{2 - 0.53 \cdot 0.46}{2} = 0.878 = 87.8\%$$

$$= \left[\frac{0.61 - 0.53 \cdot (0.14)}{0.61} = 0.878 = 87.8\% \right]$$

The bed contains 2 ft³/ft² (0.61 m³/m²) of cross-sectional area. Calculate the mass of the gel per unit of cross-sectional area.

$$M_G = 2 \cdot 41.91 = 83.82 \text{ lb/ft}^2 = [0.61 \cdot 671.2 = 409.4 \text{ kg/m}^2]$$

Calculate the moisture contained in the gel at the calculated degree of saturation.

Moisture = $83.8 \cdot 0.878 \cdot X_T = 83.8 \cdot 0.878 \cdot 0.0858 = 6.313$ lb water/ft²

$$= [409.4 \times 0.878 \times 0.858 = 30.8 \text{ kg/m}^2]$$

Calculate the moisture the air brings into the bed.

Moisture = G × humidity = 95.536 lb air/(h · ft²) · 0.00267 lb water/lb air

$$= 0.25508 \text{ lb water/(h · ft}^2) = 3.46 \times 10^{-4} \text{ kg/(m}^2 \cdot \text{s)}$$

Therefore, the breakthrough point occurs at 6.313/0.25508 h after the start or $30.8/(3.46 \times 10^{-4}) = 89{,}140$ s after the start.

Example Using Activated Carbon.   Given the following,

1. *Amount needed to be recovered.*   800 lb/h [363.6 kg/h] of ethyl acetate from a solvent-laden airstream containing 1.8 lb vapor/1000 ft³ at 90°F [32.2°C] and 14.7 lb/in² [1 atm]

2. *Carbon capacity.*   0.45 lb vapor/lb carbon [0.45 kg/kg] up to the break-through point

3. *Absorption cycle time.*   3 h

4. *Bed density.*   30 lb/ft^3 [481.5 kg/m^3]

Calculate the amount of carbon and the volume of the bed.

solution   In a 3-h period, the vapor to be absorbed is (3)(800) = 2400 lb = [(3)(363.6) = 1090.8 kg/h]. The amount of carbon required per bed is 2400/0.45 = 5330 lb = [1090.8/0.45 = 2424 kg]. The volume of carbon is 5330/30 = 177.8 ft^3 [2424/481.5 = 5.03 m^3].

13.3   Types of Adsorption Equipment

13.3.1   Fixed bed adsorption

Fixed beds are either flat or cylindrical, and in some cases the beds are in stages for space, structural, and efficiency considerations. Uniform adsorbate fluid distribution and a controlled flow rate are essential for producing predictable results. In some instances, the fixed bed is installed in metal drums for ease of replacement when breakthrough occurs. Adsorbent suppliers furnish the complete package and, if applicable, will reclaim the adsorbent for future reuse. The pressure drop is a function of the bed height, the size of the solids or adsorbent pieces, the adsorbent bulk shape, and the fluid flow rate.

13.3.2   Continuous systems (regenerative adsorbers)[5]

These systems are utilized where superheated or saturated steam is used to regenerate the gases being adsorbed. Figure 13.6 shows a typical regenerative adsorber. Typically while gases are being adsorbed in one unit, regeneration takes place on the other unit. The gases that are replenished from the steam regeneration process are cooled and condensed and reused in the process. The bed is cooled by using fresh air before being used for adsorption. Even though the cost of installation and controls is higher than that of a fixed bed, the savings accomplished in recovering the gases offset this added expense.

13.4   Basic Selection Criteria

It was previously pointed out that fixed bed adsorption is the most common process. Generally, fixed beds are used to remove small amounts of adsorbate. The selection criterion becomes one of choosing the proper adsorbent for the adsorbate to be removed. If the concentration of the adsorbate is very high, the life of the adsorbent will be

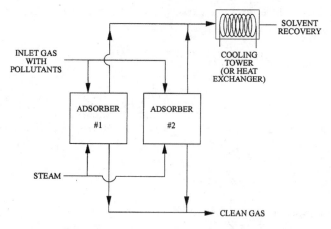

Figure 13.6 Regenerative carbon adsorbers.

shorter before regeneration and will result in higher operating costs. The following is a list of common adsorbents for the listed processes.

1. *Organic solvent removal.* Activated carbon is commonly used, but zeolites are gaining in popularity.

2. *Gaseous drying.* Silica gel is very common, and molecular sieves are sometimes used.

3. *Acid gases.* Sodium hydroxide–impregnated active carbons are most often used.

4. *Alkaline gases.* Commercially available acid-impregnated active carbons.

5. *Oxidation reactions.* Alumina impregnated with potassium permanganate is available in a variety of sizes.

6. *Indoor air quality (IAQ).* Activated carbons and synthetic zeolites are used to remove smoke and other contaminants from modern buildings. This is the largest application for these adsorbents.

7. *Water purification.* Activated carbons and zeolites.

8. *Mercury removal.* Activated carbon impregnated with sulfur or iodine.

The base cost of the adsorbent and the economic feasibility of adsorbent recovery, along with the efficiency requirement, are the overriding factors that determine the adsorbent selection. In the processes where more than trace amounts of contaminants must be removed, some form of continuous system is normally required. The fixed bed would become so large to allow for a reasonable process time prior to breakthrough that it would become impractical.

Adsorption is an exothermic process. Physical adsorption gives off much less heat than chemisorption (in most instances). The reason for the difference is the relative strengths of the bonds that are formed. Generally, physical adsorption can be thought of as having an exothermic value of the heat of sublimation of the adsorbate, whereas chemisorption gives off heat on the order of the heat of chemical reaction.

In all the descriptions of the adsorption process, the discussion was centered around a particle of the adsorbent. A particle should not be confused with the bulk or large pieces of the adsorbent. The more porous the bulk of the adsorbent, the more particles are exposed to the fluid and consequently the better the adsorption process.

13.5 Installation

The effectiveness of adsorption equipment is very sensitive to proper installation and system fluid dynamic considerations, as well as the physical layout.

13.5.1 Gas distribution

In order for the adsorbate to be removed, it must come in contact with the adsorbent. If the fluid is not uniformly distributed as it flows through the bed, the removal efficiency will be lower than predicted and breakthrough will occur more quickly than it would if the fluid flow was uniform.

13.5.2 Pluggage of the bed

Care must be taken to remove gas-borne particulate and liquid materials that can plug the adsorbent or reduce its effectiveness. Particulate and/or moisture are common examples that can lead to serious problems if not removed and yet are relatively easy to get out of the gas stream by using other gas cleaning equipment such as cyclones, baghouses, filters, and demisters.

13.5.3 Continuous systems

When the adsorbent is being carried in the fluid along with the adsorbate, there are two key factors to address. First, the gas velocity must be sufficient to keep the adsorbent airborne, and secondly the duct length must be great enough to allow time for the adsorption to be accomplished. After the adsorption has been accomplished, the adsorbent must be removed from the gas flow and replenished by using steam (see Sec. 13.3.2).

13.5.4   Face velocity and inlet pressure

The adsorbate removal efficiency is affected by both the face velocity and in some cases the inlet pressure. The permissible face velocity varies with both the adsorbate and adsorbent. The values to be used in the design must be determined from available data in order to have an efficient adsorber.

13.6   Troubleshooting

There are innumerable problems that can arise in adsorption processes. The following is a partial listing of the more common problems.

1. *Process changes.*   The biggest single problem that arises is changes in gas volume and contaminant concentrations. A volume increase affects the loading of the contaminant per unit cross-sectional area; changes the mass transfer rate; and, in the case of a fixed bed, increases the rate of the wave that describes the movement of the saturated adsorbent line in the bed, thereby reducing the useful life of the media. Minor increases in the contaminant concentration only reduce the useful life of the media. However, if the concentration increase is significant, the process calculation basis changes in that instead of having a valid assumption of an isothermic process, the exothermic nature of adsorption can raise the temperature and produce a non-steady-state condition in the bed. At times, additional contaminants can also become a part of the gas stream. If the additional contaminants are not adsorbed, the problem is not generally severe. However, if they are adsorbed, the life of the bed and the bed efficiency will be affected.

2. *Distribution of gas and liquid.*   The analysis of adsorption used illustrations for the solvent being a gas. However, the equations can also apply for solvents that are liquids. In either case, the distribution of the solvent at the entry and through the bed can affect the bed performance. In the case where the solvent is a gas, the gas discharge from the bed must also be considered. For example, if there is an abrupt change in cross-sectional area at the discharge without a transition, the effect usually begins within the bed due to the dynamics of fluid flow. Similarly, an abrupt expansion at the entry to a bed without a transition can mean that the gas has not expanded to the full bed area prior to the gas entry into the bed.

3. *Adsorbent size variations.*   The ability to predict mass transfer is dependent on being able to write an expression for the mass trans-

fer for individual particles. Obviously, if one portion of a bed has a significantly different total surface area for some of the discrete particle surfaces than another portion of the bed, the overall mass transfer for the two bed sections is different and the total mass transfer is affected. It is, therefore, necessary to have uniform adsorbent for the accurate prediction of most mass transfer processes.

4. *Media porosity variations.* The capillary action within the particle affects the individual particle mass transfer expression. If the porosity varies significantly, the capillary action will vary and the total mass transfer across the bed will vary as well.

5. *Overall mass transfer expressions.* Early research in the development of mass transfer rates for a given material did not take into account the variations in porosity and particle sizes, etc. If the adsorption device design was based on early research values, the unit will not be likely to perform as predicted.

6. *Multiple adsorbable contaminants.* Many times the effect of the adsorption of more than one contaminant is not realized, and the bed efficiency must be reanalyzed. The initial design may have ignored the additional contaminants, or the contaminants may have come about through process changes. One of the first steps in troubleshooting should be to obtain an incoming fluid constituent analysis, both as to the identification of and the concentration of each of the contaminants. Adsorption of an unwanted compound frequently reduces the effectiveness of the bed, thereby reducing the removal efficiency of the desired compound.

References

1. Dan Amuda, "Controlling VOCs and Odor," *Pollution Engineering,* July 1994, pp. 32–35.
2. Robert H. Perry and Don Green, eds., *Perry's Chemical Engineering Handbook,* 6th ed., McGraw-Hill Book Co., New York, 1984, Chapter 16.
3. Robert E. Treybal, *Mass Transfer Operations,* 2d ed., McGraw-Hill Book Co., New York, 1968, pp. 491–568.
4. Anthony J. Bunicore and Wane T. Davis, *Air Pollution Engineering Manual,* Air and Waste Management Association, Van Nostrand Reinhold, New York, 1992, p. 35.
5. Carl L. Yaws, Li Bu, and Sachin Nijhawan, "Equation Calculates Carbon Capacity for Adsorbing Pollutants," *Oil and Gas Journal,* February 13, 1995, pp. 64–69.

Thermal Oxidizers

Anthony V. Andriola

Fisher-Klosterman, Inc.

Venkatesh Subramania

Fisher-Klosterman, Inc.

14.1 Introduction

Thermal oxidizers are primarily used for the control of emissions that require destruction of a pollutant, such as toxic or hazardous gases. The thermal oxidation process is used in industries where VOCs have to be oxidized primarily to carbon dioxide and water. In applications where a halogenated hydrocarbon is to be destroyed, halogens (and acids containing halogens) are produced in addition to carbon dioxide and water. Typical applications of thermal oxidizers are throughout the chemical, printing and painting (where solvents containing hydrocarbons are used), and pharmaceutical industries.[1-3]

In the United States, awareness has increased concerning the importance of limiting these pollutants, largely a result of passage of the Clean Air Act of 1990. Industrial hydrocarbons will react with sunlight in our atmosphere to produce ozone. Ozone or photochemical smog causes respiratory problems, corrosion of bridges and buildings, and damage to vegetation. Title III of the Clean Air Act Amendment of 1990 classifies these organic hydrocarbon emissions as hazardous air pollutants (HAP).[4] A partial list of HAPs is shown in Table 14.1.

Thermal oxidation is a process where the pollutants react with oxygen under high temperature. In this process, they are changed by oxidation to a different chemical compound. The most common reaction products are water and carbon dioxide. The oxidation reaction is written as[5]

$$C_xH_y + (x + y/4)\,O_2 \rightarrow xCO_2 + (y/2)H_2O \qquad (14.1)$$

where the x and y coefficients define the hydrocarbon being oxidized. For example, consider the oxidation reaction of toluene where $x = 7$ and $y = 8$:

$$C_7H_8 + (7 + 8/4)O_2 \rightarrow 7CO_2 + (8/2)H_2O$$

or

$$C_7H_8 + 9O_2 \rightarrow 7CO_2 + 4H_2O \qquad (14.2)$$

In addition to producing carbon dioxide and water vapor during oxidation, the process also releases heat. Thermal oxidizers cause the oxidation reaction to occur in the following ways:

1. By raising the temperature of the process gas to a level where the reaction occurs spontaneously

2. By preheating the flue gas and then exposing it to a catalyst which promotes the oxidation to occur at a lower temperature (and at a much higher rate)[6]

The rate of the oxidation can be expressed by first-order reaction kinetics. This equation is based upon the premise that the rate of reaction is proportional to the amount of the reactant substance present. The amount of hydrocarbon that is consumed by oxidation in a given amount of time is estimated using the relationship[7]

$$\ln \frac{C_A}{C_{AO}} = -k \cdot t \qquad (14.3)$$

where C_A = final concentration of hydrocarbon, mol
C_{AO} = initial concentration of hydrocarbon, mol
k = rate constant, s^{-1}
t = time, s

TABLE 14.1   Hazardous Air Pollutants (HAPs)[1-4]

CAS number*	Chemical name	CAS number*	Chemical name
75070	Acetaldehyde	334883	Diazomethane
60355	Acetamide	132649	Dibenzofurans
75058	Acetonitrile	96128	1,2-Dibromo-3-chloropropane
98862	Acetophenone	84742	Dibutylphthlate
53963	2-Acetylaminofluorene	106467	1,4-Dichlorobenzene (p)
107028	Acrolein	91941	3,3-Dichlorobenzidene
79061	Acrylamide	111444	Dichloroethyl ether (Bis(2-
79107	Acrylic acid		chloroethyl) ether)
107131	Acrylonitrile	542756	1,3-Dichloropropene
107051	Allyl chloride	62737	Dichlorvos
92671	4-Aminobiphenyl	111422	Diethanolamine
92533	Aniline	121697	N,N-Diethylaniline
90040	o-Anisidine	64675	Diethyl sulfate
1332214	Asbestos	119904	3,3-Dimethoxybenzidine
71432	Benzene (including benzene	60117	Dimethyl aminoazobenzene
	from gasoline)	119937	3,3'-Dimethylbenzidine
92875	Benzidine	79447	Dimethylcarbamoyl chloride
98077	Benzotrichloride	68122	Dimethylformamide
100447	Benzyl chloride	57147	1,1-Dimethylhydrazine
92524	Biphenyl	131113	Dimethyl phthalate
117817	Bis(2-ethylhexyl)phthalate	77781	Dimethyl sulfate
	(DEHP)	534521	4,6-Dinitro-o-cresol, and salts
542881	Bis(chloromethyl)ether	51285	2,4-Dinitrophenol
75252	Bromoform	123911	1,4-Dioxane (1,4-
106990	1,3-Butadiene		Diethyleneoxide)
156627	Calcium cyanamide	122667	1,2-Diphenylhydrazine
105602	Caprolactam	106898	Epichlorohydrin (1-Chloro-2,3-
133062	Captan		epoxypropane)
63252	Carbaryl	106887	1,2-Epoxybutane
75150	Carbon disulfide	140885	Ethyl acrylate
56235	Carbon tetrachloride	100414	Ethyl benzene
463581	Carbonyl sulfide	51796	Ethyl carbamate (Urethane)
120809	Catechol	75003	Ethyl chloride (Chloroethane)
133904	Chloramben	106934	Ethylene dibromide
57749	Chlordane		(Dibromoethane)
7782505	Chlorine	107062	Ethylene dichloride (1,2-
79118	Chroacetic acid		Dichloroethane)
532274	2-Chloroacetophenone	107211	Ethylene glycol
108907	Chlorobenzene	151564	Ethylene imine (Azirdine)
510156	Chlorobenzilate	75218	Ethylene oxide
67663	Chloroform	96457	Ethylenethiourea
107302	Chloromethyl methyl ether	75343	Ethylidene dichloride (1,1-
126998	Chloroprene		Dichloroethane)
1319773	Cresols/cresylic acid (isomers	50000	Formaldehyde
	and mixture)	76448	Heptachlor
95487	o-Cresol	118747	Hexachlorobenzene
108394	m-Cresol	87683	Hexachlorobutadiene
106445	p-Cresol	77474	Hexachlorocyclopentadiene
98828	Cumene	67721	Hexachloroethane
94757	2,4-D, salts and esters	822060	Hexamethylene-1,6 diiso-
3547044	DDE		cyanate

TABLE 14.1   Hazardous Air Pollutants (HAPs)[1-4] (Continued)

CAS number*	Chemical name	CAS number*	Chemical name
680319	Hexamethylphosphoramide	85449	Phthalic anhydride
110543	Hexane	1336363	Polychlorinated biphenyls
302012	Hydrazine		(Aroclors)
7647010	Hydrochloric acid	1120714	1,3-Propane sultone
7664393	Hydrogen fluoride	57578	b-Propiolactone
	(Hydrofluoric acid)	114261	Propoxur (Baygon)
123319	Hydroquinone	78875	Propylene dichloride (1,2-
78591	Isophorone		Dichloropropane)
58899	Lindane (all isomers)	75569	Propylene oxide
108316	Maleic anhydride	75558	1,2-Propylenimine (2-Methyl
67561	Methanol		aziridine)
72435	Methoxychlor	91225	Quinoline
74839	Methyl bromide	106514	Quinone
	(Bromomethane)	100425	Styrene
74873	Methyl chloride	96093	Styrene oxide
	(Chloromethane)	1746016	2,3,7,8-Tetrachlorodibenzo-p-
71556	Methyl chloroform (1,1,1-		dioxin
	Trichloroethane)	127184	Tetrachloroethylene
78933	Methyl ethyl ketone (2-		(Perchloroethylene)
	Butanone)	7550450	Titanium tetrachloride
60344	Methyl iodide (Iodomethane)	108883	Toluene
108101	Methyl isobutyl ketone	95807	2,4-Toluene diamine
	(Hexone)	584849	2,4-Toluene diisocyanate
624839	Methyl isocyanate	95534	o-Toluidine
80626	Methyl methacrylate	8001352	Toxaphene (chlorinated cam-
1634044	Methyl tert-butyl ether		phene)
101144	4,4-Methylene (Bis(2-chloroani-	120821	1,2,4-Trichlorobenzene
	line))	79005	1,1,2-Trichloroethane
75092	Methylene chloride	79016	Trichloroethylene
	(Dichloromethane)	95954	2,4,5-Trichlorophenol
101688	Methylene diphenyl diiso-	88062	2,4,6-Trichlorophenol
	cyanate (MDI)	121448	Triethylamine
101779	4,4'-Methylenedianiline	1582098	Trifluralin
91203	Naphthalene	540841	2,2,4-Trimethylpentane
98953	Nitrobenzene	108054	Vinyl acetate
92933	4-Nitrobiphenyl	593602	Vinyl bromide
100027	4-Nitrophenol	75014	Vinyl chloride
79469	2-Nitropropane	75354	Vinylidene chloride (1,1-
684935	N-Nitroso-N-methylurea		Dichloroethylene)
62759	N-Nitrosodimethylamine	1330207	Xylenes (isomers and mixture)
59892	N-Nitrosomorpholine	95476	o-Xylenes
56382	Parathion	108383	m-Xylenes
82688	Pentachloronitrobenzene	106423	p-Xylenes
	(Quintobenzene)	0	Antimony compounds
87865	Pentachlorophenol	0	Arsenic compounds (inorganic
108952	Phenol		including arsine)
106503	p-Phenylenediamine	0	Beryllium compounds
75445	Phosgene	0	Calcium compounds
7803512	Phosphine	0	Chromium compounds
7723140	Phosphorus	0	Cobalt compounds

TABLE 14.1   Hazardous Air Pollutants (HAPs)[1-4] (Continued)

CAS number*	Chemical name	CAS number*	Chemical name
0	Coke oven emissions	0	Fine mineral fibers
0	Glycol ethers	0	Nickel compounds
0	Lead compounds	0	Polycyclic organic matter
0	Manganese compounds	0	Radionuclides (including radon)
0	Mercury compounds	0	Selenium compounds

*CAS = Chemical Abstract Service.
SOURCE: Title II, Clean Air Act Amendment, 1990.

Example   Assume toluene is to be oxidized for a period of 10 s. The rate constant $k = 0.334/s$ [see example for Eq. (14.5)]. What is the amount of hydrocarbon consumed?

solution

$$\ln \frac{C_A}{C_{AO}} = (-0.334) \cdot 10$$

$$\frac{C_A}{C_{AO}} = e^{-3.34} = 0.0354 \text{ mol/mol}$$

Therefore, for every mole of toluene in the inlet, 0.0354 mol remain after 10 s.

The destruction efficiency (DE) of the oxidizer is defined by the relation[7]

$$DE = 1 - \frac{C_A}{C_{AO}} \qquad (14.4)$$

where C_A is the final concentration of hydrocarbon, mol, and C_{AO} is the initial concentration of hydrocarbon, mol.

Example   Using the values of C_A and C_{AO} calculated from our previous example, the destruction efficiency is

$$DE = \frac{1 - 0.0354}{1} = 0.9646 = 96.46\%$$

Equation (14.2) can be used to estimate the time required to oxidize a specific fraction of a given hydrocarbon if the reaction rate constant k is known. It is important to note that in this equation, the reaction temperature is a constant. However, the heat generated by the oxidation reaction increases as the reaction progresses; therefore, the usefulness of this equation is limited to cases where the concentration of hydrocarbon is low enough to cause little change in temperature.

The rate constant k is a function of the reaction conditions including the oxygen concentration, the hydrocarbon concentration, the reaction pressure, and the temperature. The effect of temperature on the rate constant is of particular importance in thermal oxidizer design because it is usually simple to adjust the temperature at the exhaust of the oxidizer (after the oxidizer is completed).

The temperature dependence of the rate constant can be expressed as[7]

$$k = A_F{}^{(-E_A/RT)} \tag{14.5}$$

where A_F = preexponential factor, s^{-1}
$\quad\quad E_A$ = reaction activation energy, kcal/gmol
$\quad\quad R$ = universal gas constant, 1.987 cal/gmol
$\quad\quad T$ = absolute temperature, K

Example Assume toluene is to be oxidized at a temperature of 1000°C. What is the reaction rate k?

solution From Table 14.2 for toluene, $A_F = (3.73 \times 10^{16})$ s^{-1} and $E_A = 72.7$ kcal/g · mol. Therefore,

$$T = 1000 + 273 = 1273 \text{ K}$$

$$k = (3.73 \times 10^{16})^{72.7/(1.987 \cdot 1273)} = 0.334$$

Equation (14.5) is referred to as the Arrhenius equation. The frequency factor and the activation energy are a function of the type of compound being destroyed and the type of reaction (oxidation or pyrolysis). Table 14.2 gives typical values of rate constants for some VOCs.[7]

Using Eq. (14.4), the reaction constant can be obtained if the frequency factor, temperature, and reaction activation energy are known. The reaction constant can then be used to calculate the time required and the volumetric flow rate of the flue gas. As can be seen from Eqs. (14.2) and (14.4), the destruction efficiency [as defined in Eq. (14.3)] can be controlled by varying the operating temperature. Hence, the thermal oxidizer can operate at the optimum temperature

TABLE 14.2 Oxidation Data for Volatile Organic Compounds

Compound	A_F, s^{-1}	E_A, kcal/mol
Toluene	3.73×10^{16}	72.7
1,1,1-Trichloroethane	1.90×10^{8}	32.0
Xylene	4.20×10^{12}	77.6
Dichloromethane	3.00×10^{13}	64.0
Freon 113	6.62×10^{14}	69.0
Trichloroethylene	3.87×10^{5}	25.1
Tetrachloroethylene	2.60×10^{6}	33.0
Benzene	2.80×10^{6}	38.0
Chloroform	2.90×10^{12}	49.0
Carbon tetrachloride	1.26×10^{11}	50.0

needed to achieve the required hydrocarbon destruction efficiency. Therefore, we can summarize that the oxidizer design requires

1. Temperature needed for desired level of destruction
2. Retention time (i.e., the time required for desired level of oxidation)
3. Turbulence (this is an important factor that aids mixing, etc.)

Regulatory agencies normally dictate the minimum retention time and temperature for thermal oxidizers. Typically, the following parameters serve as guidelines in designing a thermal oxidizer:

1. Nominal retention time in the range of 0.2 to 2.0 s
2. Temperatures in the 1200 to 2000°F [649 to 1093°C] range for regenerative and recuperative thermal oxidizers, and 600 to 800°F [315 to 426°C] for catalytic oxidizers
3. A length-to-diameter ratio in the range of 2 to 3
4. Gas velocities in the range of 10 to 50 ft/s [3 to 15 m/s]

Typical destruction efficiencies will range between 90 and 99 percent completion of the oxidation reaction [see Eq. (14.3)].

14.2 Typical Oxidizer Configurations

Thermal oxidizers are divided into five distinct types: direct combustion (flare or afterburner), recuperative, recuperative catalytic, regenerative, and direct catalytic.

14.2.1 Direct combustion

Direct combustion oxidizers consist of a burner, a mixing zone, and a retention or combustion chamber (see Fig. 14.1). Fuel such as natural gas is premixed with combustion air (usually 7% oxygen is used for estimating the amount of combustion air) and fed into the burner to heat the incoming process gas to its oxidation temperature. The gas is homogeneously mixed with the fuel in the mixing zone where initial oxidation takes place. The final oxidation takes place in the retention zone where the gas is destroyed primarily to carbon dioxide and water. Because of the fact that virtually all the heat has to be supplied by the burner, the amount of fuel consumed may be quite large. This results in high operating costs and subsequently is used in special cases where there is

1. High inlet process gas (greater than 800°F) [427°C] so that less heat is required to achieve the oxidation temperature.

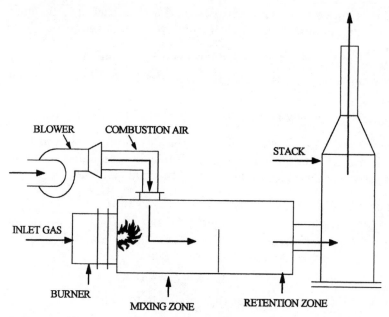

Figure 14.1 Dried flame combustion chamber. (*Illustration by Ronda White. Courtesy Fisher-Klosterman, Inc.*)

2. Very high hydrocarbon Btu content so that a design incorporating heat recovery would be in danger of sustaining damage due to self-ignition of the gas (where the burner does not have to supply a large amount of heat).

3. Low inlet volume to the system (less than 500 scfm), where savings in operating cost do not justify the added cost of the heat recovery equipment.

4. Presence of particulate in the inlet gas that can foul up the catalyst (in case of catalytic oxidation) or other heat recovery equipment used in regenerative or recuperative types of thermal oxidizers.

14.2.2 Recuperative

Recuperative thermal oxidizers (see Fig. 14.2) use an air-to-air heat exchanger to transfer heat from the clean hot outlet gas to the inlet gas stream entering the oxidizer. This allows the amount of heat required to bring the inlet gas to its oxidation temperature to be reduced substantially. Hence, less fuel is required by the burner thereby significantly reducing operating costs. Typically either a plate-type or a fin-type heat exchanger is used.

The cost of adding a heat exchanger and the additional fan horsepower required to push the air through the system makes the cost of installing and operating the recuperative thermal oxidizer higher

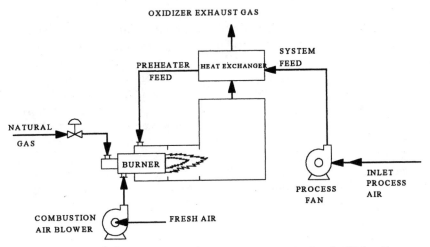

OXIDIZER EXHAUST GAS

Figure 14.2 Regenerative thermal oxidizer. (*Illustration by Ronda White. Courtesy Fisher-Klosterman, Inc.*)

than that for direct thermal oxidizers. However, the additional cost is offset by fuel savings and heat recovery that make it a viable and cost-effective alternative. In cases where the volumes are much higher, the recuperative thermal oxidizers are cheaper than direct thermal oxidizers. Care should be taken in the design of the heat exchanger to prevent excessive temperatures at the hot inlet of the heat exchanger that can cause the inlet gases to self-ignite in the heat exchanger possibly causing damage.[8]

Additionally, in the presence of corrosive elements in the inlet gas stream (like halogenated hydrocarbons), the temperature of the gases should be kept well above the dew point to prevent corrosion. Usually, the heat exchangers are made of stainless steel or nickel alloys (like Hastelloy and Inconel).

14.2.3 Recuperative catalytic[8]

Recuperative catalytic oxidizers (Fig. 14.5) are similar to recuperative thermal oxidizers except that a catalyst material is placed in the airstream. The catalyst accelerates the oxidation reaction thereby allowing spontaneous combustion to take place at much lower temperatures as compared to recuperative thermal oxidizers. Additional reduction in auxiliary fuel and lower material costs due to lower operating temperatures are achieved in comparison to recuperative thermal oxidizers. If the heating value of the fuel is sufficiently high, then use of the catalyst allows the operation of the oxidizer without any fuel. However, we must note that the catalyst has to be replaced every few years to maintain the operating efficiency of the oxidizer. Section 14.3 discusses in detail the catalytic oxidation process.

Most catalysts used to control industrial air pollution contaminants contain precious metals such as platinum or palladium. The precious metals are added to a ceramic or metal support matrix. The type and quantity of the catalysts as well as the type of support matrix will vary depending upon applications and are typically recommended by the catalyst supplier. When using catalysts, it is important not to apply them to flue gases that may contain chemicals which can poison the catalyst. A poison is defined as any substance that can inhibit the catalyst from enhancing the reaction. Some common poisons are lead, zinc, phosphorus, chlorine, silicone, and arsenic. This is an important factor since performance guarantees provided by most manufacturers have threshold limits governing poisons which if exceeded void the catalyst guarantee.

14.2.4 Regenerative

Regenerative thermal oxidizers utilize a series of beds (Fig. 14.3) made of heat-resistant material to alternately store heat from the exhaust gases exiting the combustion chamber while the second pre-heated bed releases its stored heat into the cold flue gas entering the combustion chamber. The flow of gases through the oxidizer is reversed when the bed being heated by the outlet gases becomes saturated with heat. If extremely high hydrocarbon oxidation efficiency is required, a third bed is incorporated in the system to allow purging the inlet bed before switching it to the outlet. When the regenerator roles are switched, the outlet regenerator becomes the inlet and the inlet switches to purge. The flow reversal is achieved by a series of dampers and takes place at intervals of 15 to 60 times per hour. This type of heat recovery is capable of achieving thermal efficiencies in excess of 90 percent.[8] This allows regenerative thermal oxidizers to

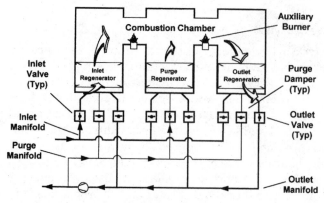

Figure 14.3 Typical regenerative thermal oxidizer. (*Illustration by Ronda White. Courtesy Fisher-Klosterman, Inc.*)

operate at minimal fuel consumption, making them particularly attractive for applications involving high air volumes and low hydrocarbon concentrations.

Because of the number of moving parts required for the airflow reversing valves, regenerative thermal oxidizers require considerable maintenance and typically have a lower reliability than recuperative oxidizers. Also, the control scheme is very elaborate and requires redundancy built into the system. In addition, the pressure drop across the regenerators is typically quite high, often double that of a recuperative system. This leads to high power consumption and high maintenance costs. Regenerative thermal oxidizers are used in applications where the gas volume is greater than 10,000 scfm and the inlet hydrocarbon concentrations are less than 1000 ppm.

14.2.5 Catalytic oxidation

An alternative to direct thermal oxidation is catalytic oxidation (Fig. 14.4) in which a catalyst is used to accelerate the oxidation reaction.[6] A catalyst is a compound that enhances the reaction without being consumed in the reaction. In catalytic oxidation, the catalysts are typically solid compounds supported on some type of high-surface-area substrate. The flue gas is heated as in direct thermal oxidation and

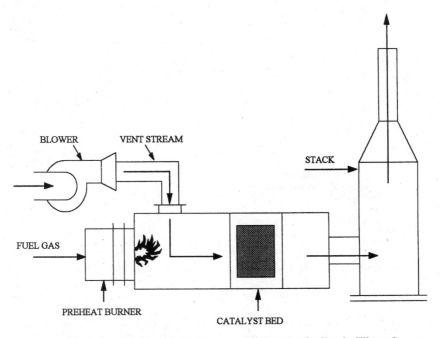

Figure 14.4 Typical catalytic oxidation process. (*Illustration by Ronda White. Courtesy Fisher-Klosterman, Inc.*)

then directed through the catalyst structure. The use of a catalyst accelerates the oxidation reaction at substantially lower oxidation temperatures thus requiring less retention time. This reduces the operating and construction cost of the system to the point where in some applications the overall cost is lower than a direct thermal oxidizer despite the added cost of the catalyst itself. The types of catalysts used are the same as those used in the recuperative catalytic oxidizers discussed in Sec. 14.2.3.

14.2.6 Selection of thermal oxidizer

Table 14.3 will serve as a general guideline for selecting a thermal oxidizer.

14.3 Catalytic Oxidation Process

Figure 14.4 describes a typical catalytic oxidation process.[5] In a typical catalytic oxidation system, the process gas passes through a preheater where it is heated by a burner to the required inlet temperature at the catalyst. Depending on the inlet concentration and removal efficiency required, an additional preburner will sometimes precede the preheater. The amount of heat required can be calculated using the relationship

$$H = M \cdot C_p \cdot \Delta T \tag{14.6}$$

TABLE 14.3 Thermal Oxidizer Selection

Item	Description	Direct flame	Recuperative	Catalytic	Regenerative
1	Low volume (<500 scfm), high particulate, high process inlet temperature >800°F (427°C)	X			
2	Inlet process temperature <200°F (93°C) and volume >10,000 scfm			X	X
3	Process temperature 200–900°F (93°C–482°C)		X		
4	Efficiency required is >99%	X	X		
5	Low hydrocarbon concentration; efficiency required is 90–99%			X	
6	Low pollutant loading				X

Note: X indicates recommended for use.

where H = heat required, Btu/h [kcal/h]
$\quad\quad M$ = mass flow rate, lb/h [kg/h]
$\quad\quad C_p$ = specific heat at constant pressure, Btu/lb · °F [kcal/kg · °C]

Example for Eq. (14.6)   Assume we have 200 lb/h [90.7 kg/h] of gas with a specific heat of 0.24 Btu/lb · °F [0.24 kcal/kg · °C] to be raised by 500°F (or 500°R) [278°C or 277.8 K]. What is the amount of heat required?

solution

$$H = 200 \cdot 0.24 \cdot 500 = 24,000 \text{ Btu/h}$$

$$= [H = 90.7 \cdot 0.24 \cdot 277.8 = 6047.2 \text{ kcal/h}]$$

Understanding the fact that the heating value h_v of natural gas is approximately 1000 Btu/lb [555.6 kcal/kg], the amount of natural gas required is

$$M_{NG} = \frac{H}{h_v} \tag{14.7}$$

where M_{NG} = amount of natural gas, lb/h [kg/h]
$\quad\quad H$ = heat required, Btu/h [kcal/h]
$\quad\quad h_v$ = heating value of natural gas, Btu/lb [kcal/kg]

Using the value for H from Eq. (14.6),

$$M_{NG} = \frac{24000}{1000} = 24 \text{ lb/h}$$

$$= \left[\frac{6047.2}{555.6} = 10.9 \text{ kg/h} \right]$$

The amount of combustion air required is determined based on the type of burner used. The oxidation process (that is coupled with increasing temperature due to heat released) takes place in the catalyst. The process gas now containing carbon dioxide, water nitrogen, and halogenated compounds (if the VOCs contain halogens) passes through the heat exchanger giving up some of the heat to increase the temperature of the inlet gas before exiting the system.

The system can be thermally more efficient for streams containing low hydrocarbon concentrations by adding gas to the heat exchanger as shown in Fig. 14.5.[7] Figure 14.5 will be used for the purpose of discussing the process calculations (see Sec. 14.4). Catalytic oxidation systems are normally designed for destruction efficiencies that range from 90 to 98 percent. Under special circumstances, destruction effi-

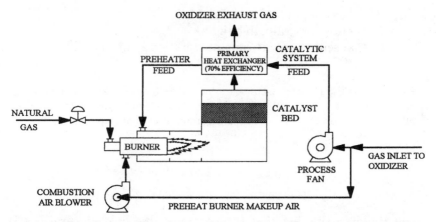

Figure 14.5 Flow diagram of catalytic oxidation system. (*Illustration by Ronda White. Courtesy Fisher-Klosterman, Inc.*)

ciencies of 99.99 percent have been demonstrated. The destruction efficiency of the catalytic oxidation process is a function of

1. Inlet and outlet temperatures at the catalyst bed
2. Volume of catalyst
3. Quantity and type of noble metal used to formulate the catalyst

Increased temperature can be achieved by either raising the inlet temperature (by increasing the catalyst volume) or increasing the noble metal content in the catalyst. While chlorinated hydrocarbons have a very low destructibility, the general rule for nonchlorinated hydrocarbons is that the higher the molecular weight, the higher the destructibility. Typical by-products of halogenated hydrocarbons are HCl, HBr, and HF. This will usually necessitate a scrubber (see Chaps. 10 and 12) after the thermal oxidizer. Unlike direct combustion processes, the catalytic oxidation process should be devoid of virtually any particulate (to remain operable) and should only be used when the oxidation does not produce any particulate. Typical catalysts contain palladium-platinum noble metals. The catalyst bed is generally made of a mesh mat, ceramic honeycomb, or ceramic matrix structure designed to increase catalyst surface area. The catalyst can also be fixed or fluidized bed configurations. Care must be taken not to increase the inlet temperature to such levels that it can decrease the activity of the catalyst. Once the catalyst is used, it can be regenerated by washing with either caustic solution or detergents. Table 14.4 gives the increase in destruction efficiencies before and after washing the catalyst.

TABLE 14.4   Typical Destruction Efficiency Before and After
Cleaning (Halohydrocarbon Destruction Catalyst)[5]

Service	Years of operation	Destruction efficiency, %	
		Before cleaning	After cleaning
Chemical	7	70	90
Flexograthic ink	1	75	99
Formaldehyde	7	99	99
Phthalic anhydride	8	92	92
Can coating	12	85	95
Fibers	3	50	99
Fibers	2	80	90
Automotive	1	91	95
Automotive	10	90	91
Automotive paint	3	92	96

14.4   Process Calculations

A typical term used in the catalytic oxidation process is *space velocity*.
This is defined as the ratio of the inlet gas volume to the catalyst and
the total volume of the catalyst. In other words

$$S_V = \frac{I_{\text{VOL}}}{T_{\text{VOL}}} \qquad (14.8)$$

where S_V = space velocity, h^{-1}
I_{VOL} = inlet gas volume, ft^3/h [m^3/h]
T_{VOL} = total volume of catalyst, ft^3 [m^3]

The space velocity changes with efficiency requirements. Usually
space velocities between 10,000 to 50,000 per hour are used. Table
14.5 gives a typical usage of space velocity and temperature based on
removal efficiency.[9]

The following example provides an explanation of the process. The
data provided is from a typical printing process[9] (see Fig. 14.6).

TABLE 14.5   Space Velocity Recommendations

Required destruction efficiency, %	Minimum temperature at catalyst bed inlet, °F	Temperature at catalyst bed outlet, °F	Space velocity, h^{-1}
90	600	1000–1200	40,000
95	600	1000–1200	30,000

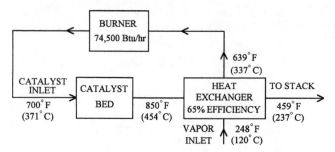

Figure 14.6 Calculated values for catalytic oxidation system.

Airflow rate = 11,000 scfm [311.485 m³/min] = 660,000 ft³/h [18,689 m³/h]

Density of air = 0.075 lb/ft³ [1.201 kg/m³]

Specific heat of air = 0.24 Btu/lb · °R [0.24 kcal/kg · K]

Mass flow rate = 11,000 · 60 · 0.075 = 49,500 lb/h [22,500 kg/h]

Removal efficiency required is 95 percent

Space velocity used is 30,000/h (see Table 14.5)

Therefore, the required catalyst volume = 660,000/30,000 = 22 ft³ [0.623 m³]. The catalyst inlet temperature = 700°F [371.1°C] and the catalyst outlet temperature = 850°F [454.4°C]. Based on a heat balance, the temperature at the outlet of the heat exchanger to the burner is

$$248 + 0.65 · (850 − 248) = 639°F = [337°C]$$

The temperature to the stack is

$$248 − 0.65 · (850 − 248) = 459°F = [237°C]$$

The heat required by the burner is

$$49,500 · 0.24 · (700 − 639) = 724,680 \text{ Btu/h} = [182,616.2 \text{ kcal/h}]$$

The above calculated temperature values are shown in Fig. 14.6. The process being solved is to calculate (1) the amount of catalyst needed and (2) heat required by burner (or called "burner capacity") in a thermal catalytic oxidizer.

14.5 Maintenance and Troubleshooting

A typical catalyst oxidizer requires the checks of the following items.

14.5.1 Burner

The burner should operate at a given burner efficiency. This is determined by the amount of natural gas and combustion air used and the pressure drop across the burner line.

14.5.2 Catalyst

Scheduled inspection of the catalyst is to be performed by either visual inspection or periodic cleaning of the catalyst. If the system operates at a higher pressure drop (than designed for) or the VOC removal efficiency is decreased, it is then time to inspect and clean the catalyst.

14.5.3 Burner ultraviolet (UV) scanner

Proper cooling air has to be injected into the UV scanner bulb to keep the bulb from failing.

14.5.4 Combustion air fans

Regular inspection and maintenance by checking the performance against the fan curves, proper balancing, and applying oil to the bearings is required to keep the combustion air fan from failing.

14.5.5 Instrumentation

The catalytic oxidation process requires intricate controls and has to operate reliably. Redundant instrumentation is incorporated into the system (due to the nature of the process) in order to increase reliability. Usually the catalytic oxidation process instrumentation is controlled through a PLC. Recommended maintenance of the instrumentation (as provided by the manufacturer) is required for proper operation.

14.5.6 Troubleshooting

Table 14.6 is a typical troubleshooting guide for the thermal-catalytic oxidation system.

References

1. L. Theodore and A. J. Buonicore, *Industrial Air Pollution Control Equipment for Particulates,* Chap. 6, CRC Press, West Palm Beach, FL. 1976.
2. M. Weiss and J. Palmisano, "Emissions Trading Gives Flexibility in Meeting Clean-Air Laws," *Power,* March 1985, pp. 55–58.

TABLE 14.6 Thermal Oxidizer Troubleshooting Guide

No.	Description	Possible causes	Recommendation
1	Low removal efficiency	1. Catalyst activity is low 2. Excessive bed outlet temperature 3. Very high space velocity 4. High heat content producing thermal runout	a. Clean or replace the catalyst b. Verify process calculations
2	High pressure drop	1. Pluggage in the line 2. Particulate in the system	a. Inspect system for pluggage and particulate
3	Uneven temperature distribution	1. Check flow distribution 2. Burner size is too high	a. Add flow vanes b. Check to see if burner size is suitable for the application
4	Burner pilot does not light up	Check natural gas lines	Check pressure at natural gas lines
5	Burner lights up and shuts off	Check for pressure switch setting for the burner	a. Decrease pressure switch setting range b. Replace pressure switch
6	Combustion fan running improperly	1. Fan not sized properly 2. Fan out of balance	Replace the fan and check fan sizing with the fan curve
7	Temperature either too high or too low at the catalyst	Heat exchanger is not working properly and the process gas is bypassing the heat exchanger	Inspect heat exchanger
8	Burner running at full capacity, but the temperature is not high enough and natural gas is consumed excessively	1. Inspect burner for flame 2. Not enough combustion air 3. Check burner linkages	a. Replace burner b. Replace or fix burner linkages
9	System does not start up properly	1. Check fresh air fan rating 2. Check instrumentation	Replace or fix fan and instrumentation. Check the PLC program
10	Low process air, low combustion air	Check belts for slippage	Replace or fix belts

3. A. J. Buonicore, "Air Pollution Control," *Chemical Engineering,* vol. 87, no. 13, June 30, 1980, pp. 81–101.
4. Clean Air Act Amendments, Public Law 101 549, Titles I–X, November 15, 1990.
5. Keith J. Herbert, "Catalyst for Volatile Organic Compound in the 1990s," Presented at 1990 Incineration Conference, San Diego, CA, May 14–18, 1990.
6. George R. Lester, "Catalytic Destruction of Hazardous Halogenated Organic Chemicals," Presented at the Air and Waste Management Association 82nd Annual Meeting and Exposition, Anaheim, CA, June 25–30, 1989.
7. Joseph L. Tessitor, John G. Pinion, and Edward DeCresie, "Thermal Destruction of Air Toxics," *Pollution Engineering,* March 1990.
8. Joseph M. Klobucar, "Choose the Best Heat Recovery Method for Thermal Oxidizers," *Chemical Engineering Series,* April 1995.
9. Michael Kususko and Carlos M. Nunez, "Destruction of Volatile Organic Compounds Using Catalytic Oxidation," *J. Waste Management Association,* vol. 140, no. 2, February 1990.

Control Technologies: Miscellaneous

Miscellaneous Technologies

William L. Heumann

Fisher-Klosterman, Inc.

15.1 Introduction

As with the prior sections of this work, Chap. 15 is divided into two general subsections: particulate separation technologies and gaseous separation technologies. The reason various technologies were included in this chapter instead of having an individual chapter within the prior text was based upon their current level of usage within the industry. In some cases, this is due to the limited applicability of the technology while in others it is based upon the relatively recent introduction of the technology into industry as of this writing. Some of the technologies covered herein will be described as new or emerging technologies. Although this wording may date this work, it will allow the reader to better understand whether or not one of these technologies may be appropriate for a given application.

15.2 Particulate Separation

In recent years most of the advances in the separation of particles and gas have occurred in the technologies covered in prior sections of this work. Nonetheless, there are numerous applications utilizing other technologies than those previously described. In most cases, these applications are for precleaners for other equipment and/or methods for enhancement of the performance of downstream equipment.

15.2.1 Settling chambers

Settling chambers (see Fig. 15.1) have been used within industry to separate relatively large particles (generally over 100 μm in diameter) from a gas stream. Settling chambers operate by allowing gravity to settle particles out of a low-velocity gas stream. The particles will drop to the bottom or the collection hopper while the clean gas continues. Care must be taken to ensure that the outlet gas velocity is increased sufficiently when it leaves the settling chamber to ensure that particulate that remains in the gas stream does not settle and cause a buildup of solids and plug horizontal sections of the ductwork.

In theory, a very large settling chamber could be designed that would allow for high-efficiency separation of small particles. Unfortunately, the cost and size of such a device would make it impractical. In practical usage, settling chambers are most commonly

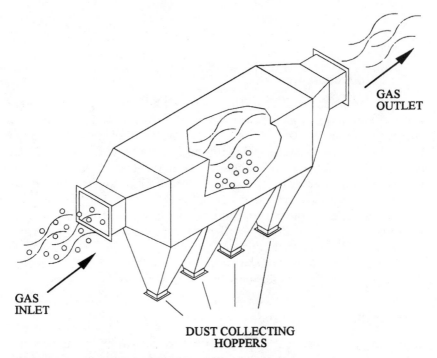

GAS OUTLET

GAS INLET

DUST COLLECTING HOPPERS

Figure 15.1 Gravity settling chamber. (*Illustration by Ronda White. Courtesy of Fisher-Klosterman, Inc.*)

utilized as precleaners for other collection equipment where there are some particles that are very large (over 100 μm in diameter) which can damage a secondary collection device. A common example would be the removal of shot and abrasive blast material from a blast cleaning booth. Although a cyclone can provide better removal efficiencies in less space, the settling chamber may be desirable due to its physical arrangement, low pressure drop, and resistance to the effects of erosion from abrasive particulate.

Settling chambers may be designed using the following equations:

$$V = t_d \cdot Q \tag{15.1}$$

where V = volume of active portion (exclusive of collection hoppers) of settling chamber, ft^3 [m^3]
t_d = time required to settle a particle of diameter d, s
Q = gas flow rate, ft^3/s [m^3/s]

and

$$t_d = \frac{h}{U_T} \tag{15.2}$$

where t_d = time required to settle a particle of diameter d, s
h = effective height of the settling chamber exclusive of hoppers, ft [m]
U_T = terminal velocity of particle to be collected, ft/s [m/s]

Table 15.1 gives some values of U_T that are useful in settling chamber design.

Example Assume a settling chamber for a 100-ft^3/s [2.8-m^3/s] gas flow is 10 ft [3.05 m] wide by 6 ft [1.83 m] high and is to be used to collect 400-μm-diameter particles. The particle density is 62.4 lb/ft^3 [1000 kg/m^3] and the gas is air at STP. What is the required length of the settling chamber?

solution From Eq. (15.2),

$$t_d = \frac{h}{U_T}$$

Since the height h = 6 ft [1.83 m] and U_T = 5.2 ft/s [1.57 m/s] (from Table 15.1),

$$t_d = \frac{6}{5.2} \left[\frac{1.83}{1.57} \right] = 1.2 \text{ s}$$

From Eq. (15.1),

$$V = t_d \cdot Q = 1.2 \cdot 100 = 120 \text{ ft}^3 = [1.2 \cdot 2.8 = 3.4 \text{ m}^3]$$

TABLE 15.1   Settling Velocity of Spherical Particles in Air[1]

Particle diameter, μm	Terminal velocity	
	ft/s	[m/s]
0.1	2.9×10^{-6}	8.7×10^{-7}
0.2	7.5×10^{-6}	2.3×10^{-6}
0.4	2.2×10^{-5}	6.8×10^{-6}
1.0	1.1×10^{-4}	3.5×10^{-5}
2.0	3.9×10^{-4}	1.19×10^{-4}
4.0	1.6×10^{-3}	5.0×10^{-4}
10.0	1.0×10^{-2}	3.00×10^{-3}
20.0	3.9×10^{-2}	1.2×10^{-2}
40.0	1.6×10^{-1}	4.8×10^{-2}
100.0	8.1×10^{-1}	2.46×10^{-1}
400.0	5.2	1.57
1000.0	1.3×10^{-1}	3.82

Note: Particle density = 62.4 lb/ft^3 [1000 kg/m^3], and the air is at a temperature of 68°F [20°C] and a pressure of 1 bar.

Since $V = L \cdot B \cdot h$, solving for L where L = chamber length, ft [m], and B = chamber width, ft [m],

$$L = \frac{V}{B \cdot h}$$

$$= \frac{120}{10 \cdot 6} = 2 \text{ ft}$$

$$= \left[\frac{3.4}{3.05 \cdot 183} = 0.61 \text{ m} \right]$$

This method of calculation is only valid for design purposes and in no way accurately predicts collection efficiency versus particle size. Furthermore, settling chamber design using these equations is only appropriate if

1. The gas flow is evenly distributed throughout the cross-sectional area.
2. The average gas velocity in the settling chamber is below the pickup velocity of material to be collected (see Table 15.2). If the pickup velocity value is unknown, use 10 ft/s [3 m/s].

15.2.2   Inertial separators

There are a wide variety of devices that use the inertia of the particulate to assist in or as the main mechanism for collection. One of the most basic examples of an inertial separator would be the addition of

TABLE 15.2   Pickup Velocities of Some
Common Materials[2]

Material	Pickup velocity	
	ft/s	[m/s]
Aluminum chips	14.1	4.3
Nonferrous foundry dust	18.7	5.7
Lead oxide	24.9	7.6
Limestone	21.0	6.4
Starch	5.9	1.8
Steel shot	15.1	4.6
Wood chips	12.8	3.9

a baffle plate to a gravity settling chamber (see Fig. 15.2). In this arrangement the collection is enhanced by directing the particulate downward. As the gas changes direction to exit the collector, the particulate will tend to continue on its original path due to inertia.

Although the primary physics allows for applying adequate inertial forces on particulate down to submicrometer sizes, it is often difficult to utilize this method exclusively for fine particle collection in industrial applications. Difficulties arise for the following reasons:

1. To generate significant levels of inertial force on small particles requires high velocities. High-velocity gas jets generate significant turbulence in which many of the flow streams are adequate to reentrain small particulate.

2. Very high velocity jets may result in particle attrition and/or severe wear on the impact surface.

3. If the particulate is separated by the gas flow by impaction in a surface where it will adhere, it may be difficult to remove (see Fig. 15.3). Further, when the particles are dislodged for cleaning of the impaction plates, they will be reentrained unless the unit is cleaned off-line.

Direct impaction of particulate can be effective for the capture of particles with diameters down to 0.5 μm but is rarely used industrially for the reasons listed above. In recent years the advance of computational fluid dynamics (CFD) models utilizing computers has allowed engineers to more accurately design inertial devices to provide more optimum separation of the flow streams from the particulate and shows some promise of providing designs that have wider industrial applicability.

The reader should understand that inertia in conjunction with other mechanisms is a very commonly utilized method of particle and

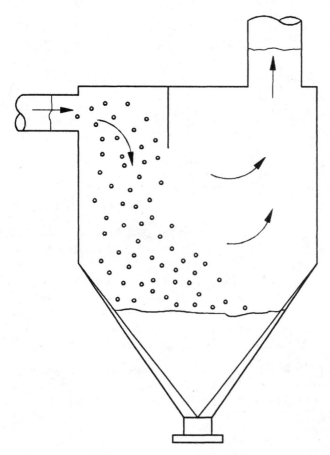

Figure 15.2 Settling chamber with inertial enhancer. (*Illustration by Ronda White. Courtesy of Fisher-Klosterman, Inc.*)

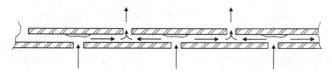

Figure 15.3 Impingement plate separator. (*Illustration by Ronda White. Courtesy of Fisher-Klosterman, Inc.*)

gas separation. Cyclones (Chap. 8) and particulate scrubbers (Chap. 10) utilize inertial forces to collect particulate. Therefore, the usage of the term *inertial* separation may be misleading. Properly speaking, cyclones and most scrubbers are forms of inertial separators but are categorized separately herein.

15.3   Gaseous Pollution Control

15.3.1   Biofiltration[3,4]

Biofiltration (see Fig. 15.4) is the name given to the process by which gaseous pollutants are removed from a process gas stream by aerobic digestion or consumption by microbes. Biofiltration is effective on certain VOCs and some inorganic compounds. It is generally most viable for treating large volumes of gas with low levels of certain appropriate contaminants. Biofiltration was originally developed for odor control but has been used for numerous other control applications since its introduction. It is a well-accepted control technology within Europe and is gaining acceptance within the United States.

A biofilter normally consists of a simple bed of material that is conducive to the support of microbe growth through which the gas passes at low velocity. To promote microbe growth and activity levels, the bed is usually kept moist and the gas is humidified before entry into the filter. Biofilters may have one or more beds of biologically active materials. The range of bed materials includes compost, peat, wood chips, soil, polystyrene, fiberglass wool, clay, and granulated activated carbon (GAC). In conjunction with GAC, biofiltration works to reactivate the adsorptive properties of the GAC. Of the purely biofiltration bed materials used, soil, peat, and fiberglass wool provide the best performance.

Biofiltration may be very effective on alcohols, ketones, and many aliphatic and aromatic hydrocarbons. Degradation rates of 50 to 100 g/(h · m³) are typical. The chlorine content in chlorinated hydrocarbons significantly reduces the rate of degradation. Successfully controlled contaminants include hydrogen sulfide, benzene, xylene, carbon disulfide, acetone, phenol, toluene, and methylmercaptan. Removal efficiencies of over 90 percent have been achieved on these contaminants.

Biofiltration beds require some maintenance and eventual replacement. Over time, mineralization will cause the bed to collapse and

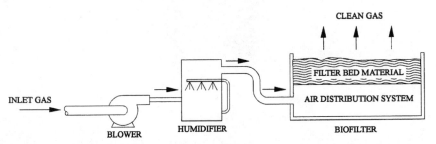

Figure 15.4 Schematic diagram of an open bed biofilter.[3] (*Illustration by Ronda White. Courtesy of Fisher-Klosterman, Inc.*)

resistance to flow (pressure drop) will increase. As a result, the bed must be turned over or replaced at that time. It is also important to note that biofiltration is dependent on the health of the microbes that are consuming the contaminant. As such, any chemicals present that poison them will impair or destroy the effectiveness of the device. Halogens (such as chlorine) and toxic metals (such as mercury) can cause losses in performance.

15.3.2   Corona reactors[3]

Although not of insignificant industrial usage as of this writing, VOCs have been successfully destroyed by exposing them to other forms of energy besides thermal energy (see Chap. 14). One such method is to pass the VOC-laden gas stream through a high-voltage, low-amperage electric field. Within this electric field or corona, free electrons are accelerated and collide with other atoms. Because of inelastic collisions between the electrons and the molecules they collide with, some or all of their energy is transferred to the target molecules. By excitation of the target molecules' atoms, the chemical bonds holding the molecules together are broken causing them to break down into simpler compounds or elements. In this manner, the VOCs (and halogenated VOCs) are destroyed by being converted most commonly into H_2O, CO_2, and CO (and halogenated acids).

The two general designs of corona reactor under development are shown in Figs. 15.5 and 15.6. The pelletized bed reactor uses a fixed

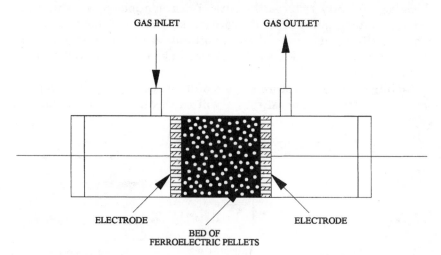

Figure 15.5 Schematic of a pelletized bed corona reactor.[3] (*Illustration by Ronda White. Courtesy of Fisher-Klosterman, Inc.*)

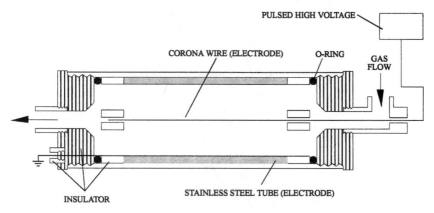

Figure 15.6 Schematic of a pulsed corona reactor.[3] (*Illustration by Ronda White. Courtesy of Fisher-Klosterman, Inc.*)

bed of ferroelectric pellets (such as barium titanate) with an alternating current. The pellets are poor conductors of electricity, and the corona fields are generated between the pellets. The pulsed corona reactor uses a very short (nanosecond) pulsed dc charge in a corona wire (electrode). The corona field is generated between the corona wire and the steel casing. In early research, it appears as if the pellet bed reactors are most efficient if limited to operating voltages below the sparking point of 5 to 8 kV/cm. In the pulsed corona reactor, operating voltages of 20 kV/cm and higher can be achieved. Since more complete VOC destruction has been associated with higher voltages, the pulsed corona reactor may have more applicability than the pelletized bed reactor although it is also more expensive. Figure 15.7 shows measured destruction efficiencies using a pellet bed corona reactor on toluene.

15.3.3 Electric arc[6]

Direct electric arcing can destroy many gaseous compounds including VOCs and some inorganic gases such as H_2S and mercaptans. In some electric arc devices, an electric arc is started at the narrow end of two diverging electrodes. The arc travels down the electrode which is also the direction of the gas flow. When the arc reaches the end of the electrodes, it breaks and immediately a new spark starts again. Tests using the particular electric arc devices have given destruction efficiencies as shown in Table 15.3.

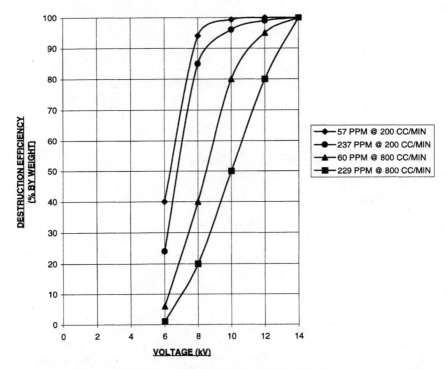

Figure 15.7 Toluene destruction efficiency from a pelletized bed corona reactor using 1-mm pellets.[3] (*Illustration by Ronda White. Courtesy of Fisher-Klosterman, Inc.*)

TABLE 15.3   Direct Electric Arc Destruction Efficiencies[6]

Gas	Destruction efficiency, %
Xylene	75
Heptane	100
Toluene	92
Methylethylketone	66
Tetrachloroethylene	100

15.3.4 Plasma treatment and other energy methods[7,8]

Another method of raising the energy state of contaminant molecules to the point at which their chemical bonds are broken, causing them to form more simple compounds and/or elements, is by putting the gas stream through plasma. Plasma is the state a gas achieves when electronically energized to the point where there are roughly the same number of positive ions and electrons and electricity is easily conducted. There are numerous plasma treatment devices under development to be used for certain gaseous emissions. Plasma treatment devices have been successful during testing, partially destroying trichloroethylene, carbon tetrachloride, and SO_2/NO_x.

This technology can effectively control certain gaseous emissions and would be classified as a *thermal treatment unit* instead of an *incinerator* for regulatory purposes. Because of the high levels of energy that may be introduced, plasma treatment processes may be found to be an economical method of removing halogenated hydrocarbons and other organic vapors with strong carbon-carbon bonds. Some other methods of raising gaseous contaminants to increased energy levels are by ultraviolet light and microwave radiation.

15.3.5 Condensers[9]

Condensation as a process is used heavily throughout industry to recover or remove certain gaseous components from a bulk gas flow. Common examples within industry include the selective distillation of various hydrocarbons in refining processes and the drying of air. The same mechanism can be used to remove a pollutant from a gas stream if the dew point of the pollutant gases is significantly higher than the noncontaminant gases. If this is the case, the total gas stream temperature can be reduced to below the dew point of the contaminant gases, which will then be separable from the remaining gas flow as a liquid.

There are several styles of condensers used industrially. They differ primarily in the method of removing heat from the gas stream. The two main mechanisms of cooling are direct and indirect. In direct cooling devices, a cooling medium (usually chilled water or air) is injected and mixed into the gas stream directly and heat is transferred from the process gas to the coolant. Indirect cooling devices separate the process gas and coolant by a barrier of some kind so that the two do not mix. Radiators are common examples of indirect cooling (or heating) devices. Indirect cooling devices primarily condense out the desired pollutant on the process side of the surface of the membrane that separates the coolant from the process gas. For this reason, these devices are also called contact condensers. Contact condensers may remove significant amounts of a contaminant even if the bulk gas temperature remains above the dew point for the given substance. The reason for this is that the temperature at the contact surface is below the dew point and condensation can form and be maintained within close proximity to this surface.

The majority of cases in which condensation is used as the primary mechanism for air pollution control utilize contact condensers, primarily of the shell and tube type of heat exchanger. Many applications within air pollution control will utilize condensation as an enhancement mechanism for collection, usually for particulate removal within scrubbers. These cases will most often utilize direct cooling methods. In addition to removal or enhancement in removal of a pollutant, these same methods are utilized frequently within industrial air pollution control to control the temperature of gases prior to entering collection devices.

Within air pollution control, condensation is successfully used as a primary collection method to remove some VOCs. Care must be taken in the design of condenser units for usage in air pollution control to allow for and/or prevent fouling of the condenser if any particulate is present in the gas stream. Condensers will perform well as collectors of fine particulate which can be an advantage if the rates of condensation and the general design of the unit result in a continual washing of the exchange surface. More commonly, however, the contact surfaces do not remain clean naturally and periodic maintenance to clean the unit is required.

Another important consideration in the design of condensers for air pollution control is that the majority of applications that are attractive candidates for condensation involve VOCs. As such, the required level of condenser performance is usually not simply one of reducing the temperature to below that for condensation to occur but to a point at which equilibrium of the vapor pressure in the gas results in a concentration of the pollutant that is below the desired level.

References

1. W. Strauss, *Industrial Gas Cleaning,* 2d ed., Pergamon Press, Oxford, England, 1975, p. 201.
2. Ibid, p. 202.
3. *The Air Pollution Consultant,* McCoy & Associates, Lakewood, CO, Nov./Dec. 1991, pp. 1.1–1.2.
4. Liu, P. K. T., R. L. Gregg, H. K. Sabol, and Barkley, "Engineering Biofilter for Removing Organic Contaminants in Air," *Air and Waste,* vol. 44, March 1994, pp. 299–303.
5. *The Air Pollution Consultant,* McCoy & Associates, Lakewood, CO, Sept./Oct. 1991, pp. 1.1–1.3.
6. A. Czernichoaski, and A. Ranalvosoloarimanana, "Zapping VOCs with a Discontinuous Electric Arc," American Chemical Society, *Chemtech,* April 1996, pp. 45–49.
7. Liu, op. cit., pp. 1.09–1.13.
8. *The Hazardous Waste Consultant,* McCoy & Associates, Lakewood, CO, Sept./Oct. 1992, pp. 1.12–1.14.
9. A. Buonicore and T. Davis (eds.), *Air Pollution Engineering Manual,* Van Nostrand Reinhold, New York, 1992, pp. 52–58.

Index

ABOUT THE EDITOR

William L. Heumann (Louisville, KY) is President and CEO of Fisher-Klosterman, an air pollution control consulting firm in operation since 1948 and a leading manufacturer of cyclones, particulate scrubbers, baghouses and cartridge collectors, gaseous scrubbers, and electrostatic precipitators. Mr. Heumann is a frequent lecturer and a contributor to industry publications such as *Chemical Engineering* magazine.